Lecture Notes of the Institute for Computer Sciences, Social Informatics and Telecommunications Engineering 422

More information about this series at https://link.springer.com/bookseries/8197

Matthias Wölfel · Johannes Bernhardt ·
Sonja Thiel (Eds.)

ArtsIT, Interactivity and Game Creation

Creative Heritage New Perspectives from Media Arts and Artificial Intelligence

10th EAI International Conference, ArtsIT 2021
Virtual Event, December 2–3, 2021
Proceedings

Springer

Editors
Matthias Wölfel
Karlsruhe University of Applied Sciences
Karlsruhe, Germany

Johannes Bernhardt
Baden State Museum
Karlsruhe, Germany

Sonja Thiel
Baden State Museum
Karlsruhe, Germany

ISSN 1867-8211 ISSN 1867-822X (electronic)
Lecture Notes of the Institute for Computer Sciences, Social Informatics and Telecommunications Engineering
ISBN 978-3-030-95530-4 ISBN 978-3-030-95531-1 (eBook)
https://doi.org/10.1007/978-3-030-95531-1

This Springer imprint is published by the registered company Springer Nature Switzerland AG
The registered company address is: Gewerbestrasse 11, 6330 Cham, Switzerland

Preface

We are delighted to introduce the proceedings of the tenth edition of the European Alliance for Innovation (EAI) International Conference on ArtsIT (ArtsIT 2021). This conference brought together researchers, practitioners, artists, and academics to present and discuss the symbiosis between art and information technology. It was intended to take place in Karlsruhe, Germany—a UNESCO Creative City of Media Arts—but finally was moved to Cyberspace due to the ongoing COVID-19 pandemic. Since 2009 ArtsIT has become a leading scientific forum for the dissemination of cutting-edge research results in the intersection between art, science, culture, performing arts, media, and technology. The role of artistic practice using digital media is also to serve as a tool for analysis and critical reflection on how technologies influence our lives, culture, and society. Therefore, ArtsIT is not only a place to discuss technological progress but also a place to reflect on the impact of art and technology on sustainability, responsibility, and human dignity.

The program of ArtsIT 2021 consisted of 31 papers selected from 57 submissions in a double-blind review process. The conference tracks were as follows: Track 1 – Theory and Reflections, Track 2 – Media Art and Virtual Reality, Track 3 – Games, Track 4 – Fusions, Track 5 – Approaches, Track 6 – Inclusion and Participation, Track 7 – Artificial Intelligence in Art, Track 8 – Artificial Intelligence in Culture, and Track 9 – Artificial Intelligence Applications. Aside from the high-quality paper presentations, the program featured the keynote "The Computable and the Uncomputable" delivered by Alexander R. Galloway, New York University, USA. Galloway addressed some lesser-known episodes from the era of digital machines, discussed how computation emerges or fails to emerge, how the digital thrives but also atrophies, and how networks interconnect while also fray and fall apart. For the publication we have restructured and concentrated the program a little.

It was a great pleasure to work with such an excellent Organizing Committee, which worked hard to organize and support the conference. In particular, the Technical Program Committee and the Publications Chair, Daniel Hepperle, helped to complete the peer-review process and produce a high-quality program. We are also grateful to the Conference Managers, Lenka Lezanska and Viltare Platzner, for their tireless support and all the authors who submitted their papers to the ArtsIT 2021 conference. We strongly believe that the ArtsIT conference provides an excellent forum for researchers, practitioners, artists, and academics to discuss all social and technological aspects that are relevant to IT-driven artistic expression. Furthermore, we expect that the future

ArtsIT conferences will be as successful and stimulating, as the papers presented in this volume demonstrate.

December 2021

Matthias Wölfel
Johannes Bernhardt
Sonja Thiel

Organization

General Chairs

Matthias Wölfel	University of Applied Sciences Karlsruhe, Germany
Johannes Bernhardt	Baden State Museum, Germany

Technical Program Committee Chairs

Elke Reinhuber	City University of Hong Kong, Hong Kong
Andres Iglesias	Universidad de Cantabria, Spain
Sarah Kenderdine	École polytechnique fédérale de Lausanne, Switzerland
Jaap Kamps	Universiteit van Amsterdam, The Netherlands
Bernd Lintermann	ZKM | Center for Art and Media, Germany

Publications Chair

Daniel Hepperle	University of Applied Sciences Karlsruhe, Germany

Web Chair

Jenia Jitsev	Forschungszentrum Jülich, Germany

Special Session Chairs

Sonja Thiel	Baden State Museum, Germany
Christiane Lindner	Baden State Museum, Germany

Technical Program Committee

Marc Engenhart	Engenhart Design Studio, Germany
Veronica Ranner	Royal College of Art, UK
Marcus Gelderie	Aalen University, Germany
Tim Schlippe	IU International University of Applied Sciences, Germany
Chris Ziegler	Arizona State University, USA
Sebastian Löwe	Mediadesign Hochschule für Design und Informatik, Germany
Thorsten Zylowski	CAS Sofware AG, Germany
Tomas Balyo	CAS Sofware AG, Germany
Andreas Sieß	Hochschule Bonn-Rhein-Sieg, Germany
Ido Iurgel	Rhine-Waal University of Applied Sciences, Germany
Mel Krokos	University of Portsmouth, UK

Sue Gollifer	University of Brighton, UK
Melinda Braun	University of Applied Sciences Karlsruhe, Germany
Andreas Reich	University of Hohenheim, Germany
Fotis Liarokapis	Cyprus University of Technology, Cyprus
Knut Hartmann	University of Applied Sciences Flensburg, Germany
Jonas Deuchler	University of Applied Sciences Karlsruhe, Germany
Andreas Siefert	PONG.Li Studios, Germany
Daniel Feige	State Academy of Fine Arts Stuttgart, Germany
Florian Arnold	State Academy of Fine Arts Stuttgart, Germany
Christian Vater	Academy of Sciences and Literature Mainz, Germany
Karsten Wendland	Karlsruhe Institute of Technology, Germany
Robert Stock	Humboldt-University Berlin, Germany
Tabea Golgath	Stiftung Niedersachsen, Germany
Michael Klipphahn	TU Dresden, Germany
Paulina Dobroć	Karlsruhe Institute of Technology, Germany

Reviewers

Anak Agung Gde Satia Utama	Universitas Airlangga, Indonesia
Andres Iglesias	University of Cantabria, Spain
Anuja Hariharan	CAS Software AG, Germany
Artur Felic	CAS Software AG, Germany
Christian Felix Purps	University of Applied Science Karlsruhe, Germany
Christian Menschik	Furtwangen University, Germany
Christine Milchram	Karlsruhe Institute of Technology, Germany
Dominik Haunß	University of Applied Science Karlsruhe, Germany
Dominik Schreiber	Karlsruhe Institute of Technology, Germany
Ilia Bagov	CAS Software AG, Germany
Ingo Stengel	University of Applied Sciences Karlsruhe, Germany
Karin Pietruska	University of Applied Sciences Karlsruhe, Germany
Katharina Glück	University of Applied Sciences Karlsruhe, Germany
Marcus Gelderie	Hochschule Aalen, Germany
Markus Iser	Karlsruhe Institute of Technology, Germany
Michael Johansson	Kristianstad University, Sweden
Noemi Christensen	CAS Software AG, Germany
Patrick Hausmann	Hochschule Bonn-Rhein-Sieg, Germany
Peter Schuller	CAS Software AG, Germany
Silke Zimmer-Merkle	Karlsruhe Institute of Technology, Germany
Sebastian Stüker	Karlsruhe Institute of Technology, Germany
Sophia Schulze-Weddige	CAS Software AG, Germany
Thorsten Zylowski	CAS Software AG, Germany
Verena Wahl	Katholische Hochschule Freiburg, Germany

Contents

Fusions

Collaboration, Inclusion and Participation

Artificial Intelligence in Art and Culture

Media Arts and Virtual Reality

Digital Art and Dissipative Structures

Sijia Tao(✉) and Alain Lioret

INREV (Images Numériques et Réalité Virtuelle), AIAC (Arts des Images et Art Contemporain), University of Paris 8, 93526 Saint Denis, France

Abstract. By briefly introducing the theory of dissipative structures and its philosophical inspiration, this paper interprets and analyzes artworks directly inspired by the theory. It illustrates on the one hand the relationship between complex systems and digital art, and on the other hand, explains the basic conditions for self-organization. The latter is one of the characteristics of complex systems. While making a distinction with the theory of autopoiesis, we try to model certain digital art creations with several features of dissipative structures. These creations incorporate different materials, with an evolutionary approach, such as interactive artworks based on living plants, and on genetic algorithms. In this way, we demonstrate the value of investigating the self-organization process of dissipative structures within both the methodological and theoretical framework of interactive digital art.

Keywords: Complex science · Dissipative structures theory · Self-organization · Digital art theory · Human-computer interaction · Genetic algorithms

1 Introduction

In the 20th century, the emergence of complexity science profoundly influenced the transformation of humanities. In the 1940s, general scientific methodologies (e.g., system theory, information theory and cybernetics) achieved different aspects and strengthened the links between different disciplines. The development of self-organization theory in the late 1960s (whose main scientific methodologies include dissipative structures theory, synergetics and hypercycle theory) revealed a concern with the whole, and with the evolution of systems in scientific research. In the 1980s, research around complexity science concepts such as nonlinearity and emergence flourished.

These complexity science studies seeped into the arts at various times, driving new media art developments, especially digital art. Not to mention, of course, the influence of complexity science on today's research on the simulation of complex systems (e.g., artificial life systems, neural networks). For example, in the 1960s, "cybernetics and art" became popular. In robotic art, we can see that Tom Shannon's *Squat* (1966) is a complex cybernetic system: a living plant is connected to a robotic sculpture and the observer controls the motors of the sculpture by touching the plant [1]. Erich Jantsch summarizes the main ideas of the self-organizing paradigm at the time in *The Self-Organizing Cosmology* (1980): "*primo*, a specific macroscopic dynamics of process systems; *secundo*,

M. Wölfel et al. (Eds.): ArtsIT 2021, LNICST 422, pp. 3–17, 2022.
https://doi.org/10.1007/978-3-030-95531-1_1

continuous exchange and thereby co-evolution with the environment, and *tertio*, self-transcendence, the evolution of evolutionary processes" [2]. By the 1970s and 1980s, the concepts of evolution and co-evolution were receiving more attention. Here, in the potential of computer development, robotic art showed an interest in telepresence, as in Eduardo Kac's 1986 robotic performance *RC Robot*, radio-controlled telerobot can talk to visitors in real time [3]. In addition, self-organization has become one of the key words in research related to the term "artificial life" coined by Christopher Langton in 1987 [4]. The combination of artificial life and bio-art has contributed to the development of generative art, such as the work of Australian artist Jon McCormack, who has been working with artificial life and evolutionary systems since the 1980s.Take for example, in *Turbulence: An Interactive Installation Exploring Artificial Life* (1994), he used genetic algorithms[1] to demonstrate virtual species and a computer perspective on nature and our relationship with it [5].

The complexity science study, which deals with complex systems, takes a non-reductionist approach in different disciplines. The physical chemist Ilya Prigogine, a Brussels School representative, made an early contribution to the field with his dissipative structures theory. Furthermore, in their 1979 book *La nouvelle alliance: métamorphose de la science*, Prigogine and Isabelle Stengers introduced the notion of "complexity science", but Prigogine did not give a clear definition of "complexity". Different schools of thought hold different ideas on the concept of "complexity". Nevertheless, complex systems usually contain the following characteristics: "Nonlinearity, Distributedness, Scale and Interaction, Multiple Levels of Observation, Self Organization, Emergence, Adaptivity, Flexible Decision Making and Feedback Loops" [6]. And dissipative structures theory describes the phenomenon of self-organization that occurs in open systems as they interact nonlinearly with their environment, acquiring macroscopically stable structures.

Here, it is important to note that self-organizing system is easily confused with the autopoietic system that appeared in the same period. According to Hermann Haken [7], a self-organizing system is one in which the internal elements of the system and the external environment interact to acquire a spatio-temporal or functional structure, provided that the external world acts in a non-specific way and is not imposed on the system. Examples include the formation of crystals and the production of lasers. An autopoietic system, as defined by Francisco Varela and Humberto Maturana [8], refers to a network that maintains itself by replicating itself. This network contains circles of production process and constituent elements, such as living cells.

On the distinction between the two, Hideo Kawamoto [9] explains it by taking the example of crystals constantly being generated in a beaker. He says that if we consider the generation of crystals as a self-organizing system (in another case, the continuously generated crystals are regarded as the self-organizing system, and the solution in the beaker is the environment), the process of generation is the object of our attention. Once crystallization has taken place, then, in the case of a self-organizing system, the

---

[1] Genetic algorithms are a type of evolutionary algorithms. According to Holland's *Adaptation in natural and artificial systems: an introductory analysis with applications to biology, control, and artificial intelligence* (1992), this is an algorithm that solves optimization problems by modelling the biological evolution process and is an adaptive probabilistic search.

same type of generative process continues to occur. For the self-organizing system as a production process, the output crystals are like factory waste. If one were to describe the self-organizing system in terms of an autopoietic system, it is only "当析出的结晶能够再次生产出产生自我的生成过程的时候，出现了生命，自生系统才开始运动" (when the precipitated crystal regenerates the self-production process that life emerges and the autopoietic system will begin to move) [9]. It follows that self-organization focuses on the production process, while autopoiesis focuses on the maintenance of a circular network of production processes. In short, both focus on the self-reference, but while self-organization emphasizes the formation of new structures, autopoiesis emphasizes self-replication.

In this paper, our focus is on the production process of new structures. Given the weakness of autopoiesis in this aspect, the paper takes a dissipative structures theory approach. On the one hand, it helps us to understand what is necessary for a self-organizing system; on the other hand we will illustrate the relationship between the complex behavior of the structure in the production process and artistic creation, and its richness for digital art theory.

2 Dissipative Structures Theory

Dissipative structures theory contains evolutionary laws in complex systems within the study of nonlinear nonequilibrium thermodynamics. The theory is based on the "Minimum Entropy Production Principle" proposed by Prigogine in 1945. It was then introduced in his paper entitled "Structure, Dissipation and Life" in 1969. Almost simultaneously with this theory came another version theory of self-organization: the theory of synergetics. Established by Haken, synergetics "is concerned with the cooperation of individual parts of a system that produces macroscopic spatial, temporal, or functional structures" [10]. Both theories provide a theoretical framework for the interconnection of living and nonliving systems that possess conflicting laws. Dissipative structures theory states that these two systems - living and nonliving - are governed by the same systemic laws under certain conditions.

Dissipative structures, according to Prigogine [11], are dynamically stable and ordered structures formed by open systems far from equilibrium through the constant exchange of matter, energy and information with the outside world. More specifically, in the dissipative motion, when the change of external conditions reaches a certain threshold, the self-organization phenomenon is generated through internal actions – such as fluctuations and mutations: the system spontaneously changes from the original disordered state to the macroscopically ordered state. For instance, a famous dissipative structure - Benard convection [12]: heating the liquid from the bottom of a pan, when the temperature gradient reaches a critical value, a regular cellular convection of the liquid occurs. This theory is also called "Self-Organization in Nonequilibrium Systems" [13], of which the focus is to put forward the irreversibility of time and the study of self-organization phenomena.

After its introduction, the theory contributed to the development of complexity science research and expanded the theory to life, ecology, brain, meteorology, and philosophy. It is to be noted that the theory has shortcomings: it uses a local equilibrium approach [14] and its application is limited. At the same time, "在远离平衡情况下,怎样合理地定义熵和温度等基本的热力学变量,仍是个很困难的问题" (how to reasonably define basic thermodynamic variables such as entropy and temperature is still a complicated problem when far from equilibrium) [15]. However, the application of this theory is still broad, and under certain conditions, it is suitable for systems in different fields.

Although Prigogine's thinking "has had little or no effect on the 'textbook science' of late twentieth century (and indeed early twenty-first century) school science curricula" [16], the theory's interdisciplinary mode of thinking has had a heuristic impact on the humanities. In the philosophical context of thinking about chaos and order, for example, Manuel de Landa [17] argues that when armies adopt decentralized tactics that are task-oriented and leave the details of execution to subordinate organizations, they act as self-organizing dissipative structures: forming some "islands of stability" and thereby leading from chaos to order.

3 Dissipative Structures Theory, Nature and Art

Dissipative structures exist widely in nature, for example, hurricanes, tornadoes [18], mineralization structures, living organisms [14]. From this perspective, living and non-living are connected. In an interview about science and art, Prigogine says, "physics, by becoming a matter of probability and emphasizing the new and a certain indetermination in nature, produces a vision that emphasizes creativity. And creativity is the most important aspect of art" [19]. He wants to try to eliminate the contradiction between science, philosophy, and art. Starting from the different concepts involved in dissipative structures theory, some artists have thought about dissipative structures in the form of installation, photography, sculpture and video[2].

For example, artist Cameron Robbins is interested in wind, air, solar energy, tides and the earth's magnetic field. Inspired by the relationship between stability and instability in dissipative structures, in 2007, he used a smoke machine to control airflow to create a vortex-like phenomenon – named "Apparition"[3] (see Fig. 1), - in his site-specific art installation "Merricks beach house". The work is like a dynamic Chinese landscape painting of the Song dynasty, incorporating a grand and abstract concept into the continuous morphological changes of the rotating smoke: motion in stillness, Space is enveloped in an abstract mood of high mountains and flowing clouds of smoke. Using the vortex as a medium, the artist is very much concerned with the connection between dissipative structures and living beings, and even the whole of nature. In addition to this work, Cameron Robbins has represented the uniqueness of vortex structure in different ways in Double Vortex (2006), and Structure of Vortices (2012).

[2] See the works of artists such as Cameron Robbins, Andrew Beck, Laura Pesce and Mattia Casalegno.

[3] In 2008, in an essay entitled "Dissipative Structures - about the Vortex", the artist introduced some phenomena and features of dissipative structures and clearly stated that he had photographed this structure for the house project. Cameron Robbins, Dissipative Structures – about the Vortex, http://cameronrobbins.com/writing/dissipative-structures-about-the-vortex/, last accessed 2021/05/29.

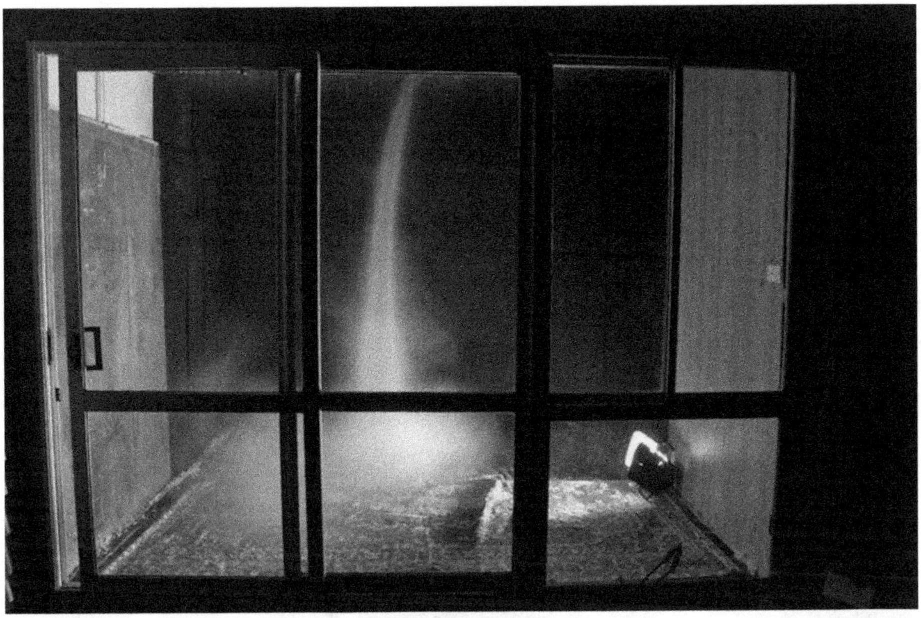

Fig. 1. Cameron Robbins, *Apparition, Smoke Room*: 26 Surf Street, Merricks beach house installation, 2007. (© Cameron Robbins.)

Dissipative structures, which are ordered on the macro level, including time, space or function, must constantly exchange matter and energy with the outside world. Cloud-streets are a type of cumulus that forms linearly with the direction of the wind [20]: clouds that were originally moving in a disorderly manner, under certain conditions, form neat columns and create a spatially ordered structure. It is therefore a common type of dissipative structure. During this process, clouds come alive. In 2018, photographer Andrew Beck used a set of photos called "Dissipative Structure I" and "Dissipative Structure II" respectively. This group of photos resembles computer-generated pictures, reminiscent of cloud-streets and cyclones' dissipative structures. The photographs highlight this "evolutionary dynamism" within the structure and suggest its existence between analog and digital.

4 Dissipative Structures and Digital Artworks

Similarly, out of this "evolution" or "between" thinking, in 2020, teamLab made an art installation called "Massless Clouds Between Sculpture and Life" (see Fig. 2). The installation is a way to think about "living" and "nonliving" from an entropy perspective, measuring the degree of chaos in a system in thermodynamics. They created self-organizing cloud masses that float in the air and gave them the ability to repair themselves.

Before the advent of dissipative structures and synergetics, nonliving systems followed the second law of thermodynamics and spontaneously changed from order to disorder. This reached a maximum entropy of the system, resulting in an irreversible process. The evolution of living beings is the opposite. Living and nonliving systems seem to be unrelated to each other. Nevertheless, Prigogine points out that living systems, unlike the conditions revealed by the second law of thermodynamics, are open and far from equilibrium, rather than isolated and in equilibrium or near-equilibrium. Under certain conditions, the system reduces entropy (emergence of a negative entropy flow) through the exchange of matter and energy with the environment. The proposed theory of dissipative structures suggests that there is no strict boundary between living and nonliving systems and that the same laws inherently exist. Erwin Schrödinger had stated in *What is life?* (1944), "What an organism feeds upon is negative entropy" [21]. In other words, through entropy reduction and self-organization to produce and maintain order to stay alive.

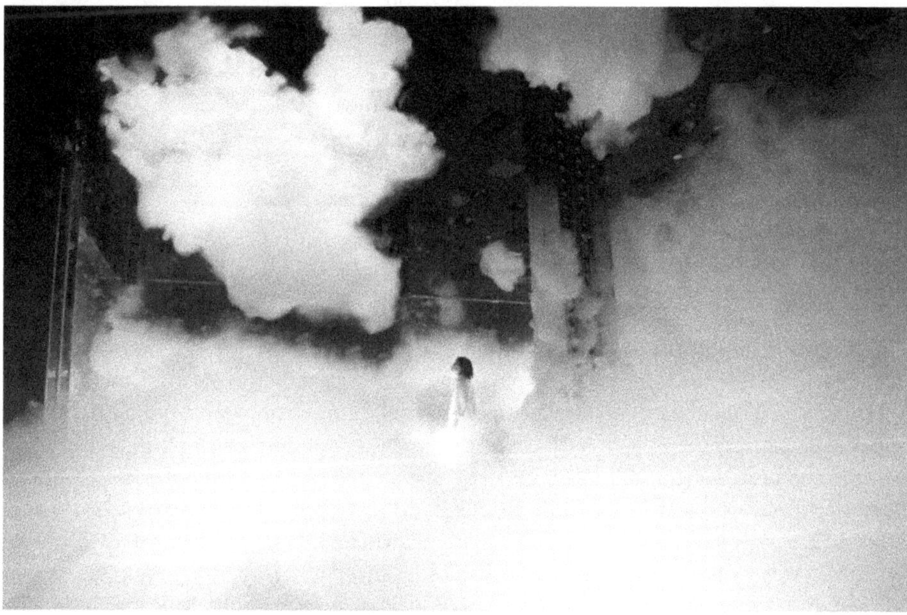

Fig. 2. TeamLab, *Massless Clouds Between Sculpture and Life*, 2020. (© TeamLab.)

TeamLab's work attempts to build a "dissipative system", with a particular emphasis on the interaction between life and the environment: "a giant white mass, floats (…) Even when people push through the sculpture and break it, it naturally repairs itself like a living thing. But, as with living things, when the cloud is destroyed beyond what it can repair, it cannot mend itself, and it collapses." [22]. Here it is necessary to give a brief explanation of the life cycle (if defined in terms of the time from condensation to dissipation) of a real cloud, using cumulus clouds as an example. The formation of cumulus clouds is based on the supply of thermals. Initially, the rise of invisible thermals produces a small wisp of cloud that is visible to an observer. Thermals then continue to flow into the cloud and out of the top into a mature cloud. For this structure to be maintained, mechanical energy needs to be dissipated and "the inflow of air from the subcloud thermal is assumed to be in balance with detrainment from the cloud into the environment" [23]. As thermals decay, the cloud begins to dry out and eventually dissipates. The formation of cumulus clouds reveals the visible presence of thermals.

In this interaction, we see that the material system in which the entire giant mass is embedded, is an open system. The space implied by the gaps between the inner masses and the observer as a representative of the external environment are the sources of information for this system. The changes in the masses caused by the involvement of the observer (traveling and or destroying) suggest the exchange of energy and information between the environment and the system. Also in this process, the degree of chaos within the floating giant mass system increases. Simultaneously, the interference of the observer intensifies the nonlinear positive feedback effects within the system, and amplifies the system's change mechanisms: the complexity within the system increases. This accelerates the system's self-organization, here manifesting as a self-healing property. For example, the giant mass system repair the parts of its organization that have been removed by the observer, bringing the system from short-term disorder to order while maintaining its "vitality". However, when the observer's interaction causes the giant mass to change beyond a certain critical range, the system's ability to self-organize will collapse and not return to its ante-interference state. This cloud system changes from order to disorder again, just like the eventual cessation of life.

Ecosystems are also common dissipative structures. When equilibrium occurs, the ecosystem also goes into death. Prigogine points out that nonequilibrium is a constant, that equilibrium is the state of a few, and nonequilibrium is the source of order. Near the threshold, a system in a nonlinear nonequilibrium state, can be subject to sudden changes in its state due to small disturbances. It occurs a bifurcation phenomenon and a re-ordered structure is formed. In advanced bifurcation phenomena, the self-organizing capacities of the sub-branches are combined, resulting in complex spatio-temporally ordered self-organization phenomena. Inspired by the questions of disorder and order, stability and instability in the theory of dissipative structures, Mattia Casalegno used the name "Strutture Dissipative"(see Fig. 3) to focus on the ecosystem from perspective of complexity theory. In the form of video projection in 2007, many particles and irregular shapes interacted with each other, evolving between symmetry and asymmetry, chaos and order. This generative artwork is a study on the combination of granular synthesis and chaotic particle systems techniques to develop live media performances and generative artworks [24]. Indeed, Philip Galanter [25] had long noted that systems combining order and disorder have long been the focus of generative artists, such as cellular automata, fractals, emergence, and other systems. As Galanter reveals, "systems are a defining aspect of generative art" [25] and suggests complexity theory as the theoretical context for systems-oriented generative art. In our understanding of the work *Strutture Dissipative*, or at least, in drawing first impressions, we also use the approach that complexity theory emphasizes for the study of systems: a combination of holistic and reductionist approaches.

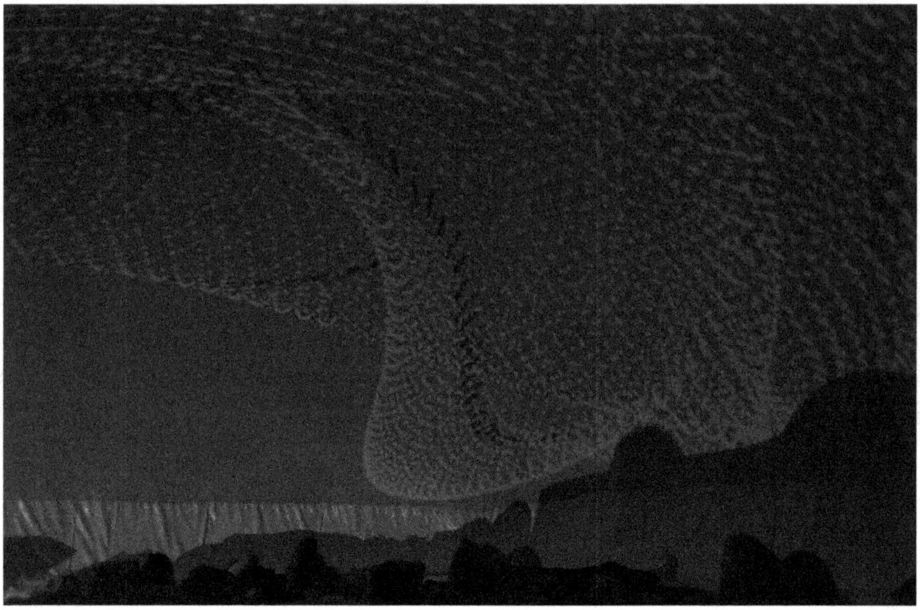

Fig. 3. Mattia Casalegno, *Strutture Dissipative*, Spherae, AxS Festival, Pasadena, CA, US, 2007. (© Mattia Casalegno.)

5 Discussion

The artists mentioned above seek to artificially create or metaphorically reproduce dissipative structures to express a concern about the evolution of nature and life. However, certain conditions need to be met in order to master this theory with restricted applications. Some of these key points are listed below:

1. The components of the system contain a large number of subsystems. Symmetry breaking[4] occurs in the system, which reflects nonlinear interactions within the system [27].
2. The system is open to the exchange of matter and energy with the outside world. In this way, the negative entropy flow injected into the system from outside, counteracts the entropy production within the system. Thus the system enters or maintains a relatively ordered state.
3. Far from equilibrium. That is, not a little bit away from equilibrium, but sufficiently away from it. It therefore needs to be distinguished from the near-equilibrium state. Because the fluctuations of the latter will be consumed and return to the average value, which makes the fluctuations seem irrelevant.
4. Bifurcations, fluctuations and mutations. Unlike evolutions of equilibrium and near-equilibrium states, there are multiple possibilities for the evolution of a system far from

[4] "Symmetry breaking", is not a term that refers to the absence of symmetry. According to Ian Stewart and Martin Golubitsky [26], it describes a certain changing relationship between an entire symmetry group and its subgroups.

equilibrium [27]. Bifurcations are one of the keys. To quote Prigogine, these "are the manifestation of an intrinsic differentiation between the parts of the system itself and the system and its environment" [28]. At the bifurcation point, random small fluctuations are amplified and produce mutations, allowing the system to acquire new macroscopic states.

Furthermore, the stability of dissipative structure is maintained by the balance of these nonlinear interactions and dissipation, constantly repeating the production process: always in a state of dynamic nonequilibrium. However, "过强的非线性相互作用或耗散作用都会使结构遭到破坏" (too strong nonlinear interactions or dissipative effects can damage the structure) [27].

When we re-examine the previously mentioned kinetic artworks in these conditions, we can see that Cameron Robbins' *Apparition* is a dynamic structure: a constant input of smoke that alters the surrounding air system, creating tiny airflows, which in turn alter the surrounding air. The fluctuations of airflows are amplified by the exquisite control of the smoke machine. Under certain invisible conditions, a vortex structure is created and its creation process is repeated. This work thus directly shows the generation of new structures and suggests their maintenance. Regarding the work of teamLab, as analyzed in more detail in the previous section, we find that this work mainly represents the maintenance of macro-structural stability and the collapse of the structure as a result of excessive nonlinear interactions. We should add here that the immersive involvement of observers and the collapse of the giant mass in this work actually reflects the dissipative structures theory's view that man's perception of nature takes place in the context of an irreversible process of natural evolution [28]. In fact, the observer's position is indispensable when talking about self-organization, as W. Ross Ashby notes that "A substantial part of the theory of organization will be concerned with properties that (…) are relational between observer and thing" [29]. In the works of Cameron Robbins and teamLab, the main components are visible to the observer. However, the production process of Mattia Casalegno's work is encapsulated in the algorithm. Thus, although it represents the evolution between disorder and order in a nonequilibrium state, it is still difficult to state, on strictly scientific terms, whether the graphs formed therein are dissipative structures or not. Through analysis we see that Cameron Robbins creates a dissipative structure and that the works of teamLab and Mattia Casalegno can only be considered to simulate, to a large extent, the self-organization of dissipative structures.

As we have emphasized repeatedly, dissipative structures theory describes a type of self-organizing phenomenon. We now return to the question at the beginning of the paper: what are the necessary conditions for a self-organizing system as learned through dissipative structures theory? According to Kawamoto's summary [9], the necessary conditions for a self-organizing system: nonequilibrium open system, changes in the system's boundary conditions, and uncertain factors within the system. The interaction with the environment that is implied by nonequilibrium open system, the uncertain nonlinear effects within the system and the fluctuations that have a triggering function, are all reflected in the works we have just reviewed. However, the boundary conditions do not seem to be deliberately emphasized in these works, but rather merge with the environment.

Gilbert Simondon [30] has made a distinction between dissipative structures as living organisms and purely physical ones. He pointed out that a purely physical dissipative structure cannot control certain external conditions and will die out as the boundary conditions disappear. Dissipative structures as organisms, on the other hand, have the ability to regulate boundary conditions and, therefore, may lead to immortality. Stuart Kauffman, in *Answering Schrödinger's "What Is Life?"*, states that "Organisms, as we shall see, *do* construct their own boundary conditions and do this by carrying out thermodynamic work to construct the very same boundary" [31]. In contrast, dissipative structures "do *not* construct their own boundary conditions" [31]. However, we need to point out that the boundary conditions still determine the shape of the dissipative structure. Kauffman explains that in Benard Convection, the shape of the pan as a container constitutes the boundary conditions. A change in the shape of the pan changes the shape and the macro pattern produced by this convection. The shape of the Benard convection can therefore be, for example, hexagonal or rolling [31].

Indeed, we can see that a strict application of these scientific conditions would make artistic creation difficult. In general, the artists mentioned above focus mainly on the philosophical inspiration inspired by the theory of dissipative structures: the intrinsic connection between the living and the nonliving, the diversity and coherence brought about by the evolution in stability and instability. This scientific theory exists here above all as a framework for the inspiration of the artists' works.

As we mentioned before, self-organization emphasizes production processes, which coincides with digital art, appealing to the nature of process. Can we then use dissipative structures theory to model certain process-oriented artistic creations, providing an interpretive tool for studying them? Or, within the framework of digital art theory and creative methods, what features of the theory are worth drawing on?

6 Dissipative Structures Theory and Digital Art Theory

The combination of complex systems and artistic creation is very common. In addition to the generative art mentioned earlier, which is closely related to complex systems, there is also the borrowing and application of autopoiesis systems in the context of art. In the framework of art theory, Niklas Luhmann, for instance, considers "the artworld" as an autopoietic social system: an operationally closed, self-referential system [32]. In artistic practice, John Mark Bishop and Mohammad Majid al-Rifaie, for example, apply the autopoiesis model to artistic creativity systems and create a drawing autopoietic artist model based on a swarm intelligence system [33].

Autopoietic systems, sometimes translated as self-producing systems. According to Kawamoto [9], the 'self' of autopoietic systems is the extent to which the system delineates its own movement, rather than being artificially specified by the observer. In contrast to the control of boundaries and the maintenance of self in autopoietic systems, in dissipative self-organization, the boundaries are constantly changing and the system's self is also continuously being re-established as the system interacts with its environment. Whether the system controls or adapts and regulates its boundaries is one of the differences between dissipative structure's self-organization and autopoiesis, and the uniqueness of the dissipative structure model in the approach to digital art creation.

If, then, the dissipative structure model is placed within the framework of artistic creation, we attempt to summarize several points:

1. Starting from the kinetic artworks above, at least two modes of creation are included. One, a self-creation method that does not require interaction with the observer. The artist outsources the creative process to the machine: the latter as a creative agent, uses the dissipative structure model as a creative framework to produce artwork with corresponding complexity features and encapsulates this process.
2. Two, the observer, participant in the work's creation, becomes the main variable and even a necessary condition of the external open environment of the object of our attention. Through the interaction with the external environment, the creation is triggered to take place. Artists not only use this interaction to contribute to the production of the work, but also make the dynamic of this production a property of the work by repeatedly embedding it in the process, so that the work exhibits a seemingly random, complex and original response to the environment. This iterative process reinforces the connection between the environment and the object of our attention, while at the same time demonstrating the object's adaptability to the environment, even as the boundaries between the two are constantly blurred, giving rise to the sense of immersion such as the observer-cloud interactions in teamLab's work.

In fact, it is easier to find similar creative modes in works of human-plant (living plant) interaction. One of the obvious reasons is that living plants, as living organisms, are typical dissipative structures. The participation of the observer, and the real environment in which the plant grows, can produce nonlinear effects within the plant. In other words, every change in environmental factors affects the growth of plants [34]. Appropriately electrifying plants is one of the most common methods for human-plant interaction. On the one hand, this affects the plant's photosynthesis, metabolic processes and other nonlinear internal interactions [34]. On the other hand, human interactions intensify the changes within the plant. In such interactions, these changes can therefore be revealed to some extent with changes in the electrical signals measured on the surface or inside the plant[5]. In some cases, plants exhibit a kind of assertion of their agency, as stated by the anthropologist Natasha Myers, mentioned by Charles Ryan John [36]: plants that appear unresponsive or delayed in their response during artistic processing. All these changes underline that living plants are not equivalent to sensors from which data measured can be easily quantified and accurately predicted. Generally speaking, in the interactive installations with living plants as the main subject, the production of a certain artistic creation caused by the plant is the object of our attention. The observer and the real environment such as the soil and air required by the plant itself together constitute the external environment for the production of this creation.

For Instance, Christa Sommerer and Laurent Mignonneau's *Interactive Plant Growing* (1992) [37] uses living plants as the medium through which the approach or touch

[5] There are variations in the methods required for different plants, or for different types of electrical signals. In plant electrophysiology, for example, to measure current changes at the cellular level, amplifiers are often needed to augment these electrical signals for artistic processing [35].

of the observer triggers the growth of virtual plants, which in turn influences the interaction of the observer, resulting in the unique qualities of the images on the screen in real time in response to the interaction. It is important to emphasize that the interaction of the observer is necessary for the existence of the work. Through the observer, the living plants extend themselves into the virtual space and change endlessly. Although the plants themselves exist here as specific materials with electronic properties, in terms of methodology, the work realizes an artistic exploration of the evolution of life through plants.

3. In his presentation of the relationship between endophysics - a kind of physics of observation from the inside out - and art, Austrian artist and new media art theorist Peter Weibel argues not only that the observer, combined with the environment in the dynamic model[6] of post-ontological art[7], constitutes a dissipative structure, but also that "Genetic algorithms that are able to separate the image from the observer-controlled context will constitute another dissipative structure" [38]. According to the previous two points, we can examine the application of evolutionary algorithms with dissipative structure features to artistic creation. Sommerer and Mignonneau's installation *A-Volve*, as an example, is based on a genetic algorithm [37]. Instead of a predetermined direction of evolution, they simulate natural biological evolution through selection, crossover and mutation. In the work, virtual creatures are created in two ways: those painted by the observers or those generated by the mating of the creatures themselves. They exist in a real environment: a glass pool filled with water. Their births, behaviours and deaths are influenced in real time by the interaction of the observer, the real pool environment and the interaction between the creatures. This community of creatures becomes a complex open system of continuous evolution. The choice and interaction of observers serve as an external environmental intervention that significantly influences the adaptation of creatures to the pool environment, amplifying the fluctuations of the community and, under the appropriate conditions, driving structural changes in the community. The work sets out that "The fittest creature will survive longest and will be able to mate and reproduce" [37]. By observing the evolution of the community of creatures, the observer is also constantly adjusting his choices and creations to select more suitable creatures or to destroy the whole community. In this work we clearly can see interactivity as an expression of the system's adaptability to its environment and as possible results that will lead to the creation of new structures.

From the points above, we are more interested in the inspiration that comes from the creations which interact with the observer (the second and third points). Because the features of the dissipative structure model are more explicitly represented in it, and enrich the form and content of the work. Although it is a challenging task to fully apply the dissipative structure model to an artistic approach, if we model artistic creation in terms of some of its features, then: in such works, the interaction of the external environment is a necessary condition for creation to take place. The environment and the work

[6] Weibel points out that in this model, the observer, the interface and the environment are covariant.

[7] Weibel mainly refers to interactive new-media art in the network.

establish a highly sensitive relationship in the development of the work: their interaction not only contributes to diversity, but also serves to maintain the life of the work. This process of creation is irreversible, irreducible to mechanical decomposition, and irreproducible. Moreover, in this type of work the object of our attention demonstrates the ability to adapt and regulate to its environment while at the same time exhibiting not only an immersion or permeability, but also a malleability and adhesion that allow for the organic integration of different types of materials. Within this dimension, Roy Ascott [39] refers to the media of bits, atoms, neurons and genes as the "Moistmedia", and this "Moist environment, located at the convergence of the digital, biological and spiritual, is essentially a dynamic environment, involving artificial and human intelligence in nonlinear processes of emergence, construction and transformation" [39]. As we can see from the previous analysis, the application of dissipative structure model to artistic creation in our hypothesis involves similar areas and processes. The prospects for the application of the model in interactive art creation will then also be explored in the framework of Moistmdia[8].

7 Conclusion

This paper is oriented towards artistic creations based on complex systems. By introducing the key ideas of dissipative structures theory, interpreting and analyzing artworks directly inspired by the theory, we briefly illustrate the influence of complex systems on digital art and explain the basic conditions for self-organization, one of the characteristics of complex systems. However, by re-examining these works on scientific terms, we find that the complete application of the dissipative structure model by the artist in the combination of science and art is a challenging task. Nevertheless, while making a distinction between self-organization and autopoiesis, we show that the emphasis on new structures and the regulation of boundary environments in the self-organization of dissipative structures imply a certain uniqueness that makes dissipative structures theory valuable to study within the methodological and theoretical framework of interactive digital art creation. To this end, in the context of case studies, we believe that some of the features of dissipative structures can be used to model certain digital art creations that have an evolutionary approach. In particular, interactive digital artworks based on living plants, and on genetic algorithms. At the same time, we point out the ability of this model to organically combine a variety of materials, and hence infer that the study of this model is suitable for the field covered by Moistmedia.

Nonetheless, this paper only presents the main ideas of dissipative structures in general, lacking a more detailed exploration of their characteristics in the scientific field and in artistic theory. Given that the paper provides only a cursory generalization of the creative approach and is limited to the context of interactive creation, the suggested conditions for modelling and the assumption of the model's applicability to the field of study are rather frivolous. In our next work, we will continue to focus on the framework

[8] On the convergence of plants, digital technology and art, Ryan [36] proposed three theoretical frameworks, which also include the field of Moistmedia. The two others are: "human-plant studies" [40], and Warwick Mules's concept of "Poiesis" [41].

of human-plant interaction, but will investigate more deeply the relationship between dissipative structures, art theory and creative methods.

References

1. Kac, E.: Foundation and development of robotic art. Art J. **56**(3), 60–67 (1997)
2. Jantsch, E.: The Self-organizing Universe: Scientific and Human Implications of the Emerging Paradigm of Evolution. Pergamon, Turkey (1980)
3. Kac, E.: Telepresence & Bio Art: Networking Humans, Rabbits, and Robots. University of Michigan Press, Ann Arbor (2005)
4. Gershenson, C., Trianni, V., Werfel, J., Sayama, H.: Self-organization and artificial life. Artif. Life **26**(3), 391–408 (2020)
5. McCormack, J.: TURBULENCE an interactive installation exploring artificial life. In: Visual Proceedings: The Art and Interdisciplinary Programs of SIGGRAPH, vol. 94, pp. 182–183 (1994)
6. Davidsson, P., Klügl, F., Verhagen, H.: Simulation of complex systems. In: Magnani, L., Bertolotti, T. (eds.) Springer Handbook of Model-Based Science. SH, pp. 783–797. Springer, Cham (2017). https://doi.org/10.1007/978-3-319-30526-4_35
7. Haken, H.: Information and self-organization A macroscopic approach to complex systems. Springer, Cham (2006). https://doi.org/10.1007/3-540-33023-2
8. Varela, F.G., Maturana, H.R., Uribe, R.: Autopoiesis: the organization of living systems, its characterization and a model. Biosystems **5**(4), 187–196 (1974)
9. Kawamoto, H.: Otopoiesisu: Daisan Sedai Sisutemu ,Seido-sha Publishers (1995) (Di san dai xi tong lun:zi sheng xi tong lun, Guo Lianyou (translator), Central Compilation & Translation Press (2016)
10. Haken, H.: Synergetics. Naturwissenschaften **67**, 121–128 (1980)
11. Prigogine, I., Stengers, I.: Order Out of Chaos: Man's New Dialogue with Nature. Verso Books, London (2018)
12. Prigogine, I.: From Being to Becoming. W. H. Freeman and Company, New York (1980)
13. Prigogine, I., Nicolis, G.: Self-organisation in nonequilibrium systems: towards a dynamics of complexity. In: Hazewinkel, M., Jurkovich, R., Paelinck, J.H.P. (eds.) Bifurcation Analysis, pp. 3–12. Springer, Dordrecht (1985). https://doi.org/10.1007/978-94-009-6239-2_1
14. Kondepudi, D., Prigogine, I.: Self-organization and dissipative structures in nature. In: Kondepudi, D., Prigogine, I. (eds.) Modern Thermodynamics: from Heat Engines to Dissipative Structures, pp. 477–486. Wiley (2014)
15. Ai, ST.: Fei ping heng tai re li xue gai lun (Di er ban). Tsinghua University Press, Beijing (2017). (Introduction to Nonequilibrium Thermodynamics (Second Edition)
16. Gough, N.: Watchmen, simultaneity, and postmodern science education : the medium and its messages. In: Graphic Novels and The(ir) World. The Graphic Novel Project: 4th Global Meeting, Dubrovnik (2015)
17. De Landa, M.: War in the Age of Intelligent Machines. Zone Books, New York (1991)
18. Prigogine, I., Nicolis, G.: Self-Organization in Non-Equilibrium Systems. Wiley, Hoboken (1977)
19. Obrist, H.U.: Science and art: a conversation with Ilya Prigogine. In: Review (Fernand Braudel Center), vol. 28, no. 2, pp. 115–128, Discussions of Knowledge (2005). Research Foundation of State University of New York for and on behalf of the Fernand Braudel Center (2005)
20. Schneider, S.H.: Encyclopedia of Climate and Weather, vol. 1. Oxford University Press, Oxford (2011)

21. Schrodinger, E.: What is Life? (1944). http://old.biovip.com/UpLoadFiles/Aaron/Files/2005051204.pdf. Accessed 29 May 2021
22. TeamLab, Massless Clouds Between Sculpture and Life (2020). https://www.teamlab.art/w/clouds/. Accessed 09 May 2021
23. Heus, T., Jonker, H.J., Van den Akker, H.E., Griffith, E.J., Koutek, M., Post, F.H.: A statistical approach to the life cycle analysis of cumulus clouds selected in a virtual reality environment. J. Geophys. Res.: Atmos.**114**(D6) (2009)
24. Casalegno, M.: Strutture dissipative exp. 2 (2007). https://vimeo.com/761810. Accessed 29 May 2021
25. Galanter, P.: What is generative art? Complexity theory as a context for art theory. In: GA2003–6th Generative Art Conference (2003)
26. Stewart, I., Golubitsky, M.: Fearful Symmetry. Dover Publications Inc., Mineola (1992)
27. Wu, J.: Kua xue ke yan jiu yu fei xian xing si wei. China Social Sciences Press, Beijing (2004). (Interdisciplinary Research and Nonlinear Thinking)
28. Prigogine, I., Stengers, I.: The End of Certainty: Time, Chaos, and the New Laws of Nature. Free Press, New York (1997)
29. Ashby, W. R.: Principles of the self-organizing system. In: Systems Research for Behavioral Sciencesystems Research, pp. 108–118. Routledge (2017)
30. Simondon, G.: L'individuation psychique et collective. Aubier, Paris (1989)
31. Kauffman, S.: Answering Schrödinger's "What is life?" Entropy **22**(8), 815 (2020)
32. Luhmann, N.: Art as a Social System. Stanford University Press, Stanford (2000).Translated by Knodt, E.M.
33. Bishop, J.M., Al-Rifaie, M.M.: Autopoiesis in creativity and art. In: Proceedings of the 3rd International Symposium on Movement and Computing, pp. 1–6 (2016)
34. Dannehl, D.: Effects of electricity on plant responses. Sci. Hortic. **234**, 382–392 (2018)
35. Sommerer, C., Mignonneau, L., Weil, F.: The art of human to plant interaction. In: The Green Thread: Dialogues with the Vegetal World, p. 233 (2015)
36. Ryan, J.C.: plant-art: the virtual and the vegetal in contemporary performance and installation art. Resil.: J. Environ. Humanit. **2**(3), 40–57 (2015)
37. Sommerer, C., Mignonneau, L.: Art as a living system: interactive computer artworks. Leonardo **32**(3), 165–173 (1999)
38. Weibel, P.: The world as interface: toward the construction of context-controlled event-worlds. In: Druckrey, T. (ed.) Electronic Culture: Technology and Visual Representation, pp. 341–351. Aperture Foundation, New York (1996)
39. Ascott, R.: Edge-life: technoetic structures and moist media. In: Ascott, R. (ed.) Art, Technology, Consciousness: Mind@ large, pp. 2–6. Intellect Books (2000)
40. Ryan, J.C.: Unbraided Lines: Essays in Environmental Thinking and Writing. Common Ground Publishing, Champaign (2013)
41. Mules, W.: With Nature: Nature Philosophy as Poetics Through Schelling, Heidegger, Benjamin and Nancy. Intellect Books, Bristol (2014)

Web-Mindscape and *REFLEXION – In Sync/Out of Sync –*: Biofeedback and Physical Computing in Inter-active New Media Art

Claudia Robles-Angel[2], Andreas Gernemann-Paulsen[1(✉)], and Uwe Seifert[1]

[1] Institute of Musicology, University of Cologne, Cologne, Germany
{andreas.gernemann,u.seifert}@uni-koeln.de,
post@claudearobles.de

[2] New Media and Audiovisual Artist, Cologne, Germany

Abstract. The article draws attention to physical computing as a method for designing interactive biofeedback systems in New Media Art. *Web-Mindscape* and *REFLEXION – In Sync/Out of Sync –*, two biofeedback installations by artist Claudia Robles-Angel serve as examples. We briefly describe the artist's intentions and give a detailed report of the technical realization of both installations in the context of physical computing. From a scientific perspective of physical computing and interactive biofeedback installations, we touch upon combining art and research in the form of art-as-science.

Keywords: New media art · Biofeedback · Physical computing · Interactive systems · Art-as-science

1 Introduction

1.1 Biofeedback and Immersion in Interactive New Media Art

Biofeedback methods developed in the 1960s by psychologist Neal E. Miller is a process that consists of measuring physiological parameters from a subject, e.g. heartbeats, brainwaves, breathing or skin resistance. Such methods were introduced in artworks (specifically in music performances) in 1965 by Alvin Lucier with his composition 'Music for Solo Performer – for Enormously Amplified Brain Waves and Percussion' [12]. For this purpose, Lucier used equipment for Electroencephalography (EEG) that belonged to the US Air Force [8]. From then on, other artists followed, e.g. Richard Teitelbaum's 'Organ Music' and 'IN TUNE', both from 1968. In these works, EEG signals control voltages in Moog synthesizers. The usage of Biofeedback in the past fifty years "has experienced substantial change, due to the rapid evolution of new technologies in the same period. The consequence of this is that nowadays we can find in the market a wide range of small and wearable interfaces for affordable prices, including also self-built and open source" [12].

M. Wölfel et al. (Eds.): ArtsIT 2021, LNICST 422, pp. 18–28, 2022.
https://doi.org/10.1007/978-3-030-95531-1_2

Physical computing can be understood as a system design method which plays an important role in the development of artistic interactive biofeedback systems.

The immersive character of the two artistic works described in this article is related to what Schacher describes as: "being submerged or enveloped, usually in water. In media arts and theory this term has been extended to mean envelopment by mediated contents, be they visual, sonic, or sometimes tactile" [18]. Furthermore, Grau conceives it as a situation of presence: "The media strategy aims at producing a high-grade feeling of immersion, of presence (an impression suggestive of 'being there')". [19]. The concept of immersion in these artistic works is related to spaces that surround visitors/performers with sound and light, which create an audiovisual environment with which they can interact with their bio-signals by measuring their inner states through respective corresponding interfaces: BCI (Brain-Computer Interface; cf. Miranda & Castet 2014 [20]) and pulse sensors, which makes them feel mentally and corporeally involved. Furthermore, such combination of bio-signals within an immersive atmosphere accentuates the feeling of "being there" [19] and accordingly, it accentuates the feedback between the audiovisual environment and the internal emotional states of performers and/or visitors. Although our two examples *WEB-Mindscape* and *REFLEXION – In Sync/Out of Sync* – implement two different types of bio-signals, they both nevertheless share a clear immersive character for audiences and/or performers alike.

1.2 Physical Computing in Interactive Biofeedback New Media Art Installations

Physical computing has revolutionized interactive media art. Based on hacking ideas [17], the Italian Arduino boards and their accessories nowadays extensively offered by third-party suppliers enable an iterative and straightforward approach to interactive New Media Art projects [5]. The artistic use of special hardware and software, which have a connection to the physical world, is a core aspect of physical computing [2, 3]. This coupling is enabled typically with microcontrollers so that sensors detect and actuators respond and change parts of the physical world. At the same time, practical artistic aspects and the simple handling of inexpensive computer technology are basic concepts. A characteristic element of physical computing is the iterative tinkering[1] process as experimental prototyping [2, 3, 5], making realizing specific components for installations possible. Also, biofeedback needs specific sensors and fastening, for example, wearable devices and modules to transmit sensor data to other devices or software interfaces. Moreover, the Arduino boards control actuators like performative lighting and sound [6]. Hence, physical computing is ideally suited for artistic biofeedback installations. In the following sections, our focus lies on the description of artistic intentions and, in particular, the technical realization of Claudia Robles Angel's above mentioned two installations in which physical computing takes a significant impact.

1.3 *WEB-Mindscape* and *REFLEXION*: Artistic Intentions and Challenges

Artistic intentions in both biofeedback installations are based on the usage of the human body as a visual and musical instrument [12], in order to enhance the body's potential

[1] Tinkering in the context of physical computing means an exploring, trying and iterative approach using IT (esp. microcontrollers) to enable and stimulate artistic creativity [2, 3, 5, 15].

by perceiving it from a different perspective, which creates a new type of relationship between performers, space, audience and media technology. Both projects bring together various interests: in addition to sound art and its combination with visual media, the interest is in the impact of combinations of human emotions and physiological parameters on digital art. Her main aim is to create a new type of relationship between sound, architecture, light and human beings in a diversity of art situations, for example, interactive installations.

The creation of interactive biofeedback installations, which are different from performances, raises a challenge and the question about which interface should be used or developed in order to invite participants to create an audiovisual environment with their emotional states. Another challenge in using bio-signals in interactive installations is that interfaces for biofeedback usually require complicated methods to attach them to the participants (for example, gel for each electrode, utmost precision in positioning electrodes, etc.). So these interfaces are difficult to implement with large audiences of participants in interactive installations [13].

2 *Web-Mindscape* and *REFLEXION – In Sync/Out of Sync*: Brain-Computer Interface, Electrocardiogram, and Electroluminiscence in Interactive New Media Art

2.1 Web-Mindscape

Web-Mindscape is a interactive installation for brainwaves, light, sound and tweets using electroluminescent (EL) wires and a BCI. This installation joins diverse aspects, such as social networks, sound, brainwaves and visual elements. It creates a site-specific immersive audiovisual environment, where sound is diffused in surround, and the visual elements consist of light produced via electroluminescent wires (EL wires). Participants are immersed in a luminous structure, surrounded by light cables and sound, the latter diffused in eight audio channels, creating an immersive audiovisual environment.

Visitors are invited (one at a time) to interact with the audiovisual environment (light and sound) by using a BCI interface, which measures their brain activity. Thereby, they are confronted with messages from a social network (Twitter) worldwide. Simultaneously, the worldwide community is invited to join an additional Twitter account. All these tweet messages are turned into an audible sound. After that, the computer measures the visitors' cerebral activity and analyses their emotional reactions to the environment and the tweets. This data is transformed into visual and audible signals, which reproduce how the inner of the subject is influenced by the outer environment, while impacting the installation's audiovisual environment.

This work was developed and firstly exhibited during an artist in residence in in 2016 thanks to a grant offered by the IK foundation in Stichting/The Netherlands. The current version was presented in May and June 2017 at Harvestworks – Digital Media Arts Center in New York City for three days, added to two full days at the ISEA 2017 – International Symposium on Electronic Art in Manizales, Colombia (Fig. 1).

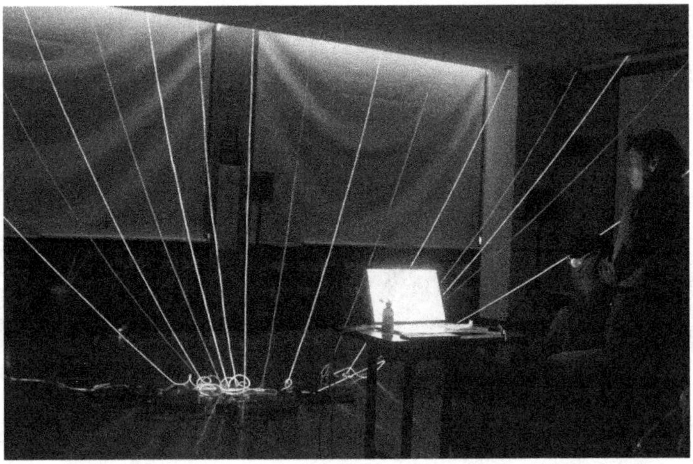

Fig. 1. *Web-Mindscape* at ISEA 2017, Manizales, Colombia. ©Claudia Robles-Angel/VG Bild und Kunst. Photo: Image Festival Manizales. See also https://vimeo.com/225400078

Web-Mindscape's Brain-Computer Interface: EMOTIV

Previous works of Claudia Robles-Angel in physical computing used the open-source EEG interface from Olimex [11], and other commercial BCI headsets were used in the work of other artists. *Web-Mindscape uses* the EEG interface EMOTIV *Insight*. This interface is developed for health and well-being purposes which consists of five polymer sensors that absorb humidity from the air, thus in opposition to the Olimex, does not require the application of gel or saline solutions and is therefore suited well for usage in art installations. On the other hand, this interface does not work as accurately as the types discussed above. However, the EMOTIV *Insight* is connected via Bluetooth to the computer and has software that helps the user check the connection of the electrodes to the scalp during the set-up process (see Fig. 2).

In order to send values from the BCI interface to MAX (an audiovisual and object-oriented software by *Cycling'74*), which is responsible for the audiovisual environment, it is necessary to find additional software to read and translate raw data from the interface. After exploring several possibilities, the software *MindYourOsc* was selected, which divides the EEG data into five basic emotions: Engagement/Boredom, Frustration, Meditation, Excitement and Excitement Long Term. Although these five emotions are used at the start of the development, test persons could not entirely understand how their brain activity influenced the audiovisual environment. For this reason, the set is reduced to only two emotions: Meditation and Excitement, inviting in this way visitors to participate by controlling the interactive space with these two opposite emotions [13].

Web-Mindscape's Visual Environment

A light structure made by electroluminescent wires (EL wires) creates the visual environment. This wired structure is set according to the characteristics of the installation space (i.e. adaptable to different venues, provided it is dark). The data of a subject's brain waves from the BCI interface turn on/off different cables and in different tempos depending on the two opposite emotional states using MAX.

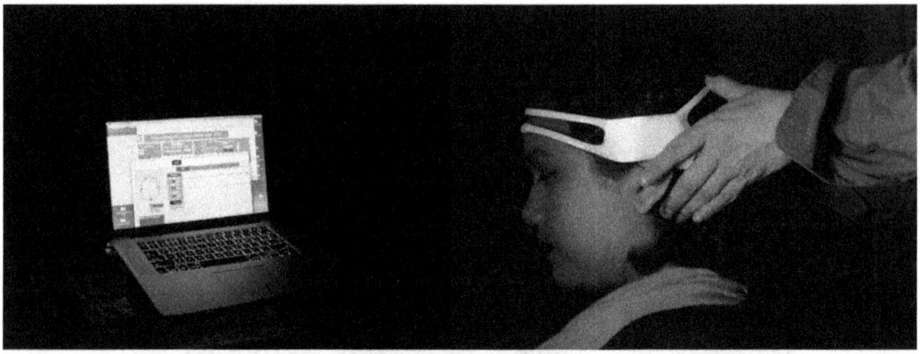

Fig. 2. EMOTIV software during the set-up process.

A ready-to-use solution for remote controlling the two 16 EL wires of 3 m each and two ELs of 25 m (34 wires and about 210 m in total) via MAX is unavailable. Thus, appropriate prototypes had to be developed in order to fulfil such artistic intention. In particular, the frequent activating of different wires, which results in various lengths of constantly glowing cables, was a challenge. A solution had to be found in an iterative tinkering process so that a realization based on Arduino boards was developed in the context of physical computing. Firstly, different serial values corresponding to the smooth dynamic states of the installation were generated in MAX and then sent via a virtual serial interface to the USB port of the host computer. Finally, two (resp. three) boxes connected to this USB port were created for the EL-control, each with a shield tower consisting of an Arduino-Uno board, two custom-made intermediate boards for adaptation and two optocouplers Escudo-Dos shields by *Sparkfun*, which drive the EL wires via triacs (see Fig. 3). Additionally, DC resp. AC to AC converters, also called as EL-inverters were used to power the EL wires with the required high AC voltage. Since there were no inverters available at the time of development that could supply such a wide range of resulting cable lengths, four of them were installed per box, so each were assigned to a group of four wires (see Fig. 4). Therefore, the Escudos had to be rebuilt to be in a position to control several inverters.[2]

***Web-Mindscape*'s Sound Environment**

The sound section of the work consists of a surround soundscape (eight independent audio channels), which changes depending on the information coming from the visitor's EEG.

There are two primary sound sources. the first is a balanced and subtle soundscape composed of frequencies from the brain waves combined with a field recording, which is activated when the participant is relaxed. The second source is the sound conversion of texts sent via Twitter, which are converted into sound by a text-to-speech algorithm inside the MAX patch. Once converted, this synthetic voice is used as sound material to which diverse sound effects are applied, such as, for example, Granular synthesis.

[2] Modern types of inverters are available that can partially handle such a bandwidth as well provide a stable output voltage. The current version of the *Web-Mindscape* boxes is equipped with these newer inverters.

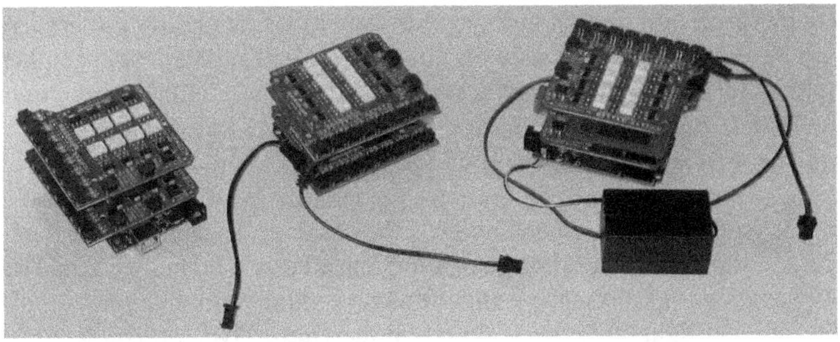

Fig. 3. Arduino Uno boards with shields used in *Web-Mindscape*.

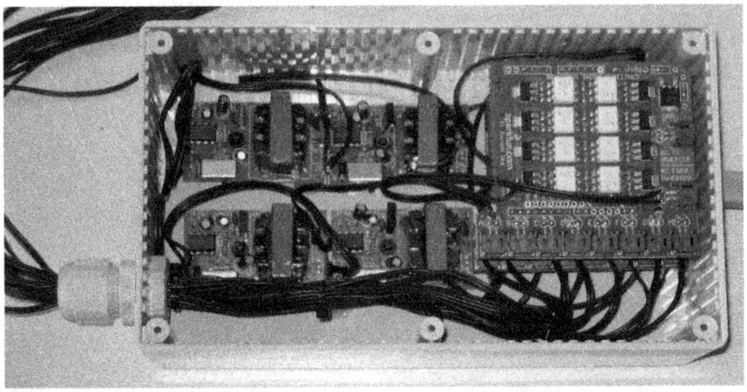

Fig. 4. One of three EL-wire control boxes with Arduino, shields and inverters.

The subject's brain activity modifies the parameters of the sound effects, and this sound is activated when the subject's relaxed condition is altered (Excitement state), which creates a sonic environment made of words as whispers, and which increases its level of complexity in a degree dependent on the data received from the brain activity.

2.2 *REFLEXION – in Sync/Out of Sync –*

REFLEXION – In Sync/Out of Sync – is an interactive installation (with an additional performance) consisting of (EL) wires and octophonic sound, in which visitors are encouraged to create together a light environment. Inspired by *Polytopes* [14] by Iannis Xenakis, the installation explores the relationship between sound, light, color and architecture with a different element, i.e. the interaction between individuals and their emotional and physiological reactions which impacts the sound and light of the installation. As with *Polytopes*, the space is fully integrated with the musical part, in which two visitors at a time are sitting next to each other are invited to create a musical and visual environment of sound and light with their inner (heartbeats) biological rhythms together.

In this manner, two visitors are invited to sit next to each other surrounded by a light structure made of EL wires steered by their heartbeat via a finger pulse sensor. The purpose of this project is, on the one hand, to make visible unconscious internal reactions that are produced in a subject in a simple situation such as sitting next to another human being; on the other hand, the project serves the purpose of inviting people to be aware of their inner-self and of the other person, as well as of their environment.

Sound and light structures are created according to the structural/architectural properties of the space: while the sound utilizes the acoustical characteristics of the space, the light structure on the other hand, reflects the shape of the space on the surface. The project does not only consist of an interactive installation combining sound and light, but it also includes a performance using the same central core principles.

The sound and light structure is based on the *Out-of-Sync/In-Sync* concept: when the two participants do not share the same rhythm of their heartbeats, the installation/performance is in an *Out-of-Sync* state; however, if the frequencies of the heart rates of the two participants run synchronously, the installation is in an *In-Sync* state and the sound and light events of the project change accordingly. This principal concept is based on research showing that our heartbeats can be synchronized by deepening the perception of others [7]. The Premiere of *REFLEXION – In Sync/Out of Sync* – took place at *Kunststation Sankt Peter* in Cologne/Germany in 2019, supported by *Innogy Stiftung* and *ON Neue Musik*. Further presentations followed in Cologne and at the *MM Gerdau Museu das Minas e do Metalin* in Belo Horizonte/Brazil in 2020 and in Bonn at *Dialograum Kreuzung an St. Helena* in 2021 (Fig. 5).

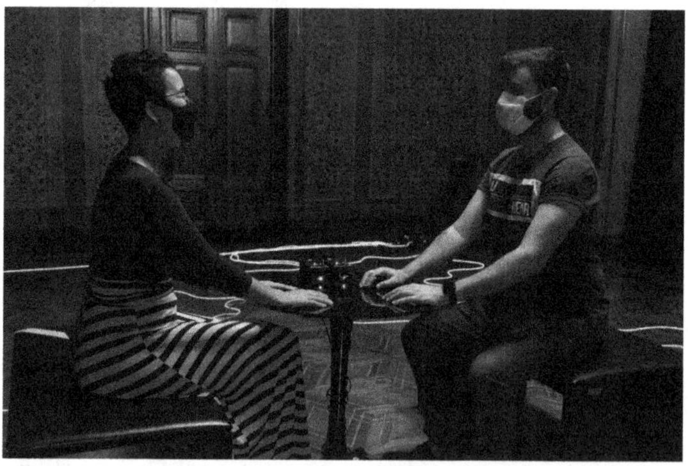

Fig. 5. *REFLEXION – In Sync/Out of Sync* 2020 –. Museum Gerdau Belo Horizonte. ©Claudia Robles-Angel/VG Bild und Kunst. Photo by Lucas D'ambrosio. See also https://vimeo.com/379450289

REFLEXION's Pulse Sensors

Because heart rates and pulse rates are identical, the suitable biosensor for this project is a finger pulse detector, which can be quickly and easily attached to each participant without much guidance avoiding the inconveniences of using an ECG (electrocardiogram) sensor, which would require to be attached to the thorax, needing constant supervision.

Hence, both the pulse sensors and the lighting control can be considered physical computing projects specially developed and assembled for this installation. This is because there are no ready-to-use commercial pulse sensors with an open standard regarding raw data, so solutions in the context of physical computing acquire relevance.

Concerning the software, the Pulse-Sensor library by Joel Murphy et al. [10] provides a basis for the Arduino sketch. The voltage coming from the sensor is analyzed, and a person's heart rate is calculated over several measurement intervals. These values are sent as serial data to the host computer and processed by different algorithms written in the MAX software, where suitable serial values are then sent to the lightning control boxes for the rhythm of the light environment. The light installation reacts, therefore, in real-time according to the participants' pulses.

Regarding the hardware, an amplifier board and fingertip by Easy Pulse (Embedded Lab) were selected for each of the sensors [4]. After a few attempts with various DIY sensors that follow the principle of photoplethysmography (PPG), the Easy Pulse sensors were convincing in terms of precision, ease of wearing and handling. In addition, Arduino Uno resp. Nano boards are used. These are connected wirelessly via Xbee modules to the host computer so that free moving space is possible for the installation and the performance.

Testing out the components, their interaction, the adaptation and the assembling was an iterative tinkering process that culminated in a first Arduino Uno version fed with a standard battery and a later, small-sized Nano version. An external mini power bank energizes this version, and it is optimized in battery life and wearability for the performance (see Fig. 6).

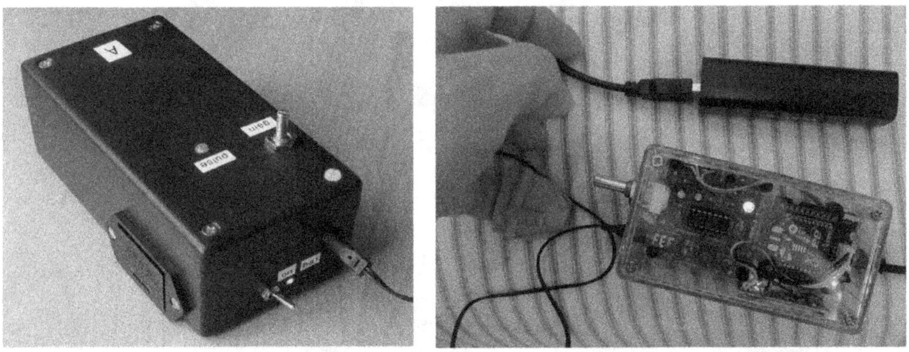

Fig. 6. Two versions of the assembled PPG sensors.

REFLEXION's Light Structure

The light structure consists of EL chasing wires, a special type of EL wires that provide an apparent uninterrupted light movement. This motion is intensified or reduced in consonance with the heartbeats when in the 'unisono' state. By this means, visitors are invited to be mindful of their surroundings: fellow human beings and the space itself. Hence, the data of the heartbeat/pulse of the two visitors are used in order to turn on/off different cables in different frequencies, creating different tempos and rhythms.

In the *Out-of-Sync* state, the asynchronous heartbeats and the inharmonic noises of the EL cables cause the light structure to flicker. When both heartbeats are, however, synchronized (In-Synch state), the light structure becomes stable (no flickering), and the roof structure of the space is reflected on the floor.

Based on the experience in *Web-Mindscape*, the physical computing components Arduino and Escudo Dos are used for the EL lightning control too. The chasing effect arises because each chasing cable consists of three thin EL wires plated together and light up one after the other. The faster the three wires are switched, the faster is the flowing/chasing impression. That means that three Escudo-Dos channels are required for each chasing cable. In this manner, for a total of 12 distributed cables, four control boxes with Arduino Uno and Escudo Dos shields are assembled. The boxes are connected with modern external inverters with a stable output of high AC voltage, which guarantees a safe operation during the installation and performance. Additionally, in practice, the JST PHR connectors of the chasing cables are very prone to failure. Therefore, all chasing cables and control boxes are equipped with stable Renk DIN plug connections suitable for continuous operation during the whole exhibition (see Fig. 7).

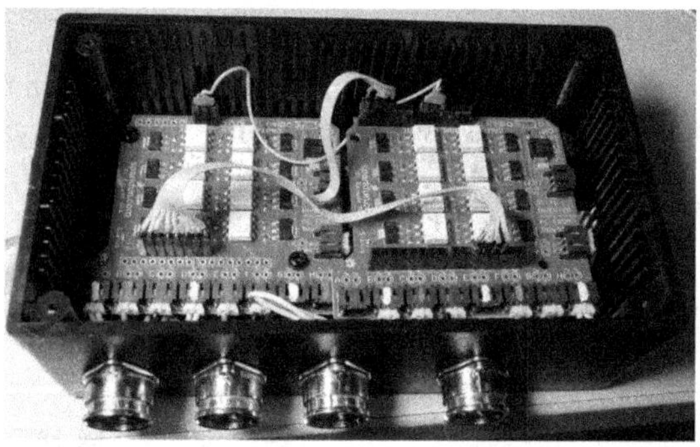

Fig. 7. One of the chasing control-boxes with Arduino Uno and two Escudo Dos boards.

***REFLEXION*'s Musical Components**

The musical components of the project consist of two sound elements:

1. After the pulse sensors read the frequencies and rhythms of the heartbeats, they are transformed using sound synthesis and sound design treatments with MAX.
2. Light cables and their circuit boards produce noise, mostly high frequencies, which were recorded and used with additional DSP functions in MAX.

When the installation is in the *Out-of-Sync* state, the sound behaves as follows: the asynchronous heart rates and the electrical noises of the light cables create a sound environment of restlessness in which inharmonic or dissonant sound constellations are dominated. However, during the *In-Sync* state heartbeats are brought into unison. So the sound and the light structure produce an environment of restlessness and harmony. Thus, harmonic and consonant sound constellations create a meditative space.

The acoustic properties of the space play, therefore, a particularly relevant role. As they are incorporated in the immersive sound conception through sound projection via eight loudspeakers, loudspeakers are distributed accordingly in the space depending on its particular acoustics.

3 Summary and Future Perspectives

We pointed out the importance of physical computing for interactive biofeedback installations in New Media Art by referring to *WEB-Mindscape* and *REFLEXION – In Sync/Out of Sync* – and by describing the artistic ideas on which both installations are based. Moreover, this article focuses on the application of physical computing by given detailed descriptions of the technology involved.

As an essential future perspective, we want to point out that physical computing represents an intersection for an art-related, expanded research term, in which "research through art" and "research in art" come into focus [5, 9, 15, 16]. At the same time, this leads to a scientific attitude that propagates a heuristic, artistic and – as typical in physical computing – iterative approach described above. This approach enables and encourages the further finding of theses and associated methodologies[3].

Especially installations with biofeedback offer the opportunity to involve them in scientific investigations as part of field studies, such as music research or research on human-machine interaction's social and artistic aspects. On the one hand, these installations are based on interactive processes, a core interest of scientific discussions in recent years. On the other hand, they can serve as a laboratory or test field for specific methods or the development of methods. Biofeedback data can be recorded and – e.g. in combination with structured observation [1, 5] – be evaluated and analyzed. According to our previous research, engagement and attention are important aspects in artistic human-machine interaction and can be observed concerning their duration, to physical expression and to biological data. For example, the presentation of one of the installations portrayed in this article included additional recordings with infrared cameras. The extensive evaluation of all the data is currently under development and will be discussed in a future publication.

To summarize, in the context of art-as-science, applied physical computing in connection with biofeedback systems is an artistic practice and an artistic extension. It invites researchers to a discerning reflection on creative new media technology and scientific exploration of embodiment, situatedness, and different kinds of interaction (e.g. Dautenhahn & Saunders 2011 [21]; Miranda 2014 [22]). Such research gains new insights into human interaction regarding this technology and the underlying psychological and mental mechanisms. In particular, a deeper understanding of the complicated processes underlying the interactive behavior patterns of humans in an artistic context will be investigated in the future [5, 16].

References

1. Bakeman, R., Quera, V.: Sequential Analysis and Observational Methods for the Behavioral Sciences. Cambridge University Press, New York (2011)

[3] E.g. empirical data acquisition and prospective data analysis.

2. Banzi, M.: Getting Started with Arduino. Make Books, Sebastopol (2008)
3. Banzi, M.: Getting started with Arduino, 2nd edn. O'Reilly, Sebastopol (2011)
4. Bhatt, R., Shahryiar, S.: EasyPulse_User_Guide (2013). http://embedded-lab.com/uploads/manuals/EasyPulse_User_Guide.pdf. Accessed July 2021
5. Gernemann-Paulsen, A.: Escapa: Eine roboterbasierte interaktive Klang-installation. Physical Computing und New Media Art in AHRI-Design und Kognitiver Musikwissenschaft. Shaker, Aachen (2018)
6. Gernemann-Paulsen, A., Robles Angel, C., Seifert, U., Schmidt, L.: Physical computing and new media art – new challenges in events, Bericht 27. Tonmeistertagung. Verband Deutscher Tonmeister, Bergisch Gladbach (2012)
7. Goldstein, P., Weissman-Fogel, I., Shamay-Tsoory, S.G.: The role of touch in regulating inter-partner physiological coupling during empathy for pain. Sci. Rep. **7**(1), 1–12 (2017)
8. Lucier, A.: Music 109: Notes on Experimental Music. Wesleyan University Press, Middletown (2012)
9. Mittelstraß, J.: Kunst und Forschung: Eine Einführung. In: Ritterman, J., Bast, G., Mittelstraß, J. (eds.) Kunst und Forschung - Können Künstler Forscher sein?, pp. 13–16. Springer, Wien (2011). https://doi.org/10.1007/978-3-7091-0753-9_2
10. Murphy, J., Gitman, Y., Needham, B.: Installing our playground for PulseSensor arduino 2019 (2018). https://pulsesensor.com/pages/installing-our-playground-for-pulsesensor-arduino. Accessed July 2021
11. Robles Angel, C.: Creating interactive multimedia works with bio-data. In: Proceedings of the International Conference on New Interfaces for Musical Expression (NIME), Oslo, pp. 421–424 (2011)
12. Robles-Angel, C.: The human body as an audiovisual instrument. In: Knight-Hill, A. (ed.) Sound and Image: Aesthetics and Practices, New York, pp. 316–330 (2020)
13. Robles-Angel, C., Scherffig, L., Birringer, J., Seifert, U.: Bio-medical signals in media art. In: Proceedings of the International Symposium on Electronic Arts (ISEA), Manizales, pp. 720–729 (2017)
14. Sterken, S.: Towards a space-time art: Iannis xenakis's polytopes. Perspect. **39**, 262–273 (2001)
15. Trogemann, G., Viehoff, J.: CodeArt. Eine elementare Einführung in die Programmierung als künstlerische Praktik. In: Ästhetik und Naturwissenschaften, Medienkultur. Springer, Wien (2005)
16. Verschure, P.F.M.J., Manzolli, J.: Computational modeling of mind and music. In: Arbib, M.A. (ed.) Language, Music, and the Brain: a Mysterious Relationship, pp. 393–414. MIT Press, Cambridge (2013)
17. Wark, M.: Das Hacker-Manifest. Beck, München (2005)
18. Grau, O.: Virtual Art: from Illusion to Immersion, p. 7. MIT Press, Cambridge (2003)
19. Schacher, J.C., Bisig, D.: Haunting space, social interaction in a large-scale media environment. In: Bernhaupt, R., Dalvi, G., Joshi, A., Balkrishan, D.K., O'Neill, J., Winckler, M. (eds.) INTERACT 2017. LNCS, vol. 10513, pp. 242–262. Springer, Cham (2017). https://doi.org/10.1007/978-3-319-67744-6_17
20. Miranda, E.R., Castet, J. (eds.): Guide to Brain-Computer Music Interfacing. Springer, London (2014). https://doi.org/10.1007/978-1-4471-6584-2
21. Dautenhahn, K., Saunders, J. (eds.): New Frontiers in Human-Robot Interaction. Benjamins, Amsterdam (2011)
22. Miranda, E.R.: Brain-computer music interfacing: interdsiciplinary research at the crossroads of music, science and biomedical engineering. In: Miranda, E.R., Castet, J. (eds.) Guide to Brain-Computer Music Interfacing, pp. 1–27. Springer, London (2014). https://doi.org/10.1007/978-1-4471-6584-2_1

NerveLoop: Visualization as Speculative Process to Explore Abstract Neuroscientific Principles Through New Media Art

Anton Dragan Maslic(✉)

School of Creative Media, City University of Hong Kong, Kowloon Tong, Hong Kong
tony.maslic@my.cityu.edu.hk

Abstract. Consciousness is a concept that is nearly impossible to visualize with conventional data visualization methods. In response to this challenge, I propose strategies for visualizing core ideas based on complex theoretical frameworks that are too abstract, incomplete, or complex to visualize through conventional data visualization methods with New Media Art playing a pivotal role in this speculative visualization process. One of those strategies targets visual speculation by introducing metaphors and allegories as a possible efficacious method. In this practice-based research paper, I present my animated video work *NerveLoop* as a case study of how New Media Art can be utilized for exploring abstract neuroscientific principles. *NerveLoop* metaphorically represents these principles by comparing the brain, the city of Hong Kong and improvisational jazz. The work is conceptually informed by theories from the scientific disciplines of neuroscience, neuropsychology, computational consciousness research and neuroanatomy. Visual and auditive metaphors are explored as potentially powerful tools for expanding research domains and enable meaning making processes that otherwise remained latent. I conclude by reflecting on the creative process and viewer response to *Nerveloop* to evaluate the effectiveness of speculative visualization as an approach.

Keywords: New media art · Speculative visualization · Consciousness · The encephalon · Neuroesthetics

1 Introduction

In 2018 I had a seizure which let me experience losing consciousness for several hours followed by the reconstruction of a conscious mind that took several months before regaining functionality. This process continues to evolve progressively and continuously. Due to my background as a visual artist, I was able to take a different perspective on exploring the pertinent questions related to neuroscience and neuroanatomy affecting my recovery. This artistic vista also provided a window into the mechanisms which are loosely based on scientific data rather than on philosophical ideas and are aspired to visualize abstract mechanism that form the core structures of how brains are processing information. Many issues, visions and different interests are affecting a collective

M. Wölfel et al. (Eds.): ArtsIT 2021, LNICST 422, pp. 29–43, 2022.
https://doi.org/10.1007/978-3-030-95531-1_3

agreement to form clear definitions, especially ones regarding consciousness, that are frequently associated with disaccord between scholars of different disciplines and this discordance is intensifying as the research areas are proliferating into multifariousness by an increasingly miscellaneous population of researchers. It is this domain of disaccord that I consider to be my working territory as an artistic researcher.

One way that I visualize and express this disaccord is through the lens of New Media Art (NMA) as a tool to initiate a subbranch of neuroimaging. This allows for an exploration process that is less precise and concentrated on visualization processes and mechanisms of complex subjects that are either too big or abstract to use conventional methods of data visualization on. This includes all aspects directly related to the functionality and the architecture of our brain, as well as the interrelationships and interconnections which I envision within my research to map diagrammatically. Consequently, I am inclined to rather use a horizontal approach to map consciousness as a model to explore among other questions, intersubjectivity, the nature of the mind, computational consciousness while seeking a plausible explanation of where general consciousness would fit in the neuroscientific approach. This approach is explored and presented through digital produced artworks that contextualizes consciousness as a framework that allows information to be processed. As consciousness itself does not have visual properties, the visualization of consciousness can only illustrate explanatory diagrammatical assumptions or conjecture of its elusive nature.

In this practice-based research paper, I present my animated video work *NerveLoop* as a case study of how New Media Art can be utilized for exploring abstract neuroscientific principles. As a foundation for understanding the work, I review a small selection of literature that deals with both consciousness to provide a neuroscientific perspective, as well as visualization techniques usually employed illustrate and exemplify existing theories. These visualization techniques are inherently subjective as they generally result in graphic representations based on simplification, abstraction, speculation, and interpretation. This subjectivity raises the question of how useful these approaches are as the visualizations are at best approximations of scientific theories of possible structural and neuroanatomical mechanisms. I then describe the process and result of creating *NerveLoop* in response to this question to claim that hypothetical projections can provide an array of multiple insights in the mechanisms of the brain and in specific the elusive nature of consciousness, which requires an unconventional approach to reformulate idiosyncrasies and purposes of its nature and even to drop the bomb by questioning its existence. Finally, I reflect upon the reactions of viewers to *NerveLoop* during its initial display during an exhibition in Hong Kong from 5–18 July, 2021.

2 Consciousness, Neuroesthetics, and Visualization

2.1 Consciousness

Consciousness has been highly enigmatic, elusive, and inspiring to thinkers, artists, and visionaries of the mind since the dawn of times. Philosophers and scientists alike have been struggling for centuries to define consciousness but failed to come up with a universally accepted description. Many theories have been posed and all fail to satisfy peer researcher from different disciplines. Each discipline has its own interpretations and definitions that works for that branch of science. In philosophy the concepts are even more diverse, which does not contribute to form an explanation that could function as a workable model of what consciousness really is. For instance, materialists and panpsychists are diametrical positioned in their concepts and these are just two of the doctrines struggling to explain it. The realm of scientists and philosophers trying to solve the questions of consciousness, also described as the 'hard problem' of consciousness [1] are rapidly gaining traction and the last decade has resulted in a radical proliferation of peer reviewed published research papers. As a result, the ideas and theories generated are widening and diverging its scope of hypotheses, making it less likely to find a unified consistently accepted theory of consciousness. At this moment of writing there are several theories that seem plausible to explain consciousness, but no accepted quantifiable system for measuring it exists. Integrated Information Theory by Tononi [2, 3] proposed a theory to quantify consciousness, but even this attempt remains controversial in its acceptance as official system to form a unit of measurement of consciousness. It therefore seems that the intangible nature of consciousness resides in the realm of arcane obscurity with conjecture as the only means of conceptualization. Another perspective would be that consciousness is progressing into its own branch of neuroscience generating a cumulative of manifold meanings and theories.

As consciousness cannot be measured through direct observation, I used a combination of phenomenology, explorations in neuroscientific theoretical concepts, and autoethnographic essays based on my seizure and recovery to develop my subjective understanding of consciousness. The work *NerveLoop* is in that sense the practical element of my research into consciousness. One of the difficulties of directly observing consciousness is due to the lack of a unified definition which propagates an array of extensively different ideas about its existence, purpose, and origins from within a multitude of different disciplines, both empirical and philosophical. Within this paper, I observe consciousness through the lens of neuroscience, neuropsychology, neuroanatomy, and the fast-emerging field of computational (artificial) consciousness research. I have omitted all other disciplines dealing with consciousness to narrow down the superabundance of interpretation and meanings that are allocated to consciousness.

Of these lenses, the most relevant to my research is the discipline of computational (artificial) consciousness which uses information generated within neuroscience and neuroanatomy to build artificial models that simulate or instantiate components that are assumed to be related and partially responsible to give rise to consciousness. Most of these models have been built in physical computational devices, using technologies like artificial intelligence and machine learning as supporting elements to developing functional prototypes. These models are usually physical computational objects that function to test assumptions and hypotheses about the neural correlates of consciousness made in neuroscience [4]. The interdiscipline as such is experimental in nature and supports neuroanatomical and neuroscientific research by feeding their findings back to neuroscientists who initially developed the hypotheses. Insights acquired through this collaborative endeavor are pivotal for increasing knowledge regarding the brain and consciousness. These neuroanatomical and mostly cognitive neuroscientific insights then inform the exploration of the origins and nature of consciousness as well as the development of the brain through the lens of evolution. Speculative visualization of this domain through NMA can contribute to producing visual metaphors that will react and provide visual feedback on the usually abstract and complex findings from this niche field of neurological research.

2.2 Visualization, Graphic Interpretation, Speculation

The purpose of any type of visualization is typically to graphically represent something complex and abstract in a simplified and codified manner that can provide easier access to the information that has been converted from a textual and often abstract format into a comprehensible infographic or animation. Visualizations are created to exemplify and clarify the information within the visual domain. It should be noted that this process of conversion of textual information into visual information generates either an approximation or a rendition that can never be precise or fully accurate. A process of interpretation leads to design decisions that speculate visual elements and motion components to conceptualize and simplify abstract data that often are devoid of any visual information. In that sense the visualization process is illustrative in codifying data through a visual redefinition of that information. This is done through a process of conceptualization and decoding research data, often generating noise as a by-product.

One can interpret this process as a specific type of visual speculation. This process is inherently characteristic to the medium and its limitations that depicts a scientific theory or component of a theory. It is worth to note that I depart from current trends in information theory and data visualization as they are developing too rapidly [5]. Traditional data visualization that uses pure scientific data is too limited for exploring issues that have less unified data output and in addition is fragmented through the various methodologies, theoretical frameworks and research disciplines that contribute to the production of both knowledge and data within the domain of cognitive neuroscience alone. In other words, consciousness is not a concept with a universal accepted definition. The approach through NMA offers a different vista by using visual metaphors[1] [6] to address these disparate information and data streams that can be processed through art visualization using principles and techniques that are usually less suitable and scarcely employed in scientific data visualization. Perhaps an exception is the field of molecular visualization, which derived its visual language chiefly through traditional animation techniques [7] Yet this extensive field has similar limitations of simplifying and abstractions to manage a level of detail that is computational possible (2014). It seems that a less defined space for interpretation allows for some flexibility to reconceptualize the various theories in a way that multiple audience groups can easier access and grasp, thereby increasing the scope of the theoretical concepts' comprehensibility and possible insights that could lead to new avenues of research.

[1] Visual metaphors can be described as visual objects that depict something representational or symbolic to elucidate something too abstract or elusive without having visual properties. Usually connecting two concepts where one has visual properties and the other does not, which develop a mental connection between the two, linking a visual quality to exemplify and simplify a complex and abstract concept. A similar explanation can be argued for auditive metaphors where sounds are conceptual representational to allude different meanings linked to specific sounds or sound patterns or even to visual or abstract information.

A potentially useful approach is that of speculative visualization which developed from speculative design [8]. This interdisciplinary approach combines artistic and design perspectives with rhetoric and scientific analysis to surpass the limits of traditional data visualization (2010). Using this approach in conjunction with NMA enhances the possibility of inspiring both scientists working in neuroscience as well informing regular audiences through artworks with unconventional characteristics.

2.3 Neuroesthetics

The exploration through NMA to explore the mechanisms of the brain and to a lesser extend consciousness have recently entered a field known as neuroestethics, a term that has been introduced from within cognitive neuroscience that centers around epistemological questions and ontological representations by the brain in a process of visualization and animation techniques, occasionally culminating in a work of art [9]. Traditionally neuroestethics focuses on processes that occur in the brain when subjects are confronted with visual art. Within this research I propose to invert this process by switching the subject from art to the brain itself. Normally, neuroestethics is employed in generating art that directly explores the brain and its mechanisms. Neuroscientific processes subsequently turn into artistic subject matter itself rather than a tool to conceptualize the reflexive inner workings of art on the brain. This particular field developed from the end of the twentieth century and evolved parallel to the technological development that provided the tools to research aspects that were only possible with the latest technological inventions in neuroimaging. These developments included technologies like computed tomography (CT), magnetic resonance imaging (MRI), functional magnetic resonance imaging (fMRI), and positron emission tomography (PET) [10]. Technological progress consequently was the driving force that instigated the field to develop.

Neuroesthetics departs from the core focus of classic aesthetics of providing definitions connected to beauty and proportional studies of aestheticized values and concentrates instead on cognitive and neural explanations mapping behavioral and social aspects in a singular approach. Neuroesthetics centers around human cognitive principles rather than abstract concepts based on culture, art history, the evolution of formalistic studies of aesthetics and so on. Furthermore, neuroestethics assumes that aesthetic cognition occurs through the interdependence of perceptual, emotional, and evaluative processes as they affect social and contextual conditions within a society. It valorizes the artworks within the domain of neuroesthetics and is thus self-referential. Within my work as an artist, I use aspects of neuroesthetics to drive the appropriate questions in the quest for exploring possible insights into mechanisms that could represent consciousness.

2.4 Evolution of the Encephalon

The advancement of the encephalon, or brain, and its functions through evolution provides a roadmap that generates insights which allows for conceptualization based on models using archaeological, neuroanatomical, neuroscientific, and cognitive neuroscientific data regarding the evolving brain. These models bypass human centric research of the brain by expanding the territory into multispecies, both extinct and alive today.

Within this expanded domain for research are highlighted both differences between and slices in time where the development of brain functions.

The neuroscientist-neuropsychologist Paul Verschure conceptualized a theory about the evolution of both the brain and consciousness in his Mind, Body Brain Nexus (MBBN) and his concept of Distributed Adaptive Control (DAC) [11]. His theory stated that during the Cambrian explosion[2] the brain was forced to develop as the competition between species became more prevalent and complex. This competition required the brain to develop a capacity to socialize but also to strategize how to survive rapidly changing environmental conditions. As a result, the capacity of the brain needed to quickly evolve and increase exponentially in computational brainpower. To account for this elevated demand of cognition, Verschure postulates that the brain starts to parallelize processes and virtualize possibilities of the world to deal with the increased complexity [12], a process that is known in psychology as 'simulation' [13]. Verschure postulates that this alteration allows the brain to simulate possible versions of the world so the entity can now impose interaction into a virtual world that enables the mind to make predictions, strategize circumstances in multiple scenarios, filling missing data, and the creation of inferences in hidden states of other agents, both alliances or enemies, anticipating behavioral and action-oriented reactions (2016).

A byproduct of virtualizing the world and its different scenarios in relation to other agents is the creation of a sense of self that leads to positioning one's self in this virtual world in relation to everything else. This position includes a sense of proprioception, which involves estimating distances measured by an entity's own dimensional awareness. This projection of a virtual world, devoid of unnecessary information, filtered out by selective focus, requires to be continuously evaluated, constantly optimized, and updated into successful interactions with other entities and to anticipate events that have not occurred yet are all increasing the chance of survival. Verschure states that the brain is constantly predicting the near future and all possible events that might happen (2016). Through the notion of self in this simulated world of probabilities is simultaneously the birth of consciousness as an epiphenomenon (2016). Consequently, the concept of reality that is happening at this specific moment in present time, perceived as now, is dominated by our unconscious states, and is only made conscious the moment it is necessary for survival and success of the species, and subsequently processed in the simulated world. This leads to believe that consciousness is always trying to catch up with reality in real-time to optimize the performance for the future (2012).

This delay of real-time has been extensively researched by Benjamin Libet, who came to fascinating results which questions our concepts of free will. Libet estimates it takes approximately 40 to 80 ms for a signal to traverse the neurological pathways towards the brain. In the brain it can take up to half a second for this information to be processed into sensory awareness [14]. Libet's experiments somehow indicated that decisions are made moments before the conscious mind intentionally does so, which questions free will of someone's rational agency as a decision maker. Verschure explains this delay as a time that is needed to rebuild changes in the constructed virtual simulated world. The

[2] The Cambrian explosion occurred approximately 541 million years ago and was a period that saw an enormous proliferation of animals and species that started to compete for survival. It is assumed that it was this period that the brain started to evolve.

more complex those changes, the longer the processing time indicates computational intensity (2016). This implies that consciousness appears within a constructed reality, created in a slightly delayed virtual world, which interprets the real-world, real-time reality, predictively to overcome this lost time and to synchronize with the real-world. Consequently, consciousness in this scenario is a truly epiphenomenal product of the encephalon.

A similar theory that deals with the construction of a virtual world has been proposed by the neuropsychologist Lisa Feldman Barret. She postulates that emotions are constructed as a response to predictions we continuously make through planned intentions within our simulated worldview. She further emphasizes that emotions are principally indistinguishable from cognition and perceptions. Barret described this as the theory of 'constructed emotion', which integrates social, psyschological, and neuroconstruction [15]. Barret's neuropsychological research supports and confirms the processes of a constructed virtual world albeit differently than Verschure's description in his concept of DAC. A question comes to mind when considering this: can humans really get in touch with the real-world in real-time as we are limited to interpret this reality filtered through our virtual model of the world in which we are undoubtedly confined? This impenetrable separation between our perceptible virtual inner world, with the external material realm is raising many philosophical questions, which are close to impossible to answer and can only be addressed through speculative reasoning, which justifies this particular methodology as not only significant, but maybe inevitable to reveal a glimpse of the world we are really living in.

2.5 Artistic Representations of the Brain and Consciousness

Santiago Ramón y Cajal, a Spanish neuroscientist, pathologist and histologist specialized in neuroanatomy and the central nervous system, created a vast body of work between 1909–1911 where he mapped and illustrated hundreds of arborizations of braincells [16]. He illustrated the complete spectrum of all different types of neurons in hundreds of detailed drawings. This body of work can be considered the first visualization study within neuroscience, and his investigations of the microscopic structure of the brain made him the pioneer of modern neuroscience. In creating his illustrations, Ramón lay the foundations for artist to work on the fringe between neuroscience and art. His work was descriptive as it depicted exactly what he observed through the microscope. This is in contrast with how 21st century artists are working with neuroscientific material using an approach that deals with different kinds of metaphors, including visual.

Christian Mio Loclair's installation work *Narciss* produced in 2018 is an example of how a phenomenon of the mind can be represented using visual metaphor and NMA. He translates human narcissism - "Mankind's self-righteous model of self-awareness, the quality of subjective findings while investigating oneself and the unequal distribution of dignity" [17] – into the phenomenon of artificial narcissism. The installation consists of a bare computer, custom made in a predominantly aesthetically black color, equipped with an embedded camera system that looks at its own reflection in a mirror positioned perpendicular opposite itself. The computer is equipped with an artificial intelligence that is constantly processing its own image. The metaphor here can be explained as a vessel to reflect constituents of a contemporary society using social media and other technology as a self-reflective narcissist mechanism of continuously carefully observing and constructing oneself through mediated representations online. He explains that the machine is displaying digital consciousness through its performance, using both Alan Turing's concept of imitation and the concept of narrative identity (2021).

A more formalistic interpretation of art that deals with the brain can be found in the work of Ralph Helmick who created in 2016 in a large-scale sculptural installation titled *Schwerpunkt*, the German term for focal point [18] The work uses sculptural representations of shiny gold leafed neurons in different sizes and scales cascading in a random array of configurations suspended in the 3-storey entrance of the McGovern Institute of Brain Research at the Massachusetts Institute of Technology (MIT). The work looks arbitrary and chaotic, but when viewed from one position on a balcony through the visual phenomenon of anamorphosis, the image forms a visual depiction of the brain. In *Schwerpunkt*, the metaphor is more general and narrates the story using loosely chaotic components which viewed from the right angle can be understood as a cohesive image. The work ultimately refers to brain research that endeavors to unravel the intricate mechanisms of how the brain functions.

Similarly, the collaborative artwork *Self Reflected* by Dr. Gregg Dunn [19] an artist and neuroscientist, and Dr. Brian Edwards, an artist and applied physicist, can be understood as a visual aesthetic representation depicting cross-sections of the brain. They developed the technique of reflective microetching, which functions as visual metaphors that show "the delicately balanced neural choreographies designed to reflect what is occurring in our minds" (2021). Dunn describes his work as "neuro art" to elucidate the nature of human consciousness. Although their work seems to take a formalistic approach, it can be argued that the concepts of observing the complexity of the human mind, which Dunn describes as "the most marvelous machine in the known universe", are represented though the complexity of the work. In *Self Reflected*, half a million neurons are animated by reflected light through the numerous micro grooves that has been generated through this process of microetching. Dunn and Edwards use combinations of hand drawings, deep neuroscience research, algorithmically simulated neural circuitry, adapted brain scan data, photolithography, gilding and strategic lighting to achieve these 2-dimensional plates. Their work bridges the aesthetic formalistic qualities of observing and thinking about a "thinking machine" thus mirroring the action of meta-analyzing our own brain.

3 NerveLoop

3.1 Overview

NerveLoop is the result of an artistic investigation to confirm the aforementioned hypotheses as described by both Verschure and Barret which state that our relation to the world is constructed in our brain in a virtual model of the world as compiled by perceived information gathered and conveyed through our senses. Conceptually, I mixed visual and auditive metaphors and conceptual metaphors to emphasize that the world we construct is reflective of the world we experience through our internalized structural scaffolding of the brain and its mechanisms, and even can be used to predict hidden or undiscovered mechanisms within the brain by researching our own creations, in this case the city of Hong Kong.

To construct this reflection, I highlighted the correlation between the transportation mechanisms of both the brain and the city of Hong Kong and focused primarily on the structural principles that both seem to share. Observing the city in comparison to the brain requires conceptualization that will help to overcome differences in spatial and temporal dimensions to synchronize between the two. A common saying across cultures is that a city is "alive" regardless of being built with mostly inanimate and non-organic materials. The dynamic characteristics of change, growth, adaptability, resilience during catastrophes, and generally progressive nature of how a city develops can all be categorized as processes that can be understood through evolution of living species and might not be so farfetched to associate with one and another. A city moves through time within a different temporality than its inhabitants. The lifespan of a city can be millennia, while that of humans is usually less than a century. This difference in temporality requires an unconventional approach in conceptualization and visualization to surpass these experiential dissimilarities. Within the film temporality is constantly shifted to experience a stretched timespan enabling to travel, point of view (POV), through the system.

3.2 Hong Kong Urban Machine Jazz and Lights

The city of Hong Kong is a living entity with a hidden capacity to generate improvisational "machine jazz." While traveling on the Mass Transit Railway (MTR) – a public transportation network of heavy rail, light rail, and buses - I noticed that the intersection between two train coaches were connected by accordion-like industrial rubber seals, which coupled with the motion and vibration of the moving train, created a sound which was vaguely reminiscent to something familiar, but which was simultaneously to abstract to identify. After recording several sound fragments, I processed the sound by slowing it down, stretching the fragments and pitching up the sound. This process released the sound which became instant recognizable as improvisation jazz. Patterns of this discovered music piece were both rhythmic and syncopated but retained characteristics of a freestyle abstract jam. Somehow the sound fragment had a visual quality which alluded to motion and speed and triggered somehow the feeling of traversing through narrow spaces like tunnels or crevices. Additionally, sounds that evoked images and a sense of tensile forces which were pulling, and pushing were interspersed throughout the track. The sound recording suggested with its intricate rich soundscape quality the first conceptual ideas of the film to travel through the brain, transposing the POV experience with the trajectory of a sequence of synaptic impulses, that travel the neurites. A new take on the work to research consciousness was to visualize a representational model of the anatomy of the brain through an animated work, that emphasized the process of thinking from a neuroscientific perspective (Fig. 1).

Duration 5 min and 33 s, 4k, Mp4. Produced and modelled in Blender, rendered in Cycles and postproduction in Da Vinci Resolve.

Hong Kong is also known for its omnipresent atmosphere of neon light. Many sci-fi books and films are directly inspired by this neon jungle and its accompanying steampunk architecture consisting of a patchwork of high-rise buildings, some gleaming new

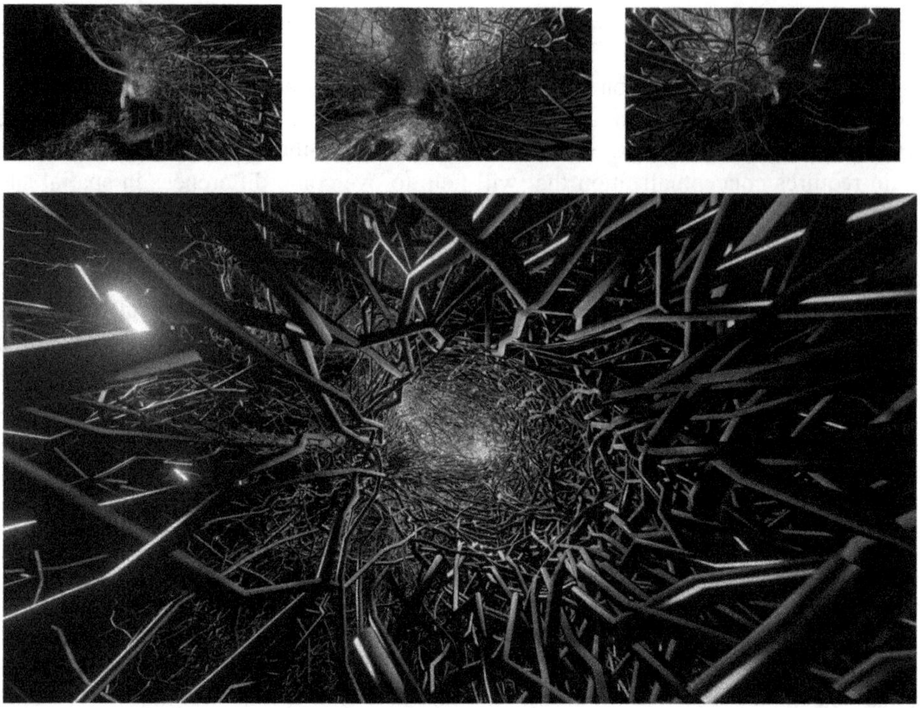

Fig. 1. Stills from NerveLoop, Hong Kong Urban Machine Jazz, 2021.

and others decrepitly dirty. Most of the external facades reveal intricate networks of external piping, wiring, bamboo scaffolding, air-condition units, or other artificial growth, which produces something locals call "AC rain", an artificial rain that descends continuously, especially in the narrow streets of the dilapidating aging districts of Kowloon. These textures, and its materiality, the sun faded colors of flakey mural paints, the stains of fungus, molds and dirt on the walls all contribute to a strictly unique atmosphere and scent. The daytime colors of the city shift at night to something reminiscent of the historical neon signs, which have been recently replaced by LED's, illuminating the night sky.

In the film I created four strong point lights, devoid from fall-off. They travel on their own elliptical trajectory in their own unique velocity. The structure of the brain is a transparent glass like material with caustics and reflections, but slightly matte, so the reflections are only light based. These lights have been colored in reference to the night city lights of Hong Kong. The motion over their elliptical orbits animates the structure of the film in a dynamic manner and contributes to experience the space as a much larger space then it has been modelled. This expanded spatial experience contributes to the capacity of the film to raises questions and to form associations and links to information in our environment. As such we could jam our habitat with a refreshed perception, which allows us to hack and discover existing elements in a rehashed manner.

3.3 Spatiality, Velocity and User Experience

The film was designed to immerse the audiences inside an experience similar to a roller-coaster ride. This was achieved by using visual effects to allow the viewer to be dragged inside this constructed digital world that represents our inner brain. As the foundation for this visualization, I modelled and simplified a tiny part of a connectome of the brain consisting of 3 clusters of neurons with simplified and limited dendrites. They are all connected through neuropathways and intertwined and are positioned in an empty space using a recent Google's AI blog for reference [20, 21]. I then selected 20 neurons from the Blakely and Januszewski model of human brain tissue which includes a dataset of 50000 cells with hundreds of millions of neurites, and 133.7 million synaptic connections (2021). This does not include the glia cells (oligodendrocytes). It should be noted that a dataset as impressive and seemingly complete as this one can only be considered an abstraction or approximation towards a comprehensible model of the real brain, as it has been achieved through various protocols, dealing with imaging, sample preparation, machine segmentation of cells, synapse detection, data storage, proofreading software and so on. Thus, even this elaborate effort to convert the brain in digital data is at best a simulacrum reflecting fragmented real-world material existence in an approximation that dives deeper towards a detailed anatomical model. Being aware of this inevitable limitation to be forced to represent a reflection rather than the real material existence a demand for visual metaphors and representational models is required. The level of necessary distancing from the provided data creates an information layer that can be explored through NMA but is in no sense scientific and at best an artistic interpretation based on the detailed dataset.

Having modeled 3 simplified clusters of neurons and dendrites, I used these as volumes to generatively grow structures with rhizome and tubular like characteristics. These structures represent the neurites and microtubules which, along with the neurons, dendrites and axions located in the brain, are potentially responsible for generating consciousness [22]. These structures can be chemically manipulated to block their function of neurotransmission which allows consciousness to be switched on or off (2003).

The modelling of these microtubules as a loosely structural element has been chosen referentially rather than as accurate representation of the architectural structure of neurons. As such the generative growth of this structure of microtubules is formed through conjecture of where consciousness could be located, but again I stress that this is an artistic decision. The specifically chosen generative method allowed the system to grow over time with a final growth period of 5 min and 22 s. This time-based generative process of evolving is representational of the new connections and neuropathways made by neurons when the brain is actively involved in adjusting its pathways. This process, known as neuronal plasticity, allows learning, restoration after injury, memory, thinking, and so on, and is a perpetual dynamic process of regenerating and restructuring the brain [23]. This neurogenesis is pivotal to the development and wellbeing of the brain throughout someone's life and has embedded some strategies for auto restorative regeneration in case of damage.

This growth period is represented by linking the different spatiality of the micro-space of the 3 neuron clusters with the macro-space of our human scaled spatial experience by radically shortening the focal length of the lens used in the POV camera. A distortion at the edge of the screen enhances the experience of having tunnel-vision. Motion blur, albeit modest was included to allow our brain to code speed and three dimensionality and to amplify the experience. Space needed to be shifted and occasionally bended, through the motion of the camera, with the ultrashort focal length lens amplifying each motion. The cinematography was determined by creating a guided path for the camera, where the camera is always facing the direction of movement. This distortion is most visible in the cases where the camera follows sharp curves in the track. The result is a slightly alienating experience of space distortion. Ultimately the space is warping around itself which suggests dimensional fluidity.

Adding to this effect is the speed of traversing through the animated space. In *NerveLoop*, the camera is radically slowed down to counterpoise and synchronize our speed of perception and motion, relative to our human scale and the speed of motion within our spatial experience. The speed of synapses to spark by impulses travelling through the neurites is approximately 40–80 μs. The time it would take for the whole film to play would make it impossible to watch, therefore I chose to convert this velocity to decelerate by a factor of approximately 5 million. This allows us to experience the speed of an impulse to travel comparable to a speed we travel through space in a subway train. Bringing back together spatial and velocity perception into our experiential comprehension.

3.4 Observations

NerveLoop was displayed for two weeks at the Jockey Club Creative Arts Centre in Sham Shui Po, Hong Kong, as part of a group exhibition featuring artistic research output. Through observations and personal conversations, it became clear that many visitors (around 22 people) were able to immediately identify that *NerveLoop* somehow referenced the brain, neurons, and dendrites without any further contextual information. A frequent comment was that watching the video felt like travelling through the brain as if one were a thought. Some people (9 people) associated the experience with rollercoasters, or an underground ride. Many visitors came quite close with the intended concepts, although sometimes expressed through rather long ruminations. Another observation was that many people were looking longer than the duration of the animation. Some individuals (7 people) stayed for 2 or 3 loops to be finished and revisited the work after seeing the rest of the exhibition. Other visitors (approximately 15 people) expressed that the work never became boring, even though it repeated in a loop and not much change was happening. They compared the sensation to looking at a campfire or at the ocean. One person mentioned that the work was very Zen. Five young children in the age between 5 and 8 years old, showed quite different reactions. Some children associated it with videogames, or sci-fi special effects and they liked to spend some time sitting with the work. One group of 6 young children were a little scared of it at first and tried to avoid looking at it and left early. The level of abstraction in combination with the propelled motion, and the absence of recognizable elements induced some discomfort in those smaller children. In overall the work was well received and made some visitors think about thinking.

4 Conclusion

Using NMA as a speculative visualization method has obvious advantages and disadvantages. There is an aspect of interpretation involved in speculation, which could arguably lead to justifiable criticism. It is therefore paramount to use visual speculation only in cases where the information is too complex, or abstract, or incomplete to use conventional data visualization methods on. In making *NerveLoop* I was able to directly see how this method affected the creative process and the reactions of the general public. As an artist, speculative visualization provided me with a new avenue by which to explore the theories and concepts regarding consciousness and the mechanisms of the brain in a new light. By incorporating this method into my practice-based research, I was able to gain new insights and knowledge regarding visualization processes and methods. I like to stress here that art and NMA do not require to have a utilizable function or purpose. I provide one option of many which in this paper encompasses scientific visualization through NMA as just one of a multitude of options and possibilities.

Another aspect that could impact the reliability of the visualization is the subjective preference of the artist to choose a specific esthetic visual language which might not be sufficient neither elucidate the scientific information. An opportunity is therefore opening whereby NMA can play a new role in contributing to science by informing and experimenting with theories that are difficult to represent through conventional data visualization. NMA may not have developed as a visualization tool to science, but it is worth noting that there is a niche possibility to do so. Of course, NMA inspired by science is a different approach and this will not impact the reliability of the scientific theory, but rather illustrate scientific ideas. Art and science in that sense can mutually benefit when the theories are reflected through art as a speculative visualization. Such speculative visualization can also inspire, bring complex scientific theories to a bigger audience, provide unconventional insights and vistas, and in conclusion can result in artworks that can be admired, enjoyed, and spark the imagination of everyone fascinated and mesmerized by it.

Conversely, the flexibility of not working empirically might be criticized but it can also be liberating to shine a light on different aspects of data as it is churned out through cognitive neuroscience. Working on this project showed me that even empirical data goes through a process of approximation, simplification, and abstraction, to overcome limitations in computational power, or the lack of proper visualization methods. I see here a space for development and NMA can be a domain that suggests a different approach, which can inspire and provide ideas to be adapted by data visualization sciences. Therefore, the contextual utilization of NMA in relation to scientific data requires clear intentions and aspirations of the parts of both artists and scientists in any form of collaboration.

Further future work will build on the ideas that have been generated through this project. Some areas that are going to be explored next will be how reality is constructed in relation to the internal virtual world as suggested by Verschure. VR as a medium will be the tool of choice to experiment with this. My personal journey of exploring the nature of consciousness would ideally result in a foreseeable future in a workable physical prototype experimenting with artificial consciousness, generating insights of how the mind and the brain inextricably exist. But more importantly as an artist I intend to provide a narrative which can inspire, provoke but most importantly provide insights that viewers could reflect on how incredible our mind and the brain really is.

References

1. Chalmers, D.J.: Facing up to the problem of consciousness. J. Conscious. Stud. **2**(3), 200–219 (1995)
2. Tononi, G.: An information integration theory of consciousness. BMC Neurosci. **5**, 1–22 (2004). https://doi.org/10.1186/1471-2202-5-42
3. Tononi, G., Boly, M., Koch, C.: Integrated information theory: from consciousness to its physical substrate. Nat. Rev. Neurosci. **17**(7), 450–461 (2016). https://doi.org/10.1038/nrn.2016.44
4. Reggia, J.A.: The rise of machine consciousness: studying consciousness with computational models. Neural Netw. **44**, 112–131 (2013). https://doi.org/10.1016/j.neunet.2013.03.011
5. Wang, C., Shen, H.W.: Information theory in scientific visualization. Entropy **13**(1), 254–273 (2011). https://doi.org/10.3390/e13010254
6. Yaman, H., Yaman, A.: Neuroesthetic: brain and art. NeuroQuantology. **17**(3), 9–14 (2016). https://doi.org/10.14704/nq.2019.17.3.1941
7. Parulek, J., Jönsson, D., Ropinski, T., Bruckner, S., Ynnerman, A., Viola, I.: Continuous levels-of-detail and visual abstraction for seamless molecular visualization. Comput. Graph. Forum **33**(6), 276–287 (2014). https://doi.org/10.1111/cgf.12349
8. Kim, T., DiSalvo, C.: Speculative visualization: a new rhetoric for communicating public concerns. In: Durling, D., Chen, L., Poldma, T., Roworth-Stokes, S., Stolterman, E. (eds.) Design Research Society International Conference, 2010: Design and Complexity, vol. 7, pp. 804–810. Design Research Society, Montreal (2010)
9. Nadal, M., Skov, M.: Neuroesthetics. In: International Encyclopedia of the Social and Behavioral Sciences, pp. 656–636. Elsevier, Amsterdam (2015)
10. Wang, J., Yang, T., Thompson, P., Ye, J.: Sparse models for imaging genetics. In: Machine Learning and Medical Imaging, pp. 129–147. Academic Press, Cambridge (2016)
11. Verschure, P.F.M.J.: Distributed adaptive control: a theory of the mind, brain body nexus. Biol. Inspired Cogn. Architect. **1**, 55–72 (2012). https://doi.org/10.1016/j.bica.2012.04.005
12. Verschure, P.F.M.J.: Synthetic consciousness: the distributed adaptive control perspective. Philos. Trans. Roy. Soc. B: Biol. Sci. **371**(1701) (2016). https://doi.org/10.1098/rstb.2015.0448
13. Barrett, L.F.: The theory of constructed emotion: an active inference account of interoception and categorization. Soc. Cogn. Affect. Neurosci. **12**(11), 1833–1833, 26 (2017). https://doi.org/10.1093/scan/nsx060
14. Libet, B.: Mind time: The Temporal Factor in Consciousness. Harvard University Press, Cambridge (2004)
15. Barrett, L.F.: How Emotions are Made: The Secret Life of the Brain, pp. 25–41. Houghton Mifflin Harcourt, Boston (2017)
16. Markram, H., et al.: Reconstruction and simulation of neocortical microcircuitry. Cell **163**(2), 456–492 (2015)
17. Loclair, C.M.: https://christianmioloclair.com/narciss
18. Helmick, R.: https://helmicksculpture.com/work/schwerpunkt
19. Dunn, G.: https://www.gregadunn.com/self-reflected
20. Blakely, T., Januszewski, M.: A Browsable Petascale Reconstruction of the Human Cortex. Google AI Blog (2021). http://ai.googleblog.com/2021/06/a-browsable-petascale-reconstruction-of.html

21. Scheffer, L.K., et al.: A connectome and analysis of the adult drosophila central brain. ELife **9**, e57443 (2020). https://doi.org/10.7554/eLife.57443
22. Hameroff, S., Penrose, R.: Conscious events as orchestrated space-time selections. Neuro-Quantology; Bornova Izmir, **1**(1) (2003). https://doi.org/10.14704/nq.2003.1.1.3
23. von Bernhardi, R., Bernhardi, L.-V., Eugenín, J.: What is neural plasticity? In: von Bernhardi, R., Eugenín, J., Muller, K.J. (eds.) The Plastic Brain. AEMB, vol. 1015, pp. 1–15. Springer, Cham (2017). https://doi.org/10.1007/978-3-319-62817-2_1

Influence of Visual Appearance of Agents on Presence, Attractiveness, and Agency in Virtual Reality

Marius Butz[1(✉)], Daniel Hepperle[1,2], and Matthias Wölfel[1,2]

[1] Faculty of Computer Science and Business Information Systems, Karlsruhe University of Applied Sciences, Karlsruhe, Germany
{buma1060,daniel.hepperle,matthias.woelfel}@h-ka.de
[2] Faculty of Business, Economics and Social Sciences, Stuttgart, Germany

Abstract. The way a system in a virtual environment interacts with the user through an agent-mediated interface could potentially influence the overall perception of the system or certain aspects of it. To investigate which type of visualization influences presence, attractiveness, and sense of agency, we designed four different visualizations of agent-mediated interfaces, namely audio-only without embodiment, as an object, as an anthropomorphic object, and with a human appearance. Our results show that presence and agency are not affected by the type of visualization of the agent, while perceived attractiveness shows differences in hedonic as well as in pragmatic quality.

Keywords: Agent-mediated interfaces · Visual appearance of avatars

1 Introduction

The way a system in a virtual environment interacts with the user through an agent-mediated interface can reduce the need to apply specific design paradigms associated with more traditional human-computer interfaces. The value of conversational user interfaces is that they provide a form of interaction and exchange in an almost natural way that simulates human-to-human verbal communication. Also for object manipulation a speech interface holds several advantages for example in regard to ease of learning and uncomplicated handling [11]. While the representation of the conversational partner is limited by physical appearance and can vary only slightly by dress, the representation of the conversational partner in virtual space is free of such constraints. The virtual representation of the interlocutor can be changed in a variety of ways including no representation at all, human-like, or in the form of an animal or object, and can be customized depending on the situation, narrative, experimental setup (e.g., for human behavior studies), or user preferences. These variations offer the interlocutor the chance to fulfill a specific role, which, in addition to the use of language, behavior, etc., is also conveyed at least in part by its visual appearance.

M. Wölfel et al. (Eds.): ArtsIT 2021, LNICST 422, pp. 44–60, 2022.
https://doi.org/10.1007/978-3-030-95531-1_4

Adapting the visual appearance of the interlocutor to the respective situation and environment can support participative perceptions, even unconsciously, including

Familiarity: sense of familiarity between an agent and a user provides a shared context that creates a mental model for the user to know what to expect from the interlocutor [3]

Competency: the appearance convey a sense of expertise (e.g. being taught about space by an astronaut) [16]

Branding: the avatar can act as a brand ambassador like Mickey Mouse or the Michelin manikin [5,15]

Acceptance: can promote the suspension of disbelief so that situations that are different from a real situation have a higher level of acceptance; e.g. nobody wonders why Superman can fly while a teacher cannot [1]

Varying the visual properties of an agent offers potential for interaction concepts that would not be possible in the real world. For example, a wizard could make objects appear out of the blue, or an anthropomorphized lamp could highlight important aspects by illuminating them [25].

The interlocutor can be either *autonomous* or *controlled by a person*, in the first case it is usually called an *agent*, in the second case an *avatar*. Avatars are common in social VR applications such as Facebook's Horizon, VRChat, AltspaceVR, or Mozilla's Hubs. A comprehensive overview of the state of the art in the visual representation of avatars in social VR applications can be found in [10]. Agents are used in many applications; for example, they can provide information, guidance, or feedback in a virtual museum or classroom context.

2 Related Work

While there are several works related to avatar and agent visual appearance in 2D environments (predominantly on video games), there is little research related to VR environments. McDonnell et al. [14] compared ten different rendering styles, ranging from a toon pencil to an actual virtual human, as well as an auditory-only rendering on a 2D monitor. They found that participants performed better on lie detection in the audio-only case than when a character was rendered. In their paper, they point out that this might be related to the participants focusing mainly on the visual appearance of the character than on what was being said. They also found that cartoon-style characters were rated as more appealing than human-style characters.

In a more dated study, also on a 2D monitor, Gulz and Haake [7] state that individuals who prefer *task-oriented* communication do not prefer a particular visual style of the avatar, but individuals who prefer *relationship-oriented* communication prefer an iconic visualization style over a realistic one.

In the study by Forlizzi et al. [6], female-looking, male-looking, and abstract agents were compared with each other with the result that human-looking agents were rated higher than abstract agents. In addition, female-looking agents were preferred over male-looking agents. They also found that gender stereotypes play a role in the expectations of an agent, even in those cases when gender cues are minimal.

While these results can provide initial guidance regarding agent and avatar appearance for immersive VR, it should be noted that results on 2D screens do not necessarily translate to VR settings or effects such as the uncanny valley are more pronounced in head-mounted VR as opposed to screen settings [9].

Bergman et al. [4] compared two different agents, one robot-like and one human-like for two types of gesture usage (with and without) in a qualitative user study. Eighty participants took part in the study, divided into four groups (two types of agents and two types of gestures). All participants started with a short video of a self-introduction of the agent, then they evaluated the first impression using a questionnaire (first measurement). In the second phase, the agent described a building in six sentences, which was again evaluated using a questionnaire (second measurement). The results indicate that the perceived interpersonal warmth is higher for the robot-like agent in the first measurement than for the human-like one. However, after the second measurement, the perceived warmth of the robot-like agent is lower than before, while it remains constant for the human-like agent. Competence is perceived significantly higher for both agent types when the agent was able to gesticulate.

Lee et al. [13] investigated whether *user performance* depended on agent appearance (actual tutor vs. 3D annotation) on three different tasks (navigating through a maze, stretching exercises, and controlling a crane). User performance was measured in execution time and task precision. They found that the 3D annotators had a higher precision in the maze task and a lower execution time in the stretching exercises. Regarding user behavior, it was found that participants in the tutoring group attempted to mimic the behavior of the virtual tutor, while participants in the annotation group attempted to meet the conditions of success.

A study by Torre et al. [24] examined differences in reliance on different emotional expressions (smiling face, positive voice modulation, or both) for a cartoonish and a photorealistic agent. The evaluation was based on behavioral data within a survival task, assessments with a questionnaire, and qualitative comments. For this purpose, a hypothetical accident scenario was created in a desert and on the moon, where participants were asked to rank six functional objects according to their importance for survival. Subsequently, the virtual agent, originally intended to serve as a navigation assistant, suggested a different (inverted) order of the user rated items, and the participants were asked to create a final order dependent on the agents suggestions. The differences between the order of the participant(s) and the agent and how many item ratings were adopted by the agent were used as the basis for behavior-based trust. The result show increased trust in an agent with congruent, neutral expression, which—according to the authors—is due to the extreme situation (stranded in the desert or on the moon), while in the opposite case the expression were deemed as sarcastic or inappropriate.

While most studies about the representation of agents and avatars have been conducted on 2D monitors, little attention has been paid to more immersive technologies such as head-mounted VR or AR. As their representation and perception might differ in VR or AR in particular on its ability of non-verbal communication (gaze direction, facial expressions, or body language) it is important to investigate the impact of different forms of visual presentation for agents and avatars [21].

3 Hypotheses

Since there has been little work on the visual representation of agents in immersive VR applications, we attempt to answer its influence on the three fundamental aspects presence, attractiveness, and sense of agency, as formulated in the following hypotheses:

H1 Presence is influenced by the type of visual appearance of the agent.
H2 Attractiveness is influenced by the type of visual appearance of the agent.
H3 Sense of agency is influenced by the type of visual appearance of the agent.

4 VR-Experience

To evaluate the hypotheses, a virtual environment is needed that provides a good match between the environment and the visual representation of the agent. A good match is given when the style and the content correspond. Our goal was to cover different interaction types and to include situations in which the agent provides information (passive user) and those in which it instructs what to do (active user). Due to the increased complexity and reduced resemblance of the test situation between participants, we decided against evaluating a two-way conversation between participant and agent.

The VR installation *Enter The Hindenburg*, which was developed by the authors in the context of the "OpenCulture-Meets-BW"-Hackathon, provides a good basis for the intended experiment. The intention of *Enter the Hindenburg* is to enable participants to experience and explore the historic airship and intangible cultural heritage "Hindenburg". Using the application, participants can explore the airship from different perspectives inside the hangar and get articulated information about the historic context. Besides this passive option, the application allows the user to *interactively* participate. The user is guided by the agent through spoken commands to control a crane and help workers build the airship, and is also given instructions on how to repair and paint the outer skin of the plane.

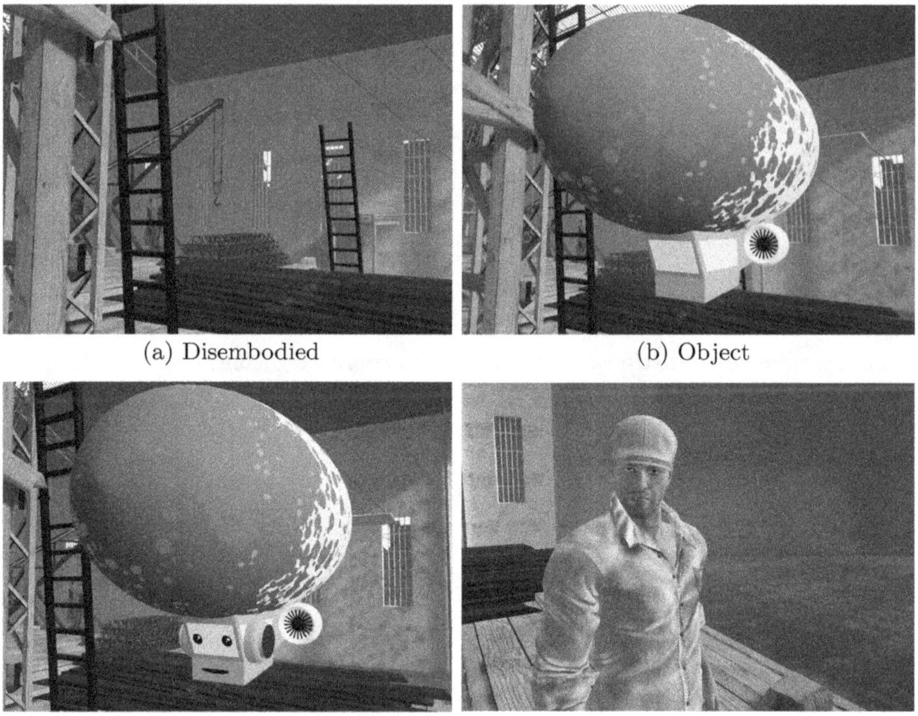

(a) Disembodied (b) Object (c) Anthropomorphic Object (d) Human

Fig. 1. Different visual agent representations. (Color figure online)

4.1 Agents

To provide an appropriate setup, we have limited our investigations to four types of agent-mediated interfaces (see Fig. 1), namely:

- disembodied (audio-only without embodiment),
- as an object,
- as an anthropomorphic object,
- and with a human appearance.

The agents only expressed themselves verbally. For better comparability, object manipulation was not demonstrated even though the human avatar would have been able to. The three different variations of agent visualizations used in our evaluation fulfill different, specific needs and support different expectations as discussed in Sect. 1. In addition to conceptual considerations such as the need for high realism in certain virtual training scenarios, technical limitations (e.g., lack of display capabilities in voice assistants such as Amazon's Alexa) and cost (e.g., high vs. low polygonal models) may also influence design decisions (Table 1).

All investigated agents share the same audio track and behavior and only differ in visual appearance and facial animations.

Table 1. Visual features of the different agents.

	Disembodied	Object	Anthropomorphic object	Human
Idle animation	✗	✓	✓	✓
3D-Audio	✗	✓	✓	✓
Gaze tracking	✗	✗	✓	✓
Lip-sync	✗	✗	✓	✓
Gestures	✗	✗	✗	✓

Disembodied (audio-only). Unlike all other agent types studied in this context, the *disembodied* agent has no visual representation and only the environment is visible (see Fig. 1a). the agent's voice is designed in such a way that no spatial orientation can be derived from it.

Object. In virtual worlds, any object can take the place of a protagonist. According to the *media equation*, it can be said that under certain circumstances objects are perceived as communicative objects [20]. Probably the most famous object protagonist is the Pixar *Luxo Jr.* mascot—it demonstrates well how expressive objects can be.

The object has to fit well with the narrative and the environment, so we decided for a shrunken, cartoon version of a Zeppelin (see Fig. 1b). To bring the protagonist to life, we added basic animations: The object moves slightly during the scene and always rotates to face the player, but with a 20° offset so that the user can also see the side of the Zeppelin. Whenever the Zeppelin is in motion (by turning or changing position), the rotors of the engines also turn, relative to speed. The voice was spatially located at the protagonist's location.

Anthropomorphic Object. By adding facial features and other body parts such as arms or legs, any object can be humanized. These types of objects, called anthropomorphic objects, potentially enhance the representation of emotions and information through gesturing or gaze direction [17]. A good example is given in the movie *Cars* by *Pixar*.

Since eyes, and the resulting eye contact, are particularly important for communication, it is therefore also of interest to evaluate a humanized version of our protagonist [2]. We got inspired by the radio character in the movie *The Brave Little Toaster* and implement a humanized version as can be seen in Fig. 1c.

In addition to the animations used in the previously described case, the agent of the anthropomorphic object blinks and the lips move in sync with what is being said.

Human. The final representation is human. To fit the narration, we chose a male person, about 40 years old, wearing a blue shirt, beige corduroy pants, and a gray beret (see Fig. 1d). The body and face were fully rigged and visemes

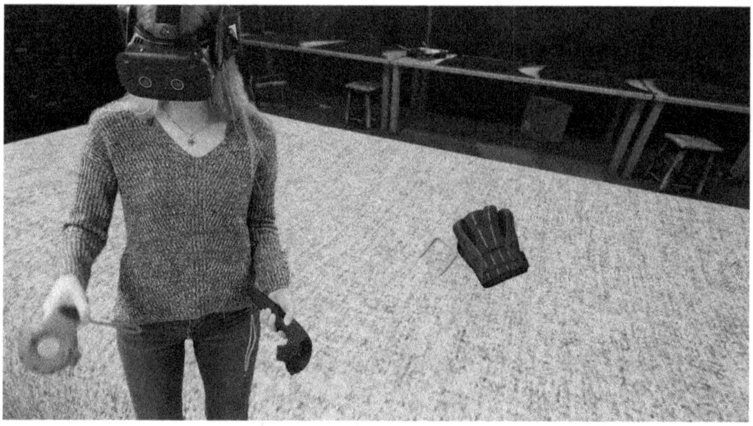

Fig. 2. Study participant during the "repairing the outer skin" task

were added to allow lip sync. Procedural animations were also added so that the agent always turns his head to have eye contact with the player. If the user would move so far that turning the character's head is not enough, the character also rotates around its axis in 90-degree increments. Additionally, the character blinks at random intervals. While speaking, the character moves with generic gestures overlaid with default animations. This allows the character to rotate, blink or point its head in the direction of the player while gesturing.

4.2 Tasks of the User

To be able to evaluate the avatars in different use cases, we implemented four tasks, one of which is passive and three are active. As a VR Headset, we used an HTC Vive Pro Eye in combination with the HTC Vive controllers (see Fig. 2).

In the *passive scene*, the user does not have a specific task, but is given information about the *Hindenburg* and its construction at three different positions (see Fig. 3). Once an agent has finished his explanation, he remains in "idle state" until the player presses the trigger button of the VR controller, which teleports the player to the next location.

In the *first interactive scene* (see Fig. 4), the user is instructed to use the levers to control a crane located on the ceiling. The levers are operated with the virtual hands of the player, which he activates by pressing the trigger button. The crane is to be used to move the aluminum struts, which are highlighted in blue, to the target position marked in yellow (see Fig. 4a). Each lever controls one axis (forward/backward, left/right or up/down) of the crane, multiple levers can be operated simultaneously. If the hook of the crane is close to the pallet, it automatically hooks and follows the position of the crane from now on. As soon as the pallet is near the target position, it automatically snaps to the correct place and the task is solved. For increased visibility even in darker areas (see Fig. 4b), two lamps, one shining towards the ceiling and the other being an

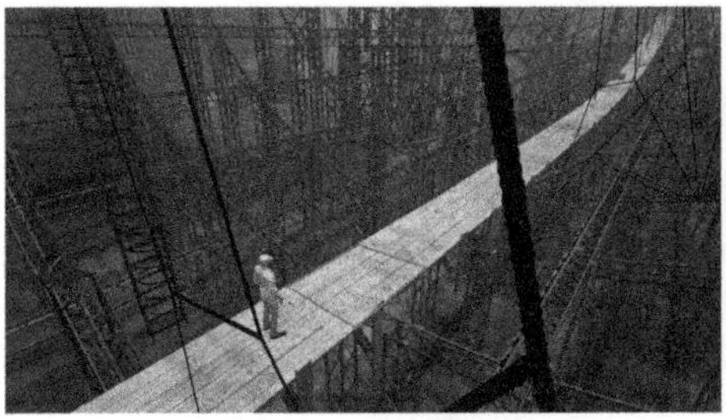

(a) First position

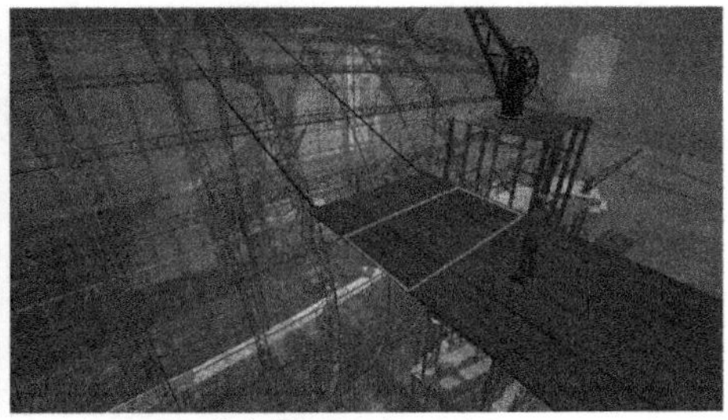

(b) Second position

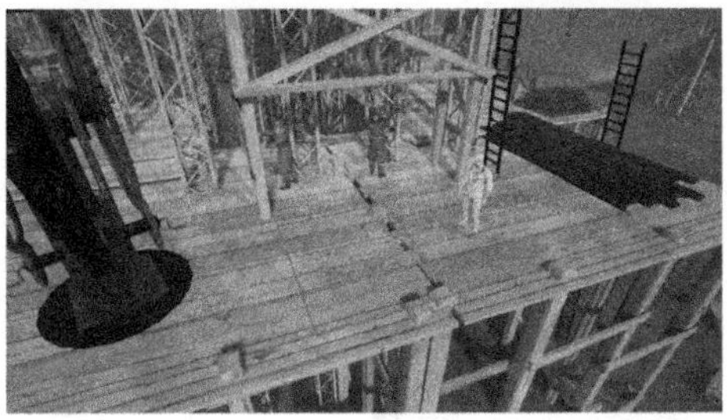

(c) Third position

Fig. 3. Positions during the *passive scene*. The walkable area is shown in red. Please note that all images show the human agent but have been evaluated for all other agent types. (Color figure online)

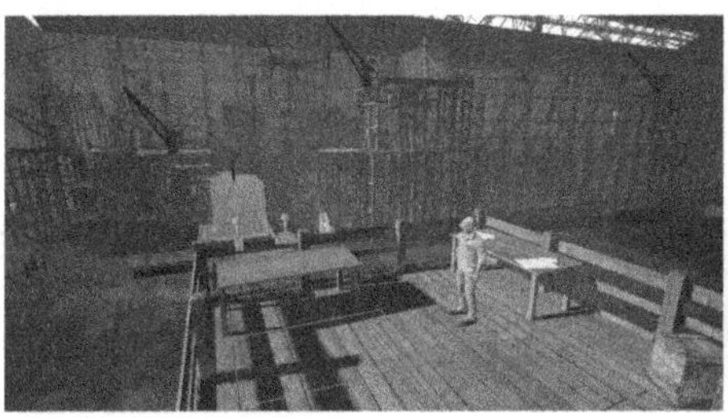

(a) Position of the user. The walkable area is shown in red.

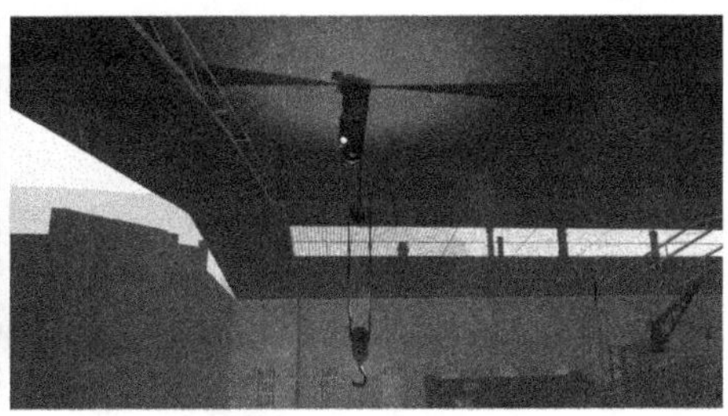

(b) Crane, which is controlled by the player

Fig. 4. *Interactive scene one*: controlling a crane in which the player has to bring the aluminum struts (highlighted in blue) to the marked target position (highlighted in yellow). (Color figure online)

omnidirectional beacon with periodic changes in brightness, are mounted on the crane's hook.

Interactive scene two requires the user to repair a part of the outer skin of the airship (see Fig. 5). The outer skin is placed on a table right in front of the user (see Fig. 5a). The tools required for the repair are available on a second table positioned to the right of the user (see Fig. 5b). A yellow light indicates

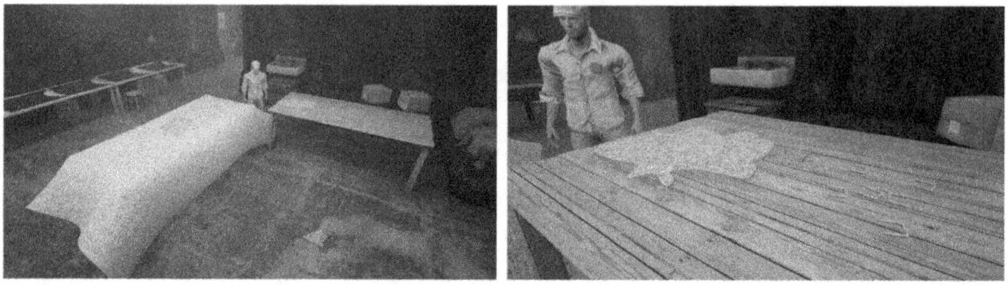

(a) Position of the user. The walkable area is shown in red.

(b) Repair utensils

Fig. 5. *Interactive scene two*: repairing the outer skin (Color figure online)

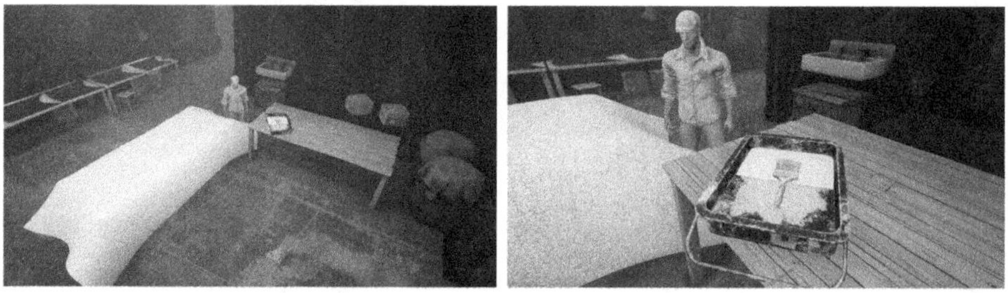

(a) Position of the user. The walkable area is shown in red.

(b) Utensils for painting

Fig. 6. *Interactive scene three*: painting the outer skin

where the repair utensils needs to be placed. Once the player is near the target position with a correct component, the placeholder location glows blue and the component can be released to snap into place. At the beginning of the scene, only the fabric is lit. After the fabric has been placed on the outer skin, the player is instructed to attach the fixing clips.

As a *third interactive scene*, the user needs to paint the Hindenburg's outer skin (see Fig. 6). After fixing the hole in the skin, it still remains on the table (see Fig. 6a) so that in can be painted accordingly. Once again, the tools needed are placed on the table right next to the participant (see Fig. 6b). In order to paint the outer skin, all the participant has to do is pass the brush over the outer skin. Once 15% of the outer skin has been painted, the task is considered complete.

5 Study Design

To obtain comparable results in the areas of agent *presence*, *attractiveness*, and *sense of agency*, we relied on established questionnaires. Each item of the used questionnaires was rated on a Likert scale between 1 and 6 and later combined as described in the subsequent subsections.

5.1 Presence

To measure the perceived *presence*, the *iGroup presence questionnaire*[1] [22] was used. This questionnaire measures *presence* based on the four components *spatial presence*, *involvement*, *experienced realism* (*realism*), and *general presence*. While *spatial presence* refers to the spatial location in the virtual world, *involvement* refers to the degree of influence the user has in the virtual world. *Realism*, on the other hand, describes to which degree the virtual world resembles the real world. In addition to the three components mentioned above, there is one item that loads on all three factors ("In the computer-generated world, I had the impression of having been there") and is therefore rated as *general presence*. Since different items load differently on a component, the value ranges of the individual components are different for the respective items. The aggregated component *spatial presence* has a value range between 4.1 and 24.5, *involvement* between 3.2 and 19.0, *realism* between 3.4 and 20.1, and *general presence* between 1 and 6. These aggregats are calculated by the sum of all products of the chosen value by the participant and the corresponding factor loading.

5.2 Attractiveness

To measure *attractiveness* the *AttrakDiff 2* questionnaire by Hassenzahl et al. [8] is used. The measurement of a product's *attractiveness* is thereby separated in three categories: *hedonic quality stimulation* (*stimulation*), *hedonic quality identity* (*identity*), and *pragmatic quality*. If a product is usable for environmental manipulation, Hassenzahl et al. label them as *pragmatic*. When an object extends the possibilities of a user through new functions, this is referred to as *hedonic quality*. While *stimulation* refers to personal development by improving one's own performance, *identity* reflects one's own personality in the product. After all, Hassenzahl et al., developed 21 items. The components were labeled according to the categories of items that loaded particularly high on them. The component *identity* has values in the interval of 5.7 and 34.1, *stimulation* has a range of values between 5.9 and 35.5, and *pragmatic quality* between 5.2 and 31.3.

5.3 Agency

To measure the *sense of agency* the *The Sense of Agency Scale* questionnaire by Tapal et al. [23] is used. It contains thirteen items. Six items relate to *sense of positive agency* and seven to *sense of negative agency*. *Sense of positive agency* describes the perceived degree to which participants felt they initiated the actions, whereas *sense of negative agency* describes the absence of this feeling. Since the items are weighted here as well, the components also have a different lower and upper bound, which is between 3.0 and 18.5 in the case of *sense of positive agency* and between 2.6 and 15.8 for *sense of negative agency*.

[1] http://www.igroup.org/pq/ipq/index.php.

5.4 Procedure

Once the participant arrived at our lab, they were given a short introduction on how to control the application by the study supervisor while calibrating the eye-tracking system. As soon as the participant was ready, the first scene was started (exemplarily shown in Fig. 2)[2]. All scenes had to be completed without the help of the supervisor, only in case of multiple misinterpretation or complete misunderstanding of the task the supervisor gave a hint to solve the problem. When the participant successfully completed the given task, they were teleported back to the neutral initial scene (blue sky, white ground). The participants were asked to take off the glasses, with the supervisor assisting them, and to answer the questionnaire on a laptop. This procedure was then repeated to a maximum of four iterations, depending on the respondent's condition, so that each respondent went through at least two and a maximum of four scenes. When selecting the possible combinations, a 4×4 matrix was used to ensure a precisely balanced distribution. Completing all 4 tasks including the onboarding and eyetracking calibration process took approximately 1 h.

5.5 Participants

34 people participated in the study, 21 female and 13 male. Age ranged from 19 to 52 years (mean $M = 29.41, sd = 8.59$). The VR experience, which was assessed participatively on a scale from 1 (no experience) to 6 (regular experience), has a mean of $m = 2.47$ and a standard deviation of $sd = 1.24$. The fact that many of the participants tested all four scenes, results in a total number of 112 observations, 28 observations per group.

6 Results

In the following, the user study is evaluated to confirm or reject the hypotheses as stated in Sect. 3. For this purpose, all components between the four groups *disembodied, object, anthropomorphic object* and *human* are analyzed for significance using the *Kruskal-Wallis Test*, since the data is not normally distributed [19]. Table 2 presents a normalized summary of all results.

6.1 Presence

As presented in Table 3 all four measures related to presence, namely *spatial presence, involvement, realism,* and *general presence* show no significant difference between the four visualizations. Therefore, the hypothesis **H1** [***Presence*** *is influenced by the type of visual appearance of the agent*] cannot be confirmed.

[2] A screen-capture of an exemplary study situation is provided on youtube. https://youtu.be/dD7RQ2inWdk.

Table 2. Since the value ranges of all comp/onents vary due to different factor loadings, the values were scaled to fit between 1 and 6. Unscaled values can be found in the corresponding sections. The given values correspond to the mean value of the results. Superscript numbers ^{1,2,3,4} mark significant differences ($p < 0.05$) between the items within a row.

	Disembodied (1)	Object (2)	anthro. Object (3)	Human (4)
Presence				
Spatial presence	5.54	5.06	5.45	5.19
Involvement	4.79	4.52	4.64	4.39
Realism	4.66	4.27	4.61	4.23
General presence	5.18	4.89	5.21	4.93
Attractiveness				
Identity	3.89^{4}	3.76^{4}	4.19	$4.39^{1,2}$
Stimulation	$3.41^{2,3}$	4.06^{1}	4.27^{1}	3.86
Pragmatic quality	4.06^{4}	$3.81^{3,4}$	4.43^{2}	$4.50^{1,2}$
Sense of agency				
Negative *agency*	2.44	2.40	2.20	2.27
Positive *agency*	5.16	4.87	5.31	4.92

Table 3. Results for *presence* components. The values shown in the table correspond to the mean value and standard deviation of the results.

	Disembodied (1)	Object (2)	anthro. Object (3)	Human (4)
Spatial presence	22.6 ± 3.57	20.6 ± 4.81	22.3 ± 3.9	21.2 ± 3.21
Involvement	15.2 ± 4.38	14.3 ± 4.82	14.7 ± 4.2	13.9 ± 3.66
Realism	15.6 ± 3.68	14.3 ± 3.79	15.5 ± 3.16	14.2 ± 3.18
General presence	5.18 ± 0.86	4.89 ± 0.87	5.21 ± 0.92	4.93 ± 0.77

6.2 Attractiveness

In Table 4 the influence on attractiveness is presented. It can be seen that there exist several differences between the agent types for all three sub-categories especially in regard to the human agent which shows significant differences with *disembodied* as well as with *object* for *identity* and *pragmatic quality.* In case of *stimulation*, the *disembodied* agent shows the largest differences to the other agents. Here we find significant differences between *disembodied* and *object* as well as *anthropomorphic object* but not for *human.* Therefore, the hypothesis **H2** [**Attractiveness** is influenced by the type of visual appearance of the agent] can be confirmed.

Table 4. Results for the *attractiveness* components. The values shown in the table correspond to the mean value and standard deviation of the results. Superscript numbers ^{1,2,3,4} mark significant differences ($p < 0.05$) between the items within a row.

	Disembodied (1)	Object (2)	anthro. object (3)	Human (4)
Identity	22.1 ± 5.03^{4}	21.4 ± 4.09^{4}	23.8 ± 3.38	$25.0 \pm 4.09^{1,2}$
Stimulation	$20.1 \pm 4.61^{2,3}$	24.0 ± 3.07^{1}	25.2 ± 4.42^{1}	22.8 ± 3.07
Pragmatic quality	21.1 ± 3.71^{4}	$19.9 \pm 3.73^{3,4}$	23.1 ± 2.49^{2}	$23.4 \pm 3.03^{1,2}$

6.3 Sense of Agency

By comparing the numbers presented in Table 5 no significant differences are present between all four agent styles: for the component *sense of positive agency*, minimally higher means show up in the groups *human* and *anthropomorphic object* compared to the groups *disembodied* and *object*. For the component *sense of negative agency* the differences are even smaller, only the component *object* has a minimally higher value than the remaining groups. It should also be noted that this component is inversely coded, as it reflects negativity. Thus, a high numerical value stands for the absence of the *sense of agency*. Therefore, the hypothesis **H3** [***Sense of agency*** *is influenced by the type of visual appearance of the agent*] cannot be confirmed.

Table 5. Results for the *sense of agency* components. The values shown in the table correspond to the mean value and standard deviation of the results

	Disembodied (1)	Object (2)	anthro. Object (3)	Human (4)
Negative *agency*	3.48 ± 4.44	3.26 ± 3.48	2.22 ± 3.33	2.57 ± 3.56
Positive *agency*	15.9 ± 3.56	15.0 ± 3.96	16.4 ± 2.98	15.1 ± 4.12

6.4 Further Notes

It should be noted that as a supervisor one could observe in some cases that the non-humanoid Zeppelin was not perceived as the origin of the voice and accordingly not as a communication partner. Other findings that came to the attention of the test supervisor during the execution of the experiments are, for example, that the non-humanized Zeppelin was often not perceived as a communication partner. According to some participants, they did not know that the voice was coming from the Zeppelin. Furthermore, the participants noted that the non-humanized Zeppelin would not fit into the overall scenery due to its comic style, whereas the Zeppelin with humanizations, on the other hand, was rather perceived as fitting, although—except for eyes, ears and mouth—the model is identical. Furthermore, it was noticed that the participants always searched for a visible figure with the *disembodied* advisor, even if the *disembodied* variant was tested first.

6.5 Scene Comparison

The question whether a specific agent type is preferable for either an *interactive scene* or a *passive scene*, was analyzed by taking a specific look at the results for each group. All scenes were compared pairwise for all agents with a *Kruskal-Wallis Test* and did not yield significant differences between any constellation. This means that no implemented agent is preferable for *interactive-* or *passive scenes.*

7 Discussion and Outlook

In this work, we evaluated the influence of agents' appearance in VR on *presence*, *attractiveness*, and the *sense of agency* using existing, established questionnaires. We implemented four different agent styles that all share the same audio-track but differ in appearance. With the presented results we can state that, while differences exist in measured *attractiveness*, no differences in *presence* and *sense of agency* between the different visualizations were found (see Table 2). Further breaking down of the results show that for *all variables measured* there is no significant difference between the *human* and the *anthropomorphic object* agent and only one significant difference between *anthropomorphic object* and *object* can be confirmed. In regard to the *anthropomorphic object* and the *object* comparison, Koda [12], found that—even though the agents' perceived *intelligence* was not influenced by whether a face was added or not— the *engagement* with the agent that had a face was significant larger. So, depending on the context, especially in entertainment and also for learning environments, adding human properties to the agent is a valuable contribution.

From this, we conclude that with current representation possibilities of virtual humans, the agent can be exchanged with an *anthropomorphic object* without losses even in cases of *attractiveness.* The other agents also do not differ in case of *presence* and *sense of agency.* This allows for several possibilities such as framing a teaching or guidance situation with the appearance of an *anthropomorphic agent* that sets up a mental model and conveys the feeling of competency (anthropomorphic saxophone guiding through music exhibition, anthropomorphic calculator giving math classes.).

There are also several arguments why a *disembodied* agent might be a valid choice since it shows no significant difference in case of *presence* as well as in *sense of agency.* For example the fact as stated by McDonnell et al. [14] that audio-only environments offer better concentration on what is spoken might be of advantage when only spoken content is of interest. It is worth mentioning that—especially with the *object* with no humanizations—no spatial localization by audio was possible, sometimes the workers (NPCs) were wrongly identified as interlocutors.

The increasing use of intelligent voice assistants (such as Amazon's Alexa) could lead to an extended acceptance of assistants in non-human appearance. Just recently, Amazon released a new version of their Alexa Echo for kids which comes with a cover to look like a panda bear or similar[3]. With regard to these findings, a non-humanoid proxy (in this case a Zeppelin) is not necessarily a worse choice than a *human*, as long as it is made clear to the user from whom he/she is receiving the information or instructions.

In case of objects, this can be accomplished by adding human features, e.g. a mouth that moves to match the speech. Such extensions can even work in the real world, as evidenced by a study by Ohmura et al. comparing an anthropomorphic version of a printer [18].

The study shows that further research in this area might be of value since for this work, only one specific cartoonish stylized *object* was humanized. We see that, with upcoming technological advances, the representation of the virtual *human* will be even closer to what a real human looks like.

References

1. Banks, J., Bowman, N.D.: Emotion, anthropomorphism, realism, control: validation of a merged metric for player–avatar interaction (PAX). Comput. Hum. Behav. **54**, 215–223 (2016). https://doi.org/10.1016/j.chb.2015.07.030, https://www.sciencedirect.com/science/article/pii/S0747563215300406
2. Bayliss, A., Tipper, S.: Predictive gaze cues and personality judgments: should eye trust you? Psychol. Sci. **17**, 514–20 (2006). https://doi.org/10.1111/j.1467-9280.2006.01737.x
3. Belanche, D., Casaló Ariño, L., Flavian, C.: Artificial intelligence in fintech: understanding robo-advisors adoption among customers. Ind. Manag. Data Syst. **119**, 1411–1430 (2019). https://doi.org/10.1108/IMDS-08-2018-0368
4. Bergmann, K., Eyssel, F., Kopp, S.: A second chance to make a first impression? How appearance and nonverbal behavior affect perceived warmth and competence of virtual agents over time. In: Nakano, Y., Neff, M., Paiva, A., Walker, M. (eds.) IVA 2012. LNCS (LNAI), vol. 7502, pp. 126–138. Springer, Heidelberg (2012). https://doi.org/10.1007/978-3-642-33197-8_13
5. Cherif, E., Lemoine, J.-F.: Human vs. synthetic recommendation agents' voice: the effects on consumer reactions. In: Rossi, P. (ed.) Marketing at the Confluence between Entertainment and Analytics. DMSPAMS, pp. 301–310. Springer, Cham (2017). https://doi.org/10.1007/978-3-319-47331-4_53
6. Forlizzi, J., Zimmerman, J., Mancuso, V., Kwak, S.: How interface agents affect interaction between humans and computers. In: Proceedings of the 2007 Conference on Designing Pleasurable Products and Interfaces, DPPI 2007, pp. 209–221. Association for Computing Machinery, New York (2007). https://doi.org/10.1145/1314161.1314180
7. Gulz, A., Haake, M.: Social and visual style in virtual pedagogical agents. In: Workshop on Adapting the Interaction Style to Affective Factors associated with the 10th International Conference on User Modeling (2005)

[3] https://www.theverge.com/2021/6/29/22554428/amazon-reading-sidekick-alexa-echo-skill-kids-voice-profiles.

8. Hassenzahl, M., Burmester, M., Koller, F.: AttrakDiff: Ein Fragebogen zur Messung wahrgenommener hedonischer und pragmatischer Qualität (2003)
9. Hepperle, D., Ödell, H., Wölfel, M.: Differences in the uncanny valley between head-mounted displays and monitors. In: 2020 International Conference on Cyberworlds (CW). IEEE (2020). https://doi.org/10.1109/cw49994.2020.00014
10. Hepperle, D., Purps, C.F., Deuchler, J., Wölfel, M.: Aspects of visual avatar appearance: self-representation, display type, and uncanny valley. Vis. Comput. 1–18 (2021). https://doi.org/10.1007/s00371-021-02151-0
11. Hepperle, D., Weiß, Y., Siess, A., Wölfel, M.: 2D, 3D or speech? A case study on which user interface is preferable for what kind of object interaction in immersive virtual reality. **82**, 321–331 (2019). https://doi.org/10.1016/j.cag.2019.06.003
12. Koda, T., Maes, P.: Agents with faces: the effect of personification. In: Proceedings 5th IEEE International Workshop on Robot and Human Communication, RO-MAN 1996 TSUKUBA, pp. 189–194. IEEE (1996). https://doi.org/10.1109/ROMAN.1996.568812
13. Lee, H., et al.: Annotation vs. Virtual tutor: comparative analysis on the effectiveness of visual instructions in immersive virtual reality. In: 2019 IEEE International Symposium on Mixed and Augmented Reality (ISMAR) (2019). https://doi.org/10.1109/ISMAR.2019.00030
14. McDonnell, R., Breidt, M., Bülthoff, H.H.: Render me real? Investigating the effect of render style on the perception of animated virtual humans. ACM Trans. Graph. **31**(4) (2012). https://doi.org/10.1145/2185520.2185587
15. Miao, F., Kozlenkova, I.V., Wang, H., Xie, T., Palmatier, R.W.: An emerging theory of avatar marketing. J. Mark. (2021). https://doi.org/10.1177/0022242921996646
16. Niewiadomski, R., Demeure, V., Pelachaud, C.: Warmth, competence, believability and virtual agents. In: Allbeck, J., Badler, N., Bickmore, T., Pelachaud, C., Safonova, A. (eds.) IVA 2010. LNCS (LNAI), vol. 6356, pp. 272–285. Springer, Heidelberg (2010). https://doi.org/10.1007/978-3-642-15892-6_29
17. Osawa, H., Ohmura, R., Imai, M.: Embodiment of an agent by anthropomorphization of a common object, vol. 10, pp. 484–490 (2008). https://doi.org/10.1109/WIIAT.2008.129
18. Osawa, H., Ohmura, R., Imai, M.: Embodiment of an agent by anthropomorphization of a common object. In: 2008 IEEE/WIC/ACM International Conference on Web Intelligence and Intelligent Agent Technology, vol. 2, pp. 484–490 (2008). https://doi.org/10.1109/WIIAT.2008.129
19. Ostertagova, E., Ostertag, O., Kováč, J.: Methodology and application of the Kruskal-Wallis test. Appl. Mech. Mater. **611**, 115–120 (2014)
20. Reeves, B., Nass, C.: The media equation: how people treat computers, television, and new media like real people and places. Bibliovault OAI Repository, the University of Chicago Press (1996)
21. Roth, D., et al.: Avatar realism and social interaction quality in virtual reality. In: 2016 IEEE Virtual Reality (VR), pp. 277–278 (2016). https://doi.org/10.1109/VR.2016.7504761
22. Schubert, T., Friedmann, F., Regenbrecht, H.: The experience of presence: factor analytic insights. Presence **10**, 266–281 (2001). https://doi.org/10.1162/105474601300343603
23. Tapal, A., Oren, E., Dar, R., Eitam, B.: The sense of agency scale: a measure of consciously perceived control over one's mind, body, and the immediate environment. Front. Psychol. **8**, 1552 (2017). https://doi.org/10.3389/fpsyg.2017.01552

24. Torre, I., Carrigan, E., McDonnell, R., Domijan, K., McCabe, K., Harte, N.: The effect of multimodal emotional expression and agent appearance on trust in human-agent interaction. In: Motion, Interaction and Games, MIG 2019. Association for Computing Machinery, New York (2019). https://doi.org/10.1145/3359566.3360065
25. Wölfel, M.: Acceptance of dynamic feedback to poor sitting habits by anthropomorphic objects. ACM (2017). https://doi.org/10.1145/3154862.3154928

Reconstructing Facial Expressions of HMD Users for Avatars in VR

Christian Felix Purps[1(✉)], Simon Janzer[1], and Matthias Wölfel[1,2]

[1] Faculty of Computer Science and Business Information Systems Karlsruhe University of Applied Sciences, Karlsruhe, Germany
christian_felix.purps@h-ka.de.com

[2] Faculty of Business, Economics and Social Sciences, Stuttgart, Germany

Abstract. Real-time recognition of human facial expressions and their transfer and use in software is now established and can be found in a variety of computer applications. Most solutions, however, do not focus on facial recognition to be used in combination with wearing a head-mounted display. In these cases, the face is partially obscured, and approaches that assume a fully visible face are not applicable. To overcome this limitation, we present a systematic approach that covers the entire pipeline from facial expression recognition using RGB images to real-time facial animation of avatars based on blendshapes for virtual reality applications. To achieve this, we (a) developed a three-stage machine learning pipeline to recognize mouth areas, extract anthropological landmarks, and detect facial muscle activations and (b) created a realistic avatar using photogrammetry, 3D modeling, and applied blendshapes that closely follow the facial action coding system (FACS). This provides an interface to our facial expression recognition system, but also allows other blendshape-oriented approaches to work with our avatar. Our facial expression recognition system performed well on common metrics and under real-time testing. Jitter and the detection or approximation of upper face facial features, however, are still an issue that needs to be addressed.

Keywords: Facial expressions · Avatars · HMD · Virtual reality

1 Introduction

Human facial expressions are a crucial aspect of nonverbal communication and a way of displaying emotions that are continuously interpreted by interlocutors [1]. As communication becomes increasingly computer-mediated for a variety of reasons (e.g., COVID-19, reduced traveling time), methods to overcome the limitations of 2D video conferencing are required. For instance, such systems cannot provide eye contact or transfer proxemic information. VR-mediated communication relying on 3D scans of the participants (e.g., as point clouds or voxels)

M. Wölfel et al. (Eds.): ArtsIT 2021, LNICST 422, pp. 61–76, 2022.
https://doi.org/10.1007/978-3-030-95531-1_5

or represented as animated avatars is able to overcome the aforementioned limitations. Both forms of representation should be able to convey human social signals of any kind in the virtual world in the same way that real people otherwise do in reality. In current applications, however, such as social VR platforms that provide a shared virtual space where people are represented by avatars or 3D scans, this goal is far from being achieved due to current limitations [2,3].

One reason for this is the lack of visualization of realistic facial expressions on an avatar's face in VR, which has several causes that could lead to inaccurate results and might cause the character to fall into the uncanny valley [2]. A key problem here is that current imaging techniques cannot correctly sense the upper face because it is occluded by the head-mounted display (HMD). Another problem lies in retargeting or mapping issues, as well as the individuality of faces, which can also greatly affect the quality of avatar facial animation.

Although several approaches to this challenge have recently emerged, RGB(D)-based real-time facial reconstruction for HMD users has not yet been made available to a wide audience. There are several solutions that address the problems on a component-by-component basis, but these components are usually not combined into a holistic system and often require complicated individual hardware installations. Hence, we present an approach that covers the entire pipeline from face tracking to sophisticated avatar face rendering and animation in HMD VR.

2 Related Work

Numerous commercial solutions[1,2,3,4] and scientific approaches [4–6] have been presented for capturing human body signals for various applications (e.g., avatar-mediated communication) including the animation of an avatars face. However, most of these approaches are not suitable for VR applications where large parts of the face are occluded by an HMD, which prevents the use of common methods, pre-trained neural networks, and commercial solutions. Moreover, these mostly machine learning-based algorithms cannot be applied only to the lower face, as they require full face recognition. Looking at the few existing sophisticated approaches that address this problem, it is clear that facial expression recognition still plays a niche role for HMD users. Only a few solutions address this challenge in general and most of them are limited to the recognition of expressions of the lower half of the face.

An early approach to facial expression tracking with retargeting in avatars comes from Wei et al. in which facial features are recognized and a 3D face model is animated using a dynamic inference algorithm and a transformation of facial motion parameters into facial animation parameters [5]. Although their

[1] https://developer.apple.com/augmented-reality/arkit/.
[2] https://developers.google.com/ar/discover.
[3] https://www.banuba.com/.
[4] https://visagetechnologies.com/facetrack/.

approach is quite old and does not address VR challenges (e.g., partially occluded facial parts), the two main system components remain basically the same: real-time image-based tracking of facial features and real-time generation of facial expressions on a 3D facial model of the avatar.

Brito and Mitchell propose a method for reusing landmark datasets for real-time face detection of HMD users and avatar animations [7]. Their method separates a given dataset into local regions for eyes and mouth, and then uses different machine learning approaches for landmark extraction. Although this system provides robust face tracking, facial regions that cannot be tracked by optical systems remain unaddressed.

Hickson et al. developed a system for classifying facial expressions in VR using eye-tracking cameras only. They used a CNN to classify 5 emotions and 10 facial action units. Their system achieved an F1 score of 0.73 for emotion and 0.68 for facial action unit detection [8]. However, their approach does not show the level of muscle activation for the facial action units because their data only included the presence or absence of activation of a specific muscle group and therefore is not suitable for continuous avatar facial animations.

Just recently, HTC released the VIVE Facial Tracker[5], one of the few commercial solutions capable of capturing facial expressions from the lower half of the face, providing an easy-to-use application for real-time animation of avatar faces. The solution uses RGBD images and is calibrated for use with a VIVE device, but also works with devices from other manufacturers. Approximate eyebrow tracking is technically possible with the build-in camera for eye-tracking (e.g., of a VIVE Pro Eye HMD), but achieves poor results in reality. Therefore, VIVE face tracking remains a solution to provide lower facial expressions.

Lou et al. presented one of the few solutions to fully reconstruct realistic facial expressions for VR HMD users [9]. They attached electromyography (EMG) sensors to the frame of an HMD, tracked muscle movements, and then used preprocessed EMG signals to reconstruct facial action units of the covered facial regions. Common imaging techniques were used to track the lower face. Their system achieved decent results by accurately assigning facial expressions to an avatar. However, the system has two major drawbacks: it requires additional hardware on each HMD frame and it cannot achieve real-time performance, which prevents its applicability for avatar-mediated communication (AMC).

A closer look at the few existing approaches reveals that there is still a need for research and development in this area to find a straightforward, real-time, and robust solution for rendering facial expressions in HMD VR.

3 System Architecture

To overcome the challenges mentioned in related work, we developed a holistic system that controls every aspect of the processing pipeline from the initial raw

[5] https://www.vive.com/de/accessory/facial-tracker/.

image acquisition until the animation of the 3D face mesh (see Fig. 1). Our main goal was to implement a solution through a simplified approach that does not require complicated or costly additional hardware. Therefore, we decided to use an ordinary RGB webcam and to keep our system flexible by loose coupling of the individual components. Therefore, the approach can be simply adjusted to different use cases and with different hardware setups. In general, our solution can be divided into two basic components as previously proposed by Wei et al. [5]. The first component (Sect. 4) is responsible for real-time recognition of facial expressions and their conversion into muscle activation values. The second component (Sect. 5) involves the creation of a rigged avatar model whose facial expressions are adjustable by morphing the facial geometry using blend-shapes. The two components are connected via an interface adapted from the facial expression recognition system (FACS) standard and thus allowing other components to be compatible (e.g. the VIVE Facial Tracker).

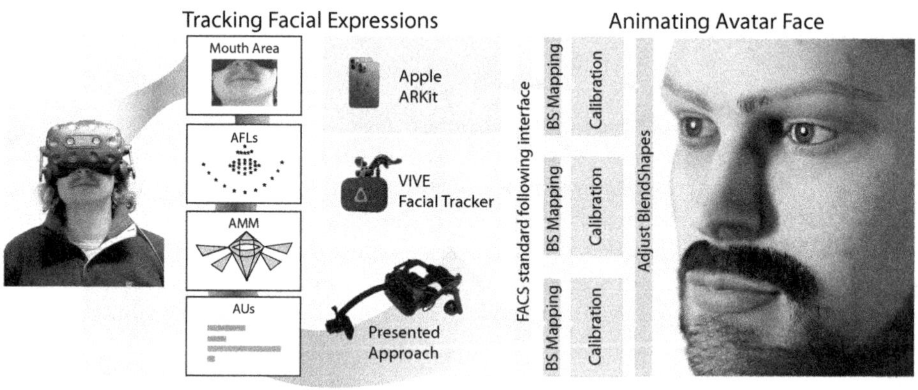

Fig. 1. Sketched architecture of the presented system. Adapted from the FACS standard an interface enables to animate our avatars face through BlendShape activation based on data coming not only from our own solution, but also from different input devices.

4 Sensing Lower Facial Expressions

Image processing and pattern recognition are typical approaches for detecting and tracking facial features. In our approach, we use a standard RGB webcam attached to the HMD with an action cam attachment. The captured RGB image stream is then processed in different stages (mouth detection, anthropological face landmark extraction, active appearance model calculation, and action unit regression, see Fig. 1) using machine learning approaches to enable facial animation of avatars.

All computations and algorithm training described in this section, as well as the live test application, interfaces, etc., were implemented using Python 3.6, with openCV 3.4.2 for image processing and rendering, numpy 1.14.2 for matrix computations, and dlib 18.18, scikit-learn 0.20 and keras 2.2.2 for artificial neural network model development and training.

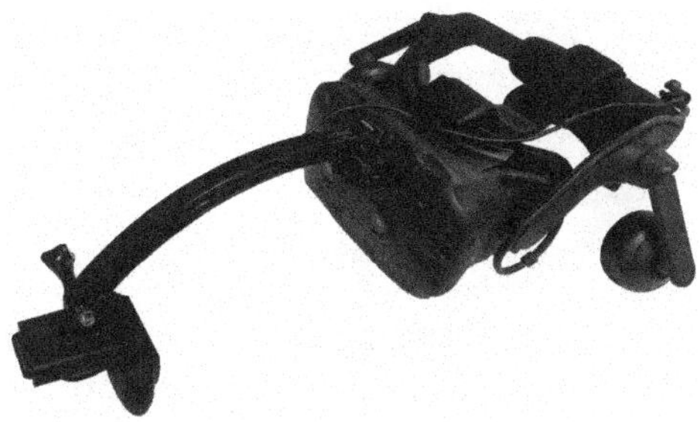

Fig. 2. HTC Vive Pro Eye with a common RGB-Webcam attached

4.1 Datasets Used

For our supervised learning-based approach to avatar facial animations for HMD users, we required multiple datasets including faces, labeled with anthropological face landmarks (AFL), FACS, and classified emotions. All the required datasets came in very different file structures, formats, and labeling styles, which meant we had to homogenize the datasets in a first pre-processing step.

Anthropological Face Landmarks are a basic notation for describing facial features in terms of x,y coordinates of relevant feature points of a face. There exist different standards with different numbers of landmarks (detail levels) that can be used to describe facial features. A common notation provides 68 landmarks and is labeled in datasets such as iBug W-300 [10] or Kaggel's Facial Keypoints (68), which we merged and used as training data.

Facial Action Units are an integral part of the FACS, that describes activation of the different facial muscle groups and is commonly used as a basic notation for labeled facial expression datasets and an implementation of avatar facial animation with minor modifications (e.g. OpenFACS) [11,12]. There are few dataset that have human portrait pictures labeled with action unit (AU) activation, such as FERA [13], DISFA [14], and FEAFA [4]. However, FERA and DISFA lack information about AUs that could be symmetrically distinguished (e.g., left/right mouth corner). In addition, AU intensity is only provided in five discrete levels which made FEAFA the favored dataset for our consideration.

Emotions can be described as a combination of the simultaneous activation of AUs. Dataset that include pictured labeled with the matching emotion are e.g. the CK+ [15] and the FACES [16] dataset. The FACES database provides a wide age range of faces with different labeled expressions, making it well suited for studying developmental and other research questions about emotions. We decided to use the FACES dataset because of its variety in individuals and better picture quality.

Although emotion recognition is not yet implemented in the current system, it is still worth mentioning that the avatar we designed already meets all the requirements for easy implementation in future work.

4.2 Augmentation of Training Data

Supervised training of a machine learning algorithm based on RGB data requires datasets of training images. To train the different algorithms, different datasets (see Table 2) were needed for the different processing stages (especially Stage 1 and Stage 2) and data augmentation. Since the learning algorithms are to be used to recognize faces or parts of faces that are partially occluded by an HMD, these HMDs must also already be included in the training images for this use case (this is not required for development/debugging purposes where no HMD is set up). Although the parts of the face that are not occluded should be detected, the recognition of the mouth, for example, is more difficult if parts of the HMD are still visible at the edge of this facial region. Thus, we used an approach similar to Suresh et al. who used a dataset (Face Mask Detection Dataset[6]) where face masks have been automatically added to non-masked faces and then jointly recognized and classified [17]. We then automatically added an overlay image showing a VIVE Pro HMD to all images from the target datasets (300-W, Kaggle FK68), and then placed, scaled, and rotated it based on the given 68 landmark notations (see Fig. 3). An excerpt from the augmentation result is depicted in Fig. 4.

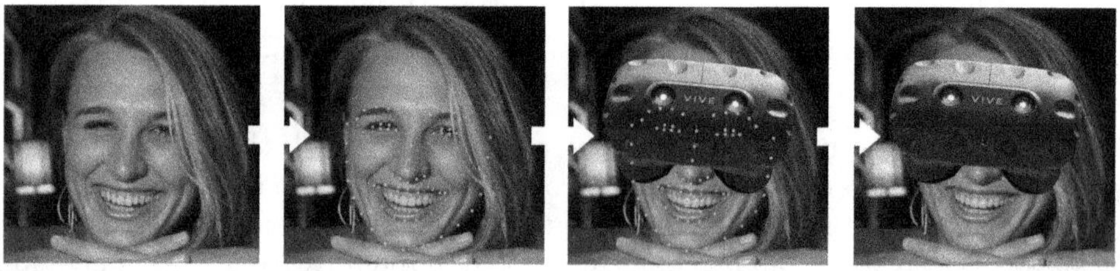

Fig. 3. Augmentation process steps from raw training image to HMD augmented picture based on 68 landmark coding

[6] https://www.kaggle.com/omkargurav/face-mask-dataset.

4.3 Mouth Detection

The recognition of the mouth area is required for all further processing and calculation steps. In particular, the accuracy of facial feature extraction (see Sect. 4.4) is highly influenced by the precision of mouth detection. To recognize the mouth within RGB images, we chose a common approach using a convolutional neural network (CNN). First introduced by Simonyan and Zisserman, the VGG16 network architecture has been very successful in large-scale image and video recognition [18]. Since it has also proven successful in similar tasks for estimating regions in images [19], we have adopted it in a slightly different configuration. We flattened the last output layer of the VGG16 network and added three dense layers and a four-unit output layer describing the normalized coordinates of the bounding box of the bottom of the face and mouth, respectively (see Fig. 5). We trained the network (VGG16 layers freezed) based on 10.300 of our preprocessed annotated images derived from the 300-W and FK68 datasets with a train/test split of 70% training, 20% validation and 10% test data using mean squared error (MSE) as loss function and adaptive moment estimation (ADAM) optimization.

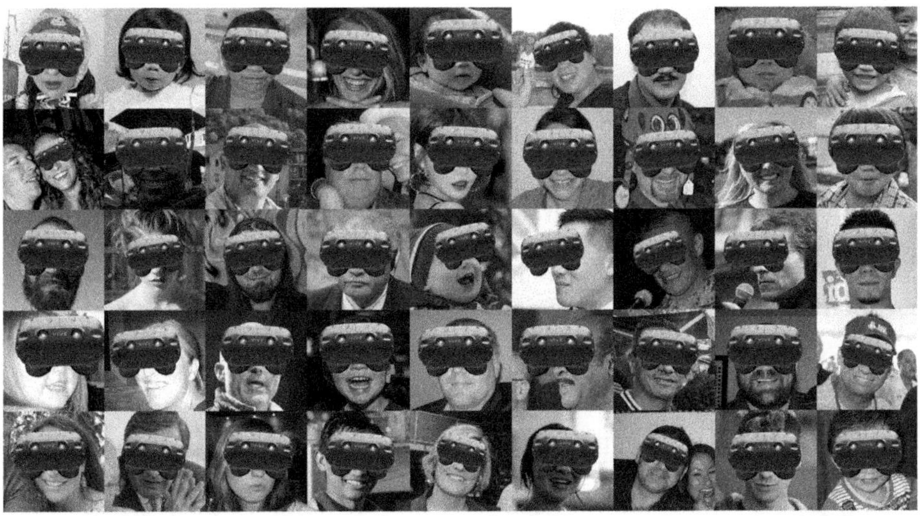

Fig. 4. Excerpt of training images augmented with HMD

4.4 Facial Landmark Extraction

Object detection based on the histogram of gradients (HoG) is a popular computer vision method for detecting semi-rigid objects and has gained acceptance in the field of face analysis. Dlib's solution, which integrates an HoG detector, can represent both appearance and shape information [20]. Dlib also offers easy implementation of predicting facial landmarks using an ensemble of regression

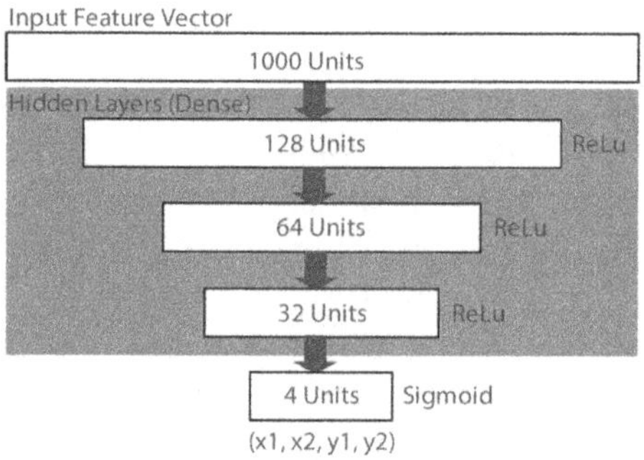

Fig. 5. Structure and parameters auf the network used for bounding box prediction of the lower facial area

trees (ERT) in an image and provides real-time high-quality predictions that perform better than other approaches, such as CNNs. Therefore, we used its shape predictor, which takes an image region containing an object and outputs a set of point positions that define the pose of the object, in our case the AFLs of the mouth, the bottom jaw, and the tip of the nose. Before training our shape predictor, we first applied our trained mouth detection algorithm to our concatenated and preprocessed training set (10,300 images) to identify the mouth region.

Moreover, it was required to crop the 68 AFLs (representing the entire face) to represent only the mouth region. For this reason, we reduced the landmarks to the lower part of the jaw (landmarks 3–14), the lower part of the nose (landmarks 30–36), and the mouth (landmarks 48–68), resulting in a total of 37 AFLs.

We then trained the shape predictor (tree depth: 5, cascade depth: 25, feature pool size: 400, oversampling amount: 5, jitter correction: 0.1) with the cropped images from the datasets 300-W and FK68.

4.5 Face Muscle Activation

There have been several approaches to animate avatar faces that estimate facial muscle activation [4,21,22]. Facial muscle activation can be normalized and retargeted to a 3D model to modify its blendshapes for facial morphological changes. Based on the detected landmarks (see Sect. 4.4), we created an active appearance model (AMM) [23]. The approach utilizing an AMM is varying from the original approach of Yan et al. who have created the FEAFA-A dataset for avatar facial animation. However, as the dataset includes faces of Asian ethnicity only it was required to find a more abstract representation to be used with faces of other ethnicities before training. Thus, we used an AMM consisting of eight polygons to describe the lower face region and its facial features (Fig. 6). Using this AMM for AU detection required us to process the FEAFA-A dataset by applying our mouth detection and facial landmark extraction algorithm on every image. Then, the new dataset used for training was created by calculating the values describing polygons of the AMM for each image. To train the algorithm, these polygons are represented as a flattened input feature vector of size 86 containing the normalized vectors representing each polygon. To measure facial muscle activation we used a different deep learning approach. For this approach, we had created our own AU mapping based on the FACS standard with slight modifications (see Table 1) as output. In addition, we had to subdivide some of the FACS AUs into separate units because it is required to be able to detect asymmetric movements of the mouth (e.g., left/right lip corners). As a network model, we used a simple 3-dense-layer fully connected artificial neural network with 86 input units (flattened polygon vectors) and 14 output units (AU activation predictions).

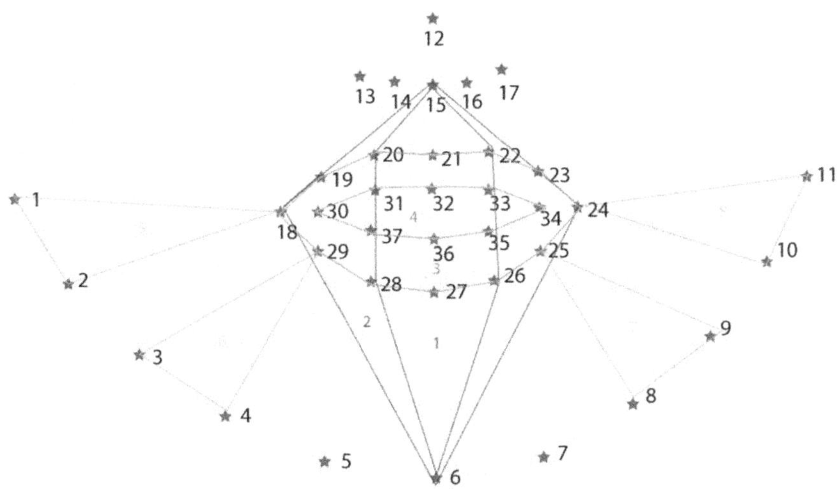

Fig. 6. Our AMM consisting of 8 polygons (43 2-dimensional vectors) created based on 37 AFLs recognized as basis for AU recognition algorithm training. The AMM was developed experimentally according to observational experience with regard on the anatomy of the facial musculature.

We trained our network based on 99,300 training examples from the preprocessed FEAFA-A dataset in a train/test split of 70% training, 20% validation and 10% test data using mean squared error (MSE) as loss function and adaptive moment estimation (ADAM) optimization.

5 Avatar Animation

Talking avatars using various facial animation techniques are ubiquitous in various media [6]. Approaches such as physics-based facial modeling and animation promise sophisticated results by taking into account potential energies and physical interaction of passive flesh, active muscles, rigid bone structure, etc., which

Table 1. Recognized action units of the lower face and the corresponding FACS AUs respectively ADs. Additionally, our AU2 - AU9 subdivide the original FACS into two distinct AUs.

AU	Our definition	Original FACS	AU	Our definition	Original FACS
1	Jaw drop	AU26 jaw drop	8	Upper lip suck	AU28 lip suck
2	Jaw slide left	AD30 jaw sideways	9	Lower lip suck	AU28 lip suck
3	Jaw slide right	AD30 jaw sideways	10	Jaw thrust	AD29 jaw thrust
4	Left lip corner pull	AU12 lip corner P.	11	Upper lip raise	AU10 upper lip raiser
5	Right lip corner pull	AU12 lip corner P.	12	Lower lip depress	AU16 lower lip Dep.
6	Left lip corner Str.	AU20 lip stretcher	13	Chin raise	AU17 chin raiser
7	Right lip corner Str.	AU20 lip stretcher	14	Lip pucker	AU18 lip pucker

offers the potential to compensate for the unnatural morphological deformations caused by HMDs [24]. However, the blendshape-based animation approach has become the most popular and is currently the leading approach for realistic facial animation [25]. This approach requires the development of a facial rig model for a 3D avatar, which is time-intensive but also provides great control over the various facial animations. For the creation of blendshapes, the FACS also provides a good standard and guidance. Even though the FACS are not equivalent to the required blendshapes for facial animation, a general reference, a standardized interface, and decoupling to the detection techniques are provided. Therefore, our approach to avatar facial animation allows the use of the methodology described in Sect. 4, but can also be used in conjunction with other, e.g., commercial systems (Fig. 1).

There are a variety of commercial rigged avatars, but our approach has specific rigging/blendshaps and interface requirements, so we created a new avatar from scratch. We used various tools and techniques for sophisticated avatar creation and implemented our final avatar including logic and interface in Unreal Engine 4.26 for rendering, applying, testing, and created a heads-up display (HUD) for monitoring and control.

5.1 Face Modeling

To create a realistic avatar face for self-representation, 3D scanning methods are crucial to achieve the most realistic results. For our face scan, we used Reality Capture[7], one of the leading photogrammetry tools that can quickly and easily create a 3D model and export both textures and polypaint. We then created a character base mesh, unwrapped the character model, and created UV maps. To match the base mesh with the created scan model, we projected the character base mesh onto the scan in Zbrush[8]. Then we adjusted the base mesh, subdivided it, and projected it again. This procedure was done until the base mesh was

[7] https://www.capturingreality.com/.

[8] http://pixologic.com/features/about-zbrush.php.

identical to the scan, or until all the details we wanted to project onto the base mesh were transferred. The resulting character base mesh consists of 32k vertices, of which almost 12k are for the head area, which is in line with today's general game engine recommendation for a character with a poly count of about 10k–100k. In this step, we also transferred the polypaint to the base mesh, which could later be used to create a colormap texture. For texturing, we used skin materials (physical-based rendering) based on the Digital Human Materials offered by Epic Games. For the creation of hair, we used an approach that adapts the method introduced by d'Eon et al., a reflection model for dielectric cylinders that has high fidelity for rough surfaces such as human hair fibers [26].

5.2 Rigging/Blendshapes

Since our approach is oriented on FACS-based blendshape, we started with Facit-BlendShapes as a basis. However, only a few of the base models BlendShapes are used for further development. The majority had to be created completely from scratch by hand. The character control rig was created using the Blender addon Auto-Rig-Pro. All weight painting had to be done by hand to achieve good deformation results. The face area itself has no weight painting, as all face movements are controlled by the BlendShapes. For detailed facial features additional shape keys have been introduced to the avatar model. The bones created for the FACS emotions are only visual bones that have no influence on the mesh itself and only activate the corresponding BlendShapes via drivers.

5.3 Calibration

Since each face to be tracked has its individual characteristics, more accurate results can be provided by calibrating the blendshape model for each user. Therefore, we have provided various modifiers to adjust the blendshape weights as well as maximum and extreme value constraints. Depending on the input device for the blendshape control, remap curves, size curves, or just manual adjustments of the float values can be used for calibration and fine-tuning.

6 Results

Under the composition of the two main components, we were able to test the overall system and its substructures individually and also the interplay in real-time. The recognition part can be quantified in numbers as well as being evaluated in real-time tests and observations. Table 2 shows the accuracy of the trained algorithms with the respective metrics. The trained network in Stage 1 for recognition of the lower face area achieved an intersection over union (IoU) of 0.91. The shape predictor trained in Stage 2 achieved an MAE of 0.52. Training of the face muscle recognition algorithm in Stage 3 resulted in an MSE of 0.015 for the network whereas AU3 Jaw Slide Right was most precise (MSE: 0.009) and AU14 Lip Pucker was most unprecise (MSE: 0.02).

Table 2. Machine learning pipeline, key figures and results

	Stage 1	Stage 2	Stage 3
Purpose	Mouth recognition	Landmark extraction	Face muscle activation
Dataset	300-W[a], Kaggle FK68[b]	300-W, Kaggle FK68	FEAFA-A[c]
Algorithm	VGG16 + BBR	HOG + ERT	FC Neural network
Input	RGB-Image	RGB-Image	AMM
Output	RGB-Image	AFLs	FACS AUs
Metrics	IoU	MAE	MSE
Result	0.91	0.52	0.015

[a] https://ibug.doc.ic.ac.uk/resources/300-W/
[b] https://www.kaggle.com/tarunkr/facial-keypoints-68-dataset
[c] https://www.iiplab.net/feafa/

The results of the avatar creation (Fig. 7), rendered in real-time by the Unreal Engine show a high degree of natural fidelity. Figure 8 shows the interaction of the tracking (incl. rendering the lower face bounding-box, AFLs, and AMM) and the avatar components and thus the corresponding facial expression of the avatar with that of the tracked person face.

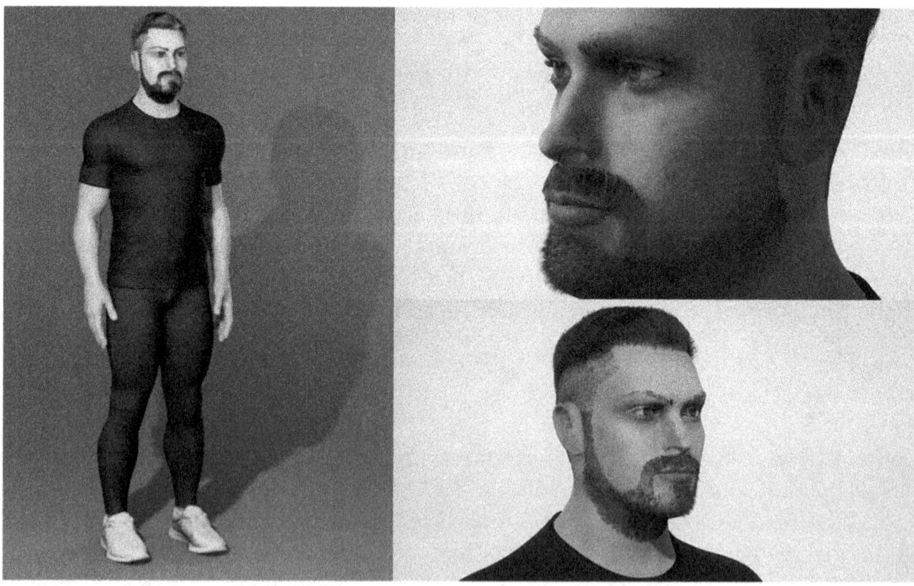

Fig. 7. "In game" screenshots of our photogrammetry scan based and rigged avatar model rendered with Unreal Engine 4.26.

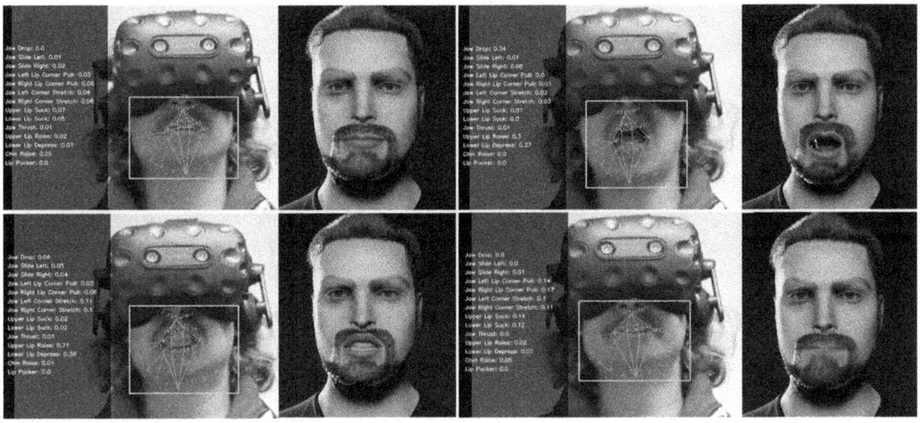

Fig. 8. Examples for FACS AUs activation (Real-time experts) of the lower face. For each of the 4 divisions it is showing: (Left) The computed AU activation as a result of our machine learning pipeling. (Middle) Real-time facial scan displaying mouth- and AFL detection and AMM. (Right) The resulting animated avatar facial expression. Considering the AU activation numbers, it is visible, that the neutral facial expression (top-left) shows almost no activation (all values close to 0). In the top-right picture the "jaw drop", "upper lip raise" and "lower lip depress" AUs are activated. The angry facial expression (bottom-left) causes only the "upper lip raise" and "lower lip depress" AUs to be activated while the bottom-right picture shows slight activation of the interacting AUs "lip corner pull/stretch", "upper lip raise" and "lower lip depress".

7 Conclusion

In this paper, we suggested another approach to visualize authentic facial expressions to be used in combination with wearing an HMD. We established a three-stage process (mouth detection, anthropological face landmark extraction, and action unit prediction) using artificial neural networks and machine learning. We created an avatar based on photogrammetry data that offers blendshape animation that has been created following the FACS standard and can be thus animated by the predicted AU values coming from our last stage of the trained neural networks in real-time via an interface. An application containing our avatar model was created using the Unreal Engine, which is loosely coupled and thus can receive data from different hardware (our own or third-party) to animate the avatar's face in real-time according to the tracked individual facial expressions.

Although the concept works in its entirety, there are some limitations. Error accumulation across the three stages of facial expression recognition often leads to a significant jitter effect and thus a loss of quality. Here, it is important to keep in mind that our AMM was created based on trials and thus was created relatively arbitrarily. Also, no hyperparameter optimization for the networks was included what could further reduce the jitter effect. Also, the implementation of a jitter correction (e.g. Kalman filter) could show improvement here.

Furthermore, it has to be considered that the FEAFA-A data set contains only Asian faces which may cause a significant bias for faces of other ethnicities. A general comparison of our approach compared with commercial ones shows still major quality differences, however, it is to mention that our model is purely based RGB data in contrast to others that also use a depth channel. Furthermore, we must state that the current version of the Unreal Engine has a known bug considering our applied hair rendering technique in VR, which is why other low-quality hair rendering techniques have to be used in this context.

However, although there are still challenges to address, key figures and the final testing results show that the concept of our approach generally works successfully and is worth being further developed.

8 Future Work

Improving the robustness and accuracy of our machine learning pipeline for facial expression recognition is a high priority in our future work, as all further progress depends on it. Thus, hyperparameter-optimization and AMM adjustments, as well as jitter reduction, have to be addressed. Furthermore, we want to take animation of the upper facial expressions into account as we already met all preconditions for emotion detection based on the mouth area. According to Blais et al. the mouth area is the most important cue for both dynamic and static facial expressions [27]. Guarnera et al. compared the ability to recognize emotions from the eye and mouth area in children and adults. Their data shows, that some basic emotions (disgust, happiness, surprise, and neutral) can be decoded just by having information about the mouth area while other emotions (anger, sadness, fear) require information about the eyes [28]. An approach to emotion recognition using imaging techniques and restricting the information to the mouth area was made by Biondi et al. [29]. They trained a CNN to classify happiness, disgust, and neutral facial expressions and achieve precise results. We want to train another artificial neural network that uses the output values (AU activation) of stage 3 (Subsect. 4.5) as an input vector to classify emotions. Additional investigation is needed to determine whether emotion recognition based on image data can produce more accurate results than based on AU or AFL input data [30]. In a natural facial expression (not a grimace), the activation of the lower and upper facial muscles is usually activated in unison, resulting in the representation of a recognizable, believable emotion. Thus, in our future work, we plan a derivation of the upper facial expression based on information about the lower which seems legitimate even if it does not represent reality. A major point missing furthermore is the classification of the missing emotions (sadness, anger, fear). An approach to this challenge can be to consider more relevant tracking data from the available sensors in a virtual world. It has been shown, that especially sadness and fear can be estimated from the body posture. As body posture data is mostly available through tracking systems in VR, this may be an approach considered to form a more holistic system. The final goal should be to detect and map the human facial expressions as realistic as possible to be even able to use this representation for interpretation in the field of human studies as no real image can be taken from participants wearing an HMD [31].

References

1. Argyle, M.: Bodily Communication, 2nd edn., pp. 1–111. Routledge, London (1986)
2. Hepperle, D., Purps, C.F., Deuchler, J., Wölfel, M.: Aspects of visual avatar appearance: self-representation, display type, and uncanny valley. Vis. Comput. (2021). https://doi.org/10.1007/s00371-021-02151-0
3. Yu, K., Gorbachev, G., Eck, U., Pankratz, F., Navab, N., Roth, D.: Avatars for teleconsultation: effects of avatar embodiment techniques on user perception in 3D asymmetric telepresence. IEEE Trans. Vis. Comput. Graph. **27**, 4129–4139 (2021)
4. Yan, Y., Lu, K., Xue, J., Gao, P., Lyu, J.: FEAFA: a well-annotated dataset for facial expression analysis and 3D facial animation, April 2019. arXiv:1904.01509 [cs, eess, stat]
5. Wei, X., Zhu, Z., Yin, L., Ji, Q.: A real time face tracking and animation system. In: 2004 Conference on Computer Vision and Pattern Recognition Workshop, pp. 71–71, June 2004
6. Zhang, J., Chen, K., Zheng, J.: Facial expression retargeting from human to avatar made easy. IEEE Trans. Vis. Comput. Graph. **28**, 1274–1287 (2020). Conference Name: IEEE Transactions on Visualization and Computer Graphics
7. Brito, C.J.D.S., Mitchell, K.: Recycling a landmark dataset for real-time facial capture and animation with low cost HMD integrated cameras. In: The 17th International Conference on Virtual-Reality Continuum and its Applications in Industry, VRCAI 2019, pp. 1–10. Association for Computing Machinery, New York (2019)
8. Hickson, S., Dufour, N., Sud, A., Kwatra, V., Essa, I.: Eyemotion: classifying facial expressions in VR using eye-tracking cameras. In: 2019 IEEE Winter Conference on Applications of Computer Vision (WACV), pp. 1626–1635 (2019). ISSN: 1550–5790
9. Lou, J., et al.: Realistic facial expression reconstruction for VR HMD users. IEEE Trans. Multimedia **22**(3), 730–743 (2020). Conference Name: IEEE Transactions on Multimedia
10. Sagonas, C., Antonakos, E., Tzimiropoulos, G., Zafeiriou, S., Pantic, M.: 300 faces in-the-wild challenge: database and results. Image Vis. Comput. **47**, 3–18 (2016)
11. Ekman, P., Rosenberg, E.L.: What the Face Reveals: Basic and Applied Studies of Spontaneous Expression Using the Facial Action Coding System (FACS). Oxford University Press, Oxford (1997). Google-Books-ID: KVmZKGZfmfEC
12. Cuculo, V., D'Amelio, A.: OpenFACS: an open source FACS-based 3D face animation system. In: Zhao, Y., Barnes, N., Chen, B., Westermann, R., Kong, X., Lin, C. (eds.) ICIG 2019. LNCS, vol. 11902, pp. 232–242. Springer, Cham (2019). https://doi.org/10.1007/978-3-030-34110-7_20
13. Valstar, M.F., et al.: FERA 2015 - second facial expression recognition and analysis challenge. In: 2015 11th IEEE International Conference and Workshops on Automatic Face and Gesture Recognition (FG), vol. 06, pp. 1–8, May 2015
14. Mavadati, M., Sanger, P., Mahoor, M.H.: Extended DISFA dataset: investigating posed and spontaneous facial expressions, pp. 1–8 (2016)
15. Lucey, P., Cohn, J.F., Kanade, T., Saragih, J., Ambadar, Z., Matthews, I.: The extended cohn-kanade dataset (CK+): a complete dataset for action unit and emotion-specified expression. In: 2010 IEEE Computer Society Conference on Computer Vision and Pattern Recognition - Workshops, pp. 94–101, June 2010. ISSN: 2160–7516

16. Ebner, N.C., Riediger, M., Lindenberger, U.: FACES-a database of facial expressions in young, middle-aged, and older women and men: development and validation. Behav. Res. Methods **42**(1), 351–362 (2010). https://doi.org/10.3758/BRM.42.1.351
17. Suresh, K., Palangappa, M., Bhuvan, S.: Face mask detection by using optimistic convolutional neural network. In: 2021 6th International Conference on Inventive Computation Technologies (ICICT), pp. 1084–1089 (2021)
18. Simonyan, K., Zisserman, A.: Very deep convolutional networks for large-scale image recognition arXiv:1409.1556, April 2015
19. Zhihong, C., Hebin, Z., Yanbo, W., Binyan, L., Yu, L.: A vision-based robotic grasping system using deep learning for garbage sorting. In: 2017 36th Chinese Control Conference (CCC), pp. 11 223–11 226, July 2017. ISSN: 1934–1768
20. King, D.E.: Dlib-ml: a machine learning toolkit. J. Mach. Learn. Res. **10**(60), 1755–1758 (2009)
21. Tian, Y.-L., Kanade, T., Cohn, J.F.: Recognizing action units for facial expression analysis. IEEE Trans. Pattern Anal. Mach. Intell. **23**(2), 19 (2001)
22. Onizuka, H., Thomas, D., Uchiyama, H., Taniguchi, R.-I.: Landmark-guided deformation transfer of template facial expressions for automatic generation of avatar blendshapes. In: 2019 IEEE/CVF International Conference on Computer Vision Workshop (ICCVW), Seoul, Korea (South), pp. 2100–2108. IEEE (2019)
23. Cootes, T., Edwards, G., Taylor, C.: Active appearance models. IEEE Trans. Pattern Anal. Mach. Intell. **23**(6), 681–685 (2001). Conference Name: IEEE Transactions on Pattern Analysis and Machine Intelligence
24. Ichim, A.-E., Kadleček, P., Kavan, L., Pauly, M.: Phace: physics-based face modeling and animation. ACM Trans. Graph. **36**(4), 153:1-153:14 (2017)
25. Lewis, J.P., Anjyo, K., Rhee, T., Zhang, M., Pighin, F., Deng, Z.: Practice and Theory of Blendshape Facial Models, p. 23 (2014)
26. d'Eon, E., Francois, G., Hill, M., Letteri, J., Aubry, J.-M.: An energy-conserving hair reflectance model. Comput. Graph. Forum **30**(4), 1181–1187 (2011)
27. Blais, C., Roy, C., Fiset, D., Arguin, M., Gosselin, F.: The eyes are not the window to basic emotions. Neuropsychologia **50**(12), 2830–2838 (2012)
28. Guarnera, M., Hichy, Z., Cascio, M., Carrubba, S., Buccheri, S.L.: Facial expressions and the ability to recognize emotions from the eyes or mouth: a comparison between children and adults. J. Genet. Psychol. **178**(6), 309–318 (2017). https://doi.org/10.1080/00221325.2017.1361377
29. Biondi, G., Franzoni, V., Gervasi, O., Perri, D.: An approach for improving automatic mouth emotion recognition. In: Misra, S., et al. (eds.) ICCSA 2019. LNCS, vol. 11619, pp. 649–664. Springer, Cham (2019). https://doi.org/10.1007/978-3-030-24289-3_48
30. Dinculescu, A.: Automatic identification of anthropological face landmarks for emotion detection. In: 2019 9th International Conference on Recent Advances in Space Technologies (RAST), pp. 585–590 (2019)
31. Wölfel, M., Hepperle, D., Purps, C.F., Deuchler, J., Hettmann, W.: Entering a new dimension in virtual reality research: an overview of existing toolkits, their features and challenges. In: International Conference on Cyberworlds (CW) (2021)

Games

Tackling Online Hate Speech?
Play Your Role!

Susana Costa(✉), Bruno Mendes da Silva, and Mirian Tavares

Research Centre for Arts and Communication, Algarve University, Faro, Portugal
{srsilva,bsilva,mtavares}@ualg.pt

Abstract. This article seeks to present and analyze methods to combat online hate speech using gamification and video games. Taking as a starting point the project "Play Your Role: Gamification Against Hate Speech", funded by the European Commission's programme, Citizenship, Rights and Justice, which is related to the ONu's goal 16, Peace, Justice and Strong Institutions, we will contextualize and present a set of complementary and interrelated tools, such as online video games, pervasive games and pedagogical itineraries to counteract online hate speech, through gaming culture and connected social spheres as a motor for the promotion of mediatic literacy and digital citizenship.

Keywords: Hate speech · Sustainable development · Serious games · Video games · Pervasive games

1 Introduction

The popular Role-Playing Games (RPG) are a genre of games in which the player assumes the role of an imaginary character, in a certain fictional world. The narrative is defined in a script and based on a system of rules. RPG originated from tabletop or pen-and-paper and evolved in the last decades to the digital and multiplayer environments, mediated by artificial intelligence. The name of the project "Play Your Role: Gamification Against Hate Speech"[1], a case study we meant to analyze in this article, quibbles with the role-playing concept - becoming someone else, somewhere else - and with the idea that events come about through consequential choices made by the player.

The starting point of the project, a multilingual initiative implemented at a European level, funded by the programme "Rights, Equality and Citizenship"[2] of the European Union, enabled the collection of data regarding interactions of young players in online games, gaming platforms and communities of gamers, which were analyzed to understand and find effective ways to prevent hate speech from proliferating in digital game

[1] Additional information at https://www.playyourrole.eu.

[2] This programme intends to contribute to the further development of an area where equality and the rights of persons, as enshrined in the Treaty, the Charter and international human rights conventions, are promoted and protected. Additional information at https://ec.europa.eu/justice/grants1/programmes-2014-2020/rec/index_en.htm.

M. Wölfel et al. (Eds.): ArtsIT 2021, LNICST 422, pp. 79–93, 2022.
https://doi.org/10.1007/978-3-030-95531-1_6

environments. For this previous study we selected students of both genders, living in Portugal, Italy, and Lithuania. The sample consisted of 572 individuals, 246 female and 291 males, divided between Italy (195), Lithuania (228) and Portugal (149). The age of the respondents varied between eleven and twenty years old, with a predominance of individuals with 12 years[3]. As a practical result of the data analysis, a set of pedagogical tools were created to instigate players and the communities to engage in and promote approaches aiming at change and learning.

The 16th global goal for sustainable development aims to promote peaceful and inclusive societies for sustainable development, provide access to justice for all and build effective, accountable and inclusive institutions at all levels. Children and youngsters are exposed to many forms of violence in the physical world. This phenomenon drifted to digital environments, where toxicity and disruptive behavior can be found, such as expression of hate speech, in the form of online blasphemies and insults. The power of words can be revealed by the influence the content has on opinions and actions, showing that violent speech can in fact have consequences outside and inside the virtual world (Hurley, 2004). According to the Sustainable Development Goals Report 2020[4], provided by the United Nations, the impact of COVID 19 on children's risk of exposure to violence due to lockdowns and associated school closures, which have affected the majority of children globally, is still yet unknown, however it is revealed that the use of the Internet for remote learning may have increase children's exposure to cyberbullying and risky online behavior.

Because they often violate the dignity of others, hate messages offer strong justifications for the need to limit them. Mechanisms that allow the authors of such messages to be silenced and banned from certain platforms for a limited time have been implemented and studied. However, despite their complexity, combating and eradicating hate speech are not the only tasks that emphasize the need to analyze and deeply understand hate speech. This research on this type of content also seeks to understand what the expression of hatred is, where it comes from, what causes it, how it rises, how it spreads on the Internet and, above all, what consequences it propagates over the network. A better understanding of the dynamics of hate speech can allow us to come up with innovative and creative responses to this problem, which allow us to go beyond certain solutions, such as repression and silencing.

In this article we intend to analyze online hate speech in the games environment, as a part of players' everyday experience and we propose counter narratives and pedagogical tools in the form of serious games, pervasive games and pedagogical itineraries to counteract the tendency of hate speech. The analysis of the state of the art reinforces that parents and educators, as well as the creation of ludic tools for educational purposes, the so-called serious games, can play a key role in the prevention and awareness of online conduct, preparing young players to deal with hate speech situations, through the promotion of empathy, as well as a safe environment of tolerance and inclusion.

---

[3] More statical information conclusions resulting from this inquiry can be found in the projects report: https://www.playyourrole.eu/wp-content/uploads/2020/07/PYR-research-report.pdf.

[4] https://unstats.un.org/sdgs/report/2020/The-Sustainable-Development-Goals-Report-2020.pdf.

2 Online Hate Speech and Toxicity

Acting out fictional dangers which, despite being controlled, seem real, can provide a certain aesthetic fascination [29]. Symbolic arenas are based on this principle and determine the general tendency of contemporary culture to use playful dynamics and mechanisms: gamification. Huizinga [15] theorized about the game as a fragment of the spatio-temporal narrative where the player acts out a parallel experience and he also underlined the essential importance of the game in the cultural construction of any society. The intimate relationship between games and new technologies leads to the emergence of new socialization networks.

The manifestation of opinion that incites hatred towards individuals or groups, giving words the power to hurt physically, characterizes hate speech. Online Hate Speech has been addressed in Europe for some time now, in the public discussion as well as at political and institutional level. The definition of hate speech online has been widely debated on a global scale and toxicity has been a topic of broad discussion in game studies since the early 2000s. Researchers on the topic describe it as a set of behaviours that one categorizes as toxic in relation to constantly renegotiated and evolving social norms [1]. According to Kwak and Blackburn [20] toxic behavior, also known as cyberbullying, griefing, or online disinhibition, “is bad behavior that violates social norms, inflicts misery, continues to cause harm after it occurs, and affects an entire community”. Despite the various definitions proposed by researchers in the recent decades, it is unanimous that the greatest difficulty in understanding and analyzing the phenomenon is its subjective character. The legal outlines of the definition vary according to cultural, political and legislative aspects evident in different countries. Ignio Gagliardone, Danit Gal, Thiago Alves and Gabriela Martinez, authors of the Countering Online Hate Speech manual published in 2015, by UNESCO [6], state that the speed and reach of the Internet, as well as the space of social media, make it difficult for governments to legislate and enforce laws regarding online hate speech, timely.

The virtual world reflects multiple tensions, the expression of conflicts between different groups within and across societies, clearly mirroring the transformative potential of the Internet, which offers both opportunities and challenges, while searching for a complex balance between fundamental rights and principles, including the freedom of expression and the defense of human equality and dignity.

Although a final, legal and global definition of hate speech is not yet consensual, according to the United Nations, it can be understood, as any type of communication or behavior that uses derogatory or discriminatory language when referring to an individual or a group, based on personal characteristics, in other words, based on religious, ethnic, racial, gender or another distinctive identity feature. The European Union, one of the institutions which pioneered in defining this type of content, also defines hate speech as a public incitement to violence or to hatred directed at groups or individuals, based on certain characteristics, including race, color, religion, descent and national or ethnic origin.

In order to prevent the spread of illegal hate speech, on 31 May 2016, the European Commission, together with Facebook, YouTube, Twitter and Microsoft, agreed a “Code of Conduct"[5] in the effort to respond to the challenge of ensuring that online platforms do not offer opportunities for illegal online hate speech to spread virally. However, the evaluation of the Code of Conduct on countering illegal online hate speech carried out by NGOs and public bodies shows a fourfold increase in the reports of hate speech. According to the Interactive Software Federation of Europe[6], in 2019 the main reason for reporting online hate speech was xenophobia (17.8%), which includes anti-migrant hatred. Xenophobia, together with anti-Muslim hatred (17.7%), as the most recurrent ground of hate speech, followed by ethnic origin (15.8%).

3 Hate Speech and Freedom of Speech

Hate speech in video games is often the result of the interactive dynamics among players, in unmoderated activities, such as team building, sharing strategies and chats, which can result in real-time conflicts. Game platforms and communities are a common medium to spread this type of discourse. Resorting to censorship as a response to these expressions of hatred can sometimes oppose freedom of expression, a pillar and an achievement of democratic societies, the foundation of self-realization, autonomy, democracy and truth [11]. Can censorship be justified to combat and react to hate speech?

The game platforms and communities usually serve as a means for the propagation of toxicity and hate speech. Discord, which allows the creation of chats and groups to unite players, already imposed its position against hate speech by banning several users linked to neo-Nazi or white supremacy ideologies and forbidding harassment or threatening messages. On the other hand, Steam, the gaming community, and store, refused to block games or content in defense of the right of decision, reaffirming itself as a game market closed to cultural disputes. Twitch and YouTube are other platforms allowing to watch live streams of almost everything, including games. The content goes live without filters, so it is impossible to predict any inappropriate actions. Live streamers can become stars, like PewDiePie, one of the most subscribed YouTuber, influencing players to act according to certain kind of attitudes.

According to Matamoros-Fernández [21] platforms are ideal spaces for racism to arise. Through a case study, the author stands that "the concept of platformed racism challenges the discourse of neutrality that characterizes social media platforms' self-representations and opens new theoretical terrain to engage with their material politics".

On the other hand, Jenni Hokka [14] analyzed PewDiePie discourse on YouTube and this platform policies regarding racist speech and she concluded that "YouTube's policy emphasise the individual's right to state his/her opinion and create content without any borders or 'gatekeepers'", contributing to the normalization of racism on social media.

European legislation, as well as some national laws, have taken important steps in combating online hate speech, for example, through criminal sanctions. On the other hand, there have been some initiatives, as mentioned above, by the most prominent technological corporations in response to this problem, through policies of use that converge on blocking users, taking the commitment to act quickly in case of complaints

[5] https://ec.europa.eu/commission/presscorner/detail/en/IP_16_1937 and https://edri.org/files/privatisedenf/euhatespeechcodeofconduct_20160531.pdf.

[6] https://www.isfe.eu.

regarding this type of abuse: "Despite initial resistance, and following public pressure, some of the companies owning these spaces have become more responsive towards tackling the problem of hate speech online, although they have not (yet) been fully incorporated into global debates" [6].

4 Methodology

An approach based on a participatory action research [13] asserts itself as an effective path to mutual understanding, common within tolerant and inclusive communities. All the actions foreseen by this project aim to promote values of democratic culture and digital citizenship in school communities, where violence has found a fertile ground for conflicts and violence.

Based on the concept of "convergence culture" [18] and considering some recent studies on media participation and media transforming [2], we propose a participatory approach in the methodology, which seems to be considered the most effective one in developing abilities and tools for change.

In this project and through video game literacy [5], the actions to be implemented propose to develop a set of skills among young people, related to critical thinking while using the media, collaboration and participation, showing how important it is to keep a certain balance between the two different forms learning, i.e. formal and informal education. The informal learning strategy, based on problem solving, simulation, evaluation and imitation, is currently considered the most effective in promoting success [12].

The methodology used in the project entails networking, particularly by fostering actions that enhance the creative and multidisciplinary power of gamification, when it comes to raising awareness and changing behaviors. We believe that this participatory methodology will contribute to digital inclusion, educational attainment, personal development and to enable an active and enlightened citizenship.

Digital literacy is one of the key elements that permeates the project - considering the culture of the game as an engine to promote democratic values and digital citizenship, promoting positive behaviors that mitigate hate speech online. Several tools were created through an approach to media literacy: 15 educational itineraries, 4 serious games and 1 pervasive game, engaging resources to combat hate speech online, to develop critical understanding of the daily practices among young people, as well as to propose alternatives to existing narratives, capable of promoting democratic values.

Games will be the methodology used to generate new and effective counter-narratives against hate speech, as well as to disseminate the results of the project in the form of urban games. Four different events will be arranged, where citizens will be invited to participate in a public game in an urban area, thus engaging actively in the results of the project.

The games were the result of an international call and a Hackathon that took place online, due to the pandemic. They are all already available on the project website and have been promoted in training sessions and workshops with students and teachers who are invited to try them out. In addition, 15 educational itineraries were also made available on the project website, which enable educators to approach the problem, using the mechanisms of the game to foster reflection on the consequences of toxic speech.

5 Serious Games

Paul Gee [9] gathered some principles that are good practices in creating serious games, guiding success as learning motors while being motivating and challenging. Also, the American Mark Prensky has been a reference for his research studies in Digital Game-Based Learning, basing his assumptions in the notion of digital natives and the need of taking the game into the classroom, while an innovative model that promotes student learning using technology [26]. Some non-governmental organizations have implemented the use of video games while working closely with several communities, looking for behavior changes, as well as educational and cultural development. Immersing a student in a virtual environment with physical world characteristics that allow him to test possibilities is one of the most effective ways of learning [10].

In many ways, video games can encourage learning, either through historical games or by depicting a historical character who teaches about the period in which he lived. As an example, let us consider "My Child Lebensborn", a nurture, survival game, based on true events, developed by Sarepta Studio AS, where driven by his own emotional drawing, the player takes care of a child from a Nazi program in the Norwegian society after the war; or "Florence", an interactive story video game developed and published by Mountains Studio, which allows the player to formulate questions about the society through a simple interactive story. The success of these games depends on the player's emotional response while interacting, the aesthetic and the design. The most important factors seem to be: awareness, the player must be sensitized by a narrative that encourages him to achieve a goal; immersion, the game must be able to shut down the player from the real world, and make him focus on the game [27]; the feeling of progress that encourages the performance [30]; the feeling of danger, when simulated with caution, can help the player focus [3]; and, finally, the feeling of conquest, able to motivate the player to continue [31].The perspective of game-based learning seems to be an important path for teaching and modeling behaviors in the era of the digital natives. Taking this into account, we can understand serious games as a tool to sensitize the player through emotional drawing, which motivates natural and fluid learning, while cumulatively avoiding boredom.

6 Four Serious Games Proposals

It was at the international hackathon that the theoretical and practical results of this study could be put into practice. With mentoring from the project team, four serious online games were developed, produced and made available through Unity platform, a leading space for creating and operating interactive, real-time 3D content. These video games are available for experimentation on the Play Your Role (PYR) project website.

Although initially intended as a face-to-face meeting, due to the pandemic situation, it was redesigned into a set of online workshops, between the project partners and the development teams, which were selected following an international call. These workshops took place between September and October 2020 and resulted in the products that we now discriminate.

6.1 Divide ET Impera

In "Divide Et Impera" the player, connected and amicable, interacts with several elements of a group. The goal is to use hate speech in a variety of ways to divide the community and instigate hostility. The player must choose the content of his speech carefully, according to the characteristics of each individual, such as nationality, sexuality, gender or religion, in order to reach the targets in the desired way and divide them.

While manipulating a small, simulated community, users are confronted with the real mechanisms used to manipulate people on social networks. This way, young people and teenagers can learn to be more critical about the sources and content of the information they find on the network (Fig. 1).

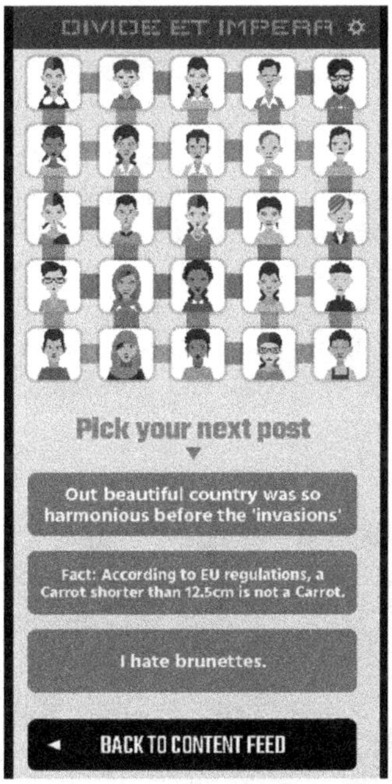

Fig. 1. Divide Et Impera

6.2 Youtuber Simulator

The player takes on the role of a Youtube streamer. The goal is to maintain a balanced life, i.e., to increase the number of subscribers to the channel and keep the discussion in the comments and chatrooms civilized, while simultaneously having to maintain his/her own mental health and social life, without being exhausted by a toxic environment or by hateful insults (Fig. 2).

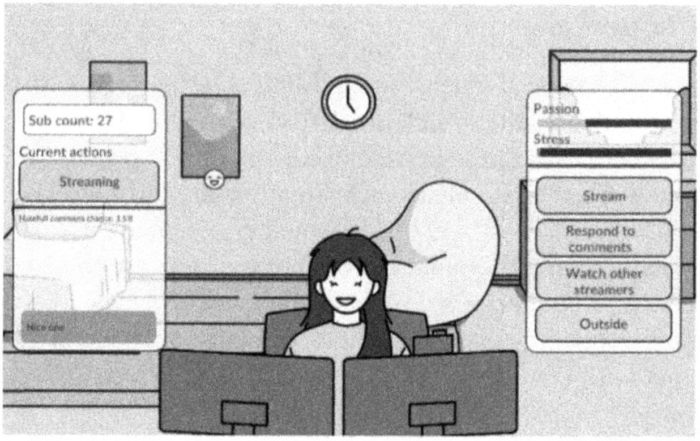

Fig. 2. Youtuber simulator

6.3 Social Threads

The game 'Social Threads' (Fig. 3) simulates social interactions that take place online and the player must react to hate speech decorously to disarm and cast away the opponent who resorts to hateful behavior.

To protect himself/herself and maintain a positive presence online, the player must select the appropriate answers from a set of hypotheses: he/she must therefore use constructive interactions to beat the opponent, and, consequently, move forward and expand his/her territory in the game.

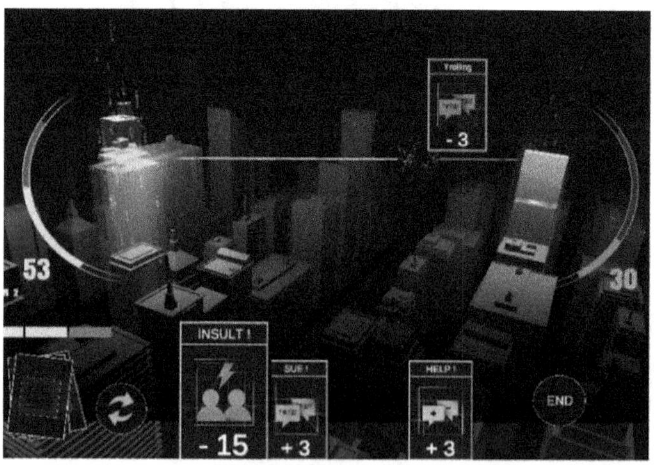

Fig. 3. Social Threads

6.4 Deplataforming

In 'Deplataforming' the player takes on the role of an activist group whose aim is to counter hate speech on multiple online platforms. The player must use the kit of available actions to be able to mitigate hate speech and demonetize and ban the users who propagate it on the platforms. Hate speech spreads quickly over the Internet, uncontrollably spreading from platform to platform; the player's mission is trying to prevent the hate speech campaign from spreading and controlling the Internet. If hate speech reaches 100% the game is over (Fig. 4).

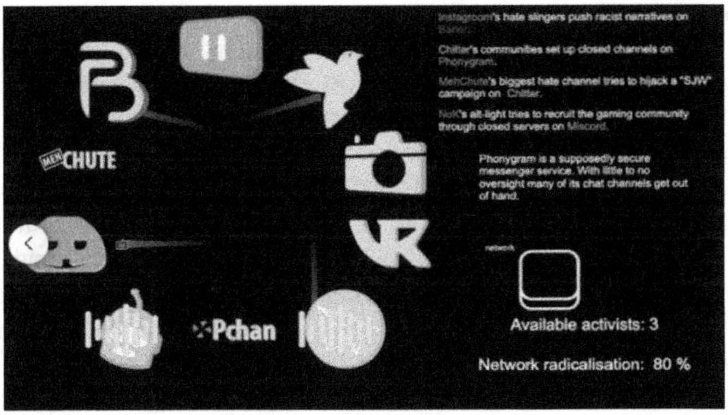

Fig. 4. Social Threads

7 Pervasive Games

Pervasive games are game situations that expand the magic circle defined by Huizinga [14] at spatial or temporal level [22]. Considered as a new form of game that escapes easy definitions, these games often include under their generic concept other forms of game, such as augmented reality games, geographic location games, urban games, hybrid reality games, among others. Adriana de Souza e Silva and Daniel M. Sutko (2009) define pervasive games as a set of ludic activities that use mobile technologies as interfaces and the physical space as a game board. Here, the game appears connected to the public space, often a city or a specific area within a city. The space dedicated to the game is always larger when compared to traditional games, since they happen on a human scale. Another feature of this type of game is the use of communication technologies, such as mobile phones, the Internet, location media, such as GPS and augmented reality, for example [22].

According to Mark Weiser [27], one of the precursors of this concept, Pervasive Computing or Ubiquitous Computing integrates information technology with everyday actions and behaviors. It is a game typology that expands the experiences of video games to the physical world, involving both physical and electronic spaces [23]. The narrative of pervasive games usually consists of finding someone or something, or avoiding being found; in some contexts, it takes the form of a treasure hunt, based on the idea of geocaching[7], for example.

Pervasive games have the potential to engage the player in contextual challenges, establishing a connection with the surrounding environment [4]: ludic and organized practices in urban environments with some type of technological/digital support and serving social purposes - i.e., with the purpose of raising awareness about specific issues.

Ferri and Coppok [7] define "Urban Games" as a specific subgroup within Pervasive Games, set in metropolitan areas, which encourage participants to move freely and interact with public spaces. According to these authors, Urban Games are often designed to create a minimum level of competition among players, emphasizing the exploration, experimentation and creative use of urban spaces instead. Jane McGonigal [19] argues that the transformation of a daily problem into a voluntary challenge activates a genuine interest, based on curiosity, motivation, effort and optimism, which would not exist otherwise. Motivation is the desire to be involved in a game that can thus acquire a new, more relevant meaning, to which the player relates.

In this project of reaction to hate speech that we have been describing, understanding the conditions that cause the expression of hatred in the interactions among players, helped determine a new goal, namely the dissemination of educational tools, in the form of serious games, such as certain ludic practices that seek, at the same time, to promote broader and more concrete effects of social awareness among players within a community, in a very wide range of urban and suburban public areas, thus transforming these spaces into a kind of "ludic interface", a term coined by Ferri and Coppok [7].

8 Prototyping a Pervasive Game

One of the proposals included in the active dissemination plan of the project is the development of urban games: a type of pervasive game, prototyped with a selected group of young people, thanks to the network of each partner: schools, youth centers or associations will be involved as key actors during the phase of dissemination and not just as target groups. Young people become game developers and use the pleasure and skills offered to video game players to activate civic participation.

The project entails the creation of four urban games in the cities of Rimini (Italy), Munich (Germany), Warsaw (Poland) and Siauliai (Lithuania), with the support of a user guide containing the main tasks, stages and descriptions, which will be made available free of charge and in the different European languages involved in the project, to spread this concept of urban game against intolerance and hate speech.

The proposed plot is cemented in a video game blog managed by two authors - a boy and a girl - who write about their gaming experiences. The narrative of the game determines that the blog will be shown as cancelled when the game starts and the only accessible page will correspond to the post-mortem of the page, thanking the fans. The

[7] In geocaching, a small box with some items and a logbook are hidden in a place the public can have access to. The GPS coordinates of that box are posted on a website. Geocachers use their portable GPS devices to find the box, from which they can retrieve an item and replace it with another object equally relevant to the challenge. The logbook is then signed and the discovery is reported on the website.

cancellation is officially declared, leading to inherent questions: What happened to the authors? Why did they decide to stop writing?

The game was designed as a treasure hunt, where each found clue allows players to gather one more piece of information about the history of the blog, as if fitting a puzzle piece into its place. The narrative unfolds in a crescendo of hatred, starting with a negative comment until it reaches a situation the authors can no longer sustain. The idea is to portray a plausible experience, which helps players reflect on the relevance of online hatred and its impact on the lives of Internet users.

Once the main narrative is concluded, users will be able to access a secret post, published after the last post they had already had the chance to read and which will be considered an epilogue. The last message will provide the narrative with a more well-rounded ending, lead to a more consistent reflection on the story and convey a message of hope.

Due to the current pandemic, the game was redesigned into a virtual format, which will be based on social networks and on the blog; however, users will be provided with tools which will allow them to move part of the treasure hunt to the physical world, the 4 cities previously mentioned.

Inspired by the mechanics of Alternate Reality Games (ARG), a manual will be developed to guide users, as well as teachers, educators and training mediators, regarding this transfer from the virtual game to the physical game, such as, for example, a note inside a book in a library, a box next to an emblematic monument, an envelope at the entrance of a theater.

The act of investigating and reconstructing a story in order to understand it, adds value to the game itself, recreating a more personal and direct connection among players, bystanders and the urban environment where the game takes place [7]. As such, participants are not "mere players", they also take charge of a small yet significant part of an investigation to discover the specific history of a certain place.

In this initiative, which we can classify as a serious urban game, we combine social interaction among players, interaction with the space, the feeling of adventure when exploring new environments or when solving the challenge; on the other hand, when reconstructing the narrative around hate speech and its consequences, it is inevitable

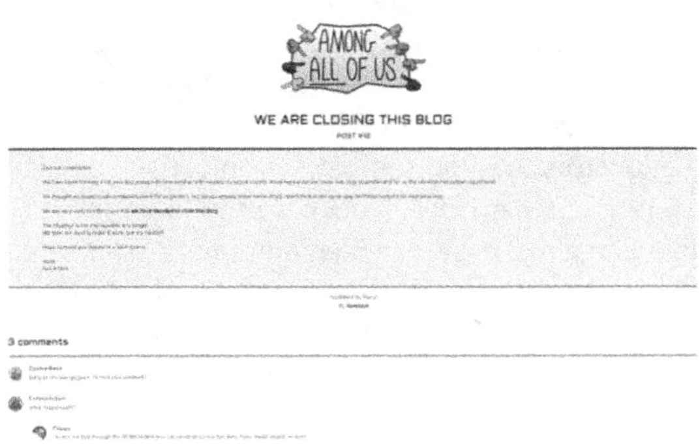

Fig. 5. Among all of us

the involvement of the player with the problem, with the social and even political consequences around an urgent issue that exists in both the physical and virtual worlds (Fig. 5).

9 Conclusions and Directions for Future Research

According to the report recently published by the team of the Project (available in May 2020[8]), it is possible to conclude that sanctions and criminalization, in response to hate speech, are relevant actions, however, they are assumed as a partial and temporary cure for a problem that needs to be studied and worked on from the roots of its cultural dimension, focusing mainly on education, literacy and the promotion of critical and active citizenship. Learning about, through and for Human Rights is essential in maintaining an active climate of empathy and respect for the other among Internet users as well.

Supranational institutions, such as UNESCO, ONU or the European Council have taken different actions with the objective of containing the online hate speech. Recent initiatives have been implemented for preventing violent extremism through education. These organizations underline that it is not enough to combat violent extremist behavior – it is necessary to avoid it through media literacy.

Our contribution in this field consists in using games imbued with a counter-narrative capable of containing online hate speech, by valuing awareness of Human Rights, through a powerful tool, i.e., the serious games.

To get to these tools, we started from the analysis of a questionnaire survey applied by the team of the project "Play Your Role" to 572 young people and teenagers, aged between 12 and 20, from three European countries (Portugal, Italy and Lithuania). This sample made it possible to analyze the interactions among young players in games and in online gaming platforms[9]. The analysis of the data showed that young people who spend more time playing online, show a greater tendency to resort to hate speech and to become victims of that same type of expression. On the other hand, it is also clear that resorting to sanctions and criminalization as a response to hate speech are relevant actions; however, and though necessary, these measures are only a partial and provisional cure for a problem that needs to be studied and addressed from the roots of its cultural dimension, focusing mainly on education, literacy and the promotion of critical citizenship. Learning about, through and for Human Rights is essential to actively generate empathy and the respect for the Other, including among Internet users.

It is intended that the experimentation with the four online games and the pervasive games, mediated by educators, has the same main objective as that of the project "Play Your Role - Gamification against Hate Speech": implementing a serious fight against online hate speech by instigating positive behavior among young people. The reinforcement of critical thinking about the sources and content of the information they find on the network can benefit several factors, such as self-esteem, self-knowledge, catharsis, deconstruction and reconstruction of real situations, as well as the construction of

[8] https://www.playyourrole.eu/wp-content/uploads/2020/07/PYR-research-report.pdf.

[9] The complete data analysis resulting from this survey can be accessed on the website of the project: www.playyourole.eu.

identity in a simulated space. After all, simulation does not represent mere objects and systems, it mainly represents models and behaviors [8].

The aim of this research was to reflect on hate speech online and suggest possible ways to combat it. The creation of a community, united by a common goal, based on the gamification of a problem and favored by spatial convergence in the form of an urban game, is a useful tool, even if just a grain of sand, a starting point in the mobilization against hate speech. The overflow between different realities and fictionality paves the way for a collective experience engaged in shared pretense, an awakening of awareness for the consequences of the toxic discourse that proliferates through the streets of the virtual world.

Finally, we would like to mention that the consortium intends to continue this fight against hate speech with the continuation of the project called Playing Against Hate. The overall objective of this new project is also to prevent and address hate speech online, enhancing video games and gamification as tools to reinforce positive behaviours in youngsters with respect to all diversities (gender, sexual orientation and gender identity, ethnic origin and religion). This is achieved by improving the capacity of teachers, educators and young people to identify and address online hate speech; promoting video games and gamification as an approach to prevent and address hate speech online in formal and non-formal education; raising the awareness of young people, educational communities and the general public through new positive narratives. Gamification, Media Education and Intersectionality are the 3 concept axes of the project, which focuses on gaming culture and connected social spheres as motor for the promotion of democratic values, critical thinking and digital citizenship. The project follows the objective of the call to promote equality and to fight against racism, xenophobia and discrimination, by promoting gaming as an approach to prevent hate speech online within formal and non-formal education (UN priority number 4).

Acknowledgments. This publication is financed by national funds through the project "UIDP/04019/2020 CIAC" of the Foundation for Science and Technology, I.P. and the PhD project UI/BD/150850/2021.

This publication was funded by the European Union's Rights, Equality and Citizenship Programme (2014–2020). The content of this publication represents the views of the author only and is his/her sole responsibility. The European Commission does not accept any responsibility for use that may be made of the information it contains.

References

1. Deslauriers, P., St-martin, L., Bonenfant, M.: Assessing toxic behaviour in dead by daylight: perceptions and factors of toxicity according to the game's official subreddit contributors. Int. J. Comput. Game Res. **20**(4) (2020). ISSN:1604-7982
2. Carpenter, N.: Media and Participation A Site of Ideological-Democratic Struggle. Intellect Ltd., Chicago (2011)
3. Chou, Y. - K.: Actionable Gamification: Beyond Points, Badges and Leaderboards. Createspace Independent Publishing Platform, Scotts Valley (2015)
4. Coelho, A., et al.: Serious pervasive games. Front. Comput. Sci. **31** (2020). https://doi.org/10.3389/fcomp.2020.00030

5. Contreras-Espinosa, R., Scolari, C.: How do teens learn to play video games? J. Inf. Lit. **13**, 45 (2019). https://doi.org/10.11645/13.1.2358
6. Gagliardone, I., Gal, D., Alves, T., Martinez, G.: Countering Online Hate Speech. United Nations Educational, Scientific and Cultural Organization, Paris (2015)
7. Ferri, G., Coppock, P.: Serious urban games. From play in the city to play for the city. In: Tosoni, S., Tarantino, M., Giaccardi, C. (eds.), Media and the City: Urbanism. Technology and Communication. Newcastle Upon Tyne, pp. 120–134. Cambridge Scholar Press (2013)
8. Frasca, G.: Ludology Meets Narratology: Similitude and Differences Between (Video) Games and Narrative. Helsinki: Parnasso 3, pp. 365–371 (1999)
9. Gee, J.P.: What Video Games Have to Teach Us About Learning and Literacy. Palgrave Macmillan, New York (2003)
10. Giasolli, V., Giasolli, M., Giasolli, R., Giasolli, A.: Serious gaming: teaching science using games. Microsc. Microanal. **12**(S02), 1698–1699 (2006). https://doi.org/10.1017/S1431927606061149
11. Greenawalt, K.: Rationales for freedom of speech. In: Moore, A.D. (ed.) Information Ethics: Privacy, Property, and Power, pp. 278–296. Washington University Press, Washington (2005)
12. Grizzle, A., Tornero, J.: Media and information literacy against online hate, radical and extremist content: some preliminary research findings in relation to youth and a research design. In: Singh, J., Kerr, P., Hamburger, E. (eds.) Media and Information Literacy: Reinforcing Human Rights, Countering Radicalization and Extremism, pp. 179–202. UNESCO, Paris (2016)
13. Gubrium, A., Harper, K.: Participatory Visual and Digital Methods. Left Coast Press, Walnut Creek, CA (2013)
14. Hokka, J.: PewDiePie, racism and youtube's neoliberalist interpretation of freedom of speech. Convergence **27**(1), 142–160 (2021). https://doi.org/10.1177/1354856520938602
15. Huizinga, J.: Homo Ludens. Lisboa, Edições 70 (1938)
16. Hurley, S.: Imitation, media violence, and freedom of speech. Philos. Stud. **117**(1/2), 165–218 (2014). https://doi.org/10.1023/B:PHIL.0000014533.94297.6b
17. ISFE. https://www.isfe.eu. Accessed 10 June 2021
18. Jenkins, H.: Cultura da Convergência. Aleph, São Paulo
19. Mcgonigal, J.: Reality Is Broken: Why Games Make Us Better and How They Can Change the World. Jonathan Cape London, London (2009)
20. Kwak, H., Blackburn, J.: Linguistic analysis of toxic behavior in an online video game. In: Aiello, L.M., McFarland, D. (eds.) SocInfo 2014. LNCS, vol. 8852, pp. 209–217. Springer, Cham (2015). https://doi.org/10.1007/978-3-319-15168-7_26
21. Matamoros-Fernández, A.: Platformed racism: the mediation and circulation of an Australian race-based controversy on Twitter Facebook and YouTube. Inf. Commun. Soc. **20**(6), 930–946 (2017). https://doi.org/10.1080/1369118X.2017.1293130
22. Montola, M.: Exploring the edge of the magic circle: defining pervasive games. In: Proceedings of Digital Arts and Culture. IT University of Copenhagen, Conpenhagen (2005)
23. Montola, M., Stenros, J., Waern, A.: Pervasive Games. Theory and Design. Experiences on the Boundary Between Life and Play. Morgan Kaufmann Publishers, Burlington (2009)
24. Perez, Ó.: Libertad de Expresión y Lenguaje del Odio como un Dilema entre Libertad e Igualdad. In: RAEIC, Revista de la Asociación Española de Investigación de la Comunicación, vol. 6, issue 12, pp. 5–34 (2019). https://doi.org/10.24137/raeic.6.12.1
25. Ponte, C., Batista, S.: EU kids online Portugal. Usos, Competências, Riscos e Mediações da Internet Reportados por Crianças e Jovens (9–17 anos). EU Kids Online e NOVA FCSH, Lisboa (2019)
26. Prensky, M.: Listen to the natives. Educ. Leadersh.: J. Dept. Superv. Curric. Dev. N.E.A **63**(4) (2006)
27. Salen, K., Zimmerman, E.: Rules of Play: Game Design Fundamentals. MIT Press, Cambridge (2003)

28. Schell, J.: The Art of Game Design: A Book of Lenses. Morgan Kaufmann Publishers, Burlington (2013)
29. Stoller, R.: Observing the Erotic Imagination. Yale University Press, London (1985)
30. Werbach, K., Hunter, D.: For the Win: How Game Thinking can Revolutionize Your Business. Wharton Digital Press, Upper Saddle River (2012)
31. Zichermann, G., Cunningham, C.: Gamification by Design: Implementing Game Mechanics in Web and Mobile Apps. O'Reilly Media, Sebastopol (2011)

Dynamic Suspense Management Through Adaptive Gameplay

Robert Levin[1], Skyler Zartman[1], and Ying Zhu[2](✉)

[1] Rochester Institute of Technology, Rochester, NY, USA
{bil1616,sz5100}@rit.edu
[2] Georgia State University, Atlanta, GA, USA
yzhu@gsu.edu

Abstract. Suspense is an important emotion for the enjoyment of games. Various methods have been proposed to manage suspense in games. Most existing works focus on managing suspense via storytelling or artifacts, such as sound effects. However, little work has been done in studying how to use gameplay to manage suspense. In this paper, we present a study in which we developed a horror-adventure game with a built-in suspense manager based on adaptive gameplay. We conducted a small user study to evaluate the effect of dynamic suspense management on game players. Our results showed that gameplay could potentially be used to manage the level of suspense experienced by players, independent of the story and artifacts in the game. The work discussed in this paper will provide game designers with new tools for suspense management in non-narrative-based games.

Keywords: Affective computing · Game design · Suspense

1 Introduction

Suspense is a feeling of anxiety or excitement about an uncertain future. It is an important emotion for the enjoyment of different types of entertainment media, such as novels, films, TV, music, sports games, and video games [11,16,17,19,25]. In this paper, we focus on managing suspense in video games. Game designers often want to manage gameplayers' emotional experience during gameplay, and suspense is an important part of that emotional experience, especially for certain game genres, such as survival horror games.

In previous works, different methods have been proposed to manage suspense in games. Most of these methods focus on manipulating stories or game artifacts such as sound effects, perhaps because similar techniques have long been studied and used in other fields such as films and TV. However, little work has been done in studying how to adjust the gameplay to manage suspense. This is a new area that does not have much to borrow from other areas. Our work is an attempt to address this gap.

M. Wölfel et al. (Eds.): ArtsIT 2021, LNICST 422, pp. 94–104, 2022.
https://doi.org/10.1007/978-3-030-95531-1_7

In this paper, we present a study in which we developed a horror-adventure game with a built-in suspense manager based on adaptive gameplay. We conducted a small user study to evaluate the effect of dynamic suspense management on game players. Our results showed promising results that gameplay could be used to manage game suspense independent of the story and artifact layer. Our work provides more tools and evidence for game designers to manage gameplayers' emotional experience through adaptive gameplay.

2 Related Work

Zhu [31] proposed a theoretical framework for managing suspense in games. Based on this framework, suspense can be elicited on three different layers: story, gameplay, and artifact. Suspense is elicited through an affective loop [31]. First, a player suspense model needs to be developed to estimate a gameplayer's level of suspense. The estimation of a player's suspense can be based on the current state of the game and, in some cases, the player's physiological inputs. Then, the estimated player suspense is fed into a suspense manager to manipulate the story, gameplay, or artifacts, which in turn elicit suspense from the player. Our demonstration game is implemented largely based on this framework.

Many previous studies have focused on manipulating stories to manage suspense [4,5,9,10,13,22–24,27]. For example, Giannatos et al. [13] and Cheong and Young [5] used plan-based models to manipulate story structures in order to manipulate the reader's suspense. O'Neill and Riedl's computational model [23] generates plans for the protagonist to avoid an impending negative outcome. These computational models were largely based on the cognitive model proposed by Gerrig and Bernardo [12]. In Szilas and Richle's model [27], tension is generated by creating paradoxical narratives. Suspense in storytelling has long been studied in arts, psychology, and economy [1,11,12,15,17,19,21,26]. There is a rich set of cognitive theories for game designers to draw from.

Artifacts can also be used to manipulate the level of suspense. A number of studies have shown that sound effects can be used to manipulate anxiety and suspense [6,14,20,28–30]. Delatorre et al. [8] suggested decorative objects could also influence the perception of suspense.

Some researchers have attempted to manage suspense by dynamically adjusting gameplay. For example, Liu et al. [18] changed the level of difficulty in a game based on the anxiety level of players. In a game developed by Vachiratamporn et al. [30], a player's level of suspense affected the timing of scary events. Bailey and Zhu [2] managed suspense by dynamically adjusting gameplayers' uncertainty, fear, and hope. For example, uncertainty was controlled by withholding information from players. Fear was controlled by adjusting the speed of the enemies and the distance between the player and enemies. Given the wide variety of gameplays in video games, there is still a lack of research on the theory and practice of using gameplay to manage suspense. This is the primary motivation for our work.

3 Managing Suspense Through Adaptive Gameplay

3.1 Game Design Goals

Our goal was to study how adaptive gameplay can be used to manipulate suspense in a video game without a story and how effective this technique might be. We first began our design process by looking at various game genres to determine which would best fit our model for suspense. We quickly found that horror-adventure would suit our goal best. Our framework relies heavily on the fear, hope, and uncertainty model of suspense proposed by Ortony et al. [21]. While games of other genres certainly still include all three, we found that horror seemed to be the most saturated in this regard.

To elaborate further on that subject, one of the issues we ran into early was giving false information to the player. Many games require that the player repeats certain mechanics over and over to become more proficient. Because our suspense manager seeks to actively change mechanics in real-time, we had to be careful when designing what our suspense manager would influence. Influencing the wrong mechanics could build a sort of mistrust in the player, who might end up feeling cheated out of learning the skills to succeed.

Another major design choice was the game's point of view. We considered virtual reality, first-person, and top-down 2D perspectives. We chose the top-down 2D perspective because it conveys more information to the player. It is easier to influence players with the mechanics worked on by the suspense manager. However, we later learned that the 2D perspective also has the significant drawback of hindering immersion and fear.

Our game, Photophobic, is a top-down 2D horror game set in a creepy apartment complex (Figs. 1, 2 and 3). The player has just woken up to find themselves alone. Now they must find the red keycard to get into the elevator and exit. In it, the player must make their way through a series of perilous rooms to find the correct keycards. Using only their fleeting flashlight, they must also eliminate and avoid the many dangers they face along the way.

An example run of a player might go as follows. After finding a blue keycard, the player moves into the blue door to enter. After moving in, noises hint that enemies may be lurking inside, so the player turns on their flashlight. Doing so will continuously drain their battery, but without the light, they will not be able to see the pair of enemies sneaking upon them. In doing so, the player stumbles upon a battery spawned in using the suspense manager. After eliminating the enemies using their light and finding a new key, the player heads out to explore another complex.

3.2 Player Suspense Model

Our suspense model creates an estimation of the player's level of suspense based on the idea that the player's level of suspense changes as they receive information from the game world. The level of suspense for the player can then be quantified based on the popularly accepted OCC suspense model [21]. In this model,

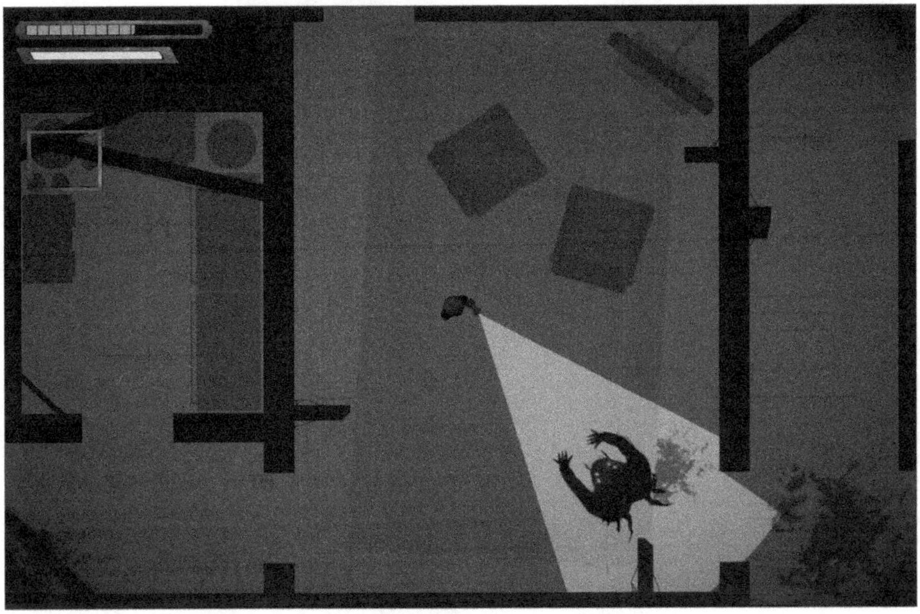

Fig. 1. Monster encounter

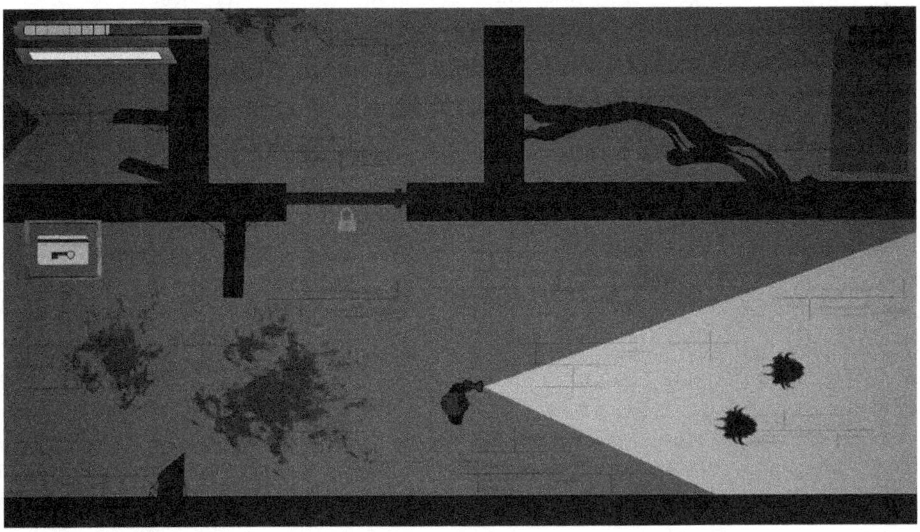

Fig. 2. Attacked in the hallway

suspense is generated by a combination of three factors: hope, fear, and uncertainty. We began breaking the various game mechanics into our three categories of hope, fear, and uncertainty and then assigned weights to each mechanic.

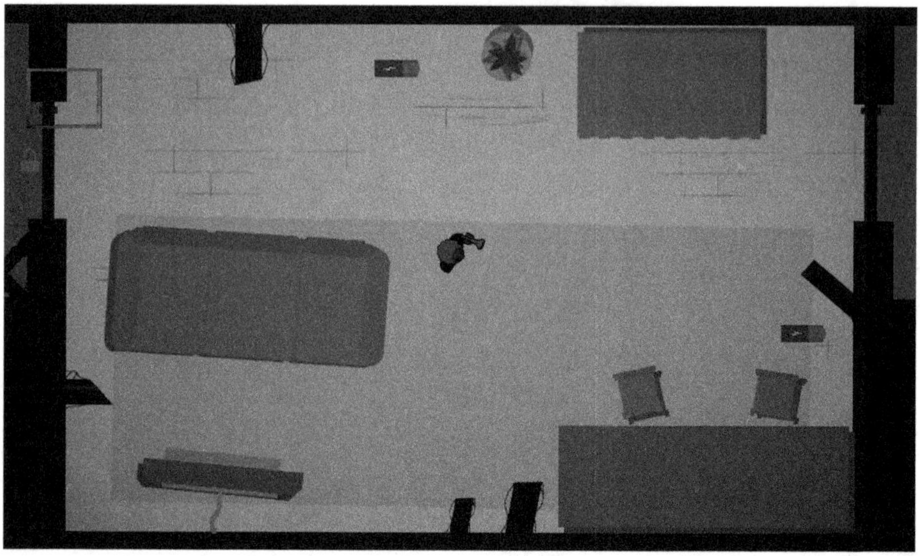

Fig. 3. The player in a saferoom

We used fuzzy logic [3] to determine the weight of each mechanic that affects each of the three categories. For example, the detection of an enemy NPC can have a high increase in fear. Not knowing the location of a battery pickup can lead to a low increase in uncertainty. Knowing the location and distance of a key or goal can have a moderate increase in hope. The specific details may vary from game to game and will be determined by game designers and developers. For our game, we had a breakdown that is shown in Fig. 4.

Once we defined each of the mechanics in terms of hope, fear, and uncertainty, we created equations to evaluate the game state and determine the player's level of suspense (see Fig. 5). We determine the value of the modifiers through the testing of our game. We specified weights based on designer expectations, and we have the system adjust those weights during play to ensure no mechanic overpowers the system. Finally, we had a value that represented the player's level of suspense during gameplay. The calculated level of suspense for the player was not necessarily the same suspense felt by the human player but served as a valid estimate of player suspense because it was based on the widely-accepted OCC cognitive model. This gave us a way to estimate the emotional experience of gameplay.

3.3 Suspense Manager

In order to run our suspense management system, we created a target suspense curve for our game based on MIT and McKinsey's work on emotional arcs [7]. The game will adjust the behavior of the game object based on the difference between the estimated suspense curve and the target suspense curve. The points

	Hope	Fear	Uncertainty
High	• Player Holding Key • Battery Amount	• Enemy location • Enemy following player • Light On/Off	• Enemy locations • Light On/Off
Med	• Player speed • Stamina Amount • Safe room proximity	• Enemy type • Battery Amount • Player Health	• Key location • Battery pickup location • Enemy status
Low	• Battery drain/recharge • Stamina drain/recharge	• Stamina Amount • Battery drain/recharge • Stamina drain/recharge • Sound effects	• Door location • Player knowledge of mechanics

Fig. 4. Breakdown of game mechanics and their mappings to hope, fear, and uncertainty. The weights are set using fuzzy logic.

$$\begin{aligned} hope =\ & holdingKey * highMod \\ & + (playerMoveSpeed + \frac{battery}{maxBattery} + \frac{Stamina}{maxStamina} + saferoomDistance) * medMod \\ & + (staminaChargeRates + batteryChargeRates) * lowMod \end{aligned}$$

$$\begin{aligned} fear =\ & (numEnemies + enemyDistance + enemyThreat + lightStatus) * highMod \\ & + (enemyTypes + \frac{maxBattery}{battery} + \frac{maxHealth}{health}) * medMod \\ & + (\frac{maxStamina}{stamina} + batteryChargeRates + staminaChargeRates + soundEffects) * lowMod \end{aligned}$$

$$\begin{aligned} uncertainty =\ & (\frac{enemiesSpotted}{TotalEnemies} + lightFlicker) * highMod \\ & + (\frac{batteriesFound}{TotalBatteries} + \frac{keysFound}{TotalKeys} + enemyMovementSet) * medMod \\ & + (\frac{doorsUnlocked}{TotalDoors} + timePlayed) * lowMod \end{aligned}$$

Fig. 5. Once broken down, we can turn the mechanics into our equations and assign the weights though testing.

of the curve were defined as specific values for suspense at key points in the level. As the player progressed through the level and reached key points, we would shift to the next stage of the curve, allowing the suspense manager to account for players progressing at wildly different paces. The manager was given control over various mechanics of the game to keep the estimated suspense level in line with the desired level. These mechanics included things such as:

- Changing battery and stamina charge and discharge rates
- Adjusting player and enemy max move speed
- Randomizing or altering enemy pathing
- Spawning or despawning enemies and battery pickups.

We chose these mechanics for the suspense manager in an effort to remain within our design goal of ensuring the player does not feel cheated by the system.

Each of these mechanics is outside of the player's ability to notice when making small adjustments and, therefore, will go unnoticed. While the adjustments themselves may not be noticed, how they interact and the cascading effects will be noticed. For example, a player may not notice their flashlight draining 1% per second vs. 1.5% per second, but they will notice when their battery charge is low, which will occur faster. Enemies that the player could easily run away from before could still escape but will remain much closer to the player and have a higher chance of catching them if the player makes a mistake. Through these changes, we can measure and affect the level of suspense a player is feeling.

We then used the established mechanics to create an AI that performs these changes in real-time based on the current and desired suspense levels. We divided the mechanics into high and low impact changes. Low impact changes are small adjustments made to the values the player cannot see, such as adjusting the rate of battery drain, which allow a fine adjustment of the suspense and can easily be reversed. High impact changes are more dramatic actions like the spawning of items or enemies, which are very noticeable and not easy to undo. Our suspense manager uses the high and low impact changes to evaluate, analyze, and adjust the estimated level of suspense in our game in order to match it to the target suspense curve.

4 User Study

We conducted a user study after finishing the game development using Unity. We had five volunteers participate in our study, undergraduate students ranging from the ages of 18 to 21. Each participant was set up with an Apple Watch in order to measure the participant's resting heart rate and their heart rate throughout the game. After finishing the game, each participant was also set up with a recording of their playthrough. Following the same time between heart rate checks, every 30 s, we also asked them to roughly estimate their levels of suspense from 0–10. This way, we would be able to analyze the difference in suspense from the tester's heart rate, personal estimation, and the estimated suspense values from our in-game player suspense model.

The results of our testing can be seen in Figs. 6, 7, and 8. In Fig. 6, each blue line shows a player's level of suspense estimated by the game AI. Each red line shows a player's heart rate. The sample size is too small to conduct meaningful statistical analysis. While there are no clear correlations between the estimated suspense and heart rate for Tester 1 and Tester 4. There are some similarities between the patterns in the estimated suspense level and the heart rate for Tester 2 and 3. Whether heart rate is a reliable measurement of suspense is still debatable, especially for low-level suspense. Our preliminary research into this area is inconclusive, and we hope that further work will yield more data for analysis.

Figures 7 and 8 show the self-rated suspense level by players and the desired suspense curve from two gameplay sessions. Here the rise and fall of the two curves show some similarities. In Fig. 7, although the desired climax (red line) is

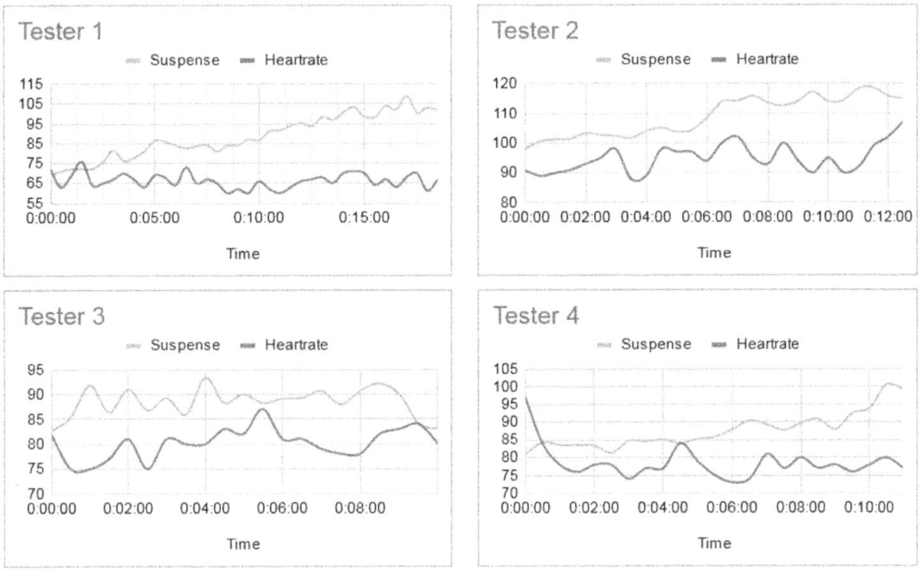

Fig. 6. Players' heart rate and estimated suspense values

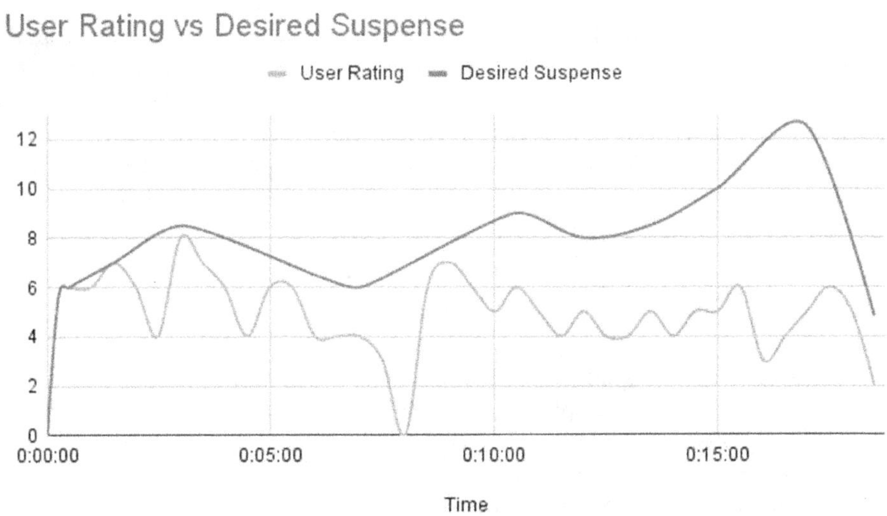

Fig. 7. Self-rated suspense levels (from 0 to 10) by players vs. desired suspense curve (Color figure online)

not manifested in the self-rated suspense curve (blue line), the "rise and fall and rise" pattern in the first half of the red line is present in the blue line. Similarly, in Fig. 8, the rise and fall patterns in the desired suspense curve (blue line) are largely present in the self-rated suspense curve (red line).

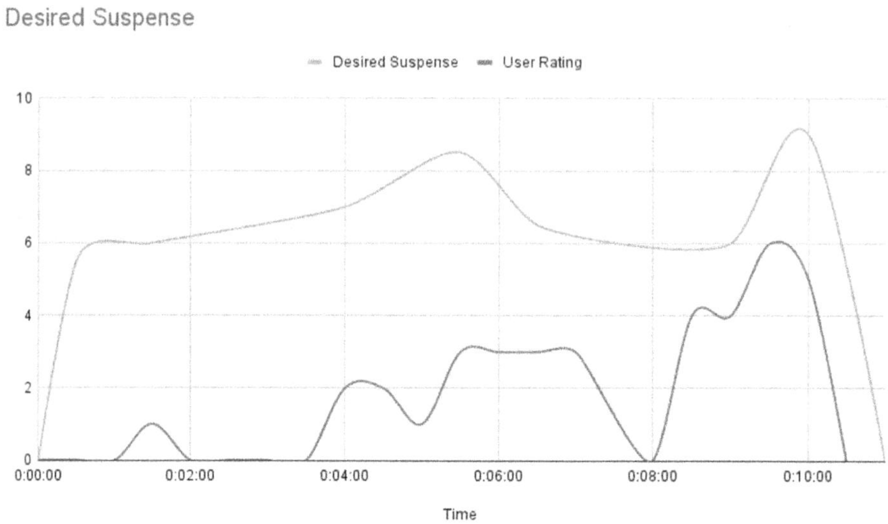

Fig. 8. Self-rated suspense levels by players vs. desired suspense curve (Color figure online)

These preliminary results demonstrate the potential of our suspense model and management system. With more user data and tweaking, the suspense curves could become more accurate estimations of suspense and anxiety. For example, the curve did not account for the drops in suspense that occur when the player ends up lost for extended periods of time. These factors could result in players experiencing different levels of anxiety than we anticipated at particular moments.

5 Conclusion and Future Work

We have presented our study in which we tested a computational model of suspense for non-narrative gameplay. Using this model, we generate and manage the user suspense using only gameplay, without the intervention from narrative and artifacts. Our small user study showed some potential, but more experiments are needed. This model can be used for non-narrative-driven games or to supplement narrative-based computational models of suspense in story-driven games. The results from our experiments show that there could be a viable solution. Using what we learned from the development process, there is an opportunity for other game developers to create a game with our framework integrated as a core feature. In the future, we plan to do more user studies and collect more user data to improve our player suspense model and collect enough data to draw meaningful conclusions about the correlation between our system and player experience.

Acknowledgement. This work was supported in part by NSF Award #1852516.

References

1. Abuhamdeh, S., Csikszentmihalyi, M., Jalal, B.: Enjoying the possibility of defeat: outcome uncertainty, suspense, and intrinsic motivation. Motiv. Emot. **39**(1), 1–10 (2014). https://doi.org/10.1007/s11031-014-9425-2
2. Bailey, E., Zhu, Y.: A computational model of suspense for non-narrative gameplay. In: Proceedings of the 24th International Conference on Information Visualisation (IV), pp. 767–770. IEEE (2020). https://doi.org/10.1109/IV51561.2020.00136
3. Capelo, D., Caminha, C., Nogueira, P.A.: Faculdade de engenharia da universidade do porto development of emotional game mechanics through the use of biometric sensors (2017)
4. Chanel, G., Rebetez, C., Bétrancourt, M., Pun, T.: Boredom, engagement and anxiety as indicators for adaptation to difficulty in games. In: Proceedings of the 12th International Conference on Entertainment and Media in the Ubiquitous Era, pp. 13–17. ACM (2008)
5. Cheong, Y.G., Young, R.M.: Suspenser: a story generation system for suspense. IEEE Trans. Comput. Intell. AI Games **7**, 39–52 (2015)
6. Chittaro, L.: Anxiety induction in virtual environments: an experimental comparison of three general techniques. Interact. Comput. **26**, 528–539 (2014)
7. Chu, E., Dunn, J., Roy, D., Sands, G., Stevens, R.: AI in storytelling: machines as cocreators. https://www.mckinsey.com/industries/technology-media-and-telecommunications/our-insights/ai-in-storytelling (2017)
8. Delatorre, P., León, C., Gervás, P., Palomo-Duarte, M.: A computational model of the cognitive impact of decorative elements on the perception of suspense. Connect. Sci. **29**, 295–331 (2017)
9. Delatorre, P., León, C., Salguero, A., Palomo-Duarte, M., Gervás, P.: Information management in interactive and non-interactive suspenseful storytelling. Connect. Sci. **31**, 82–101 (2019)
10. Doust, R., Piwek, P.: A model of suspense for narrative generation. In: Proceedings of the 10th International Conference on Natural Language Generation, pp. 178–187. Association for Computational Linguistics (2017)
11. Ely, J., Frankel, A., Kamenica, E.: Suspense and surprise. J. Polit. Econ. **123**, 215–260 (2015)
12. Gerrig, R.J., Bernardo, A.B.: Readers as problem-solvers in the experience of suspense. Poetics **22**(6), 459–472 (1994)
13. Giannatos, S., Cheong, Y.G., Nelson, M.J., Yannakakis, G.N.: Generating narrative action schemas for suspense (2012)
14. Graja, S., Lopes, P., Chanel, G.: Impact of visual and sound orchestration on physiological arousal and tension in a horror game. IEEE Trans. Games **14**, 1–13 (2020)
15. Kaspar, K., Zimmermann, D., Wilbers, A.K.: Thrilling news revisited: the role of suspense for the enjoyment of news stories. Front. Psychol. **7**, 1913 (2016)
16. Klimmt, C., Hartmann, T., Frey, A.: Effectance and control as determinants of video game enjoyment. Cyberpsychol. Behav. **10**(6), 845–848 (2007)
17. Lehne, M., Koelsch, S.: Toward a general psychological model of tension and suspense. Front. Psychol. **6**, 79 (2015)
18. Liu, C., Agrawal, P., Sarkar, N., Chen, S.: Dynamic difficulty adjustment in computer games through real-time anxiety-based affective feedback. Int. J. Hum.-Comput. Interact. **25**, 506–529 (2009)

19. Moulard, J.G., Kroff, M., Pounders, K., Ditt, C.: The role of suspense in gaming: inducing consumers' game enjoyment. J. Interact. Advert. **19**(3), 219–235 (2019). https://doi.org/10.1080/15252019.2019.1689208
20. Ogawa, S., Fujiwara, K., Kano, M.: Auditory feedback of false heart rate for video game experience improvement. IEEE Trans. Affect. Comput. 1 (2020)
21. Ortony, A., Clore, G.L., Collins, A.: The Cognitive Structure of Emotions. Cambridge University Press, Cambridge (1988)
22. O'Neill, B.: Toward a computational model of affective responses to stories for augmenting narrative generation. In: D'Mello, S., Graesser, A., Schuller, B., Martin, J.-C. (eds.) ACII 2011. LNCS, vol. 6975, pp. 256–263. Springer, Heidelberg (2011). https://doi.org/10.1007/978-3-642-24571-8_28
23. O'Neill, B., Riedl, M.: Dramatis: a computational model of suspense. In: Proceedings of the Twenty-Eighth AAAI Conference on Artificial Intelligence, pp. 944–950. AAAI Press (2014)
24. Porteous, J., Teutenberg, J., Pizzi, D., Cavazza, M.: Visual programming of plan dynamics using constraints and landmarks. In: Proceedings of the Twenty-First International Conference on International Conference on Automated Planning and Scheduling, pp. 186–193. AAAI Press (2011)
25. Shafer, D.M.: Investigating suspense as a predictor of enjoyment in sports video games. J. Broadcast. Electron. Media **58**, 272–288 (2014)
26. Smuts, A.: The desire-frustration theory of suspense. J. Aesthet. Art Criti. **66**(3), 281–290 (2008)
27. Szilas, N., Richle, U.: Towards a computational model of dramatic tension. In: Proceedings of the Workshop on Computational Models of Narrative, pp. 257–276 (2013)
28. Toprac, P., Abdel-Meguid, A.: Causing fear, suspense, and anxiety using sound design in computer games. In: Grimshaw, M. (ed.) Game Sound Technology and Player Interaction: Concepts and Developments, pp. 176–191. IGI Global (2011)
29. Vachiratamporn, V., Legaspi, R., Moriyama, K., Numao, M.: Towards the design of affective survival horror games: an investigation on player affect. In: 2013 Humaine Association Conference on Affective Computing and Intelligent Interaction, pp. 576–581 (2013)
30. Vachiratamporn, V., Moriyama, K., Fukui, K., Numao, M.: An implementation of affective adaptation in survival horror games. In: 2014 IEEE Conference on Computational Intelligence and Games, pp. 1–8 (2014)
31. Zhu, Y.: A theoretical framework for managing suspense in games. In: Proceedings of the Third IEEE Conference on Games (2021)

Toward Injury-Aware Game Design

Marinel Tinnirello[1], Ying Zhu[2(✉)], and Steven Kane[3]

[1] Marist College, Poughkeepsie, NY, USA
info@marineltinnirello.com
[2] Georgia State University, Atlanta, GA, USA
yzhu@gsu.edu
[3] Wellstar Atlanta Medical Center, Atlanta, GA, USA
Steve.Kane@wellstar.org

Abstract. As video games and esports continue to grow in popularity, gaming injuries are also on the rise. In recent years, medical professionals have placed greater emphasis on preventing and treating gaming injuries and proposed specific gaming health guidelines. However, the game industry and game research community have not done enough to address the hazards of gaming injuries or raise awareness about such hazards to players, parents, and game designers. In this paper, we propose a framework of injury-aware game design that addresses the two main causes of gaming injuries: prolonged gaming and repetitive micro-trauma. We have identified a set of injury-aware game design techniques to help raise awareness of gaming-related hazards, promote healthy gaming behavior, and optimize gameplay to prevent injuries. We believe an effective way to deliver gaming-related health information to game players is through games themselves. To demonstrate this framework, we have developed an injury-aware game and conducted a user study with players and game designers. The results from the proof-of-concept game and user study show that both players and designers have a positive reception to the idea of implementing more inclusive measures into games, with nearly all participants of the user study being interested in the idea of hand exercise recommendations.

Keywords: Game design · Gaming injuries · Injury prevention

1 Introduction

A large portion of the population has been playing video games regularly, and many people play games for long periods. According to a recent report from the Entertainment Software Association (ESA) [11], there are nearly 227 million game players across all ages in the US. Specifically, 67% of American adults and 76% of American youth are game players. Seventy-seven percent (77%) of game players play more than three hours per week, and 51% of them play over seven hours per week. Fifty-five percent (55%) of the players have played more during

M. Wölfel et al. (Eds.): ArtsIT 2021, LNICST 422, pp. 105–119, 2022.
https://doi.org/10.1007/978-3-030-95531-1_8

the pandemic. In addition, esports has been growing rapidly in recent years, and esports players play video games intensively for even longer periods, ranging from 5.5 to 10 h daily [8,10]. Top esports competitors play 12 to 14 h a day, at least six days a week [18].

Excessive video game play often leads to gaming injuries [10,16]. A study of esports players by DiFrancisco-Donoghue et al. [8] found that the most frequent complaint was eye fatigue (56%), followed by neck and back pain (42%), wrist pain (36%) and hand pain (32%). Among the players surveyed, only 2% had sought medical attention. PC gamers have a higher incidence of hand and wrist-overuse injuries such as tendinitis and carpal tunnel syndrome [6].

Many medical professionals have studied gaming injuries and published their work in medical journals or health-related media platforms. The medical community has placed greater emphasis on the identification, management, and prevention of gaming-related health hazards. A comprehensive framework and detailed guidelines to address gaming injuries have been published in the field of sports medicine [10]. However, such information does not normally reach game players since they do not regularly read medical journals.

On the other hand, the game industry and academic game research community have not done enough to address gaming injuries. For example, neither the 2021 Essential Facts about the Video Game Industry by ESA [11] and the GDC 2021 State of the Game Industry Report [12] mentions gaming injuries or health issues. The popular game engines do not include any mechanism to monitor and report excessive gameplay or hand and wrist overuse. Healthcare-related game research generally focuses on studying the potential benefits of video games for treating health issues [1,9], such as promoting exercises or helping rehabilitation. Our work is different from this type of research in that we focus on the injuries or health issues caused by gaming.

We argue that game designers and developers can do much more to address gaming injuries by introducing injury-aware mechanisms into game design and game engines. We believe the best way to deliver gaming-related health information to game players is through games themselves. In this paper, we propose a framework for injury-aware game design. This framework includes three basic injury-aware design mechanisms: feedback to game designers, feedback to game players, and injury-aware game AI. At the center of this framework is a real-time player activity monitor component that can be added to existing game engines to collect player activity data. This relatively simple and non-intrusive activity monitor can provide feedback to game designers during game design or after the game is released. Game designers can use such feedback to modify the game mechanics and level designs to alleviate the mental and physical stress of players. Game designers can also create injury-aware game AI that takes real-time feedback from the activity monitor and dynamically adjust gameplay to reduce stress to hands and wrists. In addition, a summary report of player activities can be presented to game players and/or their guardians to keep them informed of potential health risks. Finally, personalized medical advice on exercises and injury prevention can be presented to game players and their guardians. This proposed framework is a type of calm technology [4] that stays largely in the background.

As a proof of concept, we have developed an injury-aware game to demonstrate some of the features in our proposed framework. This game includes a player activity monitor that records the players' key presses and locations in the game world. Detailed finger usage data is presented to game designers to help redesign the game to reduce hand overuse. A summary report is presented to game players to raise awareness of the potential hazards of gaming injuries.

We also conducted a user study to seek feedback from players and game designers about the proposed injury-aware game design mechanism. The majority of participants (70%) had previously played the game and 60% identified as game designers or developers. Those who had played the injury-aware version leaned toward the control switching mechanic being either neutral or somewhat enjoyable. The tool's feedback for recommending hand exercises after a session was a resounding yes for a feature. However, guardians were split on if it would be beneficial for recommending how a game should be used. 83.3% of game developers also found the idea for the real-time aspect of the tool to be beneficial for reshaping level design, with the same amount noting that they would use both the temporal (the keylogger) and spatial (zoning) aspects.

We believe that injury-aware game design should be part of the normal game design process. The game design community and ultimately the game players will benefit from a comprehensive and thorough study of how games can be designed to prevent gaming injuries and keep players properly informed of such risks. Our proposed framework is a step in that direction.

The rest of the paper is organized as follows. In Sect. 2, we discuss related work from the medical community and game research. In Sect. 3, we briefly discuss gaming injuries and health issues. In Sect. 4, we describe the injury-aware game design framework. In Sects. 5 and 5.4, we discuss our proof-of-concept injury-aware game and a preliminary user study. Section 6 is our conclusion and future work.

2 Related Work

Gaming related health issues have been the subjects of some previous medical research [3,5,7,8,10,13–15,17,19–26]. For example, Emara et al. [10] classified esports related hazards into the following categories: musculoskeletal hazards, sedentary activity hazards, central neurological and psychological hazards, and infectious hazards. McGee and Ho [20] pointed out that there was still a dearth of esports-specific medical research but argued that esports competitors are subject to the kinds of repetitive loads that increase the risk for tendinopathy.

In addition, Emara et al. [10] proposed a three-point medical care framework for sports medicine providers, trainers, and coaches to care for the esports athletes. The three-point framework includes awareness and management of common musculoskeletal and health hazards, opportunities for health promotion, and recommendations for performance optimization. There is no corresponding framework for game designers and developers to tackle gaming injuries from a game design perspective. Our work is an attempt to address this gap.

Most of the research on gaming-related health hazards was conducted in the field of medicine. Relatively little work was done in the game design and development research community to address health issues caused by gaming. We reviewed the papers published in the major game design and development conferences and journals for the last three years and found no paper related to gaming injuries or gaming-related health issues. The publication venues we reviewed include Foundations of Digital Games (FDG), IEEE Conference on Games, ACM CHI PLAY, IEEE Transactions on Games, Entertainment Computing, and Games for Health Journal. For example, IEEE Transactions on Games had a special issue on Serious Games for Health [9], but the papers were about using games for rehabilitation, healthcare education, childhood obesity treatment, etc. Similarly, the papers published in Games for Health Journal [1] were primarily about studying the potential benefits of games for healthcare. Our work is different from this type of research in that we focus on the injuries or health issues caused by gaming.

3 Gaming Injuries

There are two leading causes for gaming-related musculoskeletal hazards [10]: prolonged aberrant posturing and repetitive microtrauma. Prolonged aberrant posturing can lead to neck and back pain. Repetitive microtrauma can lead to musculoskeletal illness. Game players, particularly esports players, often use rapid and repetitive hand motions in gameplay. High-intensity games may reach up to 500 to 600 moves per minute, sometimes for long periods of time. As a result, over 30% of esports players reported hand and wrist pain [8,10]. Emara et al. [10] identified 18 specific types of esports related musculoskeletal and medical hazards, including overuse shoulder tendon pathology, overuse elbow tendon pathology, cubital tunnel syndrome, overuse wrist tendon pathology, carpal tunnel syndrome, cervical pain, thoracic pain, lumbar pain, gluteal pain, ischial pain, hamstring tightness, etc.

Prolonged gaming can also cause other health hazards such as visual strain, dry eyes, headache, sleep deprivation, excess weight gain, and psychological and behavior issues [10,22,25]. For example, Pujol et al. [22] found that children who played 9 h or more of video games per week were often associated with conduct problems, peer conflicts, and reduced prosocial abilities.

4 Injury-Aware Game Design

Injury-aware game design is about crafting an enjoyable gameplay experience and informing game players and designers about the hazards of gaming, promoting healthy gaming activity, and optimizing gameplay to prevent injury. We designed our framework based on the gamer's health framework by Emara et al. [10]. However, we approach the same problem from a game design perspective, and our focus is on delivering gaming-related health information via games themselves. While the frame by Emara et al. was designed to inform sports medicine providers, trainers, and coaches, our frame is designed to inform players and game designers.

Our framework addresses the two main causes of gaming-related hazards from three perspectives: hazard awareness, health promotion, and performance optimization. Based on this general idea, we have identified the main tasks for injury-aware game design (Table 1).

Table 1. Major tasks for injury-aware game design

	Prolonged gaming	Repetitive microtrauma
Hazard awareness	Monitor and display gaming time; display warnings	Monitor and display gameplay activity; display warnings
Health promotion	Display personalized health recommendations	Create and display personalized information about injury prevention
Performance optimization	Design and develop injury-aware game mechanics to prevent prolonged gaming	Design and develop injury-aware game mechanics to prevent repetitive microtrauma

To support these tasks, an injury-aware game needs to have one or more of the following components. A player activity monitor needs to be embedded in the game to collect player activity data, such as the length of each gaming session, the frequency of keystrokes and mouse clicks, etc. Such information can be used in three different scenarios. First, the data can be presented to game designers to help them redesign their games to reduce the risk of gaming injuries. Second, game designers and developers can develop injury-aware game AI to dynamically change gameplay based on real-time player activity data to reduce the risk of repetitive microtrauma or prolonged gaming. Third, the data can be presented to game players to raise awareness of gaming injuries and promote healthy gaming activities. Game UI should include health hazard warnings and medical recommendations.

Figure 1 shows the basic components of injury-aware game design. An injury-aware game can be based on any combinations of these components.

4.1 Player Activity Monitor

The player activity monitor will collect data about keystrokes, mouse clicks, and game controller inputs at regular intervals. This information can be used to calculate the per-minute frequency of player actions. Since different keys, mouse buttons, and game controller buttons are mapped to specific figures, the detailed user input information can be mapped to specific finger activities. This can be

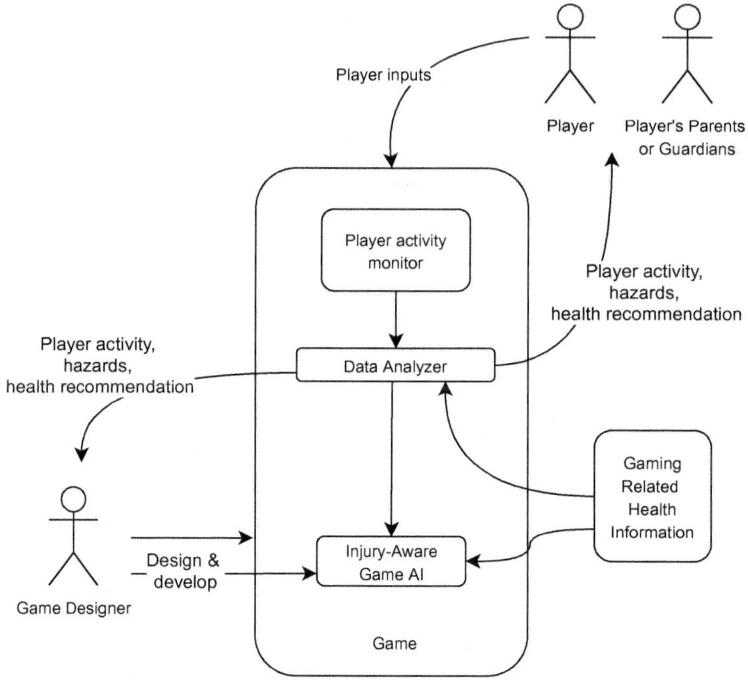

Fig. 1. Basic components of injury-aware game design

used to monitor the risk of musculoskeletal hazards associated with repetitive microtrauma. The player activity monitor can also record the length of a player's gameplay sessions, which can be used to estimate the risk of prolonged gaming.

Embedding a player activity monitor in a game is not difficult. Game engines such as Unity and Unreal have event systems for checking keyboard, mouse, and other input events. Many games may already have built-in user input loggers for user profiling and performance optimization. The data can be easily converted to gaming health-related analysis. Game engine developers can add more functions to their APIs to make it easier to monitor player input activities.

4.2 Data Analyzer

A data analyzer can be embedded in a game to apply statistical analysis to the player input data from the player activity monitor. The analyzer can generate personalized health warnings or recommendations based on the guidelines from the medical professionals.

For example, Emara et al. [10] provided detailed health guidelines and specific exercises for gamers, such as "begin with 3 to 5 min of warm-up stretch with 5 min of stretching every 2 h" and "limit continuous gaming sessions to a maximum of 1.5 to 2.0 h at a time". These guidelines and recommendations can be stored in a searchable database (Fig. 1), and user input data is used to select the most relevant health information for a player.

4.3 Feedback to Game Designers

The data analyzer can present information to game designers via a special UI during game testing. The data presented to game designers should be low-level, detailed data so that game designers can use them to design or redesign games. Three types of data can be presented to game designers: spatial, temporal, or integrated spatial-temporal data.

Spatial data shows player input activities by regions of the game world. This information may help game designers redesign the level to reduce players' hand workload for certain regions. The temporal information may include the intensity of player input activities over time so that game designers may redesign the game to reduce the intensity of repetitive hand activities. Spatial-temporal data combines both spatial and temporal data to provide a more detailed picture. Data visualization techniques such as heatmaps can be used to display spatial-temporal data.

As discussed earlier, the analyzer can use player input data (e.g., actions per minute and length of gaming session) to select relevant health information, such as musculoskeletal hazard, and present it to game designers so that game designers are aware of the potential health risks to players during game design. Between retooling a game or level to not over-rely on any given finger or to better distribute interactions across a level, some genres and mechanics are inherently more stressful than others. The feedback will allow designers to review the data to decide on the best course of action to remedy the intrinsic strains of these genres or mechanics.

4.4 Feedback to Game Players

Injury-aware games may provide feedback to game players and, in the case of young children, their guardians. The information presented to game players is less detailed than the information presented to game designers. Three types of data are presented to game players: aggregated player activity, warnings about potential hazards, and health recommendations. Again, the analyzer will use player input activities (e.g., hand actions frequency) to select the relevant health hazard information (e.g., potential hand and wrist pain) and recommendations (e.g., 5 min of stretching every 2 h).

This information can be displayed in the regular game UI during or after each gaming session. Warnings messages may be displayed if the gaming activities are deemed excessive based on the health guidelines. In some cases, the warnings may be delivered by a non-player character (NPC) in the game. The purpose of this information display is to make a player aware of the potential health hazards based on the player's personal and immediate gameplay data. The players will feel the information is more relevant because it is delivered in-game, in real-time, and based on their own personal gameplay data. This is a type of calm technology [4] that stays largely at a user's peripheral attention.

For younger children, the information may be delivered separately as a report to the parents or guardians to keep them informed. The typical parental control software can report the total amount of time of gaming but without much detail. An injury-aware game can provide parents with more specific information about hand activity, warnings on health hazards, and health recommendations.

4.5 Injury-Aware Game AI

A more advanced form of intervention is to develop game AI that can dynamically adjust the gameplay to reduce the risk of hand and wrist injury or prolonged gaming. The injury-aware game AI takes inputs from the player activity monitor and health information database. A player's activity statistics are constantly compared with the normal ranges based on the health guidelines. If a player's gaming activity goes outside the normal range, the game AI can automatically adjust the gameplay. For example, the game AI can adjust the difficulty level, reduce the frequency of enemy attacks, provide more breaks, end the game quickly, or even lock the game for a period of time.

Building injury-aware game AI is technically feasible. Adaptive gameplay, dynamic difficulty adjustment, and procedural content generation in games have been studied for many years. Many games already have a built-in dynamic difficulty adjustment mechanism. The main difference for injury-aware game AI is to incorporate gaming health guidelines into the algorithms.

5 An Example of Injury-Aware Game Design

As a proof of concept, we have developed an injury-aware game to demonstrate the mechanisms discussed earlier. This game is based on Katamari Damacy, where players roll a ball around a level to pick up items. The game consists of one level in a city setting, split into nine zones, where players roll a ball around to pick up items, according to the unlocked categories (shown by the highlighting of the item). The win condition is having a size above 2.5 m, which, on average, took players about 9 min to complete. The game is played with a keyboard and mouse. This game includes a player activity monitor, a data analyzer, feedback to game designers, and feedback to game players. We did not develop an injury-aware game AI for this proof of concept. For this study, we will be focusing on raising awareness of the Carpal Tunnel Syndrome (CTS) by tracking and analyzing keystrokes.

Our game has two versions. The first version (Fig. 2) allows for the use of both WASD and the arrow keys, with no break between them (the control group). In the second version (Fig. 3), we set the use of WASD and the arrow keys to a timer, allowing for a break on the player's hand for that given duration (the experimental group). Pauses and breaks in a game are either set in motion by players or an event happening in-game. These serve as natural checkpoints for players to catch up and reflect on what's going on in the game. For injury-aware gameplay to work while still being enjoyable, it needs to balance the subtlety of breaks without interrupting the game flow too much.

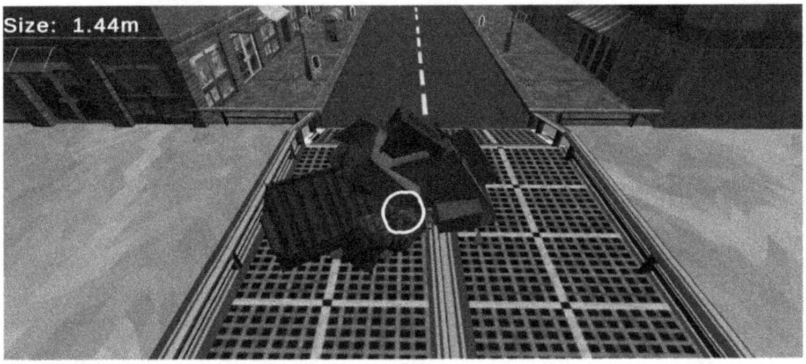

Fig. 2. The control group version.

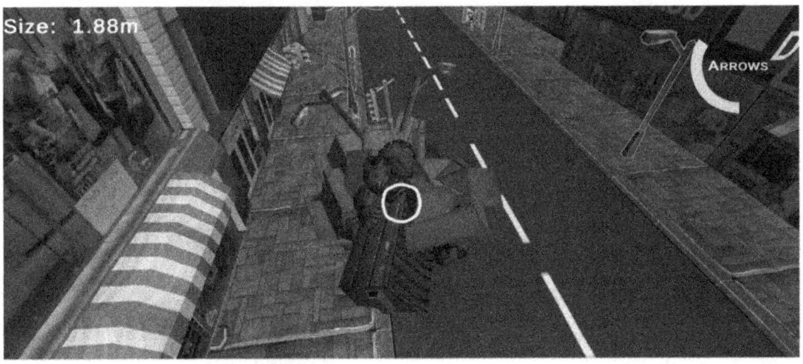

Fig. 3. The experiential group version, with a radial in the top right to indicate which controls to use.

5.1 Player Activity Monitor

Our player activity monitor features two components: a temporal aspect in the form of a keylogger and a spatial aspect in the form of area or zone IDs. The temporal aspect is a keylogger, with the designers inputting the keys they want to track, and the tool records the counts over a user-inputted period of time. The keylogger records both WASD and arrow keys at every instance in time, no matter the version. The counts will be averaged over a period of time, in the case of this game, over a period of 10 s. Because our game had only one level that took around 9 min on average to reach the win condition, we decided to set a short time span to continually export our data. As it stands, it only takes keycodes but could also be modified to handle mouse movements or controller stick movements. Built using Unity's OnGUI event system, the temporal aspect can easily be ported into any given Unity project with no dependencies or built to be its own executable.

The spatial aspect is game-specific but is meant to track the player's location. In the proof-of-concept game, the level was split into nine areas, with each item being tagged with a zone ID. Zones are a way to track the player's location at any moment of time. Each item has an assigned category and zone number. The spatial component added a zone profiling tool, which allows us to view how many objects were picked up in any given zone. If a player picked up an item, it would be added to the respective zone's overall count, showing where the player was within the last 10 s. The spatial component, the zoning tool, can be used in a variety of contexts and manners. For an open-world game, the zones could be split by the LOD chunks on a given terrain, with each object to be tracked based on a zone ID. Another example would be a rhythm game. Most beatmaps are already sectioned off into parts, so each note in said part would be tagged with a zone ID.

The two aspects work in conjunction to give an idea of how the players are using their fingers over a period of time, in relation to how objects are distributed throughout a level. All of the data is exported in real-time to Google Sheets, with a new table created for each new session. The table name records the version and a timestamp of when they started playing. Therefore, we can track the length of each gaming session.

5.2 Feedback to Game Designers

Game designers will be able to view the raw data in the Google Sheets, port the data into a data visualizer, and view both the hand usage via the keylogger, as well as the player's movement and interactions throughout a level via the zoning component.

The heatmaps (Figs. 4 and 5) display the density of finger usage to standard keyboard finger configurations. Standard left-hand configurations for the WASD keys have a setup of the left middle finger assigned to both W and S, the left ring finger assigned to A, and the left index assigned to D. Standard right-hand configurations for the arrow keys have a setup of the right middle finger assigned to both the up and down arrows, left ring finger assigned to the right arrow, and the right index assigned to the left arrow.

Once a session ends, or a designer feels as though enough data has been gathered, they can use their visualizer of choice to view the overall impact of the player's finger usage and location.

We used Plotly [2] to create heatmaps of the unaggregated key presses over a period of time. A secondary heatmap of unaggregated object interactions in relation to player location can be created out of the zoning component's data. Designers can view these heatmaps to show both the density of finger usage as well as if there's any correlation of object interaction or inaction between the two heatmaps. Levels in any given game could potentially be redesigned accordingly if certain fingers are overused.

5.3 Feedback to Game Players

Game players and their guardians do not need to see the raw, unaggregated data. Instead, a condensed summary based on the data is presented to them. In this game, players and their guardians will receive a bar chart (Fig. 6) based on

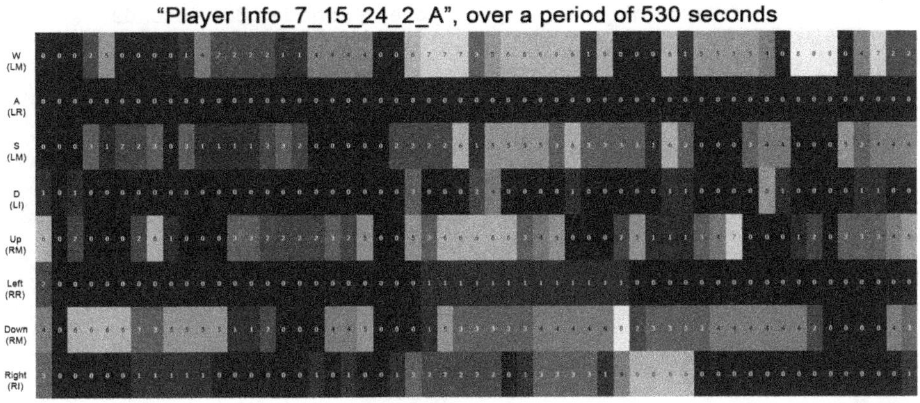

Fig. 4. The heatmap shows the key and finger usage over a period of 530 s for version A of the game. The horizontal axis is time. The vertical axis shows different keys and their finger mappings. For example, "W (LM)" means the W key is mapped to the left middle figure (LM), and "Right (RI)" means the right arrow key is mapped to the right index finger (RI). Here M, R, I mean middle, ring, and index finger, respectively.

the summed counts of each input, also marked with their corresponding finger configuration on their respective inputs. On these aggregated counts, a recommended hand exercise will be provided. These recommendations, if followed up by the players, can also aid in preventative measures against developing any gaming-related injuries.

5.4 Injury-Aware Game AI

In our proof-of-concept, we did not develop injury-aware game AI. In our game, the AI would not serve much purpose other than to potentially force a break if the player's inputs go outside the "healthy" range.

For example, in a rhythm game, a counter for the number of retries can run in the background, and after a certain threshold is reached, offer to automatically lower the difficulty of the current beatmap. For example, input-based fighting games have repetitive strains due to how much a player will be responding to an enemy's moves. In this case, a counter tagged by the moves that damage the player can be implemented. Based on the moves that hold the highest count, indicating which moves the player is struggling with the most, the probability of said moves being used by the computer can be reduced in subsequent replays of the current round or session.

5.5 User Study

Between the two studies, we gathered 29 people, 19 who played the game prior and consented to have their game session recorded, and ten who participated

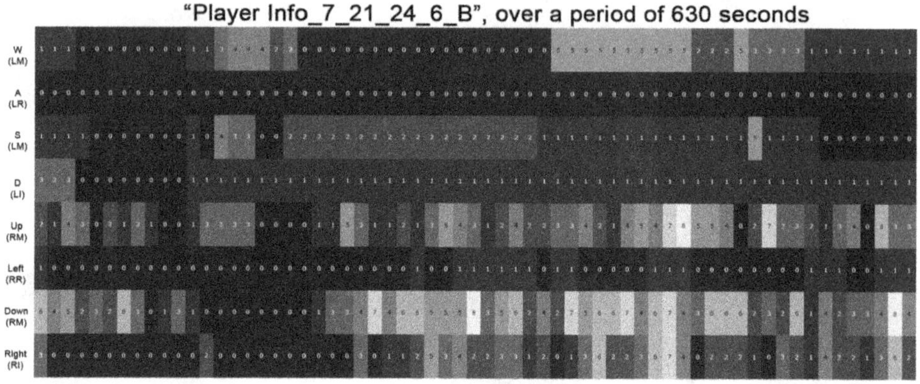

Fig. 5. The heatmap shows the key and finger usage over a period of 630 s for version B of the game. The horizontal axis is time. The vertical axis shows different keys and their finger mappings.

solely in the user study questionnaire. Participants were sent links to both studies in various Discord groups the researchers were a part of. In our preliminary user study, we gathered data both from participants who previously played the game, as well as a general populace – 60% of participants identified as game designers or developers. The questions were split into three sections: those relating to the proof-of-concept game, the relevancy of the tool's feedback to a player, and the relevancy of the tool to game designers.

About 70% of participants played the game previously versus the 30% who did not. Almost all participants stated that they relied on the WASD keys over the arrow keys throughout the game. For those that played the experimental group version, on a scale of 1 to 5 (1 being the lowest in enjoyability and 5 being the highest), participants ranked their enjoyment of the control switching mechanic between 2 and 4, with more of a lean towards the mechanic being neutral (25%), or potentially positive territory (50%). Conversely, all of the participants who played the control group version stated that a control switching mechanic would inhibit their enjoyment of the game. 62.5% of participants listed themselves as completionists, which for this game would mean remaining in a particular area to pick up as many items as they can.

The latter two sections were generalized to gauge the interest of the tool to an audience outside of this proof-of-concept, appealing to both players and designers.

All participants indicated that they would be interested in receiving hand exercise recommendations based on their key presses. While not too many participants identified as guardians or caretakers, it was an even split as to whether or not the tool's feedback would be beneficial in recommending how or when a game or games should be used.

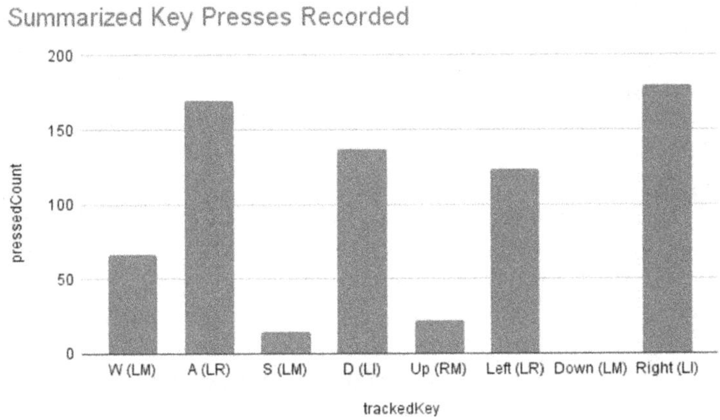

Fig. 6. This bar chart shows a summary of the key and finger usage for a gaming session. For example, "W (LM)" means the W key is mapped to the left middle figure (LM), and "Right (RI)" means the right arrow key is mapped to the right index finger (RI). Game players can see which fingers are used more frequently.

The last section focused solely on the players who identified as designers, which made up 60% of the participants, asking if the unadulterated data of the tool would be of use to them.

Eighty-three percent (83.3%) found real-time, raw data, based on players' key presses, in relation to the player's location to be beneficial in level design, whereas 16.7% did not see the need for it. The same split occurred for the next question of if they would consider using both the temporal (the keylogger) and spatial (the zones) components of this tool in any given project, with the 83.3% answering "both" and the 16.7% answering "neither."

Overall, players who experienced the injury-aware version of the game seemed not to mind or even somewhat enjoy the subtle breaks the game gave to their finger usage. While those who had not played the injury-aware version stated that the control switching mechanic would inhibit their enjoyability, all participants were interested in the idea of receiving hand exercise recommendations after play, with the majority of designers finding both aspects of the tool beneficial for any given project.

6 Conclusion and Future Work

Gaming injuries have become an important health issue as a large portion of the population plays video games regularly, and esports is growing rapidly. In recent years, the medical community has provided a specific framework and guidelines for treating and preventing gaming-related health hazards. We believe an effective way of delivering gaming-related health information to game players is to incorporate such information in game design and development via calm technology [4]. In this paper, we present a framework for injury-aware game design and identify key components for injury-aware games. As a proof of concept, we developed an injury-aware game and conducted a user study. The results showed that participants, both players and designers, would be open to exploring the avenue of injury-aware gaming, with the most positive reception of the feedback being recommendations for hand exercises. The majority of user study participants who played the injury-aware version of the game did not find their enjoyability hindered by the control switching mechanic. The majority of designers also felt both the temporal and spatial components of the tool to be beneficial to level design.

This work is the first step toward a comprehensive study of injury-aware game design. Many research questions remain to be studied. For example, there are different ways to deliver health warnings and recommendations. Their effectiveness needs to be studied and tested. Injury-aware game AI is a complicated subject, and different techniques need to be studied. Extensive user studies are needed to test the effectiveness of injury-aware game AI.

Acknowledgement. This work was supported in part by NSF Award #1852516.

References

1. Games for Health Journal. https://home.liebertpub.com/publications/games-for-health-journal/588. Accessed 25 July 2021
2. Plolty. https://plotly.com/python/. Accessed 25 July 2021
3. Ayenigbara, I.: Gaming disorder and effects of gaming on health: an overview. J. Addict. Med. Ther. Sci. **4**, 001–003 (2018)
4. Case, A.: Calm Technology: Principles and Patterns for Non-Intrusive Design. O'Reilly Media, Newton (2016)
5. Chung, T., Sum, S., Chan, M., Lai, E., Cheng, N.: Will esports result in a higher prevalence of problematic gaming? A review of the global situation. J. Behav. Addict. **8**, 384–394 (2019)
6. Cleveland Clinic: What you need to know about gaming injuries. https://health.clevelandclinic.org/what-you-need-to-know-about-gaming-injuries/ (2019). Accessed 25 July 2021
7. Columb, D., Griffiths, M.D., O'Gara, C.: Online gaming and gaming disorder: more than just a trivial pursuit. Ir. J. Psychol. Med. 1–7 (2019). https://doi.org/10.1017/ipm.2019.31. Epub ahead of print. PMID: 31366420
8. DiFrancisco-Donoghue, J., Balentine, J., Schmidt, G., Zwibel, H.: Managing the health of the eSport athlete: an integrated health management model. BMJ Open Sport Exerc. Med. **5**, 1–6 (2019)
9. Duque, D., Vilaça, J.L., Zielke, M.A., Dias, N., Rodrigues, N.F., Thawonmas, R.: Guest editorial: special issue on serious games for health. IEEE Trans. Games **12**(4), 337–340 (2020)
10. Emara, A.K., et al.: Gamer's health guide: optimizing performance, recognizing hazards, and promoting wellness in esports. Curr. Sports Med. Rep. **19**, 537–545 (2020)
11. Entertainment Software Association: 2021 essential facts about the video game industry (2021). https://www.theesa.com/wp-content/uploads/2021/07/2021-Essential-Facts-About-the-Video-Game-Industry.pdf

12. Game Developers Conference: 2021 state of the game industry report (2021)
13. Gilman, L., Cage, D.N., Horn, A., Bishop, F., Klam, W.P., Doan, A.P.: Tendon rupture associated with excessive smartphone gaming. JAMA Intern. Med. **175**, 1048–1049 (2015)
14. von der Heiden, J.M., Braun, B., Muller, K.W., Egloff, B.: The association between video gaming and psychological functioning. Front. Psychol. **10**, 1731 (2019)
15. Ince, D.C., Swearingen, C.J., Yazici, Y.: Finger and wrist pain in children using game consoles and laptops: younger children and longer time are associated with increased pain. Bull. NYU Hosp. Joint Dis. **75**, 101–104 (2017)
16. Jefferson Health: Video gaming injuries are on the rise. https://thehealthnexus.org/video-gaming-injuries-are-on-the-rise/, February 2020 (2020). Accessed 25 July 2021
17. John, N., Sharma, M.K., Kapanee, A.R.M.: Gaming- a bane or a boon-a systematic review. Asian J. Psychiatry **42**, 12–17 (2019)
18. Lajka, A.: CBS News: esports players burn out young as the grind takes mental, physical toll. https://www.cbsnews.com/news/esports-burnout-in-video-gaming-cbsn-originals/ (2018). Accessed 25 July 2021
19. McCarthey, M.: Ruptured tendon sidelines candy crush gamer after weeks of constant play. Br. Med. J. **350**, 1 p. (2015)
20. McGee, C., Ho, K.: Tendinopathies in video gaming and esports. Front. Sports Act. Living **3**, 689371 (2021). https://doi.org/10.3389/fspor.2021.689371
21. Pereira, A.M., Brito, J., Figueiredo, P., et al.: Virtual sports deserve real sports medical attention. BMJ Open Sport Exerc. Med. **5**, e000606 (2019). https://doi.org/10.1136/bmjsem-2019-000606
22. Pujol, J., et al.: Video gaming in school children: how much is enough? Ann. Neurol. **80**, 424–433 (2016)
23. Sousa, A., et al.: Physiological and cognitive functions following a discrete session of competitive esports gaming. Front. Psychol. **11**, 1030 (2020). https://doi.org/10.3389/fpsyg.2020.01030
24. Sparks, D.A., Coughlin, L.M., Chase, D.M.: Did too much Wii cause your patient's injury? J. Fam. Pract. **60**, 404–409 (2011)
25. Straker, L., Abbott, R., Collins, R., Campbell, A.: Evidence-based guidelines for wise use of electronic games by children. Ergonomics **57**, 471–489 (2014)
26. Trotter, M.G., Coulter, T.J., Davis, P.A., Poulus, D.R., Polman, R.: The association between Esports participation, health and physical activity behaviour. Int. J. Environ. Res. Public Health **17**, 1–14 (2020)

Mental Jam: A Pilot Study of Video Game Co-creation for Individuals with Lived Experiences of Depression and Anxiety

Hsiao-Wei Chen[1](✉), Jonathan Duckworth[1], and Renata Kokanovic[2]

[1] School of Design, RMIT University, Melbourne, Australia
{michelle.chen,jonathan.duckworth}@rmit.edu.au
[2] School of Global, Urban and Social Studies, RMIT University, Melbourne, Australia
renata.kokanovic@rmit.edu.au

Abstract. Mental Jam is a research project that explores how methods of video game co-creation can facilitate the participation of individuals with lived experiences of depression and anxiety to build empathy and mental health awareness among young people. Previous studies have explored the use of different artistic mediums to represent different lived experiences and to raise awareness in the community. Video games are an interactive and immersive medium which can inspire players to learn about other people's lived experiences. However, facilitating the participation of individuals with lived experience in the creation of video games is not well understood. Through a participatory action research methodology, we developed a game jam workshop designed to facilitate the co-creation of video games with participants using diverse video game design approaches, such as narrative-driven game design. We report the results from a pilot study, which comprised of narrative interviews and a game jam workshop through which a game called *Counter Attack Therapy* was produced. In conclusion, we discuss how the outcomes contribute to the field of art-based knowledge translation, as well as expand upon how game design approaches may benefit individuals with lived experiences of depression and anxiety.

Keywords: Game jam · Video game · Participatory action research · Knowledge translation · Lived experience · Depression · Anxiety

1 Introduction

Mental health is a vital part of our health and wellbeing. Mental health is defined by the World Health Organisation (WHO) as a state of wellbeing where someone can recognize their abilities, handle normal life stress, work productively, and contribute to their community [1]. In 2015, the United Nations (UN) included the promotion of mental health and wellbeing for the first time in their Sustainable Development Goals [2]. According to the WHO it is important to engage and empower people with lived experience of mental illness, by closely collaborating with them in the promotion of

M. Wölfel et al. (Eds.): ArtsIT 2021, LNICST 422, pp. 120–137, 2022.
https://doi.org/10.1007/978-3-030-95531-1_9

mental health advocacy [3]. One of the goals of the WHO's Mental Health Action Plan is to decrease stigma and discrimination by educating the public through mental health awareness campaigns. There is also a visible shift of focus from mental illness treatment to the promotion of mental health, wellbeing, and resilience [4–6].

One of the ways to promote mental health is through the knowledge translation of the lived experiences through different artistic mediums. Knowledge translation is a term used to describe how knowledge is disseminated, exchanged, and applied from a range of participants and perspectives [7]. For example, people have portrayed their first episode of psychosis through dance [8] and used drawings and digital media to express experiences of illness [9, 10]. The Big Anxiety Festival have also explored the use of arts and science to convey the people's experience of anxiety [11]. MENTAL Jam is a research project that explores the knowledge translation of young peoples' lived experiences of depression and anxiety through video games co-creation. Video games are interactive and immersive, which makes it a powerful medium for representing lived experiences and inspire players to gain a more insightful understanding [12]. Video game development is also multidisciplinary, which provide multiple ways for people with lived experience to tell their stories, such as through the narrative, art, music, and game mechanics.

In recent years, there has been an emergence of deeply personal video games about game developers' experiences of mental illness [13]. For example, *Depression Quest* and *Actual Sunlight*, are both narrative-driven games based on the game developers' lived experience of depression [14, 15]. For Zoe Quinn, developing *Depression Quest* helped her deal with her lived experience, and for her, having players experience what it feels like to live with depression is a powerful use of games as a medium [15, 16]. While for Matt Thorson, the developer of *Celeste*, a platformer game about depression and anxiety, rather than portraying representation of mental illness defined by mental health professionals, Thorson explored these themes based on his perspective [17]. Researcher Sandra Danilovic explored how lived experiences of game developers are portrayed in video games. She introduced the term autopathographical games, which are games that explore game developers' autobiographical experiences of illness as a form of self-care, understanding and therapy [13].

The existing games about the lived experiences of depression and anxiety, including Danivolic's research, are often developed in isolation by solo or small teams of game developers. For Danilovic's research, her participants were developing games on their own, which suggests they had all the skills required, including design, programming, and art [13].

In developing Mental Jam, we take a different approach that encourages young people with lived experience of depression and anxiety, to work together with game developers to co-create video games. People with lived experience are involved in the research from the very beginning and throughout the entire process to guide the research process and to ensure lived experiences are represented in every step of research.

Mental Jam is based upon game jam design workshop designs that have been used as a method by researchers to capture the whole game development process from ideation to development to release [18]. A game jam is an event where game developers can work alone or in teams, with a balance of skills and interests, to make a game based on a given theme in a short duration, which ranges from 48 h to slow jams that last a month [18]. Some events are conducted in a physical location, such as Global Game Jam, while some are conducted online, such as Ludum Dare [19]. Game jams also promote community building through a shared experience of being in the same location and a mutual interest in game development [20]. Locke et al. compared game jams to performative artworks, where video games are co-created through group participation, and the development process is as important as the games produced [21, 22]. Game jam participants are also empowered by the shared ownership of developing a game from start to finish [22].

Before the game jam workshops, we also conducted one-on-one semi-structured narrative interviews with young people with lived experiences of depression and anxiety. The participants were to give an uninterrupted account of their experience. We conducted these interviews, because participants may not be comfortable sharing their lived experiences with the group during the game jam workshops. The interview transcripts were deidentified and participants were given a pseudonym to protect their anonymity. The interviews were also thematically analysed to identify recurring themes [23], and a report of the findings was presented at the start of the game jam workshops and informed the game design.

This paper is a report from the pilot study of the research, which comprised of narrative interviews and a game jam workshop, which produced a game called *Counter Attack Therapy.*

2 Methods

To develop Mental Jam, we deployed a Participant Action Research (PAR) methodology to facilitate a collaborative process whereby the different stakeholders work together through an iterative process of reflection and action to solve a problem, and the process itself is as important as the outcome [24, 25]. PAR is a user-centred approach, where participants are the real experts of their experience [16]. They are involved in every step of the research process, from design to data gathering, analysis, and conclusions [26] which can give them a sense of empowerment [27].

This research is reviewed by an independent group of people called the Human Research Ethics Committee (HREC). This research project has been approved by the RMIT University HREC.

For this research, the participants are young people, aged 18 to 25, who were diagnosed or who self-identified with lived experience of depression and/or anxiety. Participants must also be currently, by their own account, sufficiently well to participate in research. Participants ideally should have an interest in gaming and/or in learning game development. The research also collaborates with game developers, such as programmers, artists, game designers, writers, and musicians.

As the research involves participants with mental illness, we asked participants to assess that they are sufficiently well to participate in the research. According to Roberts, people with mental illness can give informed consent [28]. Participants were the ones to determine their capacity to consent and participate in the research. The participants also must have sufficient cognitive capacity to be able to give informed consent. Each participant was also be given a Participant Information and Consent Form, in which they will give their written consent to participate.

Since the research is recruiting participants who are sufficiently well, it is unlikely that they will find this aspect of the research will be stressful or upsetting. However, reflecting on lived experiences of depression and anxiety may result in some discomfort. Before the start of each activity, we explained to all participants that their participation is voluntary and that they can discontinue or take a break at any time. During the game jam, we checked up on participants from time to time, to check on their progress, as well as check for any signs of distress. Game jams are normally low stakes environments and flexible with each participant's time and commitment. Participants at the game jam workshops are advised to maintain the confidentiality of their fellow participants. Participants were advised to only share things that they are comfortable with, and other participants are advised that any information is shared in confidence. Help-seeking information was also be provided to, which included contact numbers of mental health support services and telephone helplines.

For the pilot study, we recruited four participants through personal networks and snowballing. Eligible participants were self-identified or report receiving a diagnosis of the lived experience of depression and anxiety, who are currently, by their own account, sufficiently well to participate in research. Due to the current pandemic situation, all the interviews and game jam workshops were conducted online via Microsoft Teams, which allowed participants who are based in different countries, such as Australia, Vietnam, and the Philippines, to participate.

Prior to the commencement of the game jam workshop, we interviewed each participant between 20 min to an hour which were video recorded via Microsoft Teams. The interviews were semi-structured, and participants were invited to give an uninterrupted account of their experience with depression and/or anxiety. Participants are encouraged to talk about anything that they feel are important and as much as they are comfortable with. The interviewer asked a few follow-up questions to clarify aspects of participants' experience, as well as to ask about participants' recovery journey, and a key message that they would like to include in a video game to encourage others to seek support.

The interviews were transcribed initially using the automated transcription software, Otter [29], and manually checked and deidentified. To maintain the anonymity of research participants, they are given a pseudonym. The interviews were analysed using thematic analysis [23] to identify recurring themes. An initial coding framework was developed based on the study conducted by HealthTalk Australia on people's lived experience of depression and recovery [30]. HealthTalk Australia interviewed 39 people in Australia and they identified themes, such as "Understanding Experiences- stories of depression", "Negotiating the Health System", "Everyday Life- Support and Challenges" and "Message to Others" [30]. The interview transcripts were analysed to refine the coding framework, identify themes, and produce a report [23].

From the four participants who participated in the interviews, two of the interview participants, who were based in Vietnam were recruited for the pilot game jam workshops. The other two participants from the interviews will be recruited for another game jam workshop iteration. The participants of the game jam workshops are Helen, a recent graduate living in Hanoi; and Melisa, a student studying in Ho Chi Minh.

Due to participants availability, the game jam workshops were held online via Microsoft Teams over multiple sessions that lasted between 30 min to two hours over three weekends. The sessions were video-recorded via Microsoft Teams with the participants' consent for later analysis.

The first game jam workshop session began with a presentation about the aims of the research project and a showcase of example video games that were about depression and/or anxiety. The presentation also included the thematic analysis from the interviews and introduced some tools that would be used in the game jam workshop, such as Trello, an online post-it board [31], and Microsoft Teams.

During the second session, participants were led in a discussion about the thematic analysis report, followed by a brainstorming session about the game that they will develop. The brainstorming session was held on a Trello Board, where participants could add cards (like post-it notes) to different lists, which were labelled "Game Mechanics", "Narrative", "Art Style" and "Other Ideas" (see Fig. 1). Trello also allowed participants to attach images, links, and comments to cards, as references to the art style. The ideation session for the game jam workshop followed IDEO's design thinking and their field guide that included step-by-step instructions for ideation activities, such as brainstorming and storyboarding [32]. The brainstorming session occurred in 3 min bursts, where participants were invited to add as many ideas as possible to the Trello Board. Participants were encouraged to draw from their personal lived experiences, the thematic analysis report, and build on each other's ideas. After each burst, a brief discussion was held about the ideas added. Some bursts focused on a particular aspect, such as the narrative and building the main character of the game. The session lasted 2 h, with a clearer idea of the mechanics, narrative, and art style for the game.

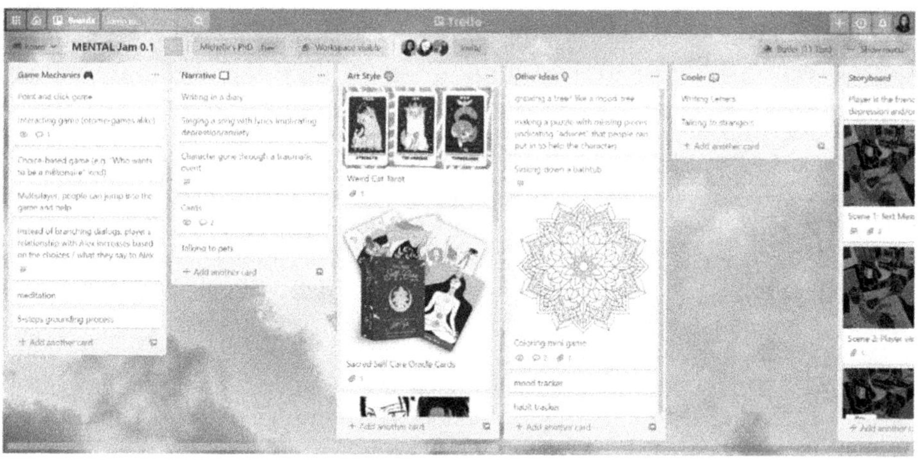

Fig. 1. Game jam workshop brainstorming session using Trello.

The Trello board was set up to track the tasks to be accomplished, with "To Do", "Doing" and "Done" lists. Trello can also add checklists and assign participants to cards. The researcher assigned tasks for the group: the participants would oversee the game design and narrative, while the researcher would oversee the art and code for the game. The researcher also scheduled another session for participants to reconvene and report on their progress. In the meantime, the researcher experimented with art styles and created the background art for the game.

The third session was scheduled two days after the first session. Unfortunately, one of the participants was unable to attend due to personal reasons. During this session, the other participant, Helen, took the lead on the narrative script of the game. The script was written in a shared Microsoft Word file on Microsoft Teams. The researcher scheduled the next session a week later.

During the week, Helen worked on the script, while the researcher developed the character art, user interface (UI) for the game. She also experimented with tools that will assist in the coding of the game. She also started developing the game using Unity, which is a free and popular game engine for game development [33], and YarnSpinner, a plugin for writing game dialogue [34]. YarnSpinner allows game developers to write the script in plain language (see Fig. 2), add options and branching dialogue to their game.

At the fourth and fifth session, which were held on the same day, both participants were able to attend and both Helen and the researcher presented their work in progress and there was a discussion on different aspects of the script, the art, and the game design. The game was further developed over the course of a week which gave the researcher time to finish development. The researcher was able to present a working version of the game at the sixth and final session for the participants to playtest. The participants were also able to give feedback on the game.

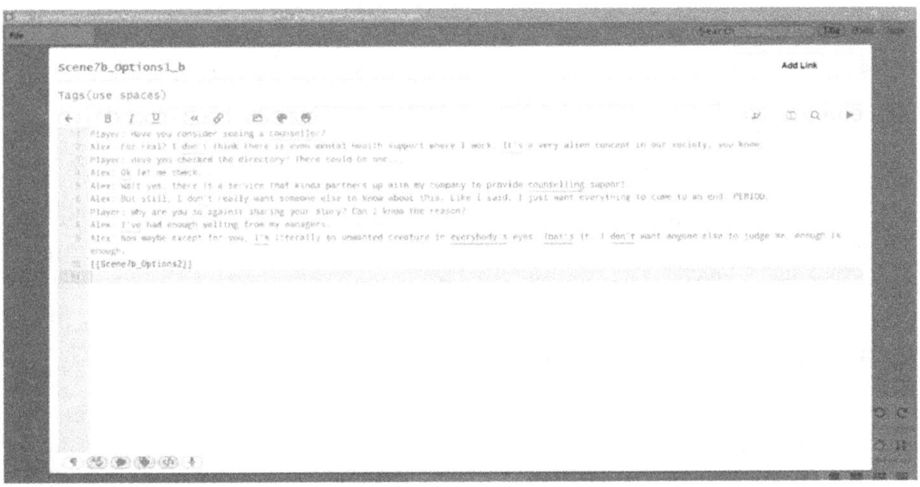

Fig. 2. The script of the game in plain language, using YarnSpinner.

3 Outcomes of the Mental Jam Online Workshop

Counter Attack Therapy is a game that was developed over three weeks with participants about their lived experiences of depression and anxiety. It is an interactive narrative game about Alex, a humanoid cat in their mid-20's. As a friend, the player will listen to Alex's story, guide them through the battle and gather useful resources to help take care of their mental health (see Fig. 3).

In the next four section we discuss the themes identified from the interviews. The narrative and design of the game were based on four main themes that were identified from the interviews: "Views about Causes of Depression and/or Anxiety", "Experiencing Depression and/or Anxiety", "Support and Challenges" and "Recovery". The game is released on itch.io, a website where independent game developers distribute their games (http://mentaljam.itch.io/cat) [35].

3.1 Theme 1: Views About Causes of Depression and/or Anxiety

Interview participants identified different reasons for the cause of their depression and/or anxiety, including isolation and traumatic events, such as bullying, sexual assault, verbal and emotionally abusive relationship, and an incident at a workplace.

Some participants felt that their depression and/or anxiety was caused by isolation, especially during the lockdown. While for another participant said her depression first started when she started living in an apartment by herself to move closer to her workplace.

For another participant, Helen, who also participated in the game jam workshops, her depression started after an incident at her workplace and a motorbike accident:

> *I made a mistake [at work], and then the, like my boss got fired instead of me, and that makes me really, like really shock and then really, really depressed. Because, like, why do they do that? Like, I couldn't really understand... I [also] had an accident. So I crashed my motorbike into another motorbike and had a twisted leg after the accident, so basically, I fell into like extreme anxiety and depression for like, half a month afterwards. Like I couldn't understand, like, Why do everything had to go wrong at the same time?... (Helen)*

Based on Helen's lived experience, for the game, the main character, Alex, also encountered some trouble at work that caused their depression and anxiety. During the game, Alex also encounters a motorbike accident and ends up with a cast on their arm.

3.2 Theme 2: Experiencing Depression and/or Anxiety

When experiencing depression and/or anxiety, some participants avoided people, slept a lot and crying, and some of them considered self-harm.

To avoid people, participants would stay in their room, attend lectures online, stop responding to emails and text messages. During the game jam workshop, Helen added:

> *at times I don't, I just don't want to talk to people. Like I have, like, I know I should tell somebody or some like, I had to reply to anyone reply to this email that message and but I just don't want to. (Helen)*

Some participants said that they spent a lot of time sleeping, one of them said they hoped that sleeping would numb the pain:

I think the general feeling was just like hoping to sleep to numb the pain. But then, waking up and realizing the pain is still there, and then you just until you just don't want to do anything anymore... I mean, what was I doing when I was at the lowest point of my depression, anxiety? Was majority spending all that time in bed? Sometimes awake, sometimes not... It's like, you know, if you're healthy, is like staying on bed for forever, you'd be so restless you like I wanna get out I wanna do something how is it that I wasted so much time staring at the ceiling and then just then suddenly like realizing what the day is gone and feeling sleepy so I sleep again like. (Jacob)

While another participant, Melisa, felt immobilized:

I feel like I'm, I'm kind of immobilized, I was kind of immobilized back in time I can sleep, like, for 14 days straight without going out my room, I cannot do anything at all, even like I cannot like doing some self care back in time. So one day, my mom just took me out my bedroom, and she decided to cut all my hair because it's just tangled into like, a big lock, and then I had to cut all of them out. So that's it. And that is like how, like my normal symptoms back in time. (Melisa)

Melisa also participated in the game jam workshop. During the brainstorming session, while designing the character Alex, she wrote: "Alex's hair is cluttered due to lack of self-care", she also included reference images to "depressed hair".

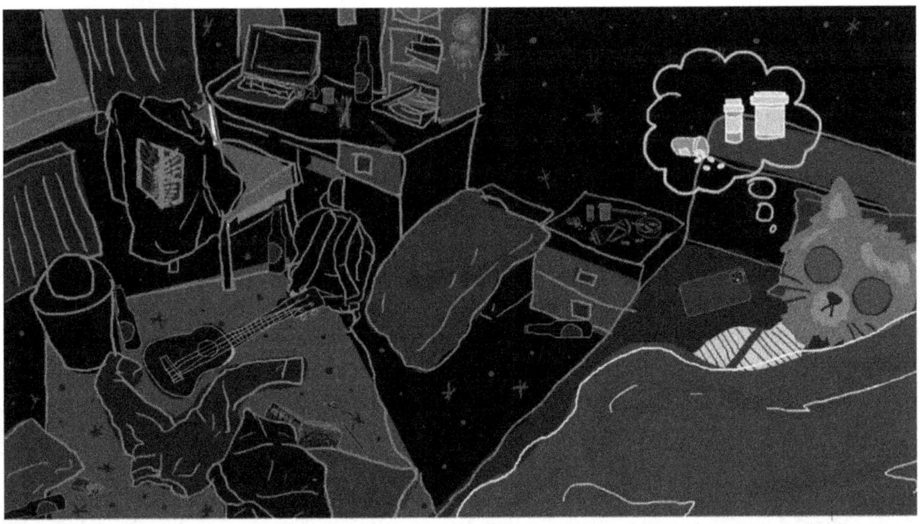

Fig. 3. Screenshot from *Counter Attack Therapy,* showing Alex sleeping, contemplating suicide.

Melisa also mentioned that some things that would remind her of a traumatic event would trigger suicidal thoughts:

Also, I I was really impulsive back in time, especially like, at the time but I feel like I I feel like when my suicidal thoughts came in, I was really impulsive. And if there was any triggers like blood or knife or any kind of news that makes that made me realize to realize to the day that I was assaulted... When I start dealing with suicidal thoughts, even I did actually suicide before. And then I got into hospital a lot. (Melisa)

During the game jam workshop, while designing Alex's room, Melisa wrote down "medicine packages are everywhere". In one of the scenes in the game, Alex went to sleep, and there is a thought bubble with pill bottles in them. According to Helen: *"But like, Alex really wishes everything would act peacefully in their sleep (see Fig. 3). So that would be indicative of like suicide by overdosing pills, you know? Yeah."* While reviewing the script for the scene, Melisa added: *"Oh, you remember like the time I was overdose in hospital? That time I sleep yeah... I was overdose so so I like when I read that. I remember that day."*

While designing Alex's room, both participants also added that there will be beer bottle and cigarettes, even though none of the participants mentioned vices during their interviews. It was only during the game jam workshop, when the researcher was showing the art of Alex's room (see Fig. 3) that Melisa pointed out the beer bottle and cigarettes and added, *"I drink a lot of beer. I used to smoke but I just stopped smoking five or six years ago"*.

3.3 Theme 3: Support and Challenges

Participants sought support from family, friends, mental health professional, and even strangers, however, some of them faced challenges such as stigma and discrimination.

Two participants felt that they were not able to talk to their parents about their depression and/or anxiety, because their parents would not understand. While another participant was quite open to tell her parents because her mother noticed her symptoms.

Most participants found that talking to their friends was useful and found their friends quite supportive. They also found that after they shared their experiences, their friends also shared their own. While another participant noted that some friends were not supportive and are quite judgmental.

All participants were seeing a mental health professional, such as a counsellor or therapist through their university or workplace. Some participants were accessing mental health services during Covid. Participants also talked about getting prescribed medication for their depression and/or anxiety. One participant noted the lack of access to mental health support in Vietnam, while another noted the cost of services in the Philippines. Melisa also talked about the culture shock she got when she moved back to Vietnam from Australia.

Some participants faced stigma and discrimination. During the game jam workshop, Melisa shared her experience with facing discrimination from her ex-boyfriend and his friends while she was living in Australia. Helen shared similar views in Vietnam:

in Vietnam, it's more viewed like the princess or Prince sickness. Because the perception is like, because they they are rich like the the like, they are rich. They don't like what like just says just something like maybe just one bit of thing went wrong, and they're already having a mental breakdown. And it's one of the stigma... (Helen)

In the game, the player is Alex's friend who offers advice. The game presents choices of dialogue that the player can tell Alex (see Fig. 4). The player also suggests to Alex to see a mental health professional, but Alex was initially resistant as portrayed in one of dialogues in the game:

Alex: For real? I don't think there is even mental health support where I work. It's a very alien concept in our society, you know. (Counter Attack Therapy)

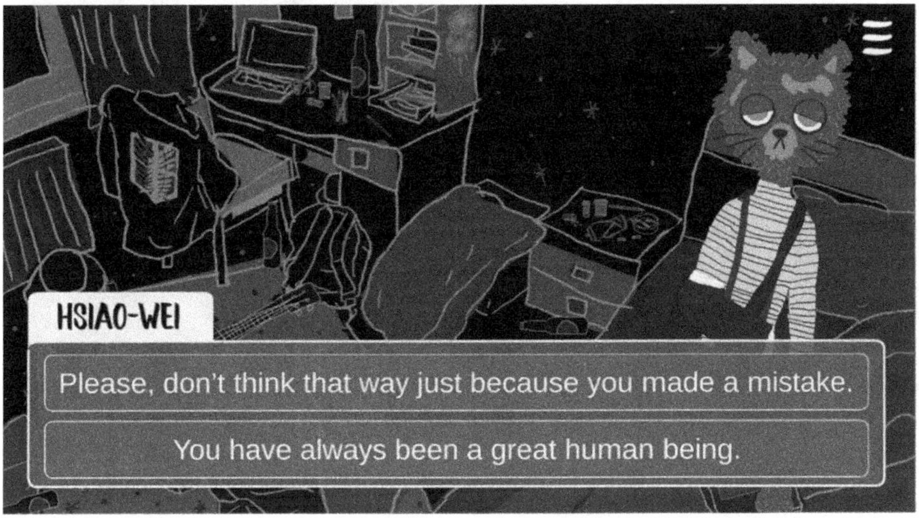

Fig. 4. Screenshot from Counter Attack Therapy, showing dialogue choices.

3.4 Theme 4: Recovery

Participants were asked about the pivot point that prompted them to seek support, they also shared their self-care and coping strategies. Participants cited that the pivot point for them to getting support is when they realised that they needed help because they cannot handle it on their own and they want to get out of the cycle of depression. Participants mentioned different coping strategies that were suggested by the mental health professional, such as breathing, meditation, gratitude journal and painting.

For the game, the participants included a breathing exercise mini-game (see Fig. 5). Participants also mentioned some coping strategies that they found useful, such as practising kendo, singing, reading books, and watching anime movie.

One participant shared that singing is one of her coping mechanisms:

Yeah, actually, singing has been helping me a lot. I've been doing cover songs that I have upload in FB. Yeah. Yeah. So that was my way to release because I think I get distracted whenever I cover songs. Because you know, I try to internalize the character, the song. Yeah, I also have to learn how to edit in GarageBand. How to mix how to set up my my mixer the microphones that I have to use etc. So you know, like keeping my mind off from overthinking and I get to do more of my creative side, also "ano ba" [I don't know] I don't know... (Rachel)

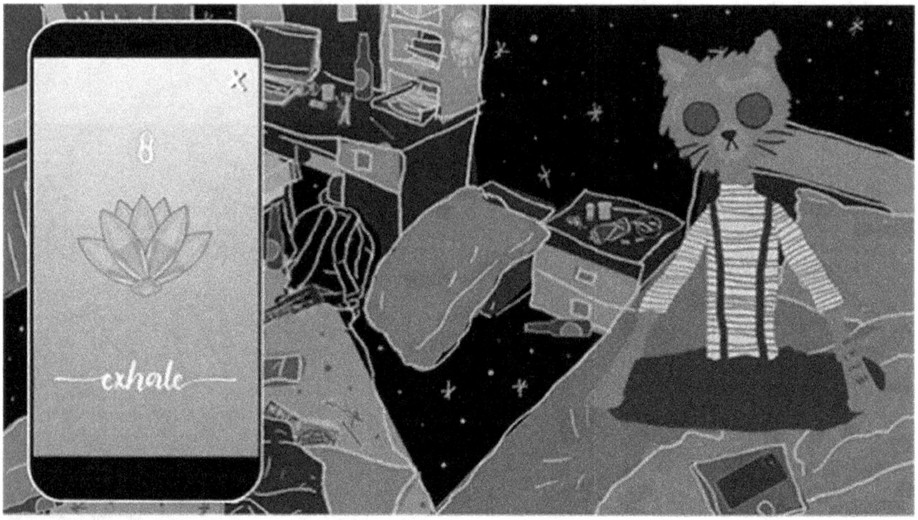

Fig. 5. Screenshot from *Counter Attack Therapy*, showing Alex practising a breathing exercise mini-game.

At the game jam workshop, the participants also discussed how some singers incorporated themes of depression and anxiety into their songs:

So I got the idea from one of my favourite singers [Jonghyun] that actually commit suicide from depression. Who was sending, like signalling help from one of his songs [Lonely], but we never realized until he passed away. (Helen)

Alex's clothes are also based on the outfit Jonghyun was wearing in his music video for the song Lonely [36]. In one the of scenes in the game, Alex plays the ukulele and sings a song with the lyrics:

I am in the middle of nothing
Wonder how did I end up like this
How did I make such a stupid mistake
To cause everything to fall
Get out of bed is even a challenge
Out to the door and there's nothing but a dark cloud
Of thoughts that I just want to scream
This tangle, when will you let me out? (Counter Attack Therapy)

Melisa also shared that watching anime movies as a coping mechanism, she cited Attack on Titan, whose main character Eren, represented a lot of what she was feeling back then:

Because it's his [Eren] action actually, like, represents my mindset. When I look at the world. I still remember one of the quotes from Mikasa is this word is cruel, but also beautiful. And I think his action also represent that quote, but also represent my mindset. Like when I'm dealing with depression as well, the motive the motive is because he is because he was paying back the world just because of his mom's death. And I think that, that trauma can lead to that kind of action. And I think it makes sense is just because, like, is it similar to when they, I don't feel that this society doesn't understand me at all, they don't understand what I'm feeling. And I want to destroy the whole world just because I don't feel like. I don't fit to this world. And I want to redo everything. And yeah, and that's, and Eren's action is actually, like, they will actually, like represent my mindset when I'm dealing with depression. So that's why it's like, even the Attack on Titan is a very violent anime. But I feel like, it's still good for me, because I feel because it represents what actually happened. (Melisa)

There are elements from Attack on Titan in Alex's room, such as the logo on the jacket and a manga on the floor. There are also references to Attack on Titan in the script of the game. The name of the game, Counter Attack Therapy is also based on the song Counterattack Mankind from Attack on Titan's soundtrack.

Helen also mentioned that one of her friends (Melisa) read tarot cards for her at the time and it helped her focus. The participants included an oracle tarot card mini-game, which will give the players and Alex advice and encouragement:

I talked to so one of my friends, like, she can read tarot cards, and I asked her how, like, how should I do? And she, she read the tarot. And then she told me to just focus on on work, because, like, by focusing on the work that could kind of help me forget about, like, stuff like that. So I did. I basically focus crazily on work and assignments and try to get that out of my head. (Helen)

In the game, Alex's appearance, background colour, and background music changes based on their current mood. As Alex starts feeling better, the background becomes lighter, and Alex's appearance becomes a lighter shade of purple, and their facial expression is happier (see Fig. 5).

4 Discussion

This paper summarizes the methods of the pilot study for Mental Jam, where four participants were interviewed and two participants participated in a series of game jam workshops to develop the game, *Counter Attack Therapy*.

We used PAR methodology to engage participants in all the phases of the research, from design to execution and dissemination [37, 38]. For this research, the participants were involved in every step of the process, from the ideation of the game design to the release and marketing of the game.

While some prior research excluded some participants, who did not feel equipped to express their experiences through academic writing [39–41], such as the participants of a research about mental health care felt that their experiential knowledge was undervalued because the reporting phase was conducted by academic researchers [40]. This research ensured that participants voices are heard throughout the process, and they had the final say of what goes into the video games and how their lived experiences are represented.

We found that participants were quite open with sharing their lived experiences of depression and anxiety during the interviews and game jam workshops. Working closely with participants in the game jam workshops over three weekends, also allowed rapport building between the participants and the researcher.

Prior research in co-design and participatory design with participants with psychosis [42] and dementia [43, 44] have found the use of a relatable fictional character allows their participants to share their lived experiences in an indirect way. Similarly, during the ideation session, as the participants are creating a composite character in the third person, a humanoid cat named Alex, they were more open to sharing personal experiences and incorporating some of the physical aspects of themselves into the character. For example, during the interview, Melisa described how her hair got so matted that her mother had to cut it off, this was translated into Alex's messy fur in the game. Halfway through the game, Alex gets into an accident, and their fur turns a darker shade of purple and messier (compare Figs. 4 and 5). Even little details, such as the logo on their jacket is based on one of Melisa's favourite anime movies.

Some information that the participants did not share during their interviews were also revealed during the ideation session, such as the beer bottles and cigarettes that the participants added to Alex's room. It was only during the game jam workshop that Melisa revealed that during her depressive episode, she used to drink and smoke a lot as a coping strategy.

The game jam workshop was originally planned to take place over 48 h on one weekend, however, due to participants availability, as well as the extended scope of the game, it took place over three weekends instead. The researcher and participants also worked on the game asynchronously during the week.

PAR methodology also encourages researchers and participants to work closely together to co-create new knowledge through iterative action and reflection [24]. After the conclusion of the game jam, we also conducted one-on-one interviews with the participants workshops to ask for their feedback about the facilitation of the game jams, and their game development process. The researcher also asked about the things that can be improved about the process.

One of the things Melisa learned from the game jam workshops is collaborating online with distant teams, which she found particularly useful during the current pandemic. Prior research found that working together as a group through a shared social experience of a game jam also fostered a sense of belonging [48]. The participants also found Trello useful in keeping track of each other's progress during the week, as well as give them a sense of a shared space even though they were based in different places.

I think like working on Trello is fine, because we have like a common platform together. Although we, you me and Helen we live in, like some places that we are very distant from each other. Right? Yeah, that by doing everything on Trello together, I think that it's really good for me to keep up with the process of how everything has gone so far... thanks to Trello... I know like how, how the ideas just like are arranged and how the process going so far... And I still like getting updated every day. And then yeah, it's still like it's kind of same to working side by side. But [sometimes] it's about internet connections. (Melisa)

Even though the participants did not have a background in game development, they found the game jam workshops rewarding because they learned new skills and developed a game for the first time. The game jam experience also challenged their notions on what game development involves:

I never thought I would be able to make a game, because I have that kind of perception that only programmers could make the game, you know. (Helen)

Prior research on game jams has also found that making is entertaining even for newcomers [45]. Game development is multidisciplinary, so participants can contribute to the game in different ways. In this pilot study, the participants contributed to the game design and narrative writing, while the researcher oversaw the art and programming. This finding concurs with another example, in The Street Arcade, game developers collaborated with a group of African American teen artists to develop video games. The teens contributed to the game design, narrative, and art for the games, while game developers were the ones who programmed them [46].

During the game jam workshops, Helen has a cast on her arm from a recent accident and participating in the game jam workshops and developing a game gave her a sense of accomplishment:

I feel like I achieved something. During my time that I thought I would never be able to do anything. Like, I was thinking like, what, what the hell can I do with a hand in the cast, and then locked down and then stuff like that, I just can't really contain the thought of being useless. But this game really gave me the chance to do something that can contribute something to the mental health issue. Like specifically from my own experience. (Helen)

In the post-game jam interview, Helen suggested that maybe during the game jam workshops, all participants can share their screens at the same time, so that they can be more engaged and provide real-time feedback. Currently, a limitation of using Microsoft Teams for the game jam workshop video calls, only one person can share their screen at one time. For future game jam workshops, we will survey alternative video conferencing platforms.

The game was released on itch.io, and so far, has over a thousand views and positive feedback from players. The participants have marketed the game on their university's social media pages, as well as getting a feature on a local lifestyle website, Urbanist Vietnam, which describes Helen's experience developing the game [47].

Through the game jam workshops and developing the game, the participants found a way to reflect on their lived experiences of depression and anxiety and share them with people in a different way:

> *By developing a game, and I think back about my story, and I think about how did I overcome every How did I overcome everything and share it with the people. It's not really like explicitly as I share about my story, but like through the game, I share the story of mine to like the audience. And I think that when I witness that the the audience like they welcome the game and they just think they were really excited and how they support it. And I thought it was I feel really relieved... because people accept my story people accept our story... and people welcome our project in a very positive manner, which is something that I've really, really treasured... (Melisa)*

> *It was absolutely like, like life changing and mind changing for me. Because it was like, because my story. It was a very negative, like a very bad memory that I somehow it's sometimes I just kind of want to forget it. And then imagine my mind is like a drawer, and I'm just gonna put it into a drawer and then lock it away and never talk about it again. But this game jam, it has made me realize that, like, not everything bad that happened in the past has to be bad. Forever, like with the right strategies, and like, with the help of team members, teammates, and with the correct like, tactics and strategies, like I can completely, turn it into something positive and then inspire other people. (Helen)*

5 Conclusion

The key findings from the pilot study are: (1) the benefits of working in groups; (2) participants were able to learn new skills; (3) a sense of belonging for the participants; (4) the research provided a venue for the participants to reflect, as well as share their lived experiences of depression and/or anxiety; (5) the use of a relatable fictional character allowed the participants to share their lived experiences in an indirect way; and (6) for future game jams, a longer and more flexible timeframe can be considered.

Working in groups and collaborating with the researcher, who is a game developer, allowed participants to develop a game even though they did not have a background in game development. Participants were also able to learn new skills, such as narrative script writing for games, and collaborating online using Trello.

The participants also reported a sense of belonging. Even though the group was based in different cities, the use of Trello to keep track of tasks and seeing each other's progress gave them a sense of shared space as if they were "working side by side".

The narrative interviews, as well as the game jam workshops, gave the participants an opportunity to share their lived experience stories. Using the literal narrative-driven approach, participants' lived experiences of depression and anxiety were translated in the narrative writing. The game included the participants' views about the causes of their depression and anxiety. This game was based on Helen's personal experiences, which was an incident she faced at work, followed by a motorbike accident. The use of the relatable fictional character also allowed the participants to create the composite character, the game's main character, Alex. Alex portrayed different symptoms that the different participants had while experiencing depression and anxiety, such as sleeping a lot, their lack of self-care, which resulted in their messy hair and room. The game also explored some of the support and challenges the participants faced, such as accessing mental health services. The game also included some mini-games, such as breathing exercises, a puzzle game and oracle tarot cards, which participants have used as coping mechanisms.

Based on the findings of the pilot study, for future game jam workshop iterations, the researcher may also consider a longer time frame, like slow jams, which last from a week to a month. This may allow participants to have more time to develop their game design and work on the game development tasks. The feedback from the pilot study will also inform the next iteration of the game jam workshop process. As the research project applies PAR, the game jam workshop process will go through iterations of planning, game jam execution, and evaluation.

The favorable and promising response from the game jam participants demonstrated that the game jam workshop was a feasible way for developing video games about the lived experiences of depression and anxiety.

References

1. World Health Organization: Promoting Mental Health: Concepts, Emerging Evidence, Practice (2004)
2. United Nations: Transforming Our World: The 2030 Agenda for Sustainable Development (2015)
3. World Health Organization: Mental Health Action Plan 2013–2020 (2013)
4. Buse, K., Hawke, S.: Health in the sustainable development goals: ready for a paradigm shift? Glob. Health **11**, 13 (2015)
5. Dybdahl, R., Lien, L.: Mental health is an integral part of the sustainable development goals. Prev. Med. Community Health **1**(1), 1–3 (2017)
6. Izutsu, T., Tsutsumi, A., Minas, H., et al.: Mental health and wellbeing in the sustainable development goals. Lancet Psychiatry **2**, 1052–1054 (2015)
7. World Health Organization: Knowledge Management and Health: News and Events (2005)
8. Boydell, K.M.: Making sense of collective events: the co-creation of a research-based dance. Forum Qual. Sozialforschung (Forum Qual. Soc. Res.) **12**(1). Art. No. 5 (2011)
9. Guillemin, M.: Understanding illness: using drawings as a research method. Qual. Health Res. **14**(2), 272–289 (2004)
10. Patel, V., Saxena, S., Lundt, C., et al.: The lancet commission on global mental health and sustainable development. Lancet **392**, 1553–98 (2018)
11. Bennett, J.: Anxiety: art and mental health. Artlink **37**, 3 (2017)
12. Solberg, D.: The problem with empathy games. https://Killscreen.Com/Articles/The-Problem-With-Empathy-Games. Accessed 21 June 2021
13. Danilovic, S.: Game design therapoetics: autopathographical game authorship as self-care, self-understanding, and therapy. PhD thesis. University of Toronto, Toronto, Canada (2018)
14. Smith, E.: 'Actual Sunlight' might be the most painfully real video game you'll ever play. https://www.Vice.Com/En_Ca/Article/4wbn9d/Actual-Sunlight-Might-Be-The-Most-Painfully-Real-Video-Game-Youll-Ever-Play-000. Accessed 21 June 2021
15. Parkin, S.: Zoe Quinn's Depression Quest. https://www.Newyorker.Com/Tech/Annals-Of-Technology/Zoe-Quinns-Depression-Quest. Accessed 21 June 2021

16. Lewis, H.: A quest for understanding. Lancet Psychiatry **1**(5), 341 (2014)
17. Grayson, N.: Celeste taught fans and its own creator to take better care of themselves. Kotaku. https://www.Kotaku.Com.Au/2018/04/Celeste-Taught-Fans-And-Its-Own-Creator-To-Take-Better-Care-Of-Themselves/. Accessed 21 June 2021
18. Foltz, A., et al.: Game developers' approaches to communicating climate change. Front. Commun. **4**, 28 (2019)
19. Kultima, A.: Defining game jam. In: Proceedings of the 10th International Conference on the Foundations of Digital Games (2015)
20. Turner, J., Thomas, L.: CoCurating game jams for community and communitas a 48 h game making challenge retrospective. In: Proceedings of the International Conference on Game Jams, Hackathons and Game Creation Events (2020)
21. Locke, R., Parker, L., Galloway, D., Sloan, R.: The game jam movement: disruption, performance and artwork. In: Proceedings of the 10th International Conference on the Foundations of Digital Games (2015)
22. Bayrak, A.T.: Jamming as a design approach. Power of jamming for creative iteration. In: Design for Next 12th EAD Conference. Sapienza University of Rome (2017)
23. Braun, V., Clarke, V.: Using thematic analysis in psychology. Qual. Res. Psychol. **3**(2), 77–101 (2006)
24. Bergold, J., Thomas, S.: Participatory research methods: a methodological approach in motion. Forum Qual. Soc. Res. **13**(1) (2012)
25. Manzo, L.C., Brightbill, N.: Toward a participatory ethics. In: Kindon, S., Pain, R., Kesby, M. (eds.), Participatory Action Research Approaches and Methods: Connecting People, Participation and Place, pp. 33–40. Routledge, London (2008)
26. Whyte, W.: Introduction. In: Whyte, W.F. (ed.), Participatory Action Research, pp. 7–18. Sage, Newbury Park, CA (1991)
27. Boote, J., Telford, R., Cooper, C.: Consumer involvement in health research: a review and research agenda. Health Policy **61**(2), 213–236 (2002)
28. Roberts, L.: Evidence-based ethics and informed consent in mental illness research. Arch. Gen. Psychiatry **57**(6), 540–542 (2000)
29. Otter. https://Otter.Ai/. Accessed 29 June 2021
30. Depression and Recovery in Australia. https://Healthtalk.Org/Experiences-Depression-And-Recovery-Australia/Overview. Accessed 21 June 2021
31. Trello Helps Teams Move Work Forward. http://Trello.com/Home. Accessed 21 June 2021
32. Brown, T.: Design thinking. Harv. Bus. Rev. **86**(6), 84–92, 141 (2008)
33. The Leading Platform for Creating Interactive, Real-Time Content. https://Unity.com/. Accessed 21 June 2021
34. Yarn Spinner the Friendly Tool for Writing Game Dialogue. https://Yarnspinner.dev/. Accessed 21 June 2021
35. About Itch.Io. https://Itch.Io/Docs/General/About. Accessed 21 June 2021
36. Smtown.: JONGHYUN 종현 'Lonely (Feat. 태연)' MV. https://www.Youtube.com/Watch?V=Nptpese9g8c. Accessed 29 June 2021
37. Vollman, A.R., Anderson, E.T., Mcfarlane, J.: Canadian Community as Partner. Lippincott Williams & Wilkins, Philadelphia (2004)
38. Smith, L., Bratini, L., Chambers, D., Jensen, R.V., Romero, L.: Between idealism and reality: meeting the challenges of participatory action research. Action Res. **8**(4), 407–425 (2010)
39. Fricker, M.: Epistemic justice as a condition of political freedom? Synthese **190**(7), 1317–1332 (2013)
40. Groot, B., Haveman, A., Abma, T.: Relational, ethically sound co-production in mental health care research: epistemic injustice and the need for an ethics of care. Crit. Public Health (2020)
41. Rose, D., Kalathil, J.: Power, privilege and knowledge: the untenable promise of co-production in mental "health." Front. Sociol. **4**, 57 (2019)

42. Nakarada-Kordic, I., Hayes, N., Reay, S.D., Corbet, C., Chan, A.: Co-designing for mental health: creative methods to engage young people experiencing psychosis. Des. Health **1**(2), 229–244 (2017)
43. Hendriks, N., Truyen, F., Duval, E.: Designing with dementia: guidelines for participatory design together with persons with dementia. In: Kotzé, P., Marsden, G., Lindgaard, G., Wesson, J., Winckler, M. (eds.) INTERACT 2013. LNCS, vol. 8117, pp. 649–666. Springer, Heidelberg (2013). https://doi.org/10.1007/978-3-642-40483-2_46
44. Tsekleves, E., Bingley, A.F., Luján Escalante, M.A., Gradinar, A.: Engaging people with dementia in designing playful and creative practices: co-design or co-creation? Dementia **19**(3), 915–931 (2020)
45. Balli, F.: Game jams to co-create respiratory health games prototypes as participatory research methodology. Forum: Qual. Soc. Res. **19**(3), Art. 35 (2018)
46. Annas, P., Groden, S.Q.: The street. Radic. Teach. **113**, 6–7 (2019)
47. Urbanist Vietnam: Nhóm Sinh Viên Ra Mắt Tựa Game Nhẹ Nhàng Đề Cao Sức Khỏe Tinh Thần. https://Urbanistvietnam.com/Hanoi-Technology/16864-Nh%C3%B3m-Sinh-Vi%C3%Aan-Ra-M%E1%BA%Aft-T%E1%BB%B1a-Game-Nh%E1%BA%B9-Nh%C3%A0ng-%C4%91%E1%BB%81-Cao-S%E1%BB%A9c-Kh%E1%BB%8Fe-Tinh-Th%E1%BA%A7n. Accessed 21 June 2021
48. Turner, J., Thomas, L., Owen, C.: Living the indie life: mapping creative teams in a 48 h game jam and playing with data. IE2013. Melbourne, Australia (2013)

Statistical Models for Predicting Results in Professional League of Legends

Robbie Jadowski[1] and Stuart Cunningham[1,2](✉)

[1] Department of Computing and Mathematics, Manchester Metropolitan University, Manchester M1 5GD, UK
robbie.jadowski@stu.mmu.ac.uk, s.cunningham@mmu.ac.uk
[2] Centre for Advanced Computational Science, Manchester Metropolitan University, Manchester M1 5GD, UK

Abstract. The esports industry has seen enormous growth in popularity. With increased viewership and revenue, further investment has been made to improve professional players' competitive strength. The modern esports team is a hierarchical business fuelled by investors and sponsorship. This paper is focused on the professional competitions in League of Legends esports. In existing real-world sports such as football or baseball, there is great attention paid to statistic driven analysis of the competition, and these stats are used to quantify player and team performance. These statistics hold significant value for competitive improvement, the gambling industry, and market influence within the esports industry.

This paper presents an analysis of data and metrics gathered from professional games during 2020 in several League of Legends international competitions. The objective was to build a predictive model through the combination of existing data analysis and machine learning that can rate team and player performance. The best performing model was able to correctly predict 67% of 306 games. Results indicate that while it is possible to predict the outcome of a competitive League of Legends game, to do so with a higher degree of accuracy would require substantially more data and contextual information.

Keywords: Esports · League of Legends · Machine learning · Regression

1 Introduction

Statistical analysis of sports, or sports analytics, has become an increasingly popular method for recruitment and strategising in modern sport and competition. The popularisation of sports analytics is often attributed to Billy Beane, who famously achieved great success as the general manager of the Oakland Athletics baseball team using a data-driven approach to evaluate and recruit players on a much lower budget than competing teams. Other teams took note of this approach and went on to achieve success through data-based decision

M. Wölfel et al. (Eds.): ArtsIT 2021, LNICST 422, pp. 138–152, 2022.
https://doi.org/10.1007/978-3-030-95531-1_10

making. This success was noticed by executives and owners of teams in other professional sports leagues, to the point where practically all modern sporting organisations now recruit analytic experts or entire departments dedicated to sports analytics [12].

The convenient nature of statistics allows managers and coaches to identify a player's strengths and weaknesses at a glance, without having to spectate each game the players compete in. The same data can used by gambling organisations to determine probability and assign odds to certain outcomes.

For example, football statistics have evolved to include automated sensing technology that can track player position, movement and other observations from fixed and mobile cameras and sensors. Several professional statistical analysis firms offer data and analysis to professional teams as a product, providing context to the data collected and helping teams make tactical decisions [2].

Since League of Legends (LoL) is a video game, an abundance of statistics can be gathered automatically as they are tracked by the game itself. The wealth of data available provides many opportunities to perform analytics on the game. Most of the existing forms of public analytics involving LoL is used by journalists and fans to make comparisons and fuel narratives. Other organisations provide LoL teams with a paid product package to enhance in-house analysis and supplement coaching.

The aim of this research is to build a statistical model using metrics from this data that can accurately rate team and player performance, with the intention of predicting the outcome of games featuring those players and teams in future games.

2 League of Legends

League of Legends was released in October 2009, and in the years since its release, it has developed a competitive infrastructure across multiple regions that rivals that of traditional sports [8]. Each region's competitive league features franchised teams that compete against each other in weekly broadcasts that regularly draw thousands of viewers and annual inter-regional championships that have drawn 44 million peak concurrent viewers during grand finals [21]. The events feature grand finals in venues such as the Staples Center, selling out the venue within 1 h of tickets being available [22], and the Beijing National Stadium, catering to live audiences in their thousands.

LoL is a team-based strategy game where two competing teams of 5 players aim to destroy their opponents base, canonically named the *Nexus*. Each game of League of Legends takes place on the same map, known as *Summoner's Rift*. Summoner's Rift is split into three lanes, commonly known as Top, middle and Bottom. These lanes form a path that leads from one team's base to the other. The two sides of Summoner's Rift, referred to as 'Blue Side' and 'Red Side' are separated by a River that runs from top lane to bottom lane, and the area in-between the lanes is known collectively as the Jungle. Blue team's base and nexus is situated in the bottom-left of the map, while red team's base and nexus is in the top-right. A representation of the map is shown in Fig. 1.

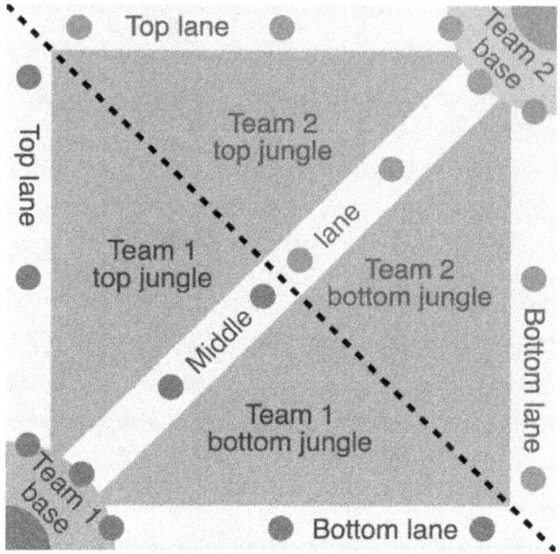

Fig. 1. Simplified version of the Summoner's Rift Map. Original PNG version by Raizin, SVG rework by sameboat licensed under CC BY-SA 3.0 [17] (Color figure online)

Players select one of over 140 champions to control in order to complete the objective and each possesses abilities that aid in combat, navigating the environment or supporting their team. Each player fulfils a different role for the team, much like the different positions in a football team. The roles featured in LoL are: Top Laner; Jungler; Mid Laner; Bot Laner; and Support. Each corresponds to the area or lane of the map that the player will operate in the opening of a game, with the Support player often partnering with the Bot Laner. These roles traditionally feature a typical character archetype, though there are exceptions and champions that buck the trend.

For a team to reach and destroy the enemy team's nexus, they must overcome a series of AI controlled structures known as Turrets. These structures are very difficult to destroy without assistance, which is usually provided by the waves of AI controlled minions that spawn periodically from a team's base. These minions will follow a lane's path to the enemy base until they run into the opposing team's champions, minions or turrets. Players must aid their minions in their advance in order to take down Turrets and reach the opposing teams base, while defending their own Turrets from the opponents.

The map also features neutral objectives, Dragon, Rift Herald and Baron Nashor. These neutral monsters can be defeated by a team to obtain permanent and temporary bonuses, ranging from additional movement speed, a percentage increase in ability power, or buffs to friendly minions to aid in sieging the opponents base.

Due to the asymmetrical nature of the map, granting the blue team easier access to the area that Baron Nashor spawns, combined with the pre-game champion draft where blue side can choose their first champion before the red team, there is a debate that blue side has an inherent advantage compared to the red team. Similar to the home advantage often seen in traditional sports. This advantage will be explored when analysing the data from competitive games and considered when making predictions if such an advantage exists.

3 Background

The use of player rankings in LoL is recognised as being an important feature of the game for individuals as well as to ensure the competitive edge of the game [11], which may arguably extend to system of team rankings and statistics. Previous work has examined the effect that the ability of LoL players working together in teams, and the presence of female gender players, has in being able to predict the competitive performance of those teams, however this relies upon individual measures being taken from players, such as measures of collective intelligence, gender, and so forth, that are not intrinsic to the LoL game statistics and so require additional information gather to take place [10]. Unsurprisingly, much existing research tends to point towards the influence that individual players, and their ability to form effective teams, can have on game outcomes [4,5]. However, in terms of win prediction, it has been shown that for other Multi-player Online Battle Arena games in *professional contexts*, accuracy rates of up to 85% are possible [9].

4 Dataset and Preparation

4.1 Dataset Source

This report is focused on seven competitive leagues in LoL: the LEC (Europe); LCS (North America); LCS Academy (North America); LCK (South Korea); PCS (Southeast Asia); CBLOL (Brazil); and TCL (Turkey). While the Chinese league is the largest and perhaps most dominant region, there is insufficient data for each individual game available, and so it is excluded from the analysis. By using the data from every competitive game played during the 2020 spring split from 24th January 2020 to 2nd March 2020, we aim to predict the outcome of games that take place in the 2020 summer split. Each training dataset was validated using 10-fold cross validation. There were a total of 306 games in the 2020 Summer Split dataset used for testing.

There are several independent analysts who create content and collect data of competitive LoL to enable community-driven analysis and discussion. The data used in this report was obtained from an independent analyst, Tim Sevenhuysen, who runs the website oracleselixir.com [19].

The training data featured 882 games of data. Each game includes 12 rows. One row for each player (10 players) and one row for each team (two teams). In order to make the data usable it was separated data, into two subsets: the raw data of per game averages of each team (Table 1); and the raw data of per game averages of each player. Player statistics comprise of: *Position; Games Played;*

Win Percentage; Counter-Pick Rate; Total Kills; Total Deaths; Total Assists; Total Kill/Death/Assist Ratio; Kill Participation; Kill Share; Average Share of Team's Deaths; First Blood Rate; Average Gold Difference at 10 min; Average Experience Difference at 10 min; Average Creep Score Difference at 10 min; Average Monsters + Minions killed per minute; Average Share of Team's Total Creep Score post-15-minutes; Average Damage to Champions per minute; Damage Share; Average Earned Gold per minute; Gold Share; Average Wards Placed per minute; and *Average Wards Cleared per minute.* The players are separated by their role in the team, since different metrics can be more important to specific roles.

Table 1. Metrics and opposite metrics in the team statistics subset

Metric	Opposite
Kills	Deaths
Gold at 15	Opponent Gold at 15
XP at 15	Opponent XP at 15
CS at 15	Opponent CS at 15
Towers	Opponent Towers
Dragons	Opponent Dragons
Vision Score per Minute	Opponent Vision Score per Minute
Kills per Minute	Opponent Kills per Minute
Damage per Minute	Opponent Damage per Minute
Barons	Opponent Barons
Heralds	Opponent Heralds
Inhibitors	Opponent Inhibitors
Wins	Losses

4.2 Performance Measures

Pythagorean Expectation. Pythagorean Expectation (PE) is used to calculate the expected total wins for a competitor over a number of games. George William 'Bill' James, known for his approach to analysing professional baseball using data and statistics, developed the formula to predict a baseball team's win percentage from the observed number of runs scored and runs allowed during a given baseball season. James is widely recognised for coining the term *Sabermetrics.* This term is a combination of the acronym SABR (Society for American Baseball Research) and the word metrics. Sabermetrics has become widely accepted as a useful baseball evaluation tool [3]. It is argued that the PE was the impetus for baseball's Sabermetricians movement, where, most notably, the Oakland Athletics adopted statistical principles that revolutionised their approach to baseball team management [12].

$$W = \frac{S^2}{S^2 + A^2} = \frac{1}{1 + (A/S)^2} \tag{1}$$

In the original formula, W is the win percentage, S is the observed number of runs scored, and A is the observed number of runs allowed. James initially used an exponent of 2, inspiring the use of Pythagorean in the formula's name. The formula has since been studied to identify the optimal exponent value for accurate predictions. Different exponents can be calculated for each team in order to more accurately predict win percentages, and methods to find those exponents, such as the Pythagenpat formula, have been developed

$$n = \frac{S + A}{G}^{0.287} \tag{2}$$

where n is the exponent, and G is the total number of games. Though originally used for baseball, the simple concept of an offensive and defensive stat forming the foundation of the PE formula means that it can be applied to other sports [13,15].

For LoL there are several metrics that can be used in an application of PE. The most obvious one would be kills and deaths. While the win condition of LoL is not having a higher margin of kills than the other team, it is an obvious metric that usually indicates the more dominant team. Another alternative would be turrets destroyed vs turrets lost. The planned model for rating teams will be calculating an overall offensive and defensive rating for each team, so these ratings can also serve as the values used in the PE formula.

Log5. Once the values of the PE formula for each team are known, we can use another formula to estimate the probability of one team beating another. James also devised Log5, a formula that uses two teams' winning percentages to calculate head-to-head match up probabilities [14].

$$pA, B = \frac{pA - pA \times pB}{pA + pB - 2 \times pA \times pB} \tag{3}$$

The Log5 formula considers the winning percentage of team A (pA) and team B (pB) and returns the percentage chance that team A beats team B. From which we can easily calculate the chance that team B beats team A. We can experiment using this formula with the values obtained from PE and compare them to predictions from logistic regression models to see if it offers better or worse performance.

Strength of Schedule. If two teams have an equal record in sports, it can be challenging to determine which one could technically be considered the better team. One way of determining this is to assess the strength of the schedule for each team. Strength of Schedule (SOS) refers to the strength of the opponents a team has faced, compared to others [6].

Calculation of SOS involves comparing the combined winning percentages of each team's opponents against their own record or adjusting statistics by adding or subtracting based on an opponent's record. Assessing a team's strength of schedule can lead to interesting insights, where a bad team who appear strong on paper, may have only played against weaker teams, and a good team with worse statistics may have only played against stronger teams. Since the LoL teams that place higher in the rankings during the spring split round robin phase progress to the spring split playoffs, they end up playing more games against tougher opponents than other teams. A team's per game average stats might be lower than a worse team, simply because they had to play more games against stronger teams.

This work will be taking SOS into account when analysing the data set, since many teams in LoL do not play against each other the same amount of times over the course of a split. This results in a method of adjusting a team's stats based on the strength of their opponents, with the goal of identifying a team's strength of schedule and building a more accurate representation of a team's overall strength.

To calculate a team's adjusted total, the metric M for a team T is

$$AdjTotal_{MT} = \sum_{i=1}^{N}(OppStat_i - AvgStat_M - SideAdv_{MT}) \tag{4}$$

where N is the number of games featuring the selected team, $OppStat_i$ is the opponent's opposite raw stat in row i, $AvgStat_M$ is the overall league average stat for metric M, and $SideAdv_{MT}$ is the average advantage/disadvantage for metric M on team T's side of the map.

The adjustment to the chosen metric is made by dividing $AdjTotal$ by the number of games a team has played and subtracting that from $RawStat$

$$AdjustedStat_{MT} = RawStat_{MT} - \frac{AdjTotal_{MT}}{TotalGames_T} \tag{5}$$

where $RawStat_{MT}$ is the raw per-game average stat for metric M for team T and $TotalGames_T$ is the total amount of games played by team T.

Using this information, one can calculate what a team's adjusted stats would be for each metric and compare them to their actual performance. If a team's adjusted stats are lower than their actual performance, this would indicate that the level of their opponents was worse in that metric and vice versa.

5 Evaluation

5.1 Team Ratings

Side Advantage. Before devising and evaluating a model, it is important to determine if the dataset is balanced or not. In this case, whether the side of the map a team starts on provides any advantage. While most physical sports feature a home advantage due to familiar locations, less travel and playing in front of their own fans, LoL takes place in a virtual environment. Therefore, no significant difference or advantage for either team should be discovered. Despite this, there are major differences between starting on either side of the map that could provide an advantage to a team.

It may be argued that the blue side of the map holds an inherent advantage due to several factors. These include the asymmetrical geometry of Summoner's Rift and the isometric point-of-view favouring the blue side of the map. Most importantly, the pick/ban phase strategy of a team is often dictated by the side of the map the team is going to playing. Data suggests that this side advantage does exist. In 2017, professional League of Legends games saw a period where blue side had a win rate of 64%. So much so that the developers of LoL, have sought to balance this advantage through various balance updates, such as making dragons a more lucrative objective.

The dataset used in this study includes 882 games, of which blue side won 477. This equates to a 54.08% win rate for blue side. A chi-square test suggests that the side of the map does have an impact on a team's chances of winning $\chi^2(1, 882) = 5.878, p = 0.015$. This infers that blue wins are expected to be more prevalent in the dataset, causing a slight imbalance.

Metric Selection. Using all available metrics in a prediction model can be detrimental to its performance and prediction accuracy. Using a point-biserial correlation coefficient (PBCC) [23] calculation for each metric, identified which metrics strongly correlate with the result of a game. Two sets of calculations were carried out on each metric, one for the true stats of each game, and one for the per game averages of each team in each game. The tables were split into calculations for blue side and red side and ordered by the highest averages PBCC (converted to absolute value).

We selected the top 8 metrics from the red and blue teams because all 8 metrics scored above 0.5 absolute true PBCC and 0.25 absolute averages PBCC.

Table 2. Offensive metrics for forming team rating

Metric	True PBCC Abs. (Blue/Red)	Average PBCC Abs. (Blue/Red)
Towers	0.887/0.891	0.319/0.309
Inhibitors	0.734/0.755	0.293/0.277
Kills	0.644/0.731	0.258/0.267
KPM	0.664/0.735	0.263/0.273

Table 3. Defensive metrics for forming team rating

Metric	True PBCC Abs. (Blue/Red)	Average PBCC Abs. (Blue/Red)
Opponent Towers	0.891/0.887	0.316/0.323
Opponent Inhibitors	0.755/0.734	0.285/0.290
Opponent Dragons	0.672/0.636	0.255/0.290
Opponent Barons	0.674/0.574	0.274/0.280

They can also be evenly split into offensive (shown in Table 2) and defensive (Table 3) metrics, which will form the basis of offensive and defensive team ratings. The coefficient values can be used to calculate a weighting for each metric when producing a team rating. Another prediction model can also be formed by using these metrics as features, meaning that the results can be compared to the prediction models using all available metrics.

Table 4. Weightings for offensive and defensive metrics

Metric	Tactic	Weight
Kills	Offensive	23.24%
KPM	Offensive	23.75%
Towers	Offensive	27.78%
Inhibitors	Offensive	25.23%
Opponent Barons	Defensive	24.75%
Opponent Dragons	Defensive	22.98%
Opponent Towers	Defensive	27.51%
Opponent Inhibitors	Defensive	24.77%

Normalization and Team Ratings. In forming team ratings, Z-score normalization was selected over min-max as Z-score does a better job at handling outliers and will grant a team a higher value if they are drastically better in a particular metric, rather than pushing all other teams to be within a smaller range of each other.

Following normalization, the next step was to determine the weight of each metric. Weighting was calculated using the PBCCs used earlier to select the most relevant features. The weights were separated into offensive and defensive and calculated by summing the mean coefficients for each metric for both blue and red team and then calculating the percentage each mean coefficient contributes, shown in Table 4.

After calculating the weights, an offensive and defensive rating were formed using the sum of each normalized metric multiplied by its weight. This creates two new metrics, the offensive rating and the defensive rating. Figure 2 displays each team in terms of their offensive and defensive ratings, creating a visualization of where a team's strengths lie in their play style. These metrics can be considered opposites, lending themselves to being used in a Pythagorean expectation formula.

Pythagorean Expectation: Exponent. To determine the most accurate PE exponent for the offensive and defensive ratings, we iteratively evaluated exponents between 0 and 10 and calculated the Mean Absolute Error (MAE) from each team's predicted y and actual x win percentages.

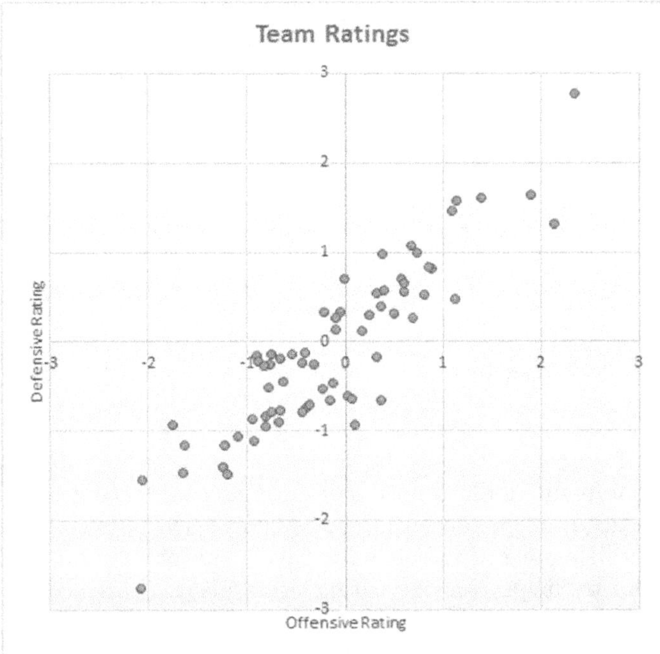

Fig. 2. Offensive and defensive ratings for all teams.

$$MAE = \frac{\sum_{i=1}^{n} |y_i - x_i|}{n} \tag{6}$$

This is done with the intention of finding the PE exponent value that minimises the MAE. The values of the defensive rating were inverted and each added to a constant of 5, since the formula relies on a lower, positive value, defensive stat being a reflection of a team's ability. We found a value of 1.82 the most accurate single exponent to use for this dataset, with MAE of 0.0397. The MAE values for this exponent range are shown in Fig. 3.

5.2 Player Ratings

Focusing on the performance of an entire team to predict results can be flawed for several reasons. In competitive LoL, teams may use substitute players to take the place of another player in a certain role. There is also the case of players transferring to a different team between each split. Teams will try to sign new players to replace under-performing ones, or more successful teams might attract the best players from lesser teams. While some teams tend to maintain a certain level of dominance despite changing their roster, this is usually down to the team's infrastructure and coaching. Most teams will notice a certain change in performance even by changing just one member of their roster.

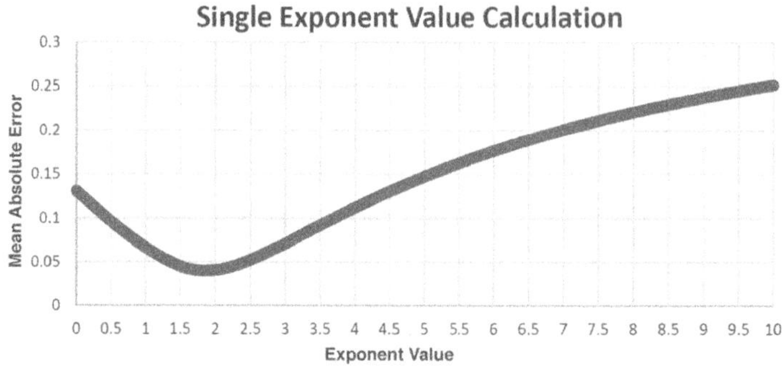

Fig. 3. Pythagorean expectation: exponent value calculation.

Predicting future results only by a team's combined results could lead to problems if that team changes its roster. In this case, rating each player may result in more accurate predictions. By assigning each player their own rating, a modular overall team rating can be formed. The process for creating player ratings is similar to the process of creating team ratings described in the previous sub-section.

Rather than choosing the same metrics for every player type in LoL, there is reason to consider the difference in each class a player can assume, and what aspects of the game are important for that role. For selection of player metrics, the same process was followed as for team metrics, namely selecting the strongest correlation coefficients and weighting them accordingly. These are displayed in Table 5, noting the acronyms: Average Dominance Factor (Dom F); average Damage dealt to champions Per Minute (DPM); average Gold Difference between a player and their opponent in their respective role at the 10-min mark (GD10); average Kills, Deaths and Assists ratio (KDA); average Creep Score Difference between a player and their opponent in their respective role at the 10-min mark (CSD10); and average experience points (XP) Difference between a player and their opponent in their respective role at the 10-min mark (XPD10).

After selecting the metrics for each role, the next task was to arrive at an overall rating. The player statistics dataset is not suitable for an offensive and defensive metric split, so each player will only have one rating based on the stated metrics. After calculating each player's rating, another model can be set up using each team's individual player ratings as a feature. Therefore, if a player is swapped out for a different one in a game, the rating will adjust to match the new player, affecting prediction outcome.

5.3 Performance Evaluation

Summary of Approaches. A total of seven sets of features and approaches were evaluated to identify the one resulting in the best prediction outcomes. These approaches were: (1) the *un-adjusted per game metrics* per team (UT); (2) the *adjusted per game metrics* per teams (AT); (3) the eight *weighted metrics*

Table 5. Player metrics and weightings

Metric	PBCC	Weight
Top Laner		
Win%	0.330	41.32%
Dom F	0.234	29.36%
Assists	0.234	29.33%
Jungle		
Win%	0.339	32.04%
Dom F	0.284	26.86%
GD10	0.161	15.20%
XPD10	0.148	14.02%
CSD10	0.126	11.88%
Mid Laner		
Win%	0.335	37.12%
Dom F	0.270	29.95%
DPM	0.149	16.49%
GD10	0.148	16.44%
Bot Laner		
Win%	0.344	29.36%
KDA	0.266	22.71%
Kills	0.232	19.82%
DPM	0.175	14.89%
GD10	0.155	13.21%
Support		
Win%	0.345	29.25%
Dom F	0.263	22.32%
Assists	0.238	20.20%
GD10	0.170	14.43%
XPD10	0.163	13.81%

selected by their PBCC scores per team (WT); (4) the calculated *offensive rating and defensive rating* per team (OD); (5) a *player rating* for each player in both teams (PR); (6) *actual win rate percentages* of each team (WP); and (7) the *expected win percentage* calculated using the Pythagorean expectation formula for both teams (PE). Approaches 1 to 5 made use of logistic regression to predict game outcomes and 6 to 7 made use of the Log5 formula for prediction.

Performance Metrics and Results. The following metrics were used to measure performance of the approaches: Classification Accuracy (CA) [18]; F1 Score

(F1) [1]; Area Under the Curve (AUC) [7]; Mathews Correlation Coefficient (MCC) [1]; Log Loss (LL) [18].

Following training of the logistic regression models and calculation of the Log5 outcomes, the results were obtained for each approach using the test data set from the 2020 Summer Split, as shown in Table 6, where the highest performing outcome for each metric is highlighted in bold.

Table 6. Evaluation results (Summer 2020)

Model	CA	F1 (Blue)	F1 (Red)	AUC	MCC	LogLoss
UT	0.618	0.670	0.545	0.649	0.217	0.544
AT	0.621	0.669	0.557	0.657	0.226	0.528
WT	0.631	0.685	0.553	0.654	0.242	0.489
OD	0.634	0.689	0.556	0.639	0.248	0.516
PR	**0.673**	**0.724**	**0.600**	**0.666**	**0.329**	**0.413**
WR	0.592	0.654	0.512	0.621	0.168	0.509
PE	0.637	0.691	0.561	0.638	0.255	0.491

The *Player Rating* model scores best in each performance metric, especially MCC, while all models suffered lower F1 scores for predicting Red Wins than predicting Blue Wins. This indicates that the models have more difficulty identifying if the red team wins, and seems resistant to predict this, despite having taken the blue side advantage into account during stat adjustments for the models. Prediction performance of wins for the *Player Rating* model is illustrated in Fig. 4.

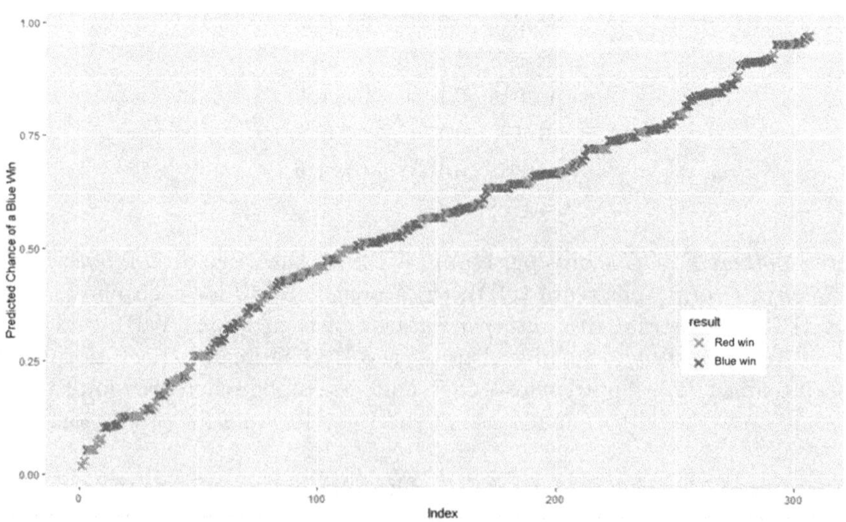

Fig. 4. Player rating model prediction results.

6 Conclusions and Future Work

The Player Rating approach achieved a significant classification accuracy of 67.3% classification when predicting 306 games from the 2020 Summer Split, which is significantly better than chance $\chi^2(1, 306) = 11.560, p < 0.001$. Compare this to results in the 2015/2016 English Football Premier league season through logistic regression, where 69.5% accuracy was achieved [16] after iterating on earlier work that achieved accuracy of 51.06% predicting the 2011/2012 season [20]. As this result is a first iteration, it stands to reason that improvements are possible. The findings may also have utility in player scouting, where a player may be performing better than their competitors, but is on a worse performing team.

The Win Rate approach scored worst in all metrics other than LogLoss. This confirms assumptions can't be made based on the previous results of teams alone, but further investigation into their actual performance in other game metrics reveals that they will better influence future results.

Since the approaches used per game averages rather than game by game data, they were unlikely to achieve a 90%+ classification accuracy, due to the inevitability of upsets. Even during the closing periods of a LoL game, the outcome can be highly volatile due to the nature of the game.

Future work should include a way to update a model after each game is played and weight more recent games higher than older when calculating a team's strength, eventually forgetting those games as they become irrelevant. For additional experimentation, a combination of team ratings and player ratings would likely be ideal. Due to the small team size in LoL, a roster change can have massive implications on the future performance of a team. There is also precedent for dominant teams falling, even without roster changes.

References

1. Chicco, D., Jurman, G.: The advantages of the Matthews correlation coefficient (MCC) over F1 score and accuracy in binary classification evaluation. BMC Genomics **21**(1), 1–13 (2020)
2. Cintia, P., Giannotti, F., Pappalardo, L., Pedreschi, D., Malvaldi, M.: The harsh rule of the goals: data-driven performance indicators for football teams. In: 2015 IEEE International Conference on Data Science and Advanced Analytics (DSAA), pp. 1–10. IEEE (2015)
3. Costa, G.B., Huber, M.R., Saccoman, J.T.: Understanding Sabermetrics: An Introduction to the Science of Baseball Statistics. McFarland, Jefferson (2019)
4. Costa, L.M., Souza, A.C.C., Souza, F.C.M.: An approach for team composition in league of legends using genetic algorithm. In: 2019 18th Brazilian Symposium on Computer Games and Digital Entertainment (SBGames), pp. 52–61. IEEE (2019)
5. Do, T.D., Dylan, S.Y., Anwer, S., Wang, S.I.: Using collaborative filtering to recommend champions in league of legends. In: 2020 IEEE Conference on Games (CoG), pp. 650–653. IEEE (2020)
6. Fearnhead, P., Taylor, B.M.: Calculating strength of schedule, and choosing teams for March Madness. Am. Stat. **64**(2), 108–115 (2010)

7. Fogarty, J., Baker, R.S., Hudson, S.E.: Case studies in the use of ROC curve analysis for sensor-based estimates in human computer interaction. In: Proceedings of Graphics Interface 2005, pp. 129–136 (2005)
8. Games, R.: League of Legends. Riot Games, Garena, Santa Monica, CA, USA (2009)
9. Hodge, V.J., Devlin, S.M., Sephton, N.J., Block, F.O., Cowling, P.I., Drachen, A.: Win prediction in multi-player esports: live professional match prediction. IEEE Trans. Games **13**, 368–379 (2019)
10. Kim, Y.J., Engel, D., Woolley, A.W., Lin, J.Y.T., McArthur, N., Malone, T.W.: What makes a strong team? Using collective intelligence to predict team performance in league of legends. In: Proceedings of the 2017 ACM Conference on Computer Supported Cooperative Work and Social Computing, pp. 2316–2329 (2017)
11. Kou, Y., Gui, X., Kow, Y.M.: Ranking practices and distinction in league of legends. In: Proceedings of the 2016 Annual Symposium on Computer-Human Interaction in Play, pp. 4–9 (2016)
12. Lewis, M.: Moneyball: The Art of Winning an Unfair Game. WW Norton & Company, New York City (2004)
13. Morey, D.: STATS basketball scoreboard, pp. 1–288 (1993)
14. Morey, L.C., Cohen, M.A.: Bias in the log5 estimation of outcome of batter/pitcher matchups, and an alternative. J. Sports Anal. **1**(1), 65–76 (2015)
15. Oliver, D.: Basketball on paper: rules and tools for performance analysis. Potomac Books, Inc., Dulles (2004)
16. Prasetio, D., et al.: Predicting football match results with logistic regression. In: 2016 International Conference On Advanced Informatics: Concepts, Theory And Application (ICAICTA), pp. 1–5. IEEE (2016)
17. Raizin, Sameboat: Simplified version of the summoner's rift map. CC BY-SA 3.0 (https://creativecommons.org/licenses/by-sa/3.0/) (2013). https://commons.wikimedia.org/w/index.php?curid=29443207
18. Saleh, H.: Machine Learning Fundamentals: Use Python and Scikit-learn to Get Up and Running with the Hottest Developments in Machine Learning. Packt Publishing Ltd., Birmingham (2018)
19. Sevenhuysen, T.: Oracle's elixir - LoL esports stats (2021). https://oracleselixir.com
20. Snyder, J.: What actually wins soccer matches: prediction of the 2011–2012 premier league for fun and profit. Thesis. University of Washington, WA: Department of Computer Science (2013)
21. Staff, L.E.: 2019 world championship hits record viewership. https://nexus.leagueoflegends.com/en-us/2019/12/2019-world-championship-hits-record-viewership/. Accessed 26 Mar 2021
22. Tassi, P.: League of Legends finals sells out LA's Staples Center in an hour. Forbes (2013)
23. Tate, R.F.: Correlation between a discrete and a continuous variable. Point-biserial correlation. Ann. Math. Stat. **25**(3), 603–607 (1954)

Fusions

Real-Time Dynamic Digital Scenography: An Electronic Opera as a Use Case

Cátia Roça(✉), Carlos Alberto Augusto, Sérgio M. Rebelo, and Pedro Martins

University of Coimbra, CISUC, DEI, Coimbra, Portugal
roca@student.dei.uc.pt, caa@carlosalbertoaugusto.org, {srebelo,pjmm}@dei.uc.pt

Abstract. In scenography, digital media has been increasingly used to design richer and more immersive experiences, for example, using video mapping or holographic projection. However, such technologies may sometimes collide with the natural (and sometimes unavoidable) improvisation moments of live interpretation. Thus, to ensure the control of these effects, large technical teams are often needed in the production of such events, which may become unfeasible in smaller size productions with smaller budgets. To test possible solutions to this problem, we endorsed the creation of an interactive low-budget scenario for the electronic opera *TMIE, Standing on the Threshold of the Outside World*, by the composer Carlos Alberto Augusto. To accomplish that, image projection techniques are combined with software to create scenarios that are capable of dynamically changing during the opera and interact with artists in real-time. This paper presents some of the experiments done so far, such as a set of interactive digital effects automatically controlled using computer vision, and simple software to manually control or time the sequence of effects during the show. Then, we discuss the potential of these techniques aimed at enabling the generation of atypical graphic solutions for live shows, as well as the practicality of having a user interface to easily manage the entry and exit times of each of the developed scenarios.

Keywords: Scenography · Computer vision · Interaction

1 Introduction

In performing and scenic arts, scenography seeks to visually organise the action's space to create a more immersive relationship between the scene and the audience [19]. The theatres' sets are traditionally built using physical and tangible materials. However, the recent technological advances promoted the democratisation of digital media, allowing the exploration of novel ways of acting. More and more, nowadays playwrights and directors begin to create, design and/or stage

C. A. Augusto—Independent Researcher.

M. Wölfel et al. (Eds.): ArtsIT 2021, LNICST 422, pp. 155–167, 2022.
https://doi.org/10.1007/978-3-030-95531-1_11

artworks exploring such technologies, making the use of digital special effects become increasingly frequent in contemporary scenic arts. The result is theatre plays capable of promoting more engaging and immersive relationships between the performers and the set, as well as between these and the audience. Most popular technologies include Video Mapping, Holography, Augmented Reality (AR), Virtual Reality, Physical Computing (PC) and Computer Vision (CV).

However, due to the natural characteristics of theatre plays, *e.g.* improvisation or chance events, the control of the special effects is, typically, a complex and time-consuming task that involves multiple technicians from various disciplinary fields, such as Sound Design, Light Design, Architecture, Graphic Design, *etc.*. These circumstances may still hinder the wider use of these techniques, namely in productions with smaller budgets. In that sense, this work seeks (I) to explore techniques from computer vision and video projection to create and apply digital special effects that autonomously react and interact with actors, and (II) to develop software to easily manage the employment of these special effects during the play.

In this project, the developed techniques are designed to be employed in the scenography of the electronic opera *TMIE, Standing on the Threshold of the Outside World*, written by the composer Carlos Alberto Augusto. In this play, the stories and realities of two deaf female characters are presented by the alternating discourses of two female singers and a male singer with the aid of an electronic soundtrack. We designed and developed a different environment of visual effects for each character, according to their characteristics and experiences in the play.

To facilitate the usage of such special effects, we developed software for controlling these effects. This software may be operated in two ways: (I) manually (*i.e.* one technician may activate/deactivate the effect); and (II) automatically (*i.e.* the effects are synchronised with certain events in an automatic manner). Also, the software is designed as a multipurpose system, allowing the development and inclusion of new effects, the blending between the existing effects, and the adaptation of the video projections. This way, it allows their use in other contexts, enabling the fulfilment of the requirements of other spaces and plays.

The remainder of this paper is organised as follows. Section 2 presents related work focusing on (I) digital effects to live shows and (II) existing software to create and handle these effects. Section 3 briefly introduces the opera understudy, *TMIE*. Section 4 describes our approach to the development of the present system and the visual effects. Finally, Sect. 5 draws the conclusions and points to the future work.

2 Related Work

Recent technological advances have allowed the exploration of new possibilities for the design of sets, which can now be fully or partially built with digital artefacts as in the virtual set of *The Jew of Malta* by Art + Com in 2002 [38] where projection is used to create a new layer of story interpretation. Since this story centres on the oscillating thoughts and postures of the various actors, this new and more readable layer of the libretto's deconstructed narration is developed through technology. For this, the actors' white costumes are used as a projection surface so that we can visualise the characters' thoughts and beliefs directly on their bodies. For example, one character's influence over another reveals itself in the fusion of their designed outfits. This is achieved with the use of an image recognition system that identifies the actors' silhouettes in real-time. From the contours of these silhouettes, virtual masks are created, later textured and projected onto the actors (see Fig. 1).

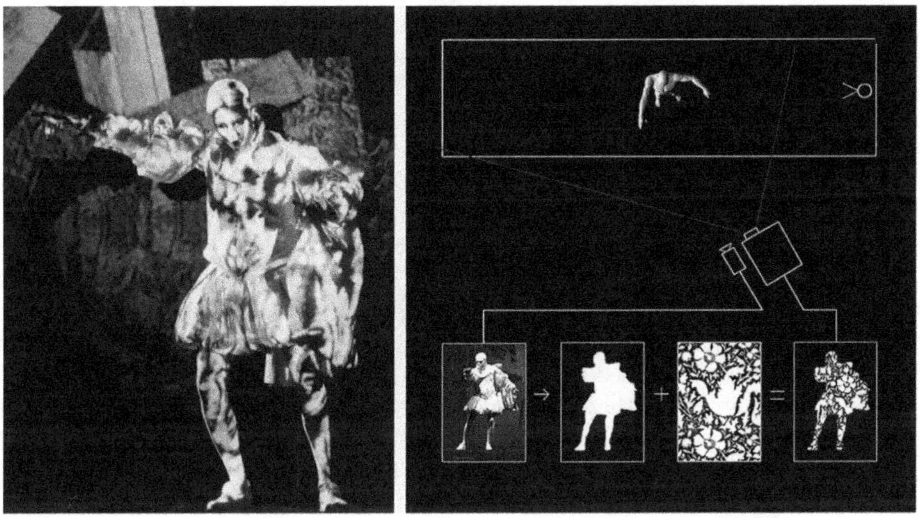

Fig. 1. Virtual Scenography *The Jew of Malta*, Art + Com, 2002. © Art + Com

The introduction of these new digital visual effects also allowed, for example, (I) the creation of more complex plots using computer simulations to visualise and test scenographic spaces [20], (II) the design of dynamic, interactive and moving scenarios through video projections or video mapping on the surrounding space, the scenic objects/props and even over the actors [32,37,42], (III) the creation of interactive and holographic scenic elements [3,32], (IV) the automatisation of the movement of physical scenarios [10], and (V) the creation of hybrid scenarios built both with real objects and virtual elements, using AR techniques [3].

More specifically in the context of Opera, we see this concept of interactivity in the work *Amazonas*, from 2010 [40], where actors interact on stage with a multi-touch table. In this case, a surface was used to influence the projected setting, therefore, the audience is able to easily follow the actors' interaction with the environment. The difference is that every live performance is different in speed, expression as well as audience's reaction. This often causes moments of improvisation or even some sections that are ignored by the actors, which makes it difficult for all the movements to be precise enough to appear real to the audience. Thus, this type of interaction becomes more realistic and natural in the eyes of the public. The use of this multi-touch table, and the interaction that comes from it, provided actors with new forms of interaction on stage. Also, they allow presenting information more naturally since the interactive exhibition is integrated with the piece.

In the context of this work, we are interested in exploring dynamic, interactive and even autonomous techniques that enhance the interaction between performers and the public and, at the same time, that may handle the unpredictability of theatrical plays. In this context, PC and CV techniques are often used to detect/recognise objects, people and sounds, and to generate audiovisual content [16] as one may see in works made using *Chordata Motion* [6], an open-source motion capture system. Also, it is possible to observe the use of sensors, such as gyroscopes, accelerometers or depth sensors (*e.g.* Microsoft Kinect), to detect and recognise the movements of performers in the three-dimensional space and, subsequently, use the gathered data to manipulate, in real-time, images projected onto the scene, like in the dance performance *Programming & Music* [13].

Several scenographic works make use of the aforementioned technologies to create more immersive performances. For example, this may be identified in the performance *8* [11], which tells a story through dance, music and dynamic scenography. In the scenario of this performance, video projections are made over mobile physical elements, in real-time. To do this, an intelligent system named *BlackTrax* [7] is used to track objects and people.

Levitation [34] is another performance that makes use of video projections and a tracking system, based on Unity Development [35], to give the illusion that the dancer is floating. To do that, the projections on the stage are automatically adapted to the dancer's movements. *Video Mapping Dance Show* [5] and *2047 Apologue I* [39] are other examples of works where artists are placed around visual projections that respond to their movements in real-time.

Furthermore, in live shows, it is still possible to complement the projected effects with other types of visual effects. Examples of this are the shows *Al Janoub Stadium* [23] and *U2 – Experience + Innocence Tour 2018* [9], which combine projections with live-action, holography, light, laser effects, sound effects and pyrotechnics, to make the environment the most immersive as possible.

With the increasing possibilities in building scenarios, there has been a growing need for more complex and capable tools. Nowadays, there are already several different tools that may be used to help the development of digital and even interactive scenarios for real-time applications, many of those without requiring the need to code. *MadMapper* [18] is an advanced tool that allows the mapping of video and light through a highly complete user interface. *Resolume* [29] is a VJing software with a modular node-based interface to create effects, mixers and video generators. *TouchDesigner* [8] consists of a visual programming system that can be applied, not only in the development of video-mapping effects but also in the creation of user interfaces, virtual reality applications, managing hardware, among other tasks. *Ventuz* [36] is a production and design environment that allows the creation of animated and interactive content using mainly, but not only, simple drag-and-drop actions. *Lumo Play* [17] allows one to create interactive floors, walls, digital signs and touchscreens by using any projector and their own hardware. Finally, *Smode* [33] allows the real-time composition and visualisation of interactive content in a simulated 3D stage. Although all these solutions may facilitate the development of digital effects, many of these tools are proprietary software and most require a considerable learning curve until the users gain the necessary ease to create ideas from scratch.

Nevertheless, for creating video mapping installations, there are also available easy-to-use and open source tools. For example, for interactivity one may use Processing [27], an easy-to-use multipurpose framework, based in Java, that was specially created for artists and designers. Regarding the mapping task, also in Processing, we refer to the SurfaceMapperGUI library [14], which allows the mapping of complex shapes but only works on an old version of Processing (1.5.1), and the Keystone library [43], which allows the use of rectangular surfaces only, but can still be a helpful tool. Furthermore, there is software such as MapMap [2] or Visution MAPIO [31], which may be useful for allowing the mapping of Processing sketches in real-time. The shortcomings are that Visution MAPIO development for macOS has been suspended and MapMap's integration with Processing seems not to be working properly, at least in macOS BigSur (used by our research team).

3 *TMIE*

*TMIE, Standing on the Threshold of the Outside World*is an electronic opera in four acts written by the Portuguese composer Carlos Alberto Augusto. A version for a sole soprano was premiered, in 2016, at O'culto da Ajuda (Lisbon, Portugal). The *libretto* is mainly based on the books *Wired for Sound* by Beverly Biderman (1998) [4] and *Miss Leavitt's Stars* by George Johnson [15], and complemented with excerpts from other texts, namely the *Fragments* by Empedocles [12] and poetry by Antero de Quental [28]. In this presentation, one single performer played all the roles with the support of a pre-recorded electronic orchestra and video sets.

The opera's plot is supported by three characters: Messier (soloist), Selena (soloist); and Coryphaeus (choir). Messier is a deaf woman who is always discovering her innermost self through the experience of listening. This character was inspired by the author of the book Wired for Sound [4], Beverly Biderman, who suffers from profound deafness. She gives a personal account of her life before and after a cochlear implant, the first effective artificial sensory organ.

Selena is a goddess who roams the skies in a silver horse-drawn cart. She was inspired by Henrietta Leavitt, the first female astronomer who also suffered from profound hearing loss. During the 19th century, Henrietta volunteered at Harvard where she developed tools that later helped Edwin Hubble calculate the distance between galaxies.

On the other hand, Coryphaeus is a philosopher studying the stories of the other two characters on the stage. This way, he has the role to mediate between them, clarifying their common ground. This character was inspired by Empedocles, a Greek philosopher who studied and created the first theory of the ear and hearing, the act of listening.

The play of this opera consists of personal reflections presented in an irregularly altered manner, by the characters. These reflections share the theme of the audition (or lack thereof). Nevertheless, the speeches are sometimes not directly related to each other, nor they are not conversations between characters.

In the context of this work, we are working with the same plot as the one presented in 2016 (see Fig. 2). However, instead of the plot being performed by one solo singer, each role will be played by a different singer.

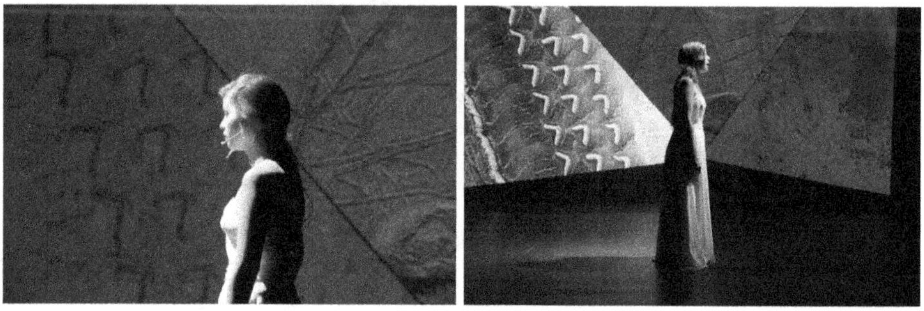

Fig. 2. Video snapshots of the premiere of *TMIE*, in 2016, at O'culto da Ajuda (Lisbon, Portugal). A full record of the opera may be visualised at https://youtu.be/3kogIlnBrfE

4 Approach

The present system plays a set of special effects to be employed in the scenography of the electronic opera *TMIE*. This project still is a work in progress, so, the opera's presentation, where this system will be introduced, is in the production stage. Nonetheless, at this moment, the system can create a set of real-time digital special effects that fulfil most of the requirements of the present opera. Also, this system can be set up without the need for large technical necessities or considerable budgets. In that sense, our current approach is focused on two developing stages: (I) the creation of real-time digital effects that automatically gather data from the stage, especially using CV techniques, and do translate these data into visuals; and (II) the development of software to set up and control the employment of effects. The following subsections will comprehensively describe each stage.

4.1 Digital Real-Time Effects

In the first stage, a set of different experiments were accomplished to read the stage environment and automatically gather data from it, and to subsequently generate visuals that translate these data into visuals. These experiments were developed by employing different technological possibilities. For instance, we have tried CV algorithms and libraries, such as PoseNet [25], U-Net [1], BodyPix [26], FaceAPI [21] or OpenCV [24] to assess the viability of using such techniques in a stage environment. Preliminary experiments using these algorithms were performed in Javascript using the *ML5* [22] and the *P5.js* [41] libraries. The final version of the system has been developed in Java using the Processing library [27]. Figure 3 displays some outputs of the referred preliminary experiments.

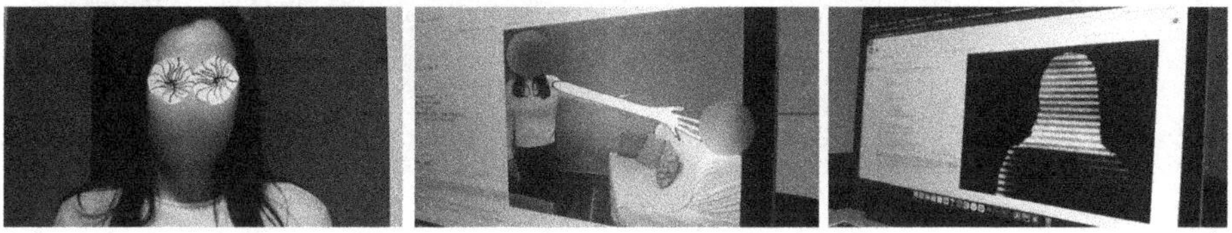

a b c

Fig. 3. Preliminary experiments on the development of real-time interactive effects to be projected on stage over artists. (a) Scenographic elements automatically following the artist's eyes/face, using the ML5 implementation of the PoseNet algorithm; (b) Scenographic elements automatically following the artist's arm, using the ML5 implementation of the PoseNet algorithm; (c) Scenographic elements automatically mapped over the artists' body, using the ML5 implementation of the U-NET algorithm.

After accessing the viability of the aforementioned technologies, and respecting the preferences of the opera's author, we started to propose scenarios/effects that would conceptually match the plot of the opera. This way, a different digital scenario environment was developed for each of the characters according to their characteristics, experiences and speech.

Selena, the astronomer, has an obvious connection with the stars. For that reason, a background pattern of stars (white circles over black background) was created. By timing the different moments and/or bypassing in the amplitude analysis of the character's miked voice (a decision left to the author), this scenario may change according to the different moments of the dialogue, varying in size, movement and colour (inverting brightness). By inverting colours and getting black stars over a white background, the scenario stops from representing a stared sky to become a representation of the negative sky scans that were part of an old technique used by astronomers to study the sky. These scans are referred to by Selena in her speech, creating a conceptual connection between the character and both the positive and negative versions of the effect (see Fig. 4). A demonstration video may be accessed at https://vimeo.com/642489858.

Messier's dialogues refer to her life experience before and after losing her hearing. To demonstrate the different moments of this evolution, a background made of dynamic luminous paths was created. These are created from moving overlapping circles that randomly change direction and which size varies with

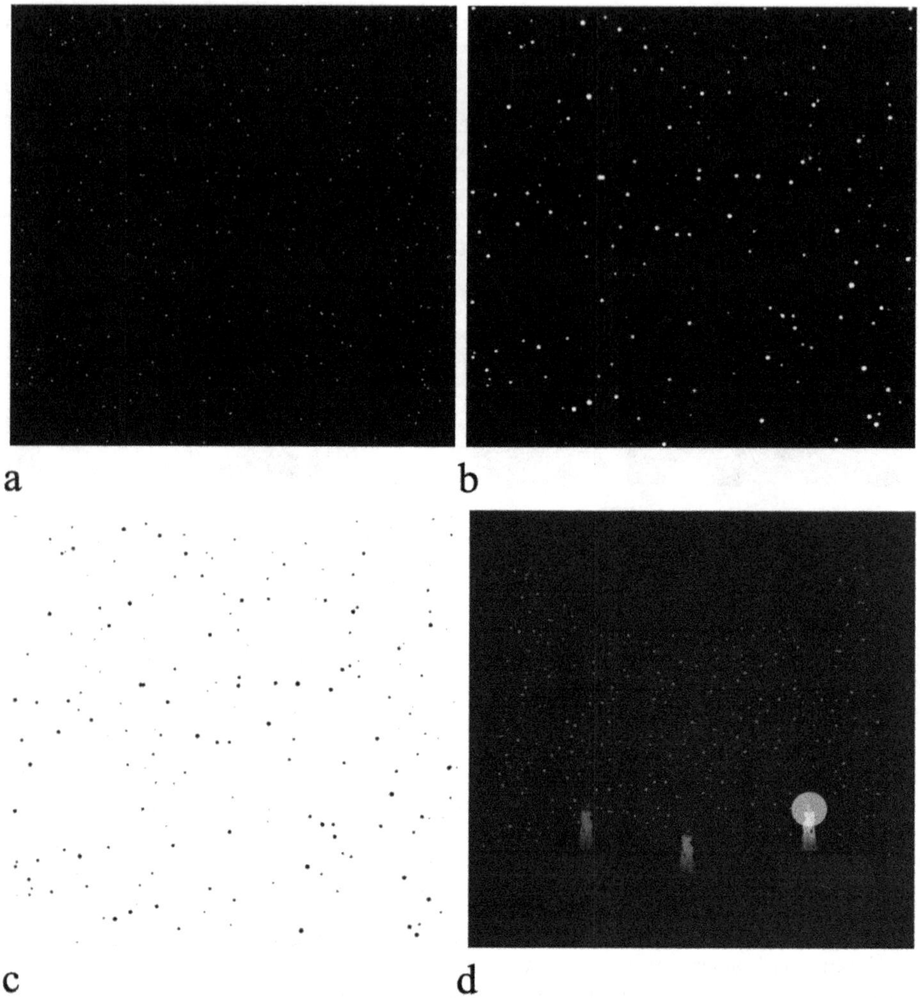

Fig. 4. Examples of different states of the environment developed for the character Selena, to be applied in the form of a background video projection: (a) small white stars over black background; (b) bigger white stars over black background; (c) black stars over white background; (d) example of application on real space.

their speed. Thus, their graphic appearance may conceptually resemble electrical impulses, neuron connections or the character's hearing connections to the world. While performing, the rays may be growing forward or backwards according to the respective moments of the characters' lives that are being referred to in the speech (see Fig. 5). To accomplish that, the two versions of this effect (forward or backwards) are to be timed in the software, yet other interactive features may be added later on. For example, speeding up the growth of the rays according to the character's movements (the more the faster), conceptually relating to her efforts/action to improve her hearing and thus understanding of the world. A demonstration video may be accessed at https://vimeo.com/642490592.

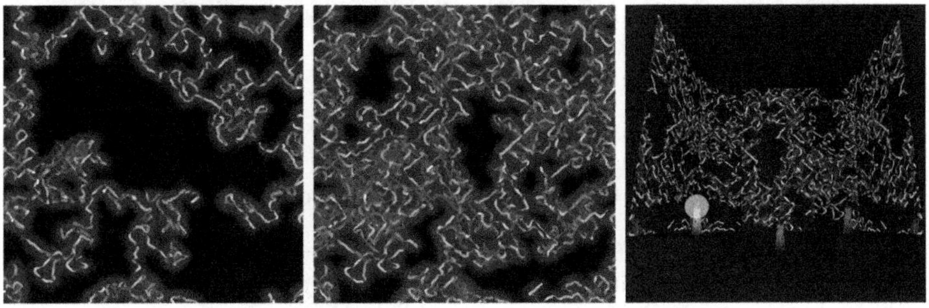

a b c

Fig. 5. (a, b) Examples of different states of the scenario/effect developed for the character Messier (white light paths over black background), to be applied in the form of a background projection; (c) example of application in a mock-up.

In addition to the environments created specifically for these two characters, all the three characters are accompanied by a light focus that follows their faces during their speeches (see Fig. 6). Coryphaeus, whose role is to clarify the connection between the remaining two, makes use of the focus effect only, which ends up being the means of connection between the three characters. This effect was ideally designed to function in the form of a hologram [30] and may take the form of different shapes or colours. Nevertheless, the effect may also be implemented as a simple projection (similarly to the aforementioned effects) or an automated mechanic light focus. To create the effect, a face detection algorithm was used to detect and follow the artist's face. When detected, a shape is drawn over the detected face with a size adapted to it. A demonstration video may be accessed at https://vimeo.com/642489151.

4.2 Controlling Software

In order to facilitate the activation and control of the developed scenarios, a controlling software was developed to allow the visual effects to be sequenced and to automatically start and stop. The fundamental objective of the proposed software is that a single technician can set up and easily control or automate a set of digital effects. For the software to be dynamic and adapt to different contexts and works, it was designed, so, it allows the introduction (or removal) of new scenarios, each implemented as a different "object" with a respective method "draw" that will draw over a given video projection display. The draw functions of the respective effects must be queued in a switch inside the draw function of the main program. Thereafter, one may automatically call the next

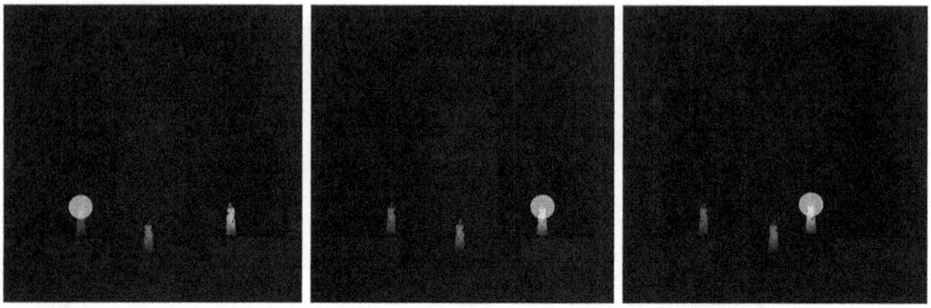

a b c

Fig. 6. Examples of different states of the scenario/effect developed for the character Coryphaeus, to be applied in the form of a hologram, a simple projection over the artists and background, or using an automated light focus.

effect in the queue by clicking the right arrow key or by setting the respective start and stop times.

The controlling software also enables the adaptation of the output to the architecture characteristics of the stage space. Thus, when necessary, the user can resize and distort the projection mask by using only the mouse and keyboard to move the vertices of the projection mask and, consequently, keystone the projection.

Lastly, the software was designed to allow simultaneous projections, in the case of being needed multiple projectors to cover all the space.

5 Discussion and Conclusions

In the design of digital scenarios and real-time scenographic effects, one may identify a growing use of digital media that has been allowing the development of richer and more immersive experiences for the audience. However, in live shows, unplanned or improvised artists actions are a constant reality, making it difficult to ensure the control of the audiovisual effects in real-time. Thus, it is often necessary to involve large technical teams to accomplish these kinds of tasks and, for this reason, in small productions with low budgets, the wider use of these techniques may become impractical.

However, based on our current observation, we believe that the use of real-time visual effects during theatre plays will enable the generation of more engaging scenarios. These effects not only enable the development of more immersive experiences but also can add a new layer of interpretation, which can aid in a better understanding of the plot.

In this paper, we have presented preliminary work on the development of solutions to surpass the problem of dealing with the real-time generation of visual effects. First, by developing a set of digital real-time dynamic and interactive visual effects and, secondly, by developing proper software to set up and easily control the sequence of effects intended. Furthermore, as a case of study, we designed and developed digital scenography for the electronic opera *TMIE*. The designs must be implemented using simple video-mapping projections. Also, as to complement, holograms [30] or light focus may be added.

Although this work is still in a preliminary phase, our experiences demonstrated to be practicable (at least so far in the implementation phase) and flexible enough to create a multitude of graphic solutions. In that sense, so far, the indicators suggest that our solution may solve the problem, bringing the desired interactivity to the studied opera without the need for a high budget or a large technical team.

Future work will focus on (I) developing a user interface to be added to the controlling software, namely to enable the user to define the entry and exit times of each visual effect, and to improve its usability, (II) developing more visual effects and scenarios environments to the opera *TMIE*; and (III) evaluating the effects and software in a real context, during the rehearsals and live shows of the studied opera.

Acknowledgments. This work is funded by national funds through the FCT - Foundation for Science and Technology, I.P., within the scope of the project CISUC - UID/CEC/00326/2020 and by European Social Fund, through the Regional Operational Program Centro 2020. The third author is funded by FCT under the grant SFRH/BD/132728/2017.

References

1. Alyafeai, Z., Lee, J.: UNET (nd). https://learn.ml5js.org/#/reference/unet. Accessed 30 July 2021
2. Audry, S., Quessy, A., Latona, M., Liaskovitis, V.: MapMap - open source video mapping software—mapmapteam.github.io (nd). https://mapmapteam.github.io/. Accessed 30 July 2021
3. Bardainne, C., Mondot A.: Mirages & miracles (2017). https://www.am-cb.net/en/projets/mirages-miracles. Accessed 30 June 2021
4. Biderman, B.: Wired for Sound: A Journey into Hearing. Trifolium Books Incorporated, Toronto, Canada (1998)
5. Carvalho, R.: Video Mapping Dance Show - Graviton (2014). https://meetgraviton.com/flv_portfolio/video-mapping-dance-show/. Accessed 30 June 2021
6. Chordata Motion: Chordata Motion - Movement made yours (2021). https://chordata.cc. Accessed 30 June 2021
7. C.G. of Companies Inc.: BlackTrax - Real Time Tracking (2021). https://blacktrax.cast-soft.com/. Accessed 30 June 2021
8. Derivative: Touchdesigner (nd). https://derivative.ca/. Accessed 30 July 2021
9. Devlin, E.: U2 - Experience + Innocence (2018). https://esdevlin.com/work/u2-experience-innocence. Accessed 30 June 2021
10. Devlin Es: Don Giovanni - ROH London (2014). https://esdevlin.com/work/don-giovanni. Accessed 30 June 2021
11. Dreamlaser: 8: A genre-bending multimedia performance (2017). https://dreamlaser.ru/en/work. Accessed 30 June 2021

12. Empedocles: Fragments of Empedocles. CreateSpace Independent Publishing Platform, Scotts Valley, CA (2017)
13. F.I.S.: Programing & Music (2020). https://fallinsession.wordpress.com/projects/. Accessed 30 June 2021
14. Jason Webb: SurfaceMapperGUI - a simple Processing interface for projection mapping (2013). https://jasonwebb.io/2013/11/surfacemappergui-a-simple-processing-interface-for-projection-mapping/#installation. Accessed 30 July 2021
15. Johnson, G.: Miss Leavitt's Stars: The Untold Story of the Woman Who Discovered How to Measure the Universe. W. W. Norton & Company, New York (2006)
16. Jung, H., Lee, J., Choi, H.J., Kim, H.: Real-time DJING + VJING with interactive elements. Contemp. Eng. Sci. **7**(24), 1321–1327 (2014)
17. Lumo Interactive: LUMOplay—LUMOplay interactive floor, wall, and touch-screen display content management software (2021). https://www.lumoplay.com/. Accessed 30 June 2021
18. MadMapper: MadMapper (2017). https://madmapper.com/. Accessed 30 June 2021
19. Maia, H.G., Muniz, E.S.: Novos caminhos para a cenografia diante da evolução tecnológica: o teatro e a realidade aumentada. Revista Tecnologia **39**(1), 1–14 (2018)
20. McKinney, J., Butterworth, P.: The Cambridge Introduction to Scenography. Cambridge University Press, Cambridge (2009)
21. Mühler, V.: face-api.js (2019). https://justadudewhohacks.github.io/face-api.js/docs/index.html. Accessed 30 June 2021
22. NYU ITP/IMA program: ml5js·Friendly Machine Learning For The Web (nd). https://ml5js.org/. Accessed 30 June 2021
23. OCUBO: Al Janoub Stadium (2019). https://www.ocubo.com/aljanub-stadium. Accessed 30 June 2021
24. OpenCV: OpenCV (nd). https://opencv.org/. Accessed 30 June 2021
25. Oved, D., Alvarado, I., Gallo, A.: Real-time human pose estimation in the browser with tensorflow. js (2018). https://blog.tensorflow.org/2018/05/real-time-human-pose-estimation-in.html. Accessed 30 July 2021
26. Oved, D., Zhu, T., Alvarado, I.: BodyPix: Real-time Person Segmentation in the Browser with TensorFlow.js (2019). https://blog.tensorflow.org/2019/11/updated-bodypix-2.html. Accessed 30 July 2021
27. Processing Foundation: Processing.org (nd). https://processing.org/. Accessed 30 June 2021
28. de Quental, A.: Poesia Completa de Antero de Quental. Abysmo, Lisbon (2018)
29. Resolume: Resolume VJ Software & Media Server - Resolume (nd). https://resolume.com/. Accessed 30 June 2021
30. Russell, S., Nelson, P., Hajicek, D.: Themed Holographic Theater (US Patent 2017/0023911A1), January 2017
31. Ryabov, I., Visution: Visution MAPIO (2021). https://visution.com/. Accessed 30 July 2021
32. Slavenski, Z.: Mapping with actors (2018). https://www.youtube.com/watch?v=juiouwvIT5k. Accessed 30 June 2021
33. Smode Tech: Real-time compositing, media server and XR - SMODE (nd). https://smode.fr/. Accessed 30 June 2021
34. Sveta, S.: Levitation (2016). https://silasveta.com/work/levitation. Accessed 30 June 2021
35. Unity Technologies: Unity Real-Time Development Platform (3D, 2D VR & AR Engine) (2021). https://unity.com/. Accessed 30 June 2021

36. Ventuz Technology: VENTUZ Realtime Graphics Software (2021). https://www.ventuz.com/. Accessed 30 June 2021
37. Wang, A.: Evolution—Mysite (2021). https://www.andidywang.com/evolution. Accessed 30 June 2021
38. Werner, A.: Marlowe: Der Jude von Malta (2002). https://www.andrewerner.eu/MJvMSeite.html. Accessed 30 July 2021
39. WHITEvoid: 2047 APOLOGUE I - Audiovisual Stage Performance (2017). https://www.whitevoid.com/2047-apologue-i/. Accessed 30 June 2021
40. Wölfel, M., Lintermann, B., Völzow, N.: Using Tangible Surfaces in Opera. In: Søndergaard, M. (ed.) Re-new - IMAC 2011 Proceedings. Aalborg University Press, Aalborg, Denmark (2011). https://www.researchgate.net/publication/259828339
41. Ye, Q., evelyn masso, McCarthy, L.L., Processing Foundation, NYU ITP: p5.js (nd). https://p5js.org/. Accessed 30 June 2021
42. Yip, T.: Storm in Emptiness (2017). https://www.timyipstudio.com/content/article/en/114. Accessed 30 July 2021
43. Zirfas, F.M., University of Applied Sciences Potsdam: Processing and Keystone - Doing Projection Mapping (2016). https://fh-potsdam.github.io/doing-projection-mapping/processing-keystone/. Accessed 30 July 2021

The Lost Film *Pontianak* (1957) as a Case Study to Evaluate Different Strategies of Performance Capture for Virtual Heritage

Benjamin Seide(✉) and Benjamin Slater

School of Art, Design and Media, Nanyang Technological University, Singapore 637458, Singapore
{bseide,baslater}@ntu.edu.sg

Abstract. The 1957 film *Pontianak* (produced by Cathay-Keris Films, directed by B.N Rao and starring Maria Menado) was a highly successful horror film and very much a product of the Malay-language film industry that was based in Singapore from the 1940s to the 1960s. However, this 'classic' film (along with one of its sequels) is now considered to be lost. For our research into the possibilities of virtual cinematic heritage, we researched scenes and sequences from the film with the aim to imaginatively 'reenact' them as a virtual reality experience. In this paper, we evaluate the creative and technical processes involved in executing such a project, demonstrating the different strategies deployed and their various outcomes. One key issue discussed is whether affordable motion capture systems are viable alternatives to capture performance for virtual heritage applications. We summarise the current status of the project and propose some possible pathways for future development.

Keywords: Virtual heritage · Motion capture · Pontianak

1 Introduction

Our initial research began in 2019 with a primary question: Can accomplishments in cultural heritage, such as the creation of virtual environments of historic sites and advancements in game development, such as inhabiting virtual environments with actors and stories, be utilised for the benefit of creating virtual 'cinematic' heritage?

We knew the outcome of the investigation would be the creation of a virtual reality application, a walk-in movie scene of a film, which the viewer can freely explore (with a headset in a room-scale VR setup), while a narrative with actors unfolds around them. We had to identify the heritage content that would be the subject of the application.

As researchers based in Singapore, we wanted to address the relatively internationally little-known Singapore and Malayan film industry and history, most notably the late colonial period (1940s to 1960s), when there was a prolific output of Malay-language cinema produced by two vertically integrated film studios - Malay Film Productions (owned and managed by The Shaw Brothers), and Cathay-Keris Studio (owned by Loke

M. Wölfel et al. (Eds.): ArtsIT 2021, LNICST 422, pp. 168–180, 2022.
https://doi.org/10.1007/978-3-030-95531-1_12

Wan Tho). Rather than pick on a frequently re-screened 'classic' film from this era, we decided to focus our research on the film *Pontianak*, made and released in 1957, by Cathay-Keris (See Fig. 1).

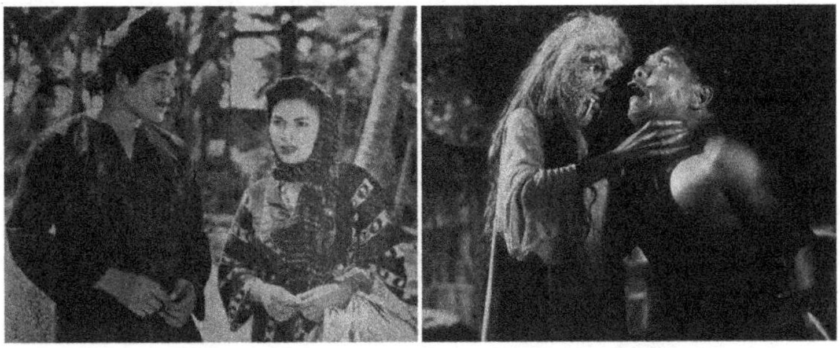

Fig. 1. Film stills from *Pontianak* (1957) and *Dendam Pontianak* (1957) © 1957, Cathay-Keris

The choice of this film as a case study was grounded in its significance as a heritage artefact. Firstly, it features representations of traditional kampongs (villages) of Malaya and Singapore from that period (and historically depicted). Our case study presents the first virtual kampong for audience exploration, which constitutes a high degree of relevance in the context of cultural and historical heritage preservation. Secondly, the key source of the film (and the series that followed) is traditional Malay mythology and folklore that was widely believed in and still is (to some extent) - in contemporary Singapore and Malaysia. Thirdly, *Pontianak* is considered to be a 'lost' film, as there are no existing prints or copies of the film in any archive, and none have been seen since at least the early 1960s [1, p. 126].

Investigating a lost film brought a new level of complexity to our work, as we were required to 'recreate' the film from scant sources, rather than 'restore' an existing (and possibly damaged or incomplete) copy. Traditional heritage approaches to cinema were thus impossible, which to some extent justified our highly non-traditional use of VR to animate a work that was otherwise impossible to experience. Our idea was to use the immersive, experiential qualities of VR to create a new work, inspired by *Pontianak*, and rather than attempt to simulate the film in an accurate way, we hoped to create an experience that imaginatively reflected the film, and our research into it, which would inspire audiences to learn more about this 'lost' piece of film heritage.

Before we began we needed to assess how similar projects were executed and what was feasible for our team to achieve. The Epic Games *Digital Human* project [2], which created a digital incarnation of actor Andy Serkis, Actor of Gollum in *The Lord of the Rings* (2001), Kong in *King Kong* (2005), Caesar in *Rise of the Planet of the Apes* (2010), and many more [3], demonstrated that convincing realism was achievable beyond big-budget film productions but in a real-time game engine environment. In 2021, Epic Games went even further and released *MetaHuman Creator* [4], a tool that enables artists to create realistic computer-generated (CG) characters to be used with their game engine Unreal. In 2018, Epic's *Digital Human* project created characters capable of believable acting by capturing the performance and facial expression of real actors utilising a professional high-end motion capture system from Vicon [5]. Since then, motion capture alternatives have become available that promise comparable results for a fraction of the cost. These significant advancements have implications beyond applications in entertainment and games, and furthermore they generate a question: Are smaller academic research teams, non-commercial projects, and artists in reaching distance of creating realistic digital humans? And how can virtual heritage applications benefit from these developments?

At time of writing, our research project, creating the virtual cinematic heritage application for the film *Pontianak*, is still ongoing; steps and processes involved in designing the virtual Malay village environment, experiments in reenacting a key scene of the film as well as findings in the history and synopses of the films have been described in a previous publication [6]. In this paper, we will provide some background into the source material and our historical research, then go on to outline different strategies of capturing performances, evaluate a low-cost motion capture system and detail further findings and knowledge gained through several iterations of capturing performances with actors for the virtual reenactments of our virtual cinematic heritage application.

2 *Pontianak* (1957) and the Snake Bite Scene

The first *Pontianak* film was released in April 1957 in Singapore. It was the first film to make prominent use of supernatural figures that were part of a common Malay folklore, originally recorded in English by anthropologists at the turn of the century [7, p. 325–326]. The portrayal of the Pontianak in the film marked a significant re-invention of the mythos, credited to the film's writer Abdul Razak in collaboration with the film's lead actress Maria Menado [8]. The films' box office success led to a second film *Dendam Pontianak* (*Revenge of the Pontianak*) being rushed into production and released that September, followed by a third film, *Sumpah Pontianak* (*Curse of the Pontianak*), released in 1958. There would be other *Pontianak* films during that period (produced by the rival studio run by the Shaw Brothers), but these three films brought together the distinct talents of Abdul Razak (as writer), the director B.N Rao (who had been 'imported' from Mumbai's burgeoning film industry), the producer Ho Ah Loke (who represented the 'Keris' part of of Cathay-Keris), and the actress Maria Menado, originally from Indonesia, who notably played three roles in the film (or rather three incarnations of the same character): the 'deformed' orphan Chomel, a 'beautiful' Chomel after she's used magic to change her appearance, and the monstrous Pontianak creature itself - a vampiric figure with elongated fangs and claw-like fingernails, wearing a white robe.

As mentioned, despite their reportedly wide distribution (throughout Malaya and Indonesia, and other Asian territories) *Pontianak* and *Dendam Pontianak*, are believed to be 'lost films'. According to sources [1, p. 126], it is said that when Ho Ah Loke broke off his production partnership with Cathay's Loke Wan Tho, the two divided the film titles between them including the remaining prints. Some years later it is said that Ho disposed of the prints, including the two *Pontianak* films, in a quarry or a lake. This oft-repeated story is flawed, mainly because there would have been hundreds of prints of each title, although little is known about film distribution in Malaya during that period and how prints were stored or destroyed.

Since *Pontianak* has not been viewed since its first period of release in 1957 and 1958 there is a scarcity of information or images available. We were able to find contemporary articles and reviews from the time of the film's release in local newspapers and magazines. One major source for stills and story information was the published synopsis of the film, which we were able to locate via private collectors of film memorabilia. Film synopses were a common form of promotional material and merchandise during this period for Malay-language films, and they would contain images from the film, behind-the-scenes photos, as well as a prose summary of the major events of the film's storyline.

We have determined, through the synopses and other secondary sources, that the basic plot of *Pontianak* is an origin story of the titular creature, who is an abandoned child, found by a bohmoh (Malay shaman), and raised as his daughter/servant, ironically given the name Chomel (meaning 'pretty' or 'cute' in Bahasa Melayu), even though she is coded as 'ugly' and 'deformed' in the narrative. When the bohmoh dies, Chomel is entrusted to burn his magic books, but instead she learns the spell to make herself beautiful. However, she is told that if she drinks human blood the spell will be broken and she will become a Pontianak. This story is in stark contrast to the commonly-known myth of the Pontianak as a ghost of a woman who died in childbirth. In the film, the beautiful Chomel, travels to a kampong where she meets and falls in love with the son of the village chief, Othman (played by M. Amin). They marry and have a daughter, Maria, and it is after this, that we reach the crucial scene when Chomel finally transforms into the Pontianak.

Another key source was the written account of A R Mustafar, an independent historian of Malay film, who reports that he watched *Pontianak* upon its release in 1957. He described the transformation scene as a crucial moment for the audience, witnessing the cinematic rendering of this infamous supernatural figure for the first time. He writes:

> Something that struck out to me was when M. Amin's calf was bitten by a snake and when he was in so much pain, Maria Menado sucked out the poison. In that moment, the cinema went absolutely silent since they knew what was going to happen next. The change from Maria Menado's beautiful face into that of the scary Pontianak shocked the audience, even causing a slight commotion for a while. When the shock died down, silence came again [9, p. 114].

In terms of narrative context, we know from the film's synopsis that Chomel has been warned after she used magic to make herself beautiful that drinking snake poison will turn her into a monster, which is something the audience would have been aware of - hence the anticipation of that moment. The synopsis goes on to describe the scene in more detail:

> (They) were having a relaxing chat alongside their daughter who was playing, Othman was suddenly bitten by a snake on his neck. Othman was moaning in pain, Chomil (sic) wanted to leave her husband to take medicine meant to fight a snake's venom, but Othman couldn't wait and asked his wife to suck out the venom that was causing him so much pain, from his neck. Othman moaned in pain again and asked his wife to suck out the snake's venom from his neck. Due to her faithfulness to her husband, Chomil held on to her husband's neck and began to suck the venom out

> of his neck...Tasting blood...it tastes good...and without realising Chomil ended up sucking up all the blood in her husband without stopping...Othman began to scream all of a sudden...Chomil started to change...her black hair became white, her skin started to wrinkle...her nails became sharp...Othman became weak due to the blood loss and...he died, falling onto the ground [10, p. 10].

Given how pivotal this scene was to the film, both in terms of the narrative and the audience response, we decided to make this the focus for our first iteration of using VR technology to recreate the sequence. The first step was to script a sequence between the two characters Othman and Chomel, in which they walked through the kampong which would build up to the moment of the snake bite and then the transformation.

We were partly inspired to have them walking as we were taking reference from a film still of the actors M. Amin and Maria Menado in character, that presents them standing, and also we wanted to create an experience in which the viewer can travel through the virtual kampong rather than be in one static location. This decision would present technical challenges described later. We wrote dialogue for the characters which was an imaginative projection rather than an attempt at speculating what the 'real' dialogue would have been. In our dialogue Othman is curious about the mysterious past of his wife, questioning her as to her origins, and revealing tensions between them. This dramatic element, was designed to function as exposition for a viewer unfamiliar with the story - it was also written the spirit of Malay-language films of the era, which tended towards being dramatically direct and expositional. However, their conversation is interrupted when the snake falls from a tree (an assumption we made about the original film given that we know that the snake targets the neck) and bites Othman, leading to him to implore Chomel to suck the venom from his wound, which she does reluctantly, and then she finally transforms into the Pontianak, which is where our sequence ends.

3 Performance Capture

To populate a virtual environment with digital humans, an artist or researcher has two basic options in regard to creating the animation of the characters. The most common approach is to use a library of actions such as idling, turning, walking, jumping etc. and then transition between these to create a flow of continuous motions. This approach is the foundation of real-time interactivity of computer games. The individual actions are created by manual key-frame animation or using motion capture performances which are then edited for short actions that can be looped. Advancements are being made regarding how seamless the transitions between actions are rendered. A second approach is to motion capture an actor's performance for the entire scene in one continuous linear action. This filmic or theatrical approach forfeits interactivity for the benefit of realism of the performance. While only workable for non-interactive background characters in games, this second approach provides an opportunity for virtual heritage applications to improve the authenticity and believability of reenactments of historical events.

As laid out in the previous chapter our main objective was to enact a scene from a film, its structure linear by design, we focused on the second approach to capture the entire performance in one continuous linear action, with our main focus on capturing the aforementioned 4-min-long Snake Bite scene. Additional shorter scenes were captured to further evaluate the two motion capture systems available to our project. The two systems are a camera-based system from Vicon, which is a permanent setup in our research facilities, and as a second system, the portable inertial sensor-based system from Rokoko, which is considered an entry-level low-cost alternative. Skogstad and Nymoen [11] analyse both concepts and conclude "If high positional precision is required, OptiTrack [a camera-based system] is preferable over Xsens [a sensor-based system], but [...] Xsens provides less noisy data without occlusion problems". The two specific systems compared here by Skogstad and Nymoen, OptiTrack and Xsens, are a fair comparison as both are considered in a similar price range; in contrast, our two systems from Vicon and Rokoko cannot be considered as such. However, the lower cost Rokoko system is promoted as an alternative to the more expensive camera-based systems and the portability feature is an advantage that must be considered, and as we were aiming to capture actors walking within a large area, the portable sensor-based system appeared more appropriate for our use case.

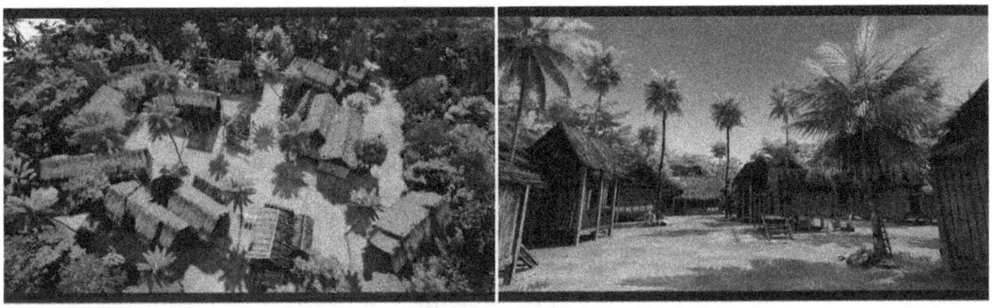

Fig. 2. Virtual kampong village, still images from VR experience, 2020, the authors

3.1 Capturing the Pontianak Snake Bite Scene

Our goal was to capture the entire 4-min-long scene with the two actors talking while walking through the virtual kampong village (See Fig. 2) in one continuous take. Following several tests, the outdoor sports field on our university campus was chosen as a capturing area as it provided the necessary open space, power outlets and, with the grass field, a relatively soft area for the actor performing as the husband to fall on. Capturing outdoors in a wide-open space (See Fig. 3) had another significant advantage, Rokoko's sensor-based suits are sensitive to electromagnetic interference, which can be minimised at an open space such as the sports field. Although the recommended distance to metallic objects is only one metre, we could not establish a fully interference-free capturing area in our school's building. However, we identified some areas in our building such as the auditorium stage with an acceptably low amount of interference.

To avoid the actors colliding with virtual buildings and trees, the walking path was translated from the virtual environment via a 2D coordinate system to the real-world capturing area, marking checkpoints and boundaries on the sports field and spanning an area of 40 by 25 m for the actors to walk through. A more sophisticated approach of

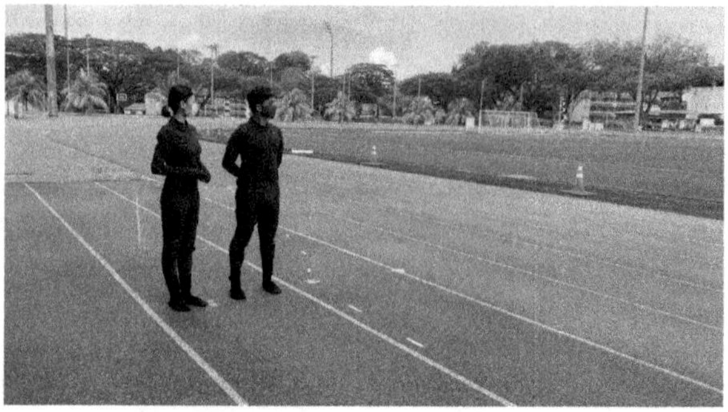

Fig. 3. Actors in sensor-based suits at outdoor capture, 2020, the authors

matching an actor's position with the virtual environment during capturing is to stream the motion capture performance to the virtual environment in real-time to create a live preview, a process considered 'virtual production'. However, sensor-based systems do not provide a reliable absolute position in 'world' space. For our particular use case with two actors walking side-by-side for minutes in a large area, the so-called 'drifting' caused the captured virtual positions of the two actors to be metres apart over time - while in reality, they were still just centimetres apart from each other. Rokoko offers a smart solution to compensate for this shortcoming by supporting SteamVR, allowing HTC Vive trackers to be mounted on actors and props. As such a setup requires several Vive base stations surrounding the capture area, it evolves into a combination of sensor and a camera-based system, neglecting some of the sensor-based system's portability advantages. Furthermore, the capture volume is limited by the base station setup, which, according to HTC, supports an area of 10 by 10 m [12]. Thus, for our application with a 40 × 25 m large outdoor area for the Snake Bite scene, adding base stations was not an option. As a result of capturing without 'absolute' position, the drift between our two characters captured simultaneously with two suits, accumulated to several metres over the entire capture time and required us to scale and reposition the data in post-production extensively to fit the layout of the virtual village.

3.2 Facial Capture

Once these positional corrections were done, a video render of the captured characters' entire walk was prepared (See Fig. 4) to support the facial capture and voice-over acting at the sound recording studio. Simultaneously to the voice-over acting, the facial data was captured using the iPhone FaceID depth map system and applied to our characters in the Reallusion iClone software.

3.3 Evaluation and Post-production

A basic post-production workflow for motion capture follows these simple steps: Review and identify the best take with the least issues and perform a clean-up of the data as necessary. The extent of the clean-up process depends on the precision of the captured data and

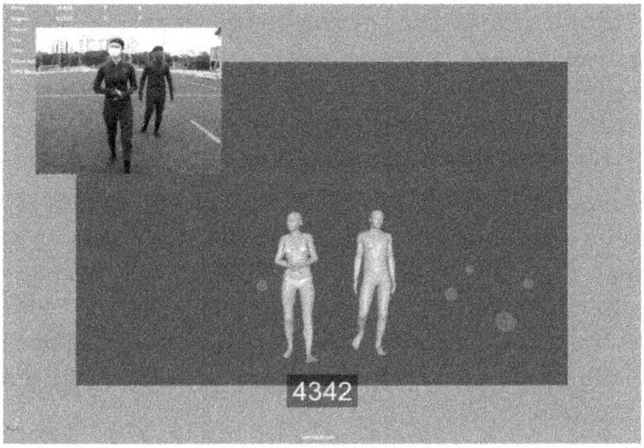

Fig. 4. Witness camera (top) and retargeted characters for voice over, 2020, the authors.

the final required quality. This manual clean-up process can only be effective if witness cameras are used to produce video references from the capturing session, commonly shot from two angles simultaneously, allowing the animator to identify discrepancies between the actor's actual movement and the captured data and to adjust accordingly. As our capturing area covered such a large area and our main witness camera being a hand-held gimbal following the actors (See Fig. 4), our reference videos were not easily usable for the clean-up stage, exposing the size of the area and the lack of static cameras as a flaw in the planning of our venture.

Evaluating the captured data and estimating the extent of how much labour-intensive clean-up would be required presented another challenge. Issues in the data which appear minor and acceptable for an animated film (for instance), might be severe and unacceptable for a virtual reality project which provides depth perception through stereopsis in the HMD. Our eyes have possibly seen countless occurrences of humans walking in real life, to a degree that, except for individuals suffering from stereoblindness, even a layman is able to identify an awkwardness in a simple walking performance of a virtual actor if the data is flawed and presented in stereoscopic 3D. Our project went through several steps of authoring such as basic clean-up, merging the facial and body data, cloth simulation, hair grooming etc. before eventually reviewing the assembled final character in VR, to only then discover that the underlying captured data possessed more severe issues than previously seen on the 2D computer display. From this experience, we concluded that every single authoring step and in particular the quality control of the motion capture data must be performed in stereoscopic 3D / virtual reality instantly and without delay.

3.4 Capturing a Second Scene with Two Systems

These findings were directly applied to the motion capture session of a second and much simpler, shorter reenactment scene, in which the Pontianak ravages a victim and, once discovered by the viewer, runs away leaving the blood-drained victim plummeting to the ground. Only requiring an area of 4 by 3 m, we were able to capture the performances with both systems available to us, the portable sensor-based suit and the studio camera-based system. To evaluate the captured data in VR, and to compare the two systems directly, we skipped previously applied intermediate steps and used simple grey-scale characters - that distinctly contrasted with the background environment, allowing us to focus precisely on potential issues in the capture data. The evaluation confirmed that similar to the Snake Bite data review, issues that appeared minor on a 2D computer display were visually amplified in stereoscopic VR. In regard to the accuracy of the motion data and perceived realism, the camera-based studio system unsurprisingly outperformed the sensor-based suit in all three - stationary, falling and running - performance actions. The data of the sensor-based suit, while still exhibiting issues, appeared most accurate for the stationary part of the performance; in contrast, the falling and running actions demonstrated severe levels of inaccuracy.

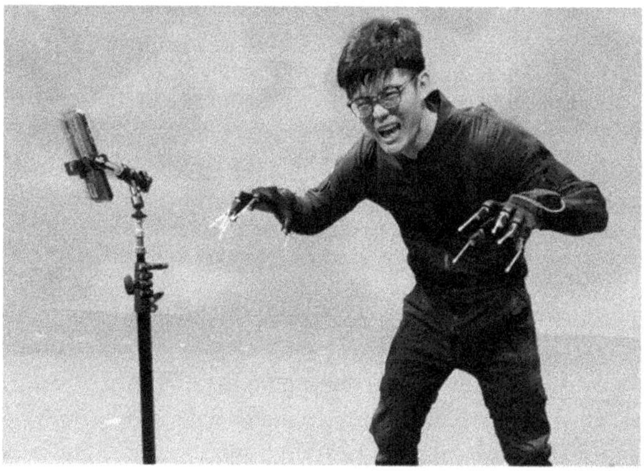

Fig. 5. Body, finger, and facial capture of the Pontianak, 2021, the authors.

3.5 Stationary Performance Capture

Based on these findings, we decided to further investigate if the sensor-based suit would produce acceptable results if the actors were kept stationary. Twenty short performances including standing, sitting, and lying were captured with a single actor playing two characters of the VR experience, and in addition to the body capture, finger and facial data were captured simultaneously (See Fig. 5). We again skipped all non-essential postproduction steps to review the data as early as possible in stereoscopic VR. The review confirmed a significant increase in accuracy in comparison to all previous attempts, allowing us to use the motion capture data with very little clean-up effort.

4 Results

At this current stage, the project has produced results beyond the films' historical findings in the form of two room-scale virtual reality applications compiled for Steam VR and viewed with an HTC Vive Pro setup.

The Pontianak Snake Bite VR Experience. The audience is invited to explore the virtual environment freely, examine the old kampong houses and the surrounding tropical vegetation. The story logic allows the user to follow Chomel and Othman on their 4-min-long stroll through the village to the jungle path location where the snake bite scene plays out and Chomel dramatically transforms into the Pontianak (See Fig. 6). As described earlier, our actors' walking path spans an area of 40 by 25 m and thus requiring the audience to use the SteamVR teleportation feature to navigate through the larger environment. Although this navigation concept works as planned, the experience of constantly teleporting to follow our actors' dialogue is overwhelming and potentially results in the user missing key moments. We therefore implemented an alternative approach which positions the user automatically at the snake bite location, allowing to uninterruptedly follow the approaching actors' conversation. Using cinematic terms (of shot sizes and camera framing) and the user representing the camera, the first approach translates to framing the actors i.e., as a medium shot, by constantly repositioning the camera location, and the second approach begins with a wide shot in which the actors are approaching and ends in a medium shot. Both approaches have their limitations and present an experimental approach to the walk-in movie idea. Among others, a lesson learned from these experiments is that continuously moving actors makes the experience 'complicated' both for the production of the work and for the user. Regarding the performance capture, as mentioned earlier, the results of the sensor-based suit from moving characters were not aesthetically or dramatically convincing, on the other hand, the sheer size of the capture area couldn't be achieved, within our resources, without the portable system.

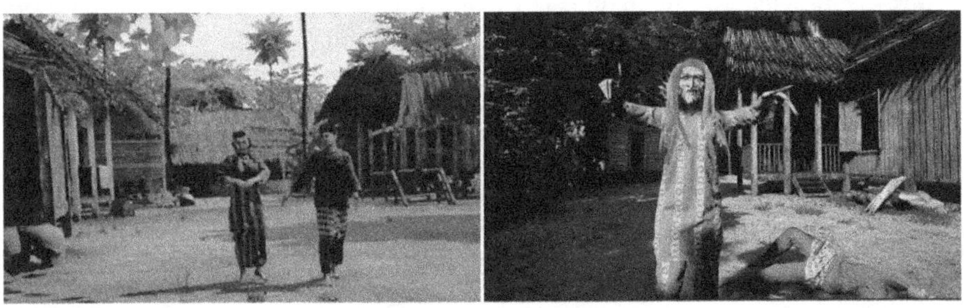

Fig. 6. Stills from *The Pontianak Snake Bite VR* Experience, 2020, the authors.

Described above, led us to create a second VR experience designed around mostly stationary performances. This 'spin-off' project is an artistic experiment and "B-movie" homage that confronts a melancholic female Stranger (an Alien/Cyborg) with the unfamiliar environment of a historic early 20th-century Malayan kampong village, and the Pontianak (See Fig. 7) [13]. The audience can explore the village freely, meander around and observe the Stranger and the Pontianak at their own leisure. There is only a minimal narrative with no dialogue or dramatic action, and although the user is required to navigate (with the teleportation feature) to the changing locations of the various scenes at which the Stranger appears, the overall experience is that of a free observer maintaining

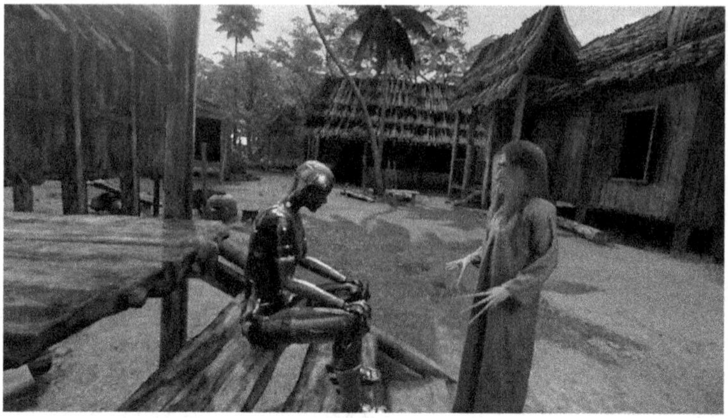

Fig. 7. Still from VR experience *The Woman Who Fell to Earth and Met the Pontianak*, 2021, the authors.

their agency. The stationary performances produced immediately usable and convincing capturing results from the low-cost system, with the finger and facial data adding significantly to the believability.

5 Discussion and Conclusion

This investigation had two main objectives, firstly to imaginatively reenact scenes of the lost *Pontianak* film as a VR experience, and secondly to evaluate if a low-cost motion capture system could support the creation of realistic acting digital humans to potentially enable virtual heritage applications to improve authenticity and believability of virtual actors.

In terms of the cinematic heritage content, we have made some progress - transforming the research into narrative scenarios that could be used in VR, but we are still grappling with the technical limitations in order to determine exactly what can be achieved in bringing these elements together (See Fig. 6). The static environmental 'set' of the Malay village possesses a high level of detail and demonstrates the exciting potential of VR as a tool for heritage, but the goal of creating realistic virtual performers, who can enact the roles from the reconstructed lost film, is still very much a work-in-progress. The script we had created for the Snake Bite scene proved to be too long and too dynamic, with its 'walking and talking' progression, revealing the many limitations of a portable low-cost motion capturing system.

In the end of this phase of production, we achieved usable results only from stationary actors, however, movements in space, such as walking and falling, produced high levels of inaccuracy which require manual clean-up to an extent of being uneconomical for a small research team. The idea of utilising the portable sensor-based suits to capture in a wide-open area to simulate actors strolling through a large virtual environment turned out to be overly ambitious in regard to the capabilities of the low-cost system. It also revealed that we did not invest enough in the importance of witness cameras to produce essential video references.

In summary, the portable low-cost system is a viable alternative for non-commercial, artistic and smaller academic research teams working on VR experiences, but only if extensive resources for manual clean-up are available or stationary actions are sufficient. Furthermore, if the primary project outcome is a stereoscopic virtual reality experience, as it is the case of our project, it is crucial to perform the review and quality control of the motion capture data directly in virtual reality. Although creating near photorealism for advanced tasks such as virtual environments and digital humans have become de facto possible and are in reaching distance, they still pose tremendous challenges for a small research team with limited resources.

This means that we had to re-evaluate what was actually possible in terms of our imaginative reenactment of scenes from *Pontianak*. While we might be able to produce key images and moments from the film (or at least our hypothesis of what happened in the film), the goal of producing a whole sequence with multiple characters is much more challenging to attain. At this stage, we are in the process of reconsidering what is possible as well as what can be stimulating and interesting to audiences interested in such a heritage project. We still believe that the VR approach to a 'lost' film is a multifaceted way to get closer to something that no longer exists, and the next stage for the project will be to present it to different audiences to gauge their reactions and feedback and to see if it is effective as a heritage or artistic experience.

Acknowledgements. The project has been kindly supported by an MOE grant in Singapore and by ADM, School of Art, Design and Media/NTU Singapore. The results would not have been possible without the diligent work of our research assistants Justin Cho, Chan Guanhua and Clemens Tan. We also express our gratitude to Arinah Bte Muhammad Sham, Toh Hung Ping, the Asian Film Archive, Dr. Rohana Said, Wong Han Ming, Hana Rosli, Fahim Fazil and Yap Wei Wen Marc.

References

1. Lim, K.T., Yiu, T.C.: Cathay 55 Years of Cinema. Landmark Books for Meileen Choo, Singapore (1991)
2. Epic Games Digital Human Project. https://docs.unrealengine.com/en-US/Resources/Showcases/DigitalHumans/index.html
3. Internet Movie Database: Andy Serkis. https://www.imdb.com/name/nm0785227/
4. Epic Games MetaHuman Creator. https://www.epicgames.com/site/en-US/news/announcing-metahuman-creator-fast-high-fidelity-digital-humans-in-unreal-engine
5. Epic Games News Blog. https://www.unrealengine.com/en-US/events/siren-at-fmx-2018-crossing-the-uncanny-valley-in-real-time
6. Seide, B., Slater, B.: Virtual cinematic heritage for the lost Singaporean film Pontianak (1957). In: Rauterberg, M. (ed.) HCII 2020. LNCS, vol. 12215, pp. 396–414. Springer, Cham (2020). https://doi.org/10.1007/978-3-030-50267-6_30
7. Skeat, W.: Malay Magic: An Introduction to the Folklore and Popular Religion of the Malay Peninsula. Macmillan and Co., London (1900). Reprint. Barnes and Noble, New York (1966)
8. The Star Malaysia. https://www.thestar.com.my/opinion/letters/2007/08/19/a-role-she-will-always-be-remembered-for

9. Mustafar, A.R.: 50 Tahun Filem Malaysia & Singapura (1930–1980). Pekan Ilmu Publications, Malaysia (2019)
10. Unknown authors: Pontianak [Promotional publication]. Harmy Press, Singapore (1957)
11. Skogstad, S., Nymoen, K., Høvin, M.: Comparing Inertial and Optical MoCap Technologies for Synthesis Control. In: SMC 2011. In: Proceedings of the 8th Sound and Music Computing Conference, pp. 421–426. Padova University Press, Padova (2011).
12. HTC Vive. https://www.vive.com/us/support/vive-pro/category_howto/minimum-and-maximum-play-area-size-for-more-than-2-base-stations.html
13. Seide, B., Slater, B.: Exploring B-movie themes in virtual reality: the woman who fell to earth and met the Pontianak. In: Proceedings of Art Machines 2, International Symposium on Machine Learning and Art 202, pp. 203–204. School of Creative Media, City University of Hong Kong, Hong Kong (2021)

Considering Authorial Liberty in Adaptive Interactive Narratives

Thomas Anthony Pedersen(✉), Tilde Hoejgaard Jensen, Vladislav Zenkevich, Henrik Schoenau-Fog, and Luis Emilio Bruni

Department of Architecture, Design and Media Technology, Aalborg University, 2450 Copenhagen, SV, Denmark
{tape,hsf,leb}@create.aau.dk, {tijens15,vzenke18}@student.aau.dk

Abstract. This article addresses the question of how much freedom an author (or a system) could be given to adapt a narrative in real-time to a potential recipient (i.e. the degrees of freedom of the author/system). While one focus in current research on adaptive storytelling is automation using artificial intelligence, we argue that the core concept of adaptive storytelling needs further development before it can be suitably implemented in an automated system. As such, we present the idea of the *Authorial Liberty Continuum*, as an authoring tool to help specify the degrees, and form, of adaptability to be provided to an Author (be it a person or a system) of a given adaptive narrative. The continuum ranges from a very limited freedom (e.g. very deterministic possibilities of change – resembling the capabilities of a drama manager), to full freedom (e.g. full control to adapt everything – resembling the power of a game master).

To explore the capabilities of this model as a framework for designing adaptive real-time interactive narratives, an exemplary system of such has been implemented, which allows a human agent (aka the *Author*) to insert elements into the experience in real-time, and thus execute small changes to the narrative. This working novel prototype showed that the perception of the events in an adaptive real-time interactive narrative varies from the real-time Author to the Recipient. This makes it difficult to foresee which elements an Author should be able to adapt, to attain a specific position on the continuum. We believe that these results warrant further exploration of the *Authorial Liberty Continuum*, in order to determine how varying points on this continuum might be classified.

Keywords: Adaptive storyworlds · Adaptive storytelling · Adaptive · Real-time · Interactive · Narrative · Digital narrative · Narratology · Narrative theory · Drama manager · Game master · Authorial liberty continuum · Events

1 Introduction

Within the field of interactive digital storytelling, a recurring theme is the idea of adaptive storytelling or adaptive storyworlds [1–4]. The concept has been approached from many different angles but is still at an incipient stage.

M. Wölfel et al. (Eds.): ArtsIT 2021, LNICST 422, pp. 181–188, 2022.
https://doi.org/10.1007/978-3-030-95531-1_13

The present article seeks to contribute to this emerging line of research, by addressing the question of how much control and/or freedom an author (or a system) could be given for seamlessly interacting with (i.e., adapting) the narrative in real-time to a potential recipient (i.e. the degrees of freedom of the author/system).

In recent years, one dimension of the research, have been toward the automation of digital storytelling through AI-driven approaches like automated virtual story generators, adaptive storytellers and intelligent narrative generators [4, 5]. In the present study we acknowledge the potential of using artificial intelligence for adaptive storytelling, as for example when addressing the combinatorial explosion in "traditional" branching structures [6]. However, we argue, that to move forward in this line of research for adaptive narrative systems, the core concept of adaptive storytelling needs further development before it can find suitable implementations in fully automated systems.

This article therefore presents the idea of *degrees of authorial freedom*, to explore the extent to which an author can adapt the narrative in real-time, as a framework for designing adaptive real-time interactive narratives. From these theoretical considerations, we have implemented an exemplary system of an adaptive interactive narrative, to explore the capabilities of the framework for designing adaptive real-time interactive narratives and illustrate the workings of what we call the *Authorial Liberty Continuum* (Fig. 1). The novel prototype affords a human agent (aka the *Author*) the capabilities to define elements of the narrative in real-time, and thus execute small changes to the storyworld, and thus, potentially, to the perceived narrative.

2 Designing for Adaptation

The rather recent advances in computational power and high-speed internet, have advanced the idea of adaptive narratives from pure theory into more mainstream practice. For example, the fairly recent Netflix series "Black Mirror: Bandersnatch" [7] can be regarded as a simple, and somewhat crude, attempt to create a cinematic interactive narrative, which adapts based on user input. These forms of narrative are nevertheless still in their infancy, and there are no clear guidelines when designing for adaptive formats. Therefore, we believe that the unique features and methods for adaptive narratives are yet to be developed.

2.1 Authorial Liberty Continuum

We suggest that the "adaptiveness" of an adaptive real-time interactive narrative, can be placed on a continuum defining the degrees of authorial freedom to adapt the narrative in real-time. We refer to this as the *Authorial Liberty Continuum* (Fig. 1). The continuum ranges from very deterministic possibilities (*Drama Manager)* to full creative freedom (*Game Master*). The two poles of the continuum are considered as two extremes (no freedom vs. total freedom) in terms of what – and how much – the Author may change (i.e. the way the Author can control the narrative, and the storyworld).

We refer to the seminal definition of *Drama Manager* by Mateas and Stern [8], who describe the drama manager as being allowed to create intractable, rich, branched stories. Drama managers (in interactive digital narratives) are usually systems that detect

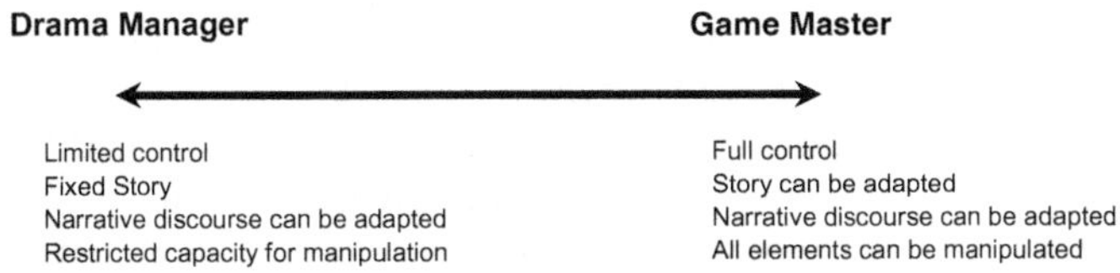

Fig. 1. *Authorial Liberty Continuum.*

happenings in a storyworld and, if needed, send a one-sided, narrow-bandwidth request to intelligent agents [8]. These requests change the behavior of the intelligent agents in order to achieve the next step in the plot. It is important to mention that in this case the communication between drama managers and agents is infrequent, allowing the latter to maintain the character-state and make *"moment-by-moment behavior decisions"* without further input from the drama manager [8]. It also needs to be noted that the term *agents* may be interpreted as both *events* as well as *characters.*

Drama managers are not given the power to change any rules in the storyworld, but work *by* the rules, and have no power to adapt the narrative outside pre-scripted scenarios. Hence, we regard *Drama Manager* as one of the extremes of the *Authorial Liberty Continuum.* By placing the Author at this extreme, the Author should only be given the ability to change the narrative discourse, but not altering the story.

At the other end of the continuum we place the *Game Master*. Tychsen et al. [9], define a set of tools and functions, similar to those of a game master in role-playing-games like "Dungeons and Dragons". They divide the functions of the game master toolkit into five groups: narrative flow, rules, engagement, environment and virtual world [9]. These groups are embracing the idea of full control of the narrative in terms of narrative beats, as well as rules of the world and the game, even how the player is engaged in the action. Based on this, we define the opposite end of the continuum as the *Game Master*. Placing an Author at this extreme, means that the Author has complete control over the narrative, and can influence all of its aspects (including discourse and the story itself).

A known issue with interactive narratives is the dilemma of providing user agency (e.g. freedom of movement and freedom of choice) while maintaining control of the narrative intelligibility [10–12]. We argue that this would still be a problem in an adaptive real-time interactive narrative, due to its interactive nature. The audience's interpretation of the author's story can go beyond what the author intended to tell, and hence, the audience can be considered as a co-author. This relates to Roland Barthes' notion of the "writerly text", as he argues that a narrative can be "co-created" by both the author and the reader, as the reader interprets the narrative discourse of the author (i.e. the words or visual representations) and constructs his own narrative from this [12, 13]. Thus, the quality of storytelling will depend on both parties - not only the author - as the audience needs to ratify what is presented [14].

Depending on the intentions of the author or designer, the different ends of the spectrum can be more or less favorable. For example, when the author has a specific message to convey, a more *didascalic* [12] (i.e. figuratively) narrative is favorable, and as such a position on the continuum towards the *Drama Manager*. However, if the intention is exploration or agency through interaction, then the user will have to rely more on their own perception of the narrative and make up their own story, creating a more abstract, or open, narrative. Hereby the Author-Audience Distance will be greater, meaning that there will be an interpretation gap between what the author intended to tell, and the narrative that the audience perceived [12]. This could allow for more freedom to adapt without breaking a specific narrative, meaning a position on the continuum towards Game Master can potentially be preferable.

As such we believe that the *Authorial Liberty Continuum* can aid designers in making decisions about the amount, and the kind, of adaptability to be provided to the author/system in order to achieve a given level of narrative intelligibility.

3 Designing with the Authorial Liberty Continuum

In order to experiment with the potential of utilizing the *Authorial Freedom Continuum* (Fig. 1) as a framework for designing an adaptive real-time interactive narrative, we designed and implemented a prototype. The prototype aimed to place the Author somewhere between the two extremes of the continuum, but tending towards the *Drama Manager* side, in order to somehow constrain the possibilities of the Author, in an attempt to guarantee some form of narrative intelligibility.

In our implementation of an adaptive real-time interactive narrative, one user is given the role as the Author and is tasked with conveying a small pre-written narrative (through environmental storytelling) using a phone application, specifically developed for this purpose (Fig. 2). The phone application gives the Author the ability to insert various 3D objects and toggle various effects in real-time. Furthermore, it shows the real-time location of the Player in the scene. The role of the Player is to explore the narrative through an interactive environment on a PC. During this, the Player experiences how the Author orchestrates the events in real-time (Fig. 3), all the while being unaware of the existence of the real-time Author and his influence on the events.

The pre-constructed setting of the narrative is a World War II bunker, hinting to the planning of the nuclear bombing of Japan. The scene is constructed to be able to communicate the narrative through environmental storytelling, without the interference of the Author, as we inserted American propaganda posters, as well as allusions to Oppenheimer through writings on a black board and other props. The Author, however, is able to insert further clues such as models of atomic bombs, blueprints, and a world map with emphasis on Japan.

As we placed the Author closer to a *Drama Manager* than to a *Game Master* on the continuum, we predefined which objects and effects the Author is granted, and we also preconstructed the general setting. However, the Author is given some degree of freedom, by being allowed to use whatever item or effect at their disposal at any time.

3.1 Evaluation of Event Perception

In order to assess whether the Recipients' perception of the nature of the available events matched the Authors' perception, we conducted a test, consisting of 10 Author/Recipient couples, randomly recruited university students through convenience sampling. After the Recipients and the Authors had tried their respective applications, they were tasked

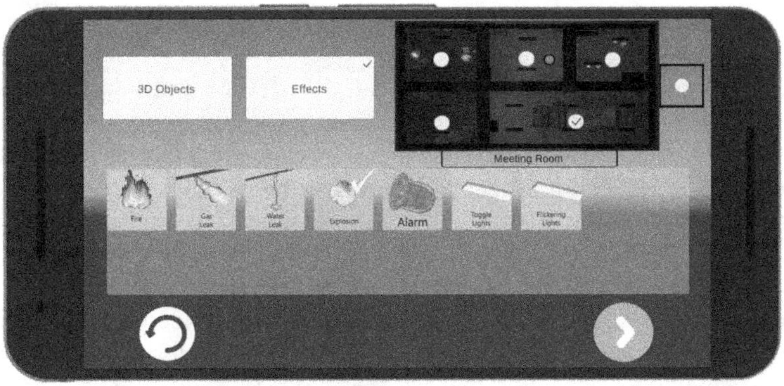

Fig. 2. User interface of Author application running on smartphone

with categorizing the different events and objects deployed by the Authors into the narrative. The categories labelled the events as either *constituent events* (i.e. affecting the understanding of the *story*) or *supplementary events* (i.e. only affecting the *narrative discourse*), following the distinctions made by Roland Barthes and Seymour Chatman [15].

The results from the tests showed that the Recipients mainly classified the events as *supplementary events* (58%), i.e. as not affecting their understanding of the story, and thus mainly affecting the narrative discourse.

The Authors on the other hand, mainly classified the events as *constituent events* (65%), i.e. affecting the understanding of the story, and thus in effect, changing the story as the events were deployed.

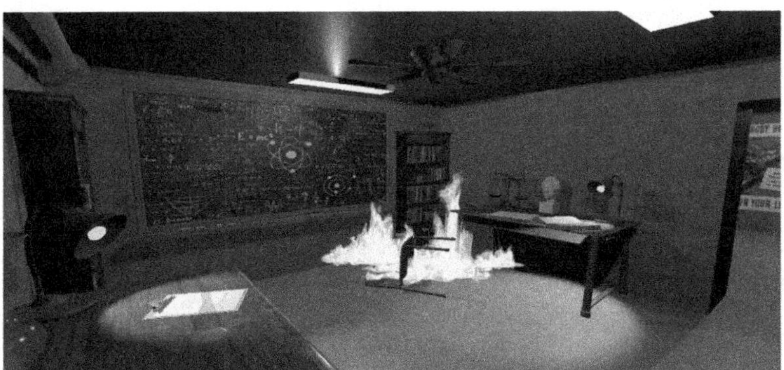

Fig. 3. Recipient's view in prototype

4 Discussion

We initially expected that the Authors would categorize events and objects as *supplementary events* (mostly changing the narrative discourse), as they would have knowledge of a story that they should try to convey, and thus concluding that the events they deployed would only change the discourse. On the flip side we believed that the Recipients would classify the events as *constituent events*. Since the Recipients would have no prior knowledge of a pre-written story, we believed they would experience all events as defining for whatever story they constructed themselves from the experience. However, our test showed the complete opposite, i.e. that Authors mainly considered the events as *constituent events* (important to the story, and thus changing the story), and the Recipients mainly considered the events as *supplementary events* (not important to the story, and only changing the narrative discourse).

The Authors' classification can potentially be linked to the fact that the Authors described themselves as omnipotent or all-knowing entities in the experience (similar to zero-focalization [16]). This could imply that the real-time aspect of the experience, gave the Authors a feeling of being a participant in the story world, rather than just creating it (i.e. being a focalizing point in the narrative).

The Recipients' perception of the events as primarily *supplementary events* could suggest that the Recipients experienced the general setting as sufficiently conveying the story through the environment, and as such not regarding the influence of the real-time Author as changing the story. However, since the Recipients were unaware of the presence of the real-time author, the classifications of the Recipients could also suggest, that the deployed elements where perceived by the Recipients as pre-programmed part of the narrative experience and thus not as elements which could be said to change anything. Some Recipients did also explain that they believed the events were simply triggered by their own movements in the virtual space.

The specific placement of the Author within the *Authorial Liberty Continuum* could also play an important role in the results, and relates to the general question of how an adaptive real-time interactive narrative could be affected by placing the Author at varying positions on the continuum.

Moving the Author towards the Drama Manager side of the *continuum* would mean less freedom (i.e. less – and more restricted – functionality for the real-time Author to alter the narrative). When designing an adaptive real-time interactive narrative with a degree of authorial freedom around this end of the continuum, the focus should then be on designing adaptive elements which solely compliments the story, but do not affect the recognition of the story, as it is intended by the designers. A simple example could be changing the look of a character based on some form of user input, without changing the role of the character, and thus still providing the same conclusion to the story. However, the results of our study open the issue of how to understand at what point an adaptation is actually changing the *story*. As we have shown, it can be difficult for a designer to foresee, which elements in an adaptive real-time interactive narrative are interpreted as *constituent* or *supplementary events*. It might even be argued that at a certain point, enough supplementary events could potentially affect the story and thus be regarded – as a whole – as constituent events.

On the other hand, placing the Author more towards the Game Master side of the continuum means more freedom (i.e. more options to change the narrative). Thus, an adaptive real-time interactive narrative around this end of the continuum should be designed to allow the Author to adapt almost any part of the narrative experience. The questionnaire showed, that the Authors in our approach, wanted to have more freedom, in the form of more objects, and additional functionality, to change in real-time, even though they felt confident that they were able to convey the story using the elements which were provided in our approach (i.e. the provided degree of freedom). However, depending on the goal of the adaptive real-time interactive narrative, we believe that this could be problematic from the perspective of the *Author-Audience Distance* [12], as more possibilities would likewise demand more of the real-time Authors and their ability to manage these added possibilities while trying to convey a coherent narrative. This, additionally, raises the question of when the degree of freedom becomes so great that it cannot be regarded as *adaptation* anymore, but merely is *creation.*

5 Conclusion/Future Works

In this paper we introduced the *Authorial Liberty Continuum*, as a way to describe the degrees of freedom given to a real-time author in an adaptive real-time interactive narrative. To explore the model, an adaptive real-time interactive narrative was implemented, in which a human real-time author was given the freedom to adapt the narrative, by deploying different pre-defined events in the scene. As such, the Author was placed towards the Drama-Manager side of the continuum. During the evaluation of the framework, Authors and Recipients were asked to classify events. The results showed that Authors tended to classify objects mostly as *constituent events*, and Recipients mostly classified the events as *supplementary events.* Taking this into consideration, another interesting topic for research could be to investigate what kind of elements the Authors would prefer to alter during the playthrough. Our approach provided only three types of tools: 3D objects, Auditory- and Visual-Effects. Also, the variation of options for the real-time Author would be an interesting topic to dig into, as it could potentially show the optimal author placement on the continuum, depending on the goal of the adaptive real-time interactive narrative.

It could also be of interest to compare adaptive real-time interactive narratives generated from both human authors and artificial intelligent (AI) solutions, in order to see how the stories would be perceived, and if they would be perceived differently by the Recipients. This comparison would be interesting as we believe that most of the pre-existing research is focused on procedurally generated narrative solutions [17] rather than human-based digital solutions.

We argue that the more we experiment with human authors at different levels of freedom on the Authorial Liberty Continuum, the more we learn of how a human author might adapt a narrative, based on the amount of freedom given. Knowledge that we perceive as imperative, if we ever want to design a credible AI for adaptive narratives.

References

1. Schoenau-Fog, H.: Adaptive storyworlds. In: Schoenau-Fog, H., Bruni, L.E., Louchart, S., Baceviciute, S. (eds.) ICIDS 2015. LNCS, vol. 9445, pp. 58–65. Springer, Cham (2015). https://doi.org/10.1007/978-3-319-27036-4_6
2. Schoenau-Fog, H., Larsen, B.A.: Creating interactive adaptive real time story worlds. In: Rouse, R., Koenitz, H., Haahr, M. (eds.) ICIDS 2018. LNCS, vol. 11318, pp. 548–551. Springer, Cham (2018). https://doi.org/10.1007/978-3-030-04028-4_64
3. Kickmeier-Rust, S.G., Albert, D.: 80 days: melding adaptive educational technology and adaptive and interactive storytelling in digital educational games. In: Proceedings of the First International Workshop on Story-Telling and Educational Games (STEG 2008), p. 8 (2018)
4. Garber-Barron, M., Si, M.: Adaptive storytelling through user understanding. In: Ninth Artificial Intelligence and Interactive Digital Entertainment Conference (2013)
5. Parag, J., Agrawal, P., Mishra, A., Sukhwani, M., Laha, A., Sankaranarayanan, K.: Story generation from sequence of independent short descriptions. arXiv preprint arXiv:1707.05501 (2017)
6. Stern, A.: Embracing the combinatorial explosion: a brief prescription for interactive story R&D. In: Spierling, U., Szilas, N. (eds.) ICIDS 2008. LNCS, vol. 5334, pp. 1–5. Springer, Heidelberg (2008). https://doi.org/10.1007/978-3-540-89454-4_1
7. John, L.: Netflix's Black Mirror: Bandersnatch is an Impressive Interactive Experience. https://interestingengineering.com/netflixs-blackmirror-bandersnatch-is-an-impressive-interactive-experience. Accessed January 2019
8. Aylett, R., Louchart, S.: Towards a narrative theory of virtual reality. In: Virtual Reality, vol. 7, pp. 2–9 (2003)
9. Louchart, S., Aylett, R.: Solving the narrative paradox in VEs – lessons from RPGs. In: Rist, T., Aylett, R.S., Ballin, D., Rickel, J. (eds.) IVA 2003. LNCS (LNAI), vol. 2792, pp. 244–248. Springer, Heidelberg (2003). https://doi.org/10.1007/978-3-540-39396-2_41
10. Mateas, M., Stern, A.: Towards integrating plot and character for interactive drama. In: Dautenhahn, K., Bond, A., Cañamero, L., Edmonds, B. (eds.) Socially Intelligent Agents. MASA, vol. 3, pp. 221–228, Springer, Boston (2002). https://doi.org/10.1007/0-306-47373-9_27
11. Tychsen, A., Hitchens, M., Brolund, T., Kavakli, M.: The game master. In: Proceedings of the Second Australasian Conference on Interactive Entertainment, pp. 215–222, Creativity & Cognition Studios Press (2005)
12. Bruni, L.E., Baceviciute, S.: Narrative intelligibility and closure in interactive systems. In: Koenitz, H., Sezen, T.I., Ferri, G., Haahr, M., Sezen, D., Çatak, G. (eds.) ICIDS 2013. LNCS, vol. 8230, pp. 13–24. Springer, Cham (2013). https://doi.org/10.1007/978-3-319-02756-2_2
13. Bensmaia, R.: Readerly, Writerly Text (Barthes). In: Herman, D., Jahn, M., Ryan, M.L. (eds.) Routledge Encyclopedia of Narrative Theory, pp. 483–484. Routledge (2010)
14. Duranti, A.: The audience as co-author: an introduction. Text-Interdiscip. J. Study Discourse **6**(3), 239–248 (1986)
15. Abbott, H.P.: The Cambridge Introduction to Narrative, 2nd edn. Cambridge University Press, Cambridge (2008)
16. Hühn, P.: Focalization. In: Multiperspectivity - The Living Handbook of Narratology. Hamburg University Press. https://wikis.sub.uni-hamburg.de/lhn/index.php/Focalization. Accessed, January 2019
17. Freiknecht, J., Effelsberg, W.: A survey on the procedural generation of virtual worlds. In: Multimodal Technologies and Interaction, vol. 1, no. 4, p. 27 (2017)

Towards Inclusive and Interactive Spaces for Breakdancing

Janica Olpindo(✉) and Doug Van Nort

DisPerSion Lab, York University, Toronto, ON M3J 1P3, Canada
{jan0,vannort}@yorku.ca
http://www.janicaolpindo.com, http://www.dispersionlab.org

Abstract. This paper provides a critical reflection on performance gestures within the context of "breaking", explored in the process of constructing an interactive system using design methodologies drawing on the framework of "defamiliarization". Through a user study conducted as exploratory dance sessions with female practitioners of breaking, we observe the relationship between movement and music generated by the Interactive Breaking Music System (IBMS). We question how practitioners embody breaking aesthetics in the gestures that emerge from their interaction with the IBMS and how this system might be leveraged to create a welcoming environment for b-girl practitioners, and possibly subvert or transform gender norms from breaking culture that manifest through movement.

Keywords: Hip-hop · Gesture · Interactive space · Feminist HCI

1 Introduction

From the perspective of embodied music cognition, listening is a full-bodied phenomenon from which a complete understanding of musical gestures and performance interactions emerge [1]. Research in this area employs multiple methodologies which are informed by critical reflection on the relationship between movement and music, including how the aesthetic of these movements or gestures change based on the music that is playing. This perspective strongly informs the research work presented in this paper.

In the world of dance, breakdancing or "breaking" is one of the four original elements of hip-hop [2]. In dialogue surrounding hip-hop culture, it is argued that it is important to understand the relationship between movement and sound in order to fully understand the medium/genre. According to Fogarty, "...the shift away from considerations of music has resulted in a lack of understanding, in both theatrical criticism and the institutionalism of breaking, of how hip hop aesthetics integrate the two" [3].

In this paper, we seek to understand the coupling of these movement/sound relationships in the context of breaking practice, and how modifications of

M. Wölfel et al. (Eds.): ArtsIT 2021, LNICST 422, pp. 189–206, 2022.
https://doi.org/10.1007/978-3-030-95531-1_14

this via an interactive system subvert or expand commonly held practices and assumptions in this genre. We ask: How do practitioners embody breaking aesthetics in the gestures that emerge from real-time interactions with a sound-generating system that varies from highly similar to highly different from the traditional music found in breaking? To this end, this research paper observes gestures and movements in different interactive breaking contexts and explores similarities in relation to established hip-hop practices. Of particular interest, in this study are practitioners who identify as b-girls.[1] A second research question we ask is: How might an interactive system be leveraged to create a welcoming environment for breakers of differing genders? We present the system design and discuss the outcome of our initial exploratory study involving b-girl practitioners. We begin with a review of literature in hip-hop scholarship and human-computer interaction, followed by a theoretical background of our methodologies, and then an outline of our system design and description of our user study.

2 Gender, Sexuality and Breaking: A Literature Review

The relationship between gestures and interactive spaces is apparent in general,[2] but how might they affect one another in the context of breaking? Even more directly, here we ask how gender norms in breaking culture, which manifest through established gestural language, might become subverted or otherwise transformed in this interactive context. To approach this, we first need to understand the forms of gendered cultural idioms related to gesture that emerge in this context. To this end, we now provide a review of literature in both hip-hop scholarship and academic discourse on technological engagements to ground the discussion.

2.1 Origins and Challenges

In "Getting off: Portrayals of Masculinity in Hip Hop Dance in Film," LaBoskey discusses the strong expression of masculinity and machismo in hip-hop culture. She mentions the influence of gangs on the rise of hip-hop as she explains the gestures used in the competition aspect of breaking:

> Dancers employ gestures through the dance that are intended to encourage a rebuttal from the opponent. Young males often grab their crotch to symbolize greater virility than their opponents. They point or stare at their rivals in direct threat, or simply move toward another individual, violating his personal space and thereby threatening him. Such taunts often trigger outbreaks of actual violence. Hip-hop dancing is, in essence, a battle. It is a translation of gang warfare into the language of dance and physical expression [4].

[1] 'B-girl' is a term referring to women specializing in the street dance style of breaking.

[2] An interactive space is defined here as both a physical space that accommodates interactions between people as well as the computational system that interprets their movements and produces media output.

LaBoskey traces the origins of hip-hop as a vernacular style of dancing in the spirit of the adolescent boys who created it and posits it as an escape from gang violence.[3] Using such gestures involving the male genitalia to demonstrate power and dominance encourages heterocentric breaking spaces to develop. Jagose states with reference from [6], "queer theory offered a way of thinking about lesbian and gay sexualities beyond the narrow rubrics of either deviance or preference, 'as forms of resistance to cultural homogenization, counteracting dominant discourses with other constructions of the subject in culture"' [7]. We seek to explore ways of promoting meaningful exchanges between gestural action and interactive spaces, in the course of advocating for a more flexible and inclusive space in the context of breaking.

2.2 Hip-Hop and Feminism

B-girls face contradictory notions of gender performativity because of the aforementioned heteromasculine structures upon which breaking culture is built. According to Johnson, "B-girls are situated between two competing notions of heterosexual femininity: one is the pornification of women in Hip Hop, and the other refers to "normal day life" expectations of polite, ladylike behaviour" [8]. Identifying as a b-girl also means identifying as an outlaw because b-girls are constantly negotiating notions of gender performativity as well as breaking's inherent qualities imposed and interpreted differently on women's bodies [9].

Morgan, Jamila, and Pough are some of the first writers to circulate the term 'hip-hop feminist'; they believe hip-hop culture and rap music, both stemming from Black culture, hold radical and liberating potential that should be tapped by the contemporary feminist movement to speak to younger feminists, particularly those of color [10]. Peoples further explores the relationship between hip-hop and feminism and the contested nature of the term "hip-hop feminism" [11]. According to Peoples, many women are hesitant of exploring feminism's possible benefits due to its racist, homophobic, classist, and xenophobic beginnings [12]. It is these challenges that inspire us to ask here if decontextualizing breaking movements and gestures outside the movement/music norms of the culture might provide opportunities to challenge and question the status quo.

2.3 "Queering the Dance Floor"

As argued by Durham et al., breaking culture operates under a heterosexist structure wherein heteromasculine engagement is normalized, devaluing performances and contributions by those marked as 'other' [13]. Gunn discusses the ability of the format known as 'all style' dance 'battles' to call into question the bodily performance organized through gender to deconstruct, disrupt and

[3] Serouj Aprahamian challenges this deeply rooted narrative in hip-hop culture to question the reliability of the source as well as to understand the implications this has in society (see [5] for more background on the street gang narrative in hip-hop culture).

expose the hierarchized distinctions in hip hop culture, or what she refers to as '*queering* the dance floor' [14]. All style dance battles are a type of event that feature specialized dancers from a variety of different street dance style backgrounds, such as waacking, popping and locking, house, krumping, and breaking, competing with and/or against one another by improvising to a diverse range of music mixed live by a deejay [15]. In all style dance battles, the significance of heteromasculine gestures to present dominance over an opponent is decontextualized. Through spaces like all style dance battles and hip-hop theatre festivals, there is great potential to innovate ways of inviting an even more diverse crowd into breaking spaces. This leads us to consider if a similar decontextualization is possible through the integration of interactive movement/music systems.

3 Methodology

Driven by the first author's personal experience in breaking culture and practice, this work is grounded in a strong conviction that this context could be made more inclusive, particularly with respect to participants coming from a broad spectrum of identities related to gender and sexuality. This paper presents results from a larger project that is grounded in five complementary areas: critical AI studies, working with machine learning techniques, interactive media development in challenging "real-world" contexts, and the use of qualitative methodologies for rigorously assessing participant experience. Our goal is to contribute a critical assessment of contemporary machine learning techniques on the one hand by testing them in this context, while simultaneously contributing to breaking practice and culture through their application, with the goal of providing new language, insights and an interactive art platform that engenders new and exciting approaches through a consideration of the long-held challenges and prospects of the breaking context in tandem with the design approach.

3.1 "Designing for Obliqueness": Tensions Between HCI and Arts/Culture

In "HCI as heterodoxy," Light explores different possibilities of designing digital tools by drawing insights on Queer Theory to challenge readings of gender as well as promote the tactic of queering in investigating resistance to the status quo [16]. The author defines 'queering' as a way of treating something obliquely: to steer into an adverse or opposite direction of the norm [17]. Further, Light alludes to Trauth et al.'s proposed criteria for conducting gender research "to engage in research flexibility; challenge the hegemonic dominance, legitimacy and appropriateness of positivist epistemologies; theorize from the margins; and problematize gender" [18]. This is in a similar vein as the concept of *Designing for obliqueness*, described as a practice (without the promotion of poor technical or interaction design) to forget, cheat, elude, and obscure the 'straightness' of machine thinking and conservative stance in the Human-Computer Interaction (HCI) discipline: the privileging of values in favour of effectiveness and efficiency, and the validation of risk averse research that stays close to the agenda of dominant interests [16].

From a movement and computing design perspective, we argue that the *Defamiliarization* approach in movement-based interaction design is an example of queering the familiar, or traditional, perspectives [19]. This approach relies on varying normal movement patterns and processes to destabilize a creative user's habitual ways of thinking about movement, to reorient their experience, and to nurture an important component of improvisation – open-ended play [20,21]. Similarly to Light's concept, Defamiliarization's goal is to avoid conforming to established design models driven by overly-proscribed gestures and movement patterns, a priori. We approach this study through the lens of defamiliarization both in terms of our interactive design decisions as well as in the user study that followed, through the introduction of an interactive system that served as a defamiliarizing element by replacing standard breaking music, subverting expected movement/sound relationships in the process.

While there are creative opportunities in integrating technology in art, Fdili-Alaoui and her collaborators in *SKIN* (a choreographic interactive dance piece) recognize that there are also tensions emerging from these opportunities. In "Making an Interactive Dance Piece," Fdili-Alaoui discusses her anti-solutionist approach in conducting research and creating an interactive dance piece [22]. She states:

> Moreover, I inscribe my work in a lineage of previous HCI approaches that go beyond the usual artificial problem solving. [...] In anti-solutionism through design fiction, Mark Blythe reveals the most influential scenarios in HCI that addresses the "monsters" of work in the twentieth century supported by technology. According to him, this is due to the fact that HCI is an applied discipline funded according to the impact and relevance of the work to industry. He argues that there are many monsters worth fighting and one can see a variation on this plot in critical design where the "monster" is the lack of informed debate [23].

Fdili-Alaoui presents this perspective as a way of navigating between standardized HCI methods or productions and creative projects, while negotiating whether technology serves or subjugates the dance in the process [24,25].

3.2 Moving and Making Strange

The Moving Making Strange (MMS) methodology, developed by Loke and Robertson, for designing movement-based interactions that are focused on the lived experience of movement and its felt, kinaesthetic experience [26]. Loke and Robertson state:

> The emphasis on understanding the felt experience of moving and keeping it alive within a human-centered design process that shifts back and forth between the multiple perspectives of mover, observer, and machine is an

effort to provide a balance to the extensive amount of existing research from a technology-centric perspective (i.e. computer vision and motion analysis) [26].

The MMS methodology is grounded in similar principles as those proposed by Light as well as Fdili-Alaoui's anti-solutionist approach. The practice of designing for obliqueness and other human-centered approaches in HCI can be seen as a design analogue to the notion of 'queering the dance floor' laid out by Gunn. These approaches attempt to mitigate the structural hegemony present in their respective fields. We import these principles from a design perspective. From an evaluation perspective, this research study follows the structure and principles of the MMS methodology by understanding the perspectives of the mover, the observer, and the machine in order to ground our focus on the process of inquiry and exploration.

4 Interactive System Development

We created a movement-to-sound system that we refer to here as the Interaction Breaking Music System (IBMS) to examine relationships between movement and gesture, as well as perceptions of engagement, agency and connectedness by b-girl participants.

4.1 Gesture and Mapping

The notion of "gesture" becomes more complex as input physical gestures undergo a process of translation during interactive sonic performances [27]. Physical gestures are processed and translated across a network of computers and sound synthesizers, having potentially multiple layers of mapping, conditioning and transformation [28]. With reference to Godøy, Donato et al. state:

> Music can invite certain gestures that are often encouraged by timbral and dynamic qualities of the sound, by mimicking the action that might have produced them, or the gestures evoked by the music which might not necessarily refer directly to the production or sound qualities [29].

This speaks to the experience of "sound tracing" higher-level attributes while listening to music. Caramiaux et al. argue that combining the use of machine-learning techniques and theoretical frameworks from an embodied music cognition perspective contextualizes the link between music perception and human actions to establish gesture-sound relationships; they defined three different mapping strategies in order to have a better understanding of these relationships:

- *Instantaneous* mapping strategies refer to the direct translation of gesture to sound features or parameters, and vice versa.
- *Temporal* mapping strategies morph the translation and adaptation of temporal profiles, timings, and event sequences between the gesture and sound data streams.

- *Metaphorical* mapping strategies refer to relationships determined by metaphorical or even semantic aspects, which do not always rely on changes between gesture and sound [30].

As noted above, people's movements – both higher level tracing and more direct action-sound gestures – convey information about perceived sonic moving forms while listening to music (see Fig. 1). For this reason, this research study employs both instantaneous and temporal mapping strategies of three types: explicit few-to-many gesture-to-sound mappings, implicit mappings learned via machine learning, and mappings of tempo changes, based on grouped data averages, with a longer temporal envelope.

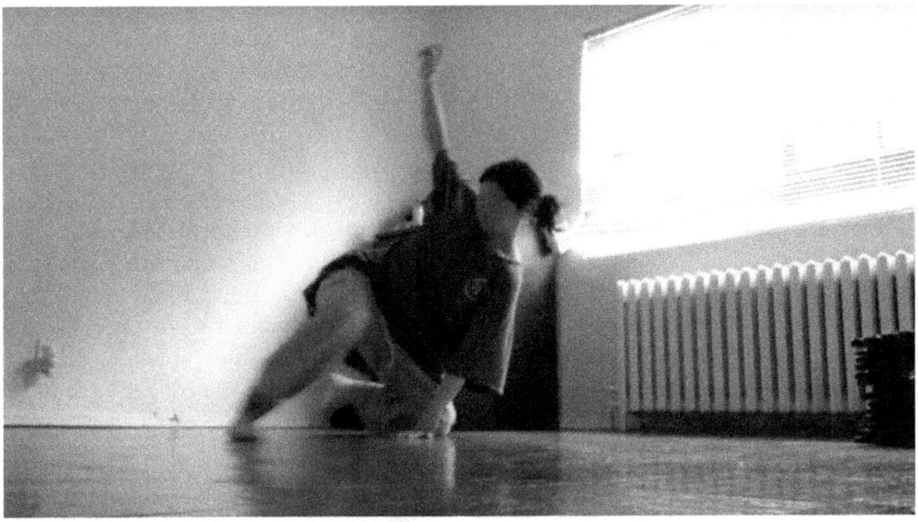

Fig. 1. Photo of a participant dancing during study session.

It is well documented that more complex parameter mappings often lead to more intuitive and expressive results [31]. The complexity of the IBMS is tied to this combination of mapping strategies, allowing intuitive use while providing a challenge for b-girl practitioners to find ways of negotiating between exploratory movements and familiarity with breaking aesthetics, i.e. the mapping choices explicitly play with the principle of defamiliarization in the breaking context.

4.2 Modeling vs. Tracking

Our process began by conducting and comparing a set of camera-tracking based experiments involving live webcam feedback and one pre-recorded dance video using both machine learning models and lower-level computer vision methods. These initial experiments provided insight into which data-capturing method is most appropriate and efficient to use in our project. We began by examining some promising contemporary machine learning models, specifically Body Pix and PoseNet. Body Pix is a real-time person segmentation, while PoseNet is a real-time human pose estimation using the TensorFlow library. We decided against these models for two reasons. First, we discovered that running them caused a heavy load on the GPU in combination with other programs to run the IBMS. With the research being conducted online, our capabilities to connect telematically with participants and record the session via Zoom would have been strained. Second, and most importantly, because Body Pix and Pose Net have been trained on bodies and/or movements that are not similar to that of breaking, the models were not able to track the dancers' movements in real-time with great accuracy. We tested this by feeding the model a breaking practice video featuring moves that transition from a standing position to the ground level. We found that, during those transitions, the Pose Net model would sporadically appear on the screen in an attempt to track the body; however, it would try to assume an upright, standing position even when the dancer is on the ground. In the case of Body Pix, We tested its real-time tracking accuracy by moving at different speeds within the camera's view. We found that body segmentation would get mixed up while turning around.

Instead, we developed the system using computer vision libraries found within Max/MSP and Jamoma [32].[4] Developing the IBMS in this way shifted the focus away from potentially normative body-models towards more open-ended data outputs that sensed a wider range of unexpected movements. In the words of Light, using this more low-level tracking method allowed more space to obscure, elude, and cheat, which inevitably provided flexibility in the resulting interaction design [16].

4.3 Feature Extraction and Mapping Design

Due to recent circumstances around the coronavirus disease (COVID-19) outbreak, the research project shifted away from the use of Kinects in a single space, and instead focused on systems that could be engaged with telematically. This involved extracting movement data from participants' webcams and applying low-pass filter thresholding and noise reduction in order to produce a simplified "essence" of a persons movements within the frame. This is represented by an all-purpose motion image generator/analyzer which is then sent to a 'motiongram module' as seen in Fig. 2. A motiongram is typically used to display motion over a period of time similar to a two-dimensional spectrogram [33]. In our explorations, we found that this transform could be exploited to provide a more reliable tracking feature than raw values in both X and Y dimensions. In particular, we independently extracted the horizontal and vertical motiongram data (producing vectors of 1×100 and 100×1 respectively) and used these to

[4] Jamoma exists as a package in Max, which is a programming language commonly used for developing interactive music systems and other systems within the interactive realm.

further extract the quantity of motion (QoM) [33] of the movement's vertical motiongram and horizontal motiongram. By capturing data this way, we move the focus away from attempting to achieve a highly accurate model of human body using machine learning methods, and instead focus on understanding the more low-level dynamics of movement with relation to sound. We then apply *this* data to machine learning methods during the mapping process.

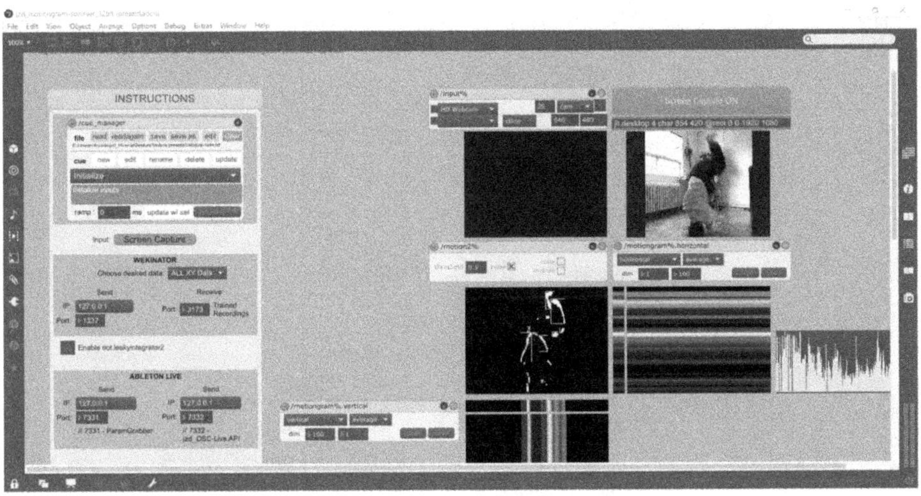

Fig. 2. Screenshot of the project's Max patch.

This processed X/Y QoM data is then parsed out into three distinct streams: smoothed using an exponential smoothing method (X_s/Y_s), instantaneous velocity (X_v/Y_v), and instantaneous acceleration (X_a/Y_a), values. These values are used directly, and are further used to train a continuous neural network input/output mapping using Wekinator [34], which subsequently maps these values into an Ableton Live set as Open Sound Control Messages (OSC) through Max for Live (M4L) tools (ParamGrabbr by Showsync and a simple M4L instrument created for this project that allow for communication between Max and Ableton). The IBMS features a complex combination of implicit mapping using machine learning and explicit 'one-to-many' mapping strategies using these parameters, such that one variable is mapped to multiple device parameters through both means at any given moment, as depicted in Fig. 3. For example, the variable 'wek-xv' (Wekinator output of x_v) is mapped to parameters including reverb, probability of variation, decay, delay time, distance and filter cutoff across multiple tracks within the Live set.

5 User Study

This research employs one-on-one and group dance sessions with 8 b-girls, involving improvised movements to traditional breaking music – these are highly rhythmic songs that incorporate elements of funk and hip hop, typically those with a fast bpm and long drum breaks/solos – as well as the interactive breaking music system (IBMS) via Zoom. These study sessions are followed by individual or group interviews to reflect on participants' experience of the session as well as their experience as a b-girl and/or their experience of breaking culture in general.

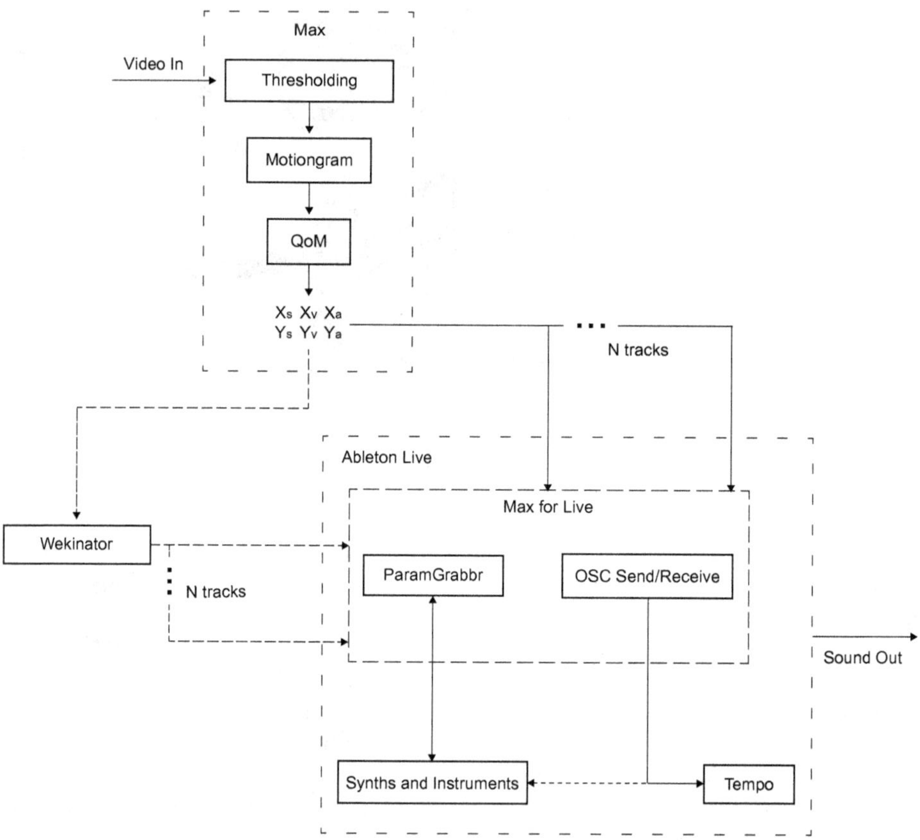

Fig. 3. Interactive breaking music system's mapping control structure.

5.1 Study Structure

In this study, participants were invited by the first author from within their personal network of dancers in the Toronto area – all of whom were informed about the interactive nature of the music system and gave their written informed consent to participate in the study. Each dancer had at least 5 years of breaking experience and were between the ages of 20 and 40. They were asked to improvise to (A) traditional breaking music – which included "Give It Up Or Turnit a Loose" by James Brown, and "Apache" by Incredible Bongo Band – and (B) music generated by the IBMS with different tempo settings: B1) unchanging tempo at 118 bpm, B2) accumulated/averaged movements which, after a threshold value, triggered a tempo of either 118 bpm or 90 bpm, and B3) tempo changes continuously with dancer movement in the range of 20 bpm to 118 bpm (see Table 1). Individual dance sessions were conducted with the following structure: 10 min of dancing to A, and 10 min of dancing to B2. This was followed by two group sessions of size 4 and 3 (unfortunately one of our participants was not able to attend the following session). This group dance session followed a similar structure: 10 min of dancing to A, 10 min of dancing to B1, and 10 min of dancing to B3. These sessions were held online to allow dancers to participate from wherever they felt the safest during the pandemic.

Table 1. Study structure

	Individual	Group
A - Traditional breaking music	10 mins.	10 mins.
B - Music from IBMS		
1 - unchanging tempo (118 bpm)		10 mins.
2 - trigger tempo (90 bpm/118 bpm)	10 mins.	
3 - continuous tempo change (20 bpm–118 bpm)		10 mins.

The tempo settings were changed to B1 and B3 for the group dance sessions to observe how dancers would react to music without a stable tempo, and to gradually introduce them to B3, B1 was used as an intermediary setting between the music they were familiar with and an extreme setting of the IBMS. Each dancer was initially asked to raise their hand before the start of their round to avoid everyone dancing at once; however, an interesting emergent phenomena occurred: dancers instead took turns by ending their round with a gesture that the next dancer would emulate at the start of their round.

As discussed earlier, we entered the study with two main questions:

RQ1 Would practitioners embody breaking aesthetics in gestures that emerged from interactions with IBMS, and how would this vary across states?

RQ2 How might the IBMS be leveraged to create a welcoming environment for b-girl practitioners, and possibly subvert or transform gender norms from breaking culture that manifest through movement?

Because breaking music typically employs steady rhythms at a fast tempo, we hypothesized that non-breaking music with a sudden change to the tempo (B2) or an "inverted" tempo that is driven by dancer movements (B3) would encourage a defamiliarizing experience for the participants. All three of the non-traditional interaction states were presented to the dancer participants.

Following principles of qualitative methodologies including Grounded Theory that seek to allow emergent narratives to arise organically [35], general questions were asked during the individual sessions that were then updated for the group sessions based on trends found within participant responses. The initial one-on-one sessions focused on general questions of participant experience, comparing to "normal" breaking spaces, and comfortability with both using existing breaking moves and trying new moves in this context. The group sessions updated the questions to address emergent trends, which we further discuss in the next section.

5.2 Results

With respect to RQ1, the first author, as a practitioner in the field, observed that there were moves that would only be present in breaking in general – these included six-steps, hooks, back-rock variations, and threading. People seemed more on "autopilot" with state A in terms of producing standard breaking moves, meaning that they were more likely to perform movements from muscle memory rather than trying new ones. This was supported by participant responses, including the following:

> *For the traditional music, because those are songs I've literally practiced to or like battled to, my rounds are more traditional breaking like this top rock, footwork, freeze. But for the experimental music, [...] I felt like I was experimenting with different movements and like of my qualities and just kind of like going into the void. Just putting my body and places are carrying through with the momentum and just going somewhere like unfamiliar....*

This implied that state A did not present an environment where the participant felt free to explore new movements, whereas the B states – or what they called "experimental music" – did. Because this research explicitly foregrounded breaking, participants were more inclined to at least attempt breaking movements, and thus it is unsurprising that all B states seemed similar in terms of initial presence of breaking gestures. In reaction to the experimental music, another participant noted:

> *I used one sound reference to kind of like, locate myself in the music, and in my movement. Like there was that constant beat [recurring sound] kind of going through the whole track.*

Nonetheless, they seemed to be mediated by or driven by the music that was produced by the system. The explicitness of breaking in this study, however, posed an interesting trade-off. One participant noted:

> *This is a specific study about breaking so I kind of had to bring myself back to that.... So, although it was nice to break to a non-traditional breaking sound, it still felt like I was overthinking, or I was thinking a little more*

> *for the experimental sound. Whereas for the traditional sound, I was like okay yeah I know how to dance to this. So...it didn't make me think more than I did when I was dancing [with] the experimental sound.*

Although the nature of the study forced the participant to be conscious of their movements in order to fit what they deemed as breaking, which made it so that they inherently embodied breaking aesthetics in their gestures initially, they were also forced to be more aware of their thought-process on the movements they were making while interacting with the IBMS. One mentioned about their interaction with the IBMS during B3:

> *When the experimental music was on, at one point, I was like maybe I'll pretend I'm water, and then as soon as I started [...] contemporary dance starts coming out. I was like 'Okay stop that. No water today.' When traditional breaking music is on, I'll never think "Oh, what are some experimental things I can implement right now. It's more like what moves could I do. Can I try this move, maybe?"*

This participant tried new ways of exploring movements by attempting to follow a different form of moving when dancing to what they determined to be experimental music. This meant trying more fluid motions, which made their experience in contemporary dance to unintentionally influence their movements and forced them to negotiate between dance aesthetics of breaking and contemporary dance. These results lead us to believe that an initial answer to RQ1 is really more about how practitioners negotiate with breaking aesthetics, taken as an initial starting point, via gestures that emerge only from their interactions with the IBMS. The process of "inverting" the tempo by making this an interactive variable, in the tempo ranges we examined, does indeed seem to be an important element that links breaking with the "experimental" spaces entered into by participants.

With respect to RQ2, the IBMS was seen to foster a supportive and non-competitive space through which participants were able to play freely. This parallels broader reflections on the benefits of an all b-girl environment for breaking. One participant recalled encountering the emphasis on competition in Toronto's breaking scene during their first few years being involved with the community:

> *When I came to Toronto, I was like, you know, everyone's just practicing for competition. I was like, is there even room to, like, vibe and have fun? [...] I found the b-girl community was just so supportive right away. And the b-boy community, it took me four years to, like, penetrate it.*

They felt that the competitive environment made it harder for them to integrate with the overall breaking community, as opposed to b-girl spaces where they felt accepted immediately, especially when their intention was not to participate at every competition. During the group dance sessions with both B states, 5 out of 7 participants noted they felt encouraged to play. One participant explained:

> *I felt like with the experimental music I want to play more with stuff [...] because it's not like a break [beat], I don't feel that competitive, or that battle kind of energy so I'm not like trying to do my hard [moves]. I just want to kind of play.*

The lack of competitiveness in the space allowed them freedom to play and explore movements while interacting with the IBMS, and therefore provided a space where participants were not judged on the execution and technicality of movements. In short, participants noted that an all b-girl space in general increased a sense of support and inclusion, and the IBMS was seen as providing a similar function through its linking between breaking-appropriate musical context and invitation to explore new "experimental" moves.

5.3 Discussion

Our research questions were aimed at investigating the potential of the IBMS to subvert gender norms of breaking movements in order to facilitate more inclusive breaking spaces. The tempo settings of B2 and B3 both encouraged movement-making processes that differ from the dancers' typical practice routines. The tempo setting of B2 presented sudden changes to the generated interactive breaking music, forcing participants to adopt to the new tempo. Further, responses suggest that B3 provided a more defamiliarizing experience due to a direct mapping from dancer movement to (smooth) continuous changes in tempo, in the range of 20 bpm to 118 bpm. Because breaking music traditionally has a fast and stable tempo, these settings posed a new challenge for dancers to move and create movements outside of what they are used to, while smooth transitions allowed dancers to make continuous adjustments in the moment. In light of participant responses, we believe that the root of a complete answer to RQ1 lies in considering the process of negotiation between breaking aesthetics in gestures that emerged from interactions with the IBMS, and gestures that are "brought" into the session a priori as a starting point. Participants' experience of the sessions can be seen as defamiliarizing, based on subject feedback that the IBMS made them become more aware of their own movement-making process – something that they normally would not have paid attention to at traditional breaking spaces. The IBMS thus encouraged them to be more conscious and introspective of their movements, yet (almost paradoxically) more comfortable in their own movements than they might feel at times in traditional breaking spaces. As a starting point to a fuller answer to RQ2, this is significant because the movements that breaking practitioners are familiar with were developed in a heteromasculine space. An interactive breaking space mediated by something like IBMS could provide a context where a practitioner feels they are properly "in" a breaking space, yet are naturally reflective of the dominant movement vocabulary while being inspired to try new movements that deviate from this. In further support of the need for such a space, the following were comments made by a participant reflecting on their experience in male vs. female dominated breaking spaces:

> *I feel like as a female, it's like it's really good practice to learn how to take up space, like through going to male dominated spaces. I've learned how to be more confident in myself and like how to take up space and, like, be grounded in my intention because of that, I can actually show up anywhere now.... I think I prefer b-girl spaces, but there aren't many b-girls dominated females spaces for breaking.*

This highlights that female-dominated breaking spaces immediately felt more supportive because of the collective intention of improving their dance. Because the IBMS dance sessions focused on their experience as a group and in relation to one another, i.e. the broader interactive space that included framing context as well as the interactive system, the b-girls involved felt safe to telematically dance around one another. The lack of competitiveness allowed for more room to play and disorient their typical auditory and kinesthetic senses. The interactive breaking space can be seen as consisting of three distinct parts: the framing of the breaking context, the focus on "defamiliarizing" aspects of breaking movement from highly similar to highly experimental, and the system designed around complex mapping strategies and tempo changes that had this in mind. We believe that the synthesis of these three aspects shows the true potential for how a system like the IBMS could be leveraged to foster a more welcoming environment for b-girl practitioners.

6 Conclusion

Integrating technology in art interactive spaces that foster human-machine relationships opens up the possibility for dancers to explore a wide spectrum of movements they have never done (or possibly seen) before [36]. According to Berman and James:

> ...the tight coupling between the participants' behaviours may lead to interestingly unpredictable outcomes. For example, if the dancer imitates or extends the avatar's movements, her expressiveness and other aspects of her movement will influence the avatar's subsequent output, resulting in a feedback loop. The complexity of the interaction makes the outcome difficult to predict.... [36]

Similarly, women are able to make more space for themselves in hip-hop culture with the help of digital technology. Johnson argues:

> The Internet becomes an especially powerful medium through which to redefine this discursive terrain, and to assert the interests and concerns of female Hip Hop practitioners. Jessica Pabon, a performance studies scholar writing about female graffiti artists, calls the Internet a system of visibility and communication for these women [37].

B-girls are able to sustain their identities in hip-hop culture through web videos and other specialized programming on the Internet, such as the webseries, "Strictly B-Girl," that featured interviews with b-girls from across North America [37].

This study begins to explore this possibility at the intersection of these two worlds. The results are a promising first step that we intend to build upon in future iterations examining both in-person and telematic/networked contexts, towards building a data library of breaking movements for gesture recognition purposes, facilitating integrated IBMS and basic breaking classes/workshops over a period of time, and exploring more explicit and implicit mapping strategies using other low-level tracking systems that might amplify the defamiliarizing potential that we have initially observed in this study.

References

1. Caramiaux, B., Françoise, J., Schnell, N., Bevilacqua, F.: Mapping through listening. Comput. Music J. **38**(3), 34 (2014). https://doi.org/10.1162/COMJ_a_00255
2. Hoch, D.: Toward a hip-hop aesthetic: a manifesto for the hip-hop arts movement. In: Chang, J. (ed.) Total Chaos: The Art and Aesthetics of Hip-Hop, pp. 351–353. BasicCivitas Books (2006). https://hdl-handle-net.ezproxy.library.yorku.ca/2027/heb.32663
3. Fogarty, M.E.: Dance to the Drummer's Beat: Competing Tastes in International B-Boy/B-Girl Culture, 188 (2011). https://era.ed.ac.uk/handle/1842/5889
4. LaBoskey, S.: Getting off: portrayals of masculinity in hip hop dance in film. Dance Res. J. **33**(2), 114 (2001). https://doi.org/10.2307/1477808
5. Aprahamian, S.: Hip-hop, gangs, and the criminalization of African American culture: a critical appraisal of yes yes y'all. J. Black Stud. **50**(3), 298–315 (2019). https://doi.org/10.1177/0021934719833396
6. De Lauretis, T.: Queer theory: lesbian and gay sexualities: an introduction. Differ. J. Feminist Cult. Stud. **3**(2), iii–xvii (1991)
7. Jagose, A.: Feminism's queer theory. Feminism Psychol. **19**(2), 157 (2009). https://doi.org/10.1177/0959353509102152
8. Johnson, I.K.: From blues women to b-girls: performing badass femininity. Women Perform. **24**(1), 15–28 (2014). https://doi.org/10.1080/0740770X.2014.902649
9. Johnson, I.K.: From blues women to b-girls: performing badass femininity. Women Perform. **24**(1), 16 (2014). https://doi.org/10.1080/0740770X.2014.902649
10. Peoples, W.A.: 'Under construction': identifying foundations of hip-hop feminism and exploring bridges between black second-wave and hip-hop feminisms. Meridians **8**(1), 20 (2008). www.jstor.org/stable/40338910
11. Peoples, W.A.: 'Under Construction': identifying foundations of hip-hop feminism and exploring bridges between black second-wave and hip-hop feminisms. Meridians **8**(1), 19–52 (2008). www.jstor.org/stable/40338910
12. Peoples, W.A.: 'Under construction': identifying foundations of hip-hop feminism and exploring bridges between black second-wave and hip-hop feminisms. Meridians **8**(1), 27 (2008). www.jstor.org/stable/40338910
13. Durham, A., Cooper, B., Morris, S.: The stage hip-hop feminism built: a new directions essay. Signs J. Women Cult. Soc. **38**(3), 15 (Spring 2013). https://doi.org/10.1086/668843

14. Gunn, R.: Dancing away distinction: queering hip hop culture through all style battles. Queer Stud. Media Pop Cult. **4**(1), 23 (2019). https://doi.org/10.1386/qsmpc_00002_1
15. Gunn, R.: Dancing away distinction: queering hip hop culture through all style battles. Queer Stud. Media Pop Cult. **4**(1), 13 (2019). https://doi.org/10.1386/qsmpc_00002_1
16. Light, A.: HCI as heterodoxy: technologies of identity and the queering of interaction with computers. Interact. Comput. **23**(5), 430–438 (2011). https://doi.org/10.1016/j.intcom.2011.02.002
17. Light, A.: HCI as heterodoxy: technologies of identity and the queering of interaction with computers. Interact. Comput. **23**(5), 432 (2011). https://doi.org/10.1016/j.intcom.2011.02.002
18. Light, A.: HCI as heterodoxy: technologies of identity and the queering of interaction with computers. Interact. Comput. **23**(5), 431 (2011). https://doi.org/10.1016/j.intcom.2011.02.002
19. Carlson, K., Fdili-Alaoui, S., Corness, G., Schiphorst, T.: Shifting spaces: using defamiliarization to design choreographic technologies that support co-creation. In: Proceedings of the 6th International Conference on Movement and Computing, MOCO 2019, pp. 1–8. Association for Computing Machinery, Tempe (2019). https://doi.org/10.1145/3347122.3347140
20. Loke, L., Robertson, T.: Moving and making strange: an embodied approach to movement-based interaction design. ACM Trans. Comput.-Hum. Interact. **20**(1), 1–25 (2013). https://doi.org/10.1145/2442106.2442113
21. Essex, J.A.: Moov: Scaffolding Motion-Based, Paired Play Creation. Masters, OCAD University (2017). http://openresearch.ocadu.ca/id/eprint/1988/
22. Fdili Alaoui, S.: Making an interactive dance piece: tensions in integrating technology in art. In: Proceedings of the 2019 on Designing Interactive Systems Conference, DIS 2019, pp. 1195–1208. Association for Computing Machinery, New York (2019). https://doi.org/10.1145/3322276.3322289
23. Fdili Alaoui, S.: Making an interactive dance piece: tensions in integrating technology in art. In: Proceedings of the 2019 on Designing Interactive Systems Conference, DIS 2019, pp. 1196–1197. Association for Computing Machinery, New York (2019). https://doi.org/10.1145/3322276.3322289
24. Fdili Alaoui, S.: Making an interactive dance piece: tensions in integrating technology in art. In: Proceedings of the 2019 on Designing Interactive Systems Conference, DIS 2019, p. 1196. Association for Computing Machinery, New York (2019). https://doi.org/10.1145/3322276.3322289
25. Fdili Alaoui, S.: Making an interactive dance piece: tensions in integrating technology in art. In: Proceedings of the 2019 on Designing Interactive Systems Conference, DIS 2019, p. 1204. Association for Computing Machinery, New York (2019). https://doi.org/10.1145/3322276.3322289
26. Loke, L., Robertson, T.: Moving and making strange: an embodied approach to movement-based interaction design. ACM Trans. Comput.-Hum. Interact. **20**(1), 2 (2013). https://doi.org/10.1145/2442106.2442113
27. Jarvis, I., Van Nort, D.: Posthuman gesture. In: Proceedings of the 5th International Conference on Movement and Computing, MOCO 2018. Association for Computing Machinery, New York (2018). https://doi.org/10.1145/3212721.3212807
28. Van Nort, D., Wanderley, M., Depalle, P.: Mapping control structures for sound synthesis: functional and topological perspectives. Comput. Music J. **38**(3), 6–22 (2014). https://doi.org/10.1162/COMJ_a_00253

29. Donato, B.D., Dewey, C., Michailidis, T.: Human-sound interaction: towards a human-centred sonic interaction design approach. In: Proceedings of the 7th International Conference on Movement and Computing (2020). https://doi.org/10.1145/3401956.3404233
30. Caramiaux, B., et al.: Mapping through listening. Comput. Music J. **38**(3), 44 (2014)
31. Hunt, A., Wanderley, M.: Mapping performer parameters to synthesis engines. Org. Sound **7**(2), 97–108 (2002)
32. Place, T., Lossius, T.: Jamoma: A Modular Standard for Structuring Patches in Max, 1 January 2006
33. Jensenius, A.: Motion-sound interaction using sonification based on motiongrams. In: ACHI 2012–5th International Conference on Advances in Computer-Human Interactions (2012)
34. Fiebrink, R., Cook, P.: The Wekinator: a system for real-time, interactive machine learning in music. In: Proceedings of The Eleventh International Society for Music Information Retrieval Conference (ISMIR 2010), 1 January 2010
35. Charmaz, K.: Constructing Grounded Theory: A Practical Guide Through Qualitative Analysis. Sage Publications, Thousand Oaks (2006)
36. Berman, A., James, V.: Kinetic dialogues: enhancing creativity in dance. In: Proceedings of the 2nd International Workshop on Movement and Computing - MOCO 2015, p. 82. ACM Press, Vancouver (2015). https://doi.org/10.1145/2790994.2791018
37. Johnson, I.K.: From blues women to b-girls: performing badass femininity. Women Perform. **24**(1), 24 (2014). https://doi.org/10.1080/0740770X.2014.902649

Collaboration, Inclusion and Participation

Creative Collaboration with the "Brain" of a Search Engine: Effects on Cognitive Stimulation and Evaluation Apprehension

Mélanie Gozzo[1(✉)], Michiel Koelink Woldendorp[2,3,4], and Alwin de Rooij[1,5]

[1] Department of Communication and Cognition, Tilburg School of Humanities and Digital Sciences, Tilburg University, Warandelaan 2, 5037 AB Tilburg, The Netherlands

[2] ArtechLAB Amsterdam, Amsterdam University of the Arts, Overhoeksplein 2, 1031 KS Amsterdam, The Netherlands

[3] Master of Education in Arts, Hanze University of Applied Sciences, Praediniussingel 59, 9711 AG Groningen, The Netherlands

[4] Master of Education in Arts, NHL Stenden University of Applied Sciences, Rengerslaan 8, 8917 DD Leeuwarden, The Netherlands

[5] Situated Art and Design Research Group, St. Joost School of Art and Design, AVANS University of Applied Sciences, Parallelweg 21, 5223 AL 's-Hertogenbosch, The Netherlands

Abstract. Artificial Intelligence (AI) is rapidly becoming part of how we do creative work. This to the extent that commonly used AI-powered systems, such as search engines, are already routinely used to support our day-to-day creative tasks. However, surprisingly little is known about how creative collaboration with the "brain" of a typical search engine compares to creative collaboration with other people. We propose that exploring this requires a cognitive and a social perspective. Firstly, the output of a search engine might influence cognitive stimulation differently than human collaborative forms, i.e. the degree to which output by another inspires more and more novel associations. Secondly, evaluation apprehension, i.e. not sharing all your ideas due to a fear of being evaluated negatively, might be reduced when collaborating with such AI-powered systems due to their limited perceived social agency. Thirdly, a user's attitude towards AI might moderate this effect, e.g., due to fears about what such systems do with the user's data. An experiment (n = 139) was conducted where participants were instructed to collaborate with an AI powered by a search engine, or with another person, during a divergent thinking task (in reality these collaborations were scripted). The results indicated that 1) collaborating with another person increased cognitive stimulation, 2) collaborating with the AI decreased evaluation apprehension, and 3) people's general attitude towards AI did not moderate its effect on evaluation apprehension. Herewith, the study contributes to an emerging body of work on creative collaboration with AI.

Keywords: Co-creative AI · Cognitive stimulation · Creativity · Divergent thinking · Evaluation apprehension · Search engines

M. Wölfel et al. (Eds.): ArtsIT 2021, LNICST 422, pp. 209–223, 2022.
https://doi.org/10.1007/978-3-030-95531-1_15

1 Introduction

Artificial intelligence (AI), as science, aims to create artifacts that exhibit some form of intelligence, with achieving human-level creativity as one of its hallmark challenges [1]. Along with other recently achieved AI milestones, such as DeepMind's AlphaGo beating the Go world champion Lee Sedol [2] people professionally engaged in domains that are historically associated with creativity are now starting to compete with AI algorithms [3]. Generated by a neural network and developed by artist Collective Obvious AI & Art, the Portrait of Edmond de Belamy sold for $432,500 at the world-renowned auction house Christie's on October 25th, 2018 [4]. Alongside this ongoing development of (collaborative) creative AI systems, one could state that AI has already permeated our day-to-day creative activities via another route.

As AI-driven hardware and software have become omnipresent, the possibilities of running excessively large trained neural network algorithms have increased. Neural networks mimic the semantic network of the human brain to better reason, classify, and understand received input, and to produce suitable output [5]. This has not only sparked the further development of creative AI [1, 3] but has also led to the further optimization of search engines used on the web, for example Google's reverse image search [6]. The "brains" of these search engines are the semantic networks that emerge from the search engine's neural networks, vectorization techniques, and other algorithms (e.g., content analysis, meta-data, and ranking models), from which it draws its associations. Recent work suggests that creatives, from laypersons to professional artists, routinely rely on these search engines to provide them with input to support their creative activities [7]. For example, a fine artist might search for images to inspire new ideas, or a layperson might seek inspiration for what to cook for dinner. Therefore, one could claim that AI has already permeated our day-to-day creative work, via our reliance on search engines to support our creative thinking. Despite this, relatively little is known about how our collaborations with the "brains" of these search engines affect creative task performance [7].

In light of these developments, we propose that taking both a cognitive and a social perspective could provide a useful starting point for further investigation. A cognitive perspective is relevant because differences in the semantic networks of search engines and human collaborators [8] may directly affect cognitive stimulation, i.e. the degree to which output by another person or system inspires more and more novel associations [9]. Specifically, it is an open question whether the output of search engines increase or decrease cognitive stimulation when compared to the output of other human beings [5]. A social perspective is relevant because of a common tendency to anthropomorphize AI systems as a collaborator or teammate [5]. This raises novel questions about whether effects on creative task performance might be explained by the mitigation of issues that commonly arise during creative collaborations among people, such as evaluation apprehension [10], i.e. not sharing ideas due to a fear of being evaluated negatively [11]. This due to the technology's limited perceived social agency [12]; or alternatively whether a user's attitude towards AI technologies might also elicit technology-specific forms of evaluation apprehension, e.g., due to fears about what such systems do with their data [13].

The study presented in this paper aims to shed more light on these conjectures by answering the following research question: *How does creative collaboration with the "brain" of a search engine affect creative task performance?* The paper is structured as follows: first, the rationale introduced above is developed in more detail, based on which three hypotheses are conjectured. Second, the methodological details of an experiment (n = 139) that was developed and conducted to test the hypotheses are presented. Third, the results of the experiment are explained. Fourth, the results and key limitations are discussed and future work is proposed.

2 Creative Collaboration with the "Brain" of a Search Engine

Creativity can be defined as the creation of novel yet useful ideas, problem solutions, or products [14]. Whether it is laypeople or professional artists, the general process by which they arrive at a creative outcome tends to be similar [15]: People undertake activities to understand a problem, generate ideas, evaluate these ideas, and (iteratively) revise and test these ideas to arrive at a revised version of their idea, problem solution, or product [16]. *Divergent thinking*, the ability to produce variation [17] contributes to the creative process at various stages [18]. The assumption is that unrestricted quantity in some parts of the creative process will ultimately lead to quality in other parts [19]. This depends in part on the organization of a person's semantic memory [20] and the ease and semantic distance which associations can be retrieved [21, 22]. Faster retrieval of associations from semantic memory enables people to generate more options to develop a creative solution from within a limited time frame, whereas the semantic distance of the retrieved associations correlates with the likelihood that these associations enable novel outcomes of the creative process [20–22]. For example, having many novel associations can benefit the early stages of understanding a problem, by generating a diverse set of perspectives on a problem [23]; and during idea generation, generating many novel candidate solutions increases the chance of developing a truly creative revised idea, solution, or product in the remainder of the creative process [24]. As such, divergent thinking can be viewed as an indicator of creative task performance and divergent thinking tests are often used to assess creative potential [17, 18].

It is well known that collaboration with other people can benefit divergent thinking due to *cognitive stimulation* [9]. Output by others may contain semantic categories that enable an individual to make new associations faster, which would otherwise require them to engage in an increasingly time-consuming search in their semantic memory [25]. Output by others can also contain semantic categories with a more semantic distance than the categories that are immediately accessible by a person [26] due to the idiosyncrasies of how an individual's semantic memories are organized [27]. A person can therefore benefit from others' output by increasing the number and semantic distance of the associations they are able to make themselves, beyond what they are capable of alone, which positively influences divergent thinking [26, 27]. However, it is common that the semantic categories contained in the output from a collaborator might not be so different from the associations that an individual would make on their own [26] or that the categories contained stimulate having common associations or verge towards the useful at the cost of novelty [27] with a negative influence on divergent thinking.

Therefore, the quality of the output during collaboration can affect cognitive stimulation positively or negatively, and by extension divergent thinking. Extending this *cognitive perspective* to our day-to-day reliance on search engines for creative work [7] suggests that creative task performance is affected in the same way. However, it is not known whether the quality of the output generated by a typical search engine in 2021 causes more or less cognitive stimulation, than say, an averagely creative human being.

The literature appears to be ambiguous on this topic. On the one hand, AI systems in general learn and organize their semantic networks differently than humans do [8]. In theory, these systems could retrieve more, more efficient, and more apt associations from its semantic network than people can, due to the unimaginably extensive database they could be based on [28]. The information retrieved by the AI is different from what another person is likely to provide, sometimes to the extent that an AI's retrieved information violates human expectations and is characterized as weird [29]. Possibly, weird stimuli entail novelty by prompting the generation of semantically distant associations [30] thereby positively influencing divergent thinking [26, 27]. Thus, one could be tempted to conclude that the output of search engines might be more cognitively stimulating than the output of an averagely creative human being. On the other hand, researchers have also voiced concerns about the limits of search engines in particular in this regard, citing the argument that search engines are often designed to retrieve data based on similarity [5]. This would suggest that our ubiquitous but everyday reliance on search engines [7] may negatively influence cognitive stimulation, and subsequent divergent thinking [26, 27]. As such, the available literature suggests that cognitive stimulation is likely to be affected, but it is not clear whether this effect is positive or negative. Therefore, the following non-directional hypothesis is proposed:

H1: Creative collaboration with the "brain" of a search engine, compared to creative collaboration with an averagely creative person, influences divergent thinking due to its effect on cognitive stimulation.

A *social perspective* might provide additional insight into how creative collaboration with the "brain" of a search engine compares to creative collaboration with another person. *Evaluation apprehension*, i.e. not sharing ideas due to a fear of being negatively evaluated [10] is a key example of how social interactions among people may affect creative task performance negatively [11]. A reduced willingness to share ideas directly affects the amount and diversity of information shared with others due to self-imposed constraints about what is "safe" to share or not. A direct consequence can be a reduction in the number and diversity of responses shared between collaborators and possibly also generated during the ideation process. Although evaluation apprehension can have several causes [9] it is well known that often social anxiety underlies evaluation apprehension [31]. People regularly do not share ideas because they fear the negative social consequences they might incur from others in response to the information they share. Past experimental research, for example, suggests that a fear of being evaluated negatively reduces the number of ideas when interacting with other people. This effect is mitigated when working alone [31]. We propose that creative collaborations with AI systems in general might reduce evaluation apprehension due to their limited social agency and could consequentially positively influence divergent thinking.

Even though there is a common tendency to anthropomorphize AI systems as a collaborator or teammate [5] this does not mean that people attribute the same level of social agency to creative AI systems as they do to other people [12]. Even when AI systems are specifically designed to increase (the illusion of) social agency, e.g., by endowing them with the ability to detect and send social signals [32] people do not attribute the same social abilities to these technologies they also attribute to other people [12]. Social robots, for example, tend to be seen at best as in between an anthropomorphic being and a technological object. Previous work suggests that interacting with these technologies can reduce social anxiety (a common cause of evaluation apprehension), when compared to interacting with people, due to the limited social agency users attribute to these technologies [33, 34] (but see [10] for an alternative finding). Therefore, we conjecture that people are less likely to expect social repercussions from any kind of AI system than from another person. In terms of social repercussions there is little to fear from an AI. Thus, another way in which creative collaboration with the "brain" of a search engine might affect creative task performance, is social in nature: Collaborating with an AI might reduce evaluation apprehension and its negative effects on divergent thinking. Based on these conjectures, we propose the second hypothesis:

H2: Creative collaboration with the "brain" of a search engine, compared to creative collaboration with an averagely creative person, positively influences divergent thinking due to a negative effect on evaluation apprehension.

As previously suggested, evaluation apprehension can have many causes [9]. When AI technologies are cast into an anthropomorphic framework of collaborators and teammates [5] we cannot expect that people forgo any mistrust they might have about the type of technology they are interacting with, simply because these technologies allude to something more human-like than other similar technologies [13]. Despite the positives that come with using search engines [6] and AI more broadly [8] people are often concerned about whether they can trust using the output of an AI [35] or have privacy or other concerns about how search engines use their data [36]. Thus, speculatively, when people do not view AI systems positively [13] it is possible that this might introduce another cause of evaluation apprehension that is specific to creative collaboration with these kinds of technologies. A user's *attitude towards AI* might thus moderate the expected effects on evaluation apprehension and divergent thinking expressed in H2 because it introduces another cause of evaluation apprehension. Based on these speculations, the third and final hypothesis is proposed:

H3: The effects of creative collaboration with the "brain" of a search engine, compared to creative collaboration with an averagely creative person, on divergent thinking via evaluation apprehension, is moderated by a person's attitude towards AI.

3 Method

To test the hypotheses an online experiment was conducted with a between-subject design.

3.1 Participants

A total of one hundred forty-one participants were recruited. One participant did not sign the informed consent and one did not finish the experiment. The data from these two participants were therefore removed from the data set. Data from the remaining one hundred thirty-nine participants ($M_{age} = 22.54$, $SD_{age} = 3.70$) were used in the analysis. Eighty-four of these participants self-identified as females and fifty-five participants self-identified as male. The participants were recruited by convenience sampling using the researcher's network (n = 52) and the human subjects pool (n = 87) of the Department of Communication and Cognition, Tilburg University. All participants were previously or currently engaged in a higher education program. The Research Ethics and Data Management Committee of the Tilburg School of Humanities and Digital Sciences approved the study.

3.2 Materials and Measurements

3.2.1 Experimental Manipulations

Participants were randomly instructed to collaborate on a divergent thinking task [25] with either an AI (referred to as AI collaborator, for the sake of brevity in the following sections), which was powered by a search engine, or with another person (referred to as human collaborator). In both conditions participants were asked to generate as many creative associations with either a depiction of a flower or a rocket as they could (also randomly assigned; Fig. 1). Two objects were chosen to reduce the chance that any results could be attributed to the specifics of one object. Collaboration entailed that while the participants were generating and typing in associations, they saw the associations generated by the other in real-time while thinking that the other also read their associations (Fig. 2). This was emphasized by asking participants to press the ENTER key after each association to "share" their associations with the collaborator. In reality, however, these collaborations were scripted. Using a scripted collaboration made the data collection feasible. The scripted collaborations entailed that participants were presented with associations that were previously collected. The associations by the collaborator were presented in a simple text output window in both conditions. This was done to avoid confounding variables that might be introduced by differences in, e.g., seeing another person while collaborating versus using the GUI of a search engine. The AI collaborator's associations were collected by uploading the rocket and flower image into Google's reverse image search. The first fifty images retrieved were manually translated into a word and stored. Google search was chosen because it is one of the search engines that is routinely used in everyday creative tasks [7]. As Fig. 3 shows, the related associations given by Google's reverse image search were also images. Therefore, a human translation of these images into words was necessary as the associations were later presented to the participants via a text-output field in both conditions. The human collaborator's associations were collected by asking eight people from the researcher's network to do the divergent thinking task for the flower and the rocket image. The associations were pooled and redundant associations were removed. Then, fifty of their responses were randomly selected for the rocket and flower image and stored. These eight people represented an "averagely creative person" as indicated by their scores on Runco's

Ideational Behavior scale ($M = 3.10$, $SD = 0.27$) [37]. In the AI collaborator condition, each participant was presented over time with 21 out of 50 randomly selected associations obtained from the search engine. In the human collaborator condition, participants were presented with 21 of 50 randomly selected associations obtained from the collected human associations. To realize a sense of real-time collaboration, the interval at which associations were presented mimicked the human divergent thinking process (0–30 s= 3.33 sec interval, 31–60 s = 5 sec interval, 61–90 s = 7.5 sec interval, 91–120 s = 15 sec interval) [25].

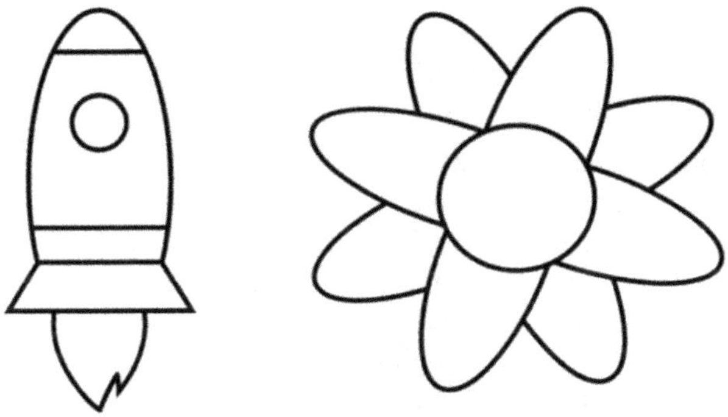

Fig. 1. The rocket and flower stimuli used in the divergent thinking task.

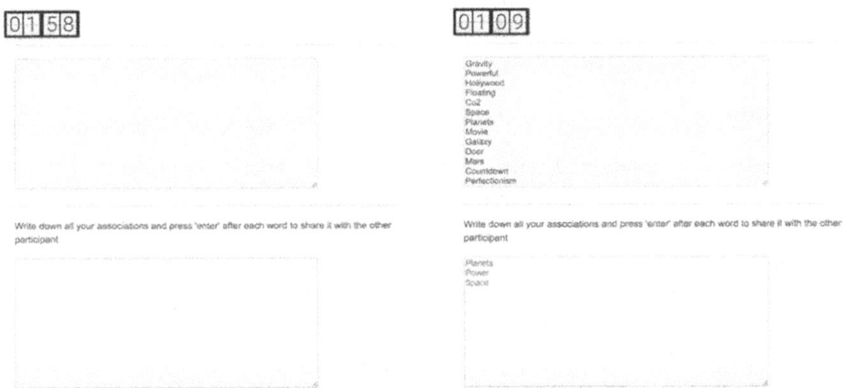

Fig. 2. Textual interface through which associations were shared by the collaborator (top), and where the associations were entered by the participant (bottom).

3.2.2 Assessing Divergent Thinking

Divergent thinking performance was assessed with 1) fluency, i.e., the number of associations produced, counted by the researchers [25] and 2) the average semantic distance between the associations and the rocket or flower, calculated with the SemDis software tool [38]. SemDis was previously shown to have good criterion validity as a divergent thinking performance measure [38].

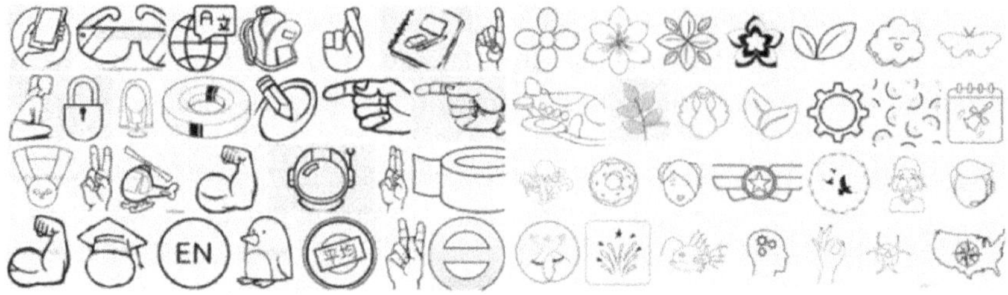

Fig. 3. Google's reverse image search associations with the rocket (left) and flower (right).

3.2.3 Assessing Cognitive Stimulation, Evaluation Apprehension, and General Attitudes Towards AI

Cognitive stimulation was assessed with a self-developed six-item five-point Likert scale (1 = strongly disagree, 5 = strongly agree). The items asked participants about whether the collaboration simulated their divergent thinking (e.g., "The input of the collaborator helped me to think creatively"). Cronbach alpha suggested acceptable reliability, $\alpha =$ .74. Evaluation apprehension was assessed with a seven-item five-point Likert scale (1 = strongly disagree, 5 = strongly agree) developed by [39]. The scale was adapted to suit the present study better, e.g., "I felt nervous to share my ideas with the group" was rephrased as "I felt nervous to share my associations with my collaborator". Cronbach alpha suggested good reliability, $\alpha = .85$. The participants' general attitude towards AI was assessed using a twenty-item five-point Likert scale (1 = strongly disagree, 5 = strongly agree) developed by [13]. Cronbach alpha suggested good reliability, $\alpha = .83$.

3.3 Procedure

The experiment was conducted online using Qualtrics. There, participants were asked to read the study information, to sign informed consent, and were randomly assigned to one of the conditions. Information that could reveal the deceptions in the experiment was withheld at this point such as participants assuming that collaboration was done in real-time. The participants were asked to fill in demographic information and the general attitude towards AI scale. After this, they received the divergent thinking task instructions, and were presented with an example to aid in their understanding: "If the illustration depicts a 'Cow' you could answer with: 'Milk, 'Grass', ..."). They were randomly assigned to either the instruction that they would be collaborating with an AI, which was powered by Google's reverse image search, or with another person. Subsequently, they were presented with their stimulus and started the divergent thinking task. Important to note is that even though the experiment was held in English, participants were allowed to respond in their native language (Dutch) whenever they experienced a language barrier to allow for fluency of the associations spilled by the participants. Participants were instructed to write down all their associations for the next two minutes, to press ENTER after every association in order to share their association with the collaborator, to use the received input from the collaborator to think of other associations related to the concept, to answer in single words, and to answer either in English or

Dutch. After finishing the task, the participants filled in the cognitive stimulation and evaluation apprehension scales, were fully debriefed, and thanked.

4 Results

To provide insight into the general characteristics of the data the descriptive statistics and correlations were calculated. Visual inspection of the histograms suggested that the data distribution of the variables evaluation apprehension and fluency deviated from normality. Therefore, the non-parametric Kendall's tau-b correlation coefficients were reported. These are presented in Table 1.

Table 1. Means and standard deviations (between parentheses) and Kendall's tau-b correlations (two-tailed). Note. † $p < .100$, * $p < .050$, ** $p < .010$.

	Experimental manipulations		Correlations				
	AI collaborator	Human collaborator	1	2	3	4	5
1. Cognitive stimulation	3.00 (.65)	3.50 (.57)	–				
2. Evaluation apprehension	1.89 (.57)	2.14 (.68)	−.144†	–			
3. Fluency	16.67 (.6.14)	15.88 (5.69)	.137	−.072	–		
4. Semantic distance	.78 (.05)	.76 (.07)	.085	−.067	.283**	–	
5. General attitude towards AI	3.51 (.51)	3.59 (.37)	.269**	−.344**	.093	.103	–

To test whether creative collaboration with the "brain" of a search engine, compared to creative collaboration with an averagely creative person, positively influences divergent thinking due to its effect on cognitive stimulation (hypothesis 1), two mediation analyses were conducted using Hayes' bootstrapping method [40]. This method is robust against deviations from normality. The model terms were both specified with collaboration type as the independent variable (human collaborator coded: 0, AI collaborator coded: 1) and with self-reported cognitive stimulation as the mediator. Model 1 was specified with fluency as the dependent variable, and model 2 with semantic distance as the dependent variable. Assumption checks suggested heteroskedasticity in both models, which was tested by visually inspecting the distribution of the studentized residuals plotted against the standardized predictor values [41]. Therefore, Huber-White heteroscedasticity consistent standard errors were used to calculate the test statistics for both models [42]. The models and unstandardized coefficients are presented visually in Figs. 4a (model 1) and 4b (model 2), whereas the indirect and direct effects are presented in the text below.

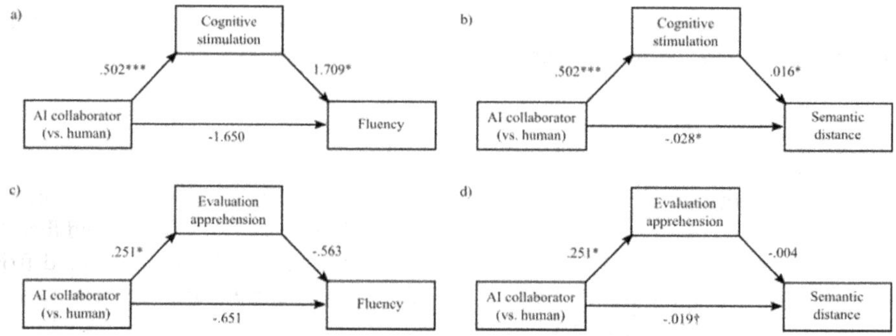

Fig. 4. a) Mediation analysis of the effects of collaborating with the AI on a) cognitive stimulation and subsequent fluency (model 1), and b) semantic distance (model 2), and c) mediation analysis of collaborating with the AI on evaluation apprehension on subsequent fluency (model 3), and d) semantic distance (model 4). Data are unstandardized coefficients. † $p < .100$, * $p < .050$, ** $p < .010$, *** $p < .001$.

These mediation analyses showed a significant indirect effect of human collaborator, compared to the AI collaborator, on the relationship of cognitive stimulation with fluency, $b = -.858$, $se = .487$, 95% CI[.040, 1.976] and with semantic distance, $b = .008$, $se = .005$, 95% CI[.001, .018]. These indirect effects were characterized by a significant positive effect of collaborating with a human on cognitive stimulation, $b = .502$, $se = .104$, 95% CI[.296, .707] and of cognitive stimulation on fluency, $b = 1.709$, $se = .786$, $p = .031$, 95% CI[.155, 3.264] and semantic distance, $b = .016$, se $= .008$, $p = .034$, 95% CI[.001, .031]. In contrast no significant direct effect was found on fluency, $b = -1.659$, $se = 1.116$, $p = .142$, 95% CI[−3.857, .558]. However, the results did show a significant negative direct effect on semantic distance, $b = -.028$, $se = .011$, $p = .013$, 95% CI[−.049, −.006]. This finding suggests that creative collaboration with the "brain" of a search engine, compared to creative collaboration with an averagely creative person, negatively influences divergent thinking due to its effect on cognitive stimulation. As such, this finding confirms hypothesis 1 in the sense that there is an influence, and it adds that this influence is negative.

To test whether creative collaboration with the "brain" of a search engine, compared to creative collaboration with an averagely creative person, positively influences divergent thinking due to a negative effect on evaluation apprehension (hypothesis 2), two more mediation analyses were conducted using Hayes' bootstrapping method [40]. Again, the model terms were both specified with collaboration type as the independent variable (human collaborator coded: 0, AI collaborator coded: 1), but now with self-reported evaluation apprehension as the mediator. Model 3 was specified with fluency, and model 4 with semantic distance, as the dependent variable. Assumption checks suggested heteroskedasticity in model 4 [41]. Therefore, Huber-White heteroscedasticity consistent standard errors were used to calculate the test statistics for model 4 [42]. No further corrections were thus applied to model 3. The models and unstandardized coefficients are presented in Figs. 4c (model 3) and 4d (model 4). The indirect and direct effects are presented in the text below.

The results of these tests showed no significant indirect effect of the human collaborator, compared to the AI collaborator, on fluency, $b = -.141$, $se = .237$, 95% CI[−.656 .329] nor on semantic distance, $b = -.001$, $se = .002$, 95% CI[−.006, .004] that was mediated by its effects on evaluation apprehension. Furthermore, no significant direct effects were found of the human collaborator, compared to the AI collaborator, on fluency, $b = -.651$, $se = .972$, $p = .504$, 95% CI[−2.573 1.272] nor on semantic distance, $b = -.019$, $se = .011$, $p = .099$, 95% CI[−.041 .004]. Note, however, that the results did show a significant positive effect of the human collaborator, compared to the AI collaborator, on evaluation apprehension in model 3, $b = .251$, $se = .108$, $p = .021$, 95% CI[.038 .464] and in model 4, $b = .251$, $se = .105$, $p = .019$, 95% CI[.043, .460]. These findings suggest creative collaboration with the "brain" of a search engine, compared to creative collaboration with an averagely creative person, negatively affects evaluation apprehension, as expected. However, there is no subsequent effect on divergent thinking. As such, these results only partially confirm hypothesis 2.

The results from models 3 and 4 also suggest no significant moderation of a person's general attitude towards AI of the effects human collaboration, compared AI collaboration, on the fluency and semantic distance of the associations produced by the participants, that was mediated by evaluation apprehension (hypothesis 3). That is, given that no mediation effect was found, there is no effect to moderate. However, because the results from models 3 and 4 did suggest an effect of AI collaboration, compared to human collaboration, on evaluation apprehension, we can test whether this effect is moderated by a person's general attitude towards AI. To this end, a regression model was calculated with collaboration type, the general attitude towards AI, and the product of these two variables (interaction) as the independent variables, and self-reported evaluation apprehension as the dependent variable. The results showed no interaction effect of collaboration type and general attitude towards AI on evaluation apprehension, $b = .200$, $se = .236$, $p = .398$, 95% CI[−.267 .667]. These findings suggest that the effects of creative collaboration with the "brain" of a search engine, compared to creative collaboration with an averagely creative person, on divergent thinking via evaluation apprehension, is not moderated by a person's attitude towards AI. As such, these results do not confirm hypothesis 3.

5 Discussion

The presented study was conducted to take a first look at how creative collaboration with the "brain" of a search engine affects creative task performance in comparison to creative collaboration with the averagely creative person.

The results suggested that creative collaboration with the "brain" of a search engine, compared to creative collaboration with an averagely creative person, influenced divergent thinking due to its effect on cognitive stimulation (hypothesis 1). Specifically, the results indicate that this is a negative effect, meaning that participants who interacted with the AI collaborator, experienced less cognitive stimulation, and produced fewer associations, with a lower average semantic distance, when compared to participants who interacted with the human collaborator. Speculatively, general search engines, such as Google's reverse image search, may retrieve information that is different or weird "in the wrong way" [29] or perhaps just too similar [5]. What stands out, however, is that our routine reliance on AI-powered search engines [6] for our day-to-day creative tasks [7] in 2021, may negatively affect cognitive stimulation and subsequent divergent thinking when compared to creative collaboration with an averagely creative person [26, 27]. Creatives, from laypersons to professional artists, might therefore need to be careful when considering the source of their inspirations. Choosing these types of AI technologies over people for creative collaboration may thus harm creative task performance. At least, from a cognitive perspective.

The results also suggested that creative collaboration with the "brain" of a search engine, compared to creative collaboration with an averagely creative person, negatively influences evaluation apprehension. However, these effects did not subsequently enhance divergent thinking (hypothesis 2). Despite people's common anthropomorphization of AI technologies as collaborators and teammates [5] it was conjectured that AI collaboration would reduce evaluation apprehension. The underlying reasoning was that the limited perceived social agency of these types of technologies would mitigate a common cause of evaluation apprehension, social anxiety [11]. Although confirmation of this particular mechanism is outside the scope of this paper, the results, for now, suggest this might be the case. One possible explanation could be that AI-powered systems might serve as psychological safety net that helps people to be less socially pressured [43–45]. Thus, from a social perspective, these types of human-technology interactions might benefit creative task performance via a reduction of evaluation apprehension. Though note that the former could not be confirmed.

Furthermore, the results did not show that effects of creative collaboration with the "brain" of a search engine, compared to creative collaboration with an averagely creative person, on divergent thinking via evaluation apprehension, was moderated by a person's attitude towards AI (hypothesis 3). Also, further testing confirmed that the effect of collaboration type on evaluation apprehension was not moderated by a person's general attitude towards AI. Thus, in the present study, we could not confirm our speculation that a person's general attitude towards AI introduces a different cause of evaluation apprehension.

The study leaves several unanswered questions that merit future research, partly due to the study's limitations. The basic form of collaboration with the "brain" of Google's reverse image search, e.g., normally happens via its graphical user interface (GUI) [6] whereas we presented its retrieved associations via a text-output field. Interacting with Google's AI via its GUI may affect divergent thinking differently [7]. For example, the associations generated by Google's reverse image search might have been biased in some way as the images were manually translated into words to avoid confounding variables in a later state of the experiment. Yet this might have influenced the original process of interpreting the AI-generated associations by participants. Additionally, the effects on evaluation apprehension might differ from situations that are socially richer than the present study. Although low social richness helped to reduce confounds, because the associations could be presented similarly in both experimental conditions, it also removed the (non-)verbal expressions of others that may worsen evaluation apprehension [11, 31]. Indeed, the average scores on the questionnaire suggested low evaluation apprehension, possibly too low to affect divergent thinking [10]. Future work could therefore focus on the effects of social cues on evaluation apprehension and subsequent divergent thinking, by comparing face-to-face creative collaborations between people and socially rich AI systems such as social robots [10]. Finally, the positive direct effect of creative human-AI collaboration on semantic distance observed in model 2 (Fig. 4b) requires further investigation. It may be that the quality of the associations of information that general purpose AI systems retrieve [8] can stimulate divergent thinking but is not perceived as cognitively stimulating, or its effects may be best explained by other key psychological mechanisms that affect creative collaboration between people, such as social loafing or social disinhibition [9]. This should also be the subject of future research.

Herewith, the present study contributes to an emerging body of work on the efficacy of creative human-AI collaboration by showing that creative collaboration with the "brain" of a search engine, compared to collaboration with an averagely creative person, reduces cognitive stimulation but also evaluation apprehension, and that a person's general attitude towards AI does not introduce a novel form of evaluation apprehension.

References

1. Du Sautony, M.: The Creativity Code: Art and Innovation in the Age of AI. Harvard University Press, Cambridge (2020)
2. Google. https://blog.google/technology/ai/alphagos-ultimate-challenge/. Accessed 2 Mar 2021
3. Miller, A.I.: The Artist in the Machine: The World of AI-Powered Creativity. MIT Press, Cambridge (2019)
4. Christies. https://christies.com/features/A-collaboration-between-two-artists-one-human-one-a-machine-933201.aspx. Accessed 15 June 2021
5. Seeber, I., et al.: Machines as teammates: a research agenda on AI in team collaboration. Inf. Manag. **57**(2), 103174 (2020)
6. Wired. https://www.wired.com/2016/02/ai-is-changing-the-technology-behind-google-searches/. Accessed 2 Mar 2021
7. Zhang, L., Capra, R.: Understanding how people use search to support their everyday creative tasks. In: Conference 2019 on Human Information and Retrieval, pp. 153–162 (2019)
8. Zador, A.M.: A critique of pure learning and what artificial neural networks can learn from animal brains. Nat. Commun. **10**(1), 1–7 (2019)
9. Sawyer, K.R.: Explaining Creativity: The Science of Human Innovation. Oxford University Press, Oxford (2011)
10. Geerts, J., de Wit, J., de Rooij, A.: Brainstorming with a social robot facilitator: better than human facilitation due to reduced evaluation apprehension? Front. Robot. AI **8**. Article 156
11. Diehl, M., Stroebe, W.: Productivity loss in brainstorming groups: toward the solution of a riddle. J. Pers. Soc. Psychol. **53**(3), 497–509 (1987)
12. Scassellati, B., Heny, A., Matarić, M.: Robots for use in autism research. Ann. Rev. Biomed. Eng. **14**(1), 275–294 (2021)
13. Schepman, A., Rodway, P.: Initial validation of the general attitudes towards artificial intelligence scale. Comput. Hum. Behav. Rep. **1**, 100014 (2020)
14. Runco, M.A., Jaeger, G.J.: The standard definition of creativity. Creat. Res. J. **24**(1), 92–96 (2021)
15. Glaveanu, V., Lubart, T., Bonnardel, N., Botella, M., Biaisi, P.D., Desainte-Catherine, M., Zenasni, F.: Creativity as action: findings from five creative domains. Front. Psychol. **4**, 176 (2013)

16. Lubart, T.I.: Models of the creative process: past, present and future. Creat. Res. J. **13**(3–4), 295–308 (2001)
17. Wreen, M.: Creativity. Philosophia **43**(3), 891–913 (2015). https://doi.org/10.1007/s11406-015-9607-5
18. Runco, M.A., Acar, S.: Divergent thinking as an indicator of creative potential. Creat. Res. J. **24**(1), 66–75 (2012)
19. Osborn, A.F.: Applied Imagination. Revised ed. Scribner (1957)
20. Benedek, M., Kenett, Y.N., Umdasch, K., Anaki, D., Faust, M., Neubauer, A.C.: How semantic memory structure and intelligence contribute to creative thought: a network science approach. Think. Reason. **23**(2), 158–183 (2017)
21. Beaty, R.E., Silvia, P.J., Nusbaum, E.C., Jauk, E., Benedek, M.: The roles of associative and executive processes in creative cognition. Mem. Cognit. **42**(7), 1186–1197 (2014). https://doi.org/10.3758/s13421-014-0428-8
22. Benedek, M., Könen, T., Neubauer, A.C.: Associative abilities underlying creativity. Psychol. Aesthetics Creat. Arts **6**(3), 273 (2012)
23. Reiter-Palmon, R.: The role of problem construction in creative production. J. Creat. Behav. **51**(4), 323–326 (2017)
24. Isaksen, S.G., Dorval, K.B., Treffinger, D.J.: Creative Approaches to Problem Solving: A Framework for Innovation and Change. Sage Publications, Thousand Oaks (2010)
25. Benedek, M., Neubauer, A.C.: Revisiting Mednick's model on creativity-related differences in associative hierarchies. Evidence for a common path to uncommon thought. J. Creat. Behav. **47**(4), 273–289 (2013)
26. Kohn, N.W., Smith, S.M.: Collaborative fixation: effects of others' ideas on brainstorming. Appl. Cogn. Psychol. **25**(3), 359–371 (2011)
27. Nijstad, B.A., Stroebe, W.: How the group affects the mind: a cognitive model of idea generation in groups. Pers. Soc. Psychol. Rev. **10**(3), 186–213 (2006)
28. Gallant, S.I.: Neural network learning and expert systems. 3rd edn. MIT Press (1995)
29. Norton, D., Heath, D., Ventura, D.: Finding creativity in an artificial artist. J. Creat. Behav. **47**(2), 106–124 (2013)
30. Gibbert, M., Hampton, J.A., Estes, Z., Mazursky, D.: The curious case of the Refrigerator-TV: similarity and hybridization. Cogn. Sci. **36**(6), 992–1018 (2012)
31. Camacho, L.M., Paulus, P.B.: The role of social anxiousness in group brainstorming. J. Pers. Soc. Psychol. **68**(6), 1071–1080 (1995)
32. Yan, H., Ang, M.H., Poo, A.N.: A survey on perception methods for human–robot interaction in social robots. Int. J. Soc. Robot. **6**(1), 85–119 (2014)
33. Hwang, A.H.C., Won, A.S.: IdeaBot: investigating social facilitation in human-machine team creativity. In: Proceedings of the 2021 CHI Conference on Human Factors in Computing Systems, pp. 1–16 (2021)
34. Nomura, T., Kanda, T., Suzuki, T., Yamada, S.: Do people with social anxiety feel anxious about interacting with a robot? AI Soc. **35**(2), 381–90 (2019)
35. Shin, D.: The effects of explainability and causability on perception, trust, and acceptance: implications for explainable AI. Int. J. Hum.-Comput. Stud. **146**, 102551 (2021)
36. Morris, M.R.: AI and accessibility. Commun. ACM **63**(6), 35–37 (2020)
37. Runco, M.A., Plucker, J.A., Lim, W.: Development and psychometric integrity of a measure of ideational behavior. Creat. Res. J. **13**(4), 393–400 (2001)
38. Beaty, R.E., Johnson, D.R.: Automating creativity assessment with SemDis: an open platform for computing semantic distance. In: Behavior Research Methods 2020, pp. 1–24 (2020)
39. Bolin, A.U., Neuman, G.A.: Personality, process, and performance in interactive brainstorming groups. J. Bus. Psychol. **20**(4), 565–585 (2006)
40. Hayes, A.F.: Introduction to Mediation, Moderation, and Conditional Process Analysis: A Regression-Based Approach. Guilford Publications (2017)

41. Daryanto, A.: Tutorial on Heteroskedasticity using HeteroskedasticityV3 SPSS macro. Quant. Methods Psychol. **16**(5), 8–20 (2020)
42. Long, J.S., Ervin, L.H.: Using heteroscedasticity consistent standard errors in the linear regression model. Am. Stat. **54**(3), 217–224 (2000)
43. Suh, M., Youngblom, E., Terry, M., Cai, C.J.: AI as social glue: uncovering the roles of deep generative AI during social music composition. In: Proceedings of the 2021 CHI Conference on Human Factors in Computing Systems (2021)
44. de Rooij, A., Corr, P.J., Jones, S.: Emotion and creativity: hacking into cognitive appraisal processes to augment creative ideation. In: Proceedings of the 2015 ACM SIGCHI Conference on Creativity and Cognition, pp. 265–274 (2015)
45. de Rooij, A., Corr, P.J., Jones, S.: Creativity and emotion: enhancing creative thinking by the manipulation of computational feedback to determine emotional intensity. In: Proceedings of the 2017 ACM SIGCHI Conference on Creativity and Cognition, pp. 148–157 (2017)

Designing Mobile Tasks to Improve Art Description Accessibility for People with Visual Impairments

Megan Corbett[1], Jeehan Malik[1], Vero Rose Smith[2], and Kyle Rector[1](✉)

[1] University of Iowa, Iowa City, IA 52245, USA
kyle-rector@uiowa.edu
[2] Greenfields Academy, Chicago, IL 60618, USA

Abstract. All people should be able to experience museums, but there are barriers for people with visual impairments (VIs) including few museums that have accessibility accommodations and having to plan their visit. There are museum and technical efforts to supply accessible experiences, but they require curation by experts, making it difficult for these solutions to scale. To address this problem, we used the Art Beyond Sight (ABS) Accessibility Guidelines as a framework to develop mobile tasks to guide laypeople in composing accessible artwork descriptions. We compared the ratings of 31 people with VIs and four docents on curations from Amazon's Mechanical Turk between two approaches: 1) baseline tasks inspired from prior museum HCI research, and 2) our designed tasks. Both people with VIs and docents rated the second descriptions higher than the first in understandability and adherence to the ABS Accessibility Guidelines. The second descriptions vivid details and orientation information. Our work shows the potential to bring these tasks to a museum space.

Keywords: Art descriptions · Blind · Visually impaired · Crowdsourcing · Docents

1 Introduction

All people should be able to experience museums to engage with art, culture, and history. However, there are barriers for people with visual impairments (VIs) including a lack of museums with accessibility accommodations [10, 16]. While the number of museums with accessibility accommodations is increasing, People with VIs have to plan their visit to guarantee the accessible experience [23, 25, 28, 29].

Technology efforts aim to make museum spaces accessible, such as smartphone apps with art descriptions (e.g., [11, 19]). Bluetooth beacons can sense a person's location [3, 26, 30] or depth cameras can sense one's distance [24] to play relevant artwork descriptions. However, these experiences require experts to compose the artwork descriptions. Audio guides are costly to implement (both in cost and staff time). Audio guides that do exist might not conform to best practices as outlined by the Art Beyond Sight (ABS)

M. Wölfel et al. (Eds.): ArtsIT 2021, LNICST 422, pp. 224–247, 2022.
https://doi.org/10.1007/978-3-030-95531-1_16

Accessibility Guidelines [6] because they make the assumption that the person can see the art. For example, an audio description might solely focus on the artist biography and give no information about what the art looks like. There are accessible audio tours, but for a limited set of museums. There are unanswered questions for how to curate these descriptions from people other than curators who are already occupied – for example, laypeople who are already visiting the museum.

Our research investigated how to curate audio guide-worthy content without the need for expert composition by guiding laypeople in composing accessible artwork descriptions. Our multidisciplinary team with Human-Computer Interaction (HCI) researchers and an Associate Curator at an art museum used the ABS Accessibility Guidelines [6] as a framework. We created four tasks (or short text assignments) inspired by HCI research (Baseline Approach) and four tasks inspired by the established ABS Accessibility Guidelines (ABS Approach). ABS is from Art Education for the Blind, which leads a multidisciplinary collaborative of sighted and blind professionals and advisors [7]. ABS Guidelines were developed from theory and research by sighted and blind scholars, professionals, and artists [8]. Our work is the first step toward curating accessible descriptions based on these guidelines.

We included different stakeholders in our research. To understand the feasibility of using artwork descriptions written by museum patrons, we analyzed artwork descriptions written by Mechanical Turk workers (MTurkers). Four docents evaluated each artwork description per the ABS Accessibility Guidelines. 31 people with VIs evaluated the sets of contributions from both approaches in terms of how well they understood each artwork's contents. Both people with VIs and docents rated the descriptions from ABS Approach higher than the Baseline Approach in understandability and per the ABS Guidelines, respectively. People with VIs appreciated the ABS Approach descriptions because they highlighted prominent elements, described layout of the artwork, and made the artwork come alive. The ABS Approach shows potential – by having patrons respond tasks, work by museum employees is reduced from composition to vetting. We show the feasibility of gathering accessible descriptions of artworks through a multidisciplinary process. We make three contributions.

1. We describe the multidisciplinary team's design process of the ABS Approach.
2. We designed and developed tasks through collaborative work between experts in art and HCI to ensure the crowdsourced descriptions were accessible to people with VIs.
3. We conducted an empirical study that compares ratings from 31 people with VIs and four docents of artwork descriptions from MTurkers. The second set of descriptions had higher ratings of accessibility from docents and understandability from people with VIs. They had vivid details and orientation information.

2 Background and Related Work

2.1 Gathering Textual Contributions in Museum Spaces

HCI research has investigated museumgoer engagement by gathering textual responses to artworks. Alelis et al. [4] studied visitor engagement through a task that requested emotional responses to artworks, and found that people were motivated to find personal connections with the artwork. Clarke et al. [15] deployed MyRun, a "participatory platform" with 13 touchscreens as a part of a 3-month exhibition about a famous half marathon. MyRun asked visitors to give stories about the half marathon and collected ~ 13,000 contributions. Cosley et al. [17] deployed MobiTags, a mobile system to improve visitor interaction with exhibits. They allowed users to view and "place" tags on objects throughout an exhibit. They found that people used the tags to form impressions of objects and as navigational tools. Cosley et al. [18] deployed ArtLinks, a standalone computer with keyboard, mouse, and display at a museum exhibit to foster social awareness and reflections. ArtLinks asked users to provide words and short phrases while reflecting on an artwork. Participants liked the social aspects of the interaction and being part of the museum system. Though this research engaged people and collected artwork information, it is unclear whether the descriptions are accessible. Our Baseline Approach is based on this prior research.

2.2 Art Beyond Sight

Art Beyond Sight (ABS) is from Art Education for the Blind who leads a multidisciplinary collaborative of sighted and blind professionals and advisors. ABS is a collaborative of community-based groups, museums, schools, advocacy groups, and people with VIs. Art Education for the Blind was founded in 1987 and their mission is to "make art, art history, and visual culture accessible to people who are blind or visually impaired." They create accessible art programming, materials, advance knowledge in the field, and encourage visually impaired artists. They also partner with museums around the world to increase accessibility [7].

The ABS Guidelines [6] are 16 guidelines spanning from museum information, context of the artwork within the museum, explaining the artworks, to allowing people to interact with the artworks. They were developed from theory and research by sighted and blind scholars, professionals, and artists [8]. We did not focus on "Standard Information" because we provided it (e.g., artist, title, see Fig. 1). We excluded "Focus on the Style" and "Provide Information on the Historical and Social Context" because the associate museum curator determined these guidelines require formal education (e.g., brushwork, history of the artwork). Because the study was electronic, it would be inappropriate to have people "Describe the Importance of the Technique or Medium" because it would involve seeing the art in person. Finally, our study is museum agnostic, so we excluded "Indicate Where the Curators Have Installed a Work" because the artworks were presented as an image in a survey.

We excluded another 4 guidelines because we restricted artwork descriptions to be text: "Incorporate Sound in Creative Ways," "Allow People to Touch Artworks," "Alternative Touchable Materials," and "Tactile Illustrations of Artworks." We worked with the remaining 7 guidelines, possible to address by laypeople:

1. General Overview: Subject, Form, and Color: A general overview of the painting's subject, form, and color is given by presenting visual information in a sequence.
2. Orient the Viewer with Directions: The viewer is oriented with directions using specific and concrete information on the location of objects or figures in the image.

3. Use Specific Words: The description uses specific words and includes clear and precise language that can be taken literally.
4. Provide Vivid Details: Vivid details of different parts of the painting are provided.
5. Refer to Other Senses as Analogues for Vision: Visual experiences are translated into other senses.
6. Explain Intangible Concepts with Analogies: Difficult to describe visual phenomena are explained by using analogies that compare the phenomena to objects or experiences from everyone's common experience.
7. Encourage Understanding through Reenactment: Instructions to mimic a depicted figure's pose are given.

2.3 Crowdsourced Descriptions of Images for People with VIs

Crowdsourcing is a low-cost solution to quickly source verbal descriptions of images for people with VIs. Several research efforts that involve VizWiz [12–14] enable people with VIs to receive information about their environment through crowdsourced responses to photo questions about objects. Burton et al. [14] explored subjective responses to fashion accessibility for people with VIs. This study used volunteers, none of whom were fashion experts. People with VIs trusted the volunteers' responses. UCap is an android application that allows blind consumers to request product descriptions based on images [20]. While these works are successful at obtaining visual information, our contribution is developing tasks with a museum expert to curate art-related accessible information.

2.4 Art and Museum Accessibility for People with VIs

People with VIs do not experience the same level of access to museums as sighted people. While they want to experience museums, planning trips is time consuming [25] and there are limited availability of quality accessible materials [9]. While The Museum of Modern Art [23] and Smithsonian American Art Museum [28, 29] offer accessible tours, people must make appointments or attend on a bimonthly schedule.

Further, several museums do not provide audio descriptions or accessible information on their website. VocalEye's *"State of Museum Access Report of 2018"* [16] studied museum accessibility across the United Kingdom and found that most museums fail to offer adequate online information about accessible services; for example, only 3% of museum websites mentioned *"audio-descriptive guides,"* or audio guides with accessible information. As of April 2020, the American Council of the Blind curated ~ 100 museums, parks, and exhibits with audio description across the US [10].

For public art institutions, there is a lack of funding to implement these solutions. Free platforms could meet accessibility needs, but there is a risk of platforms monetizing or cutting access to content. It is hard to predict long-term costs, complicating budgets. Another barrier is staff time and training, where museums stretch curators with other responsibilities (e.g., fundraising, teaching).

Several research efforts use technology to make museum spaces accessible. Rector et al. [24] created and deployed Eyes-Free Art, which allowed people with VIs to independently explore, immerse in, and engage with art. The researchers behind NavCog [3, 26] and the creators of the Andy Warhol Museum's Out Loud audio guide app [30] used Bluetooth beacons to supply people with VIs with navigation instructions paired with audio descriptions. The Museum of Contemporary Art Chicago developed Coyote, open-source software to curate accessible descriptions of artwork [11, 19]. There are opportunities for technological solutions that do not require experts to compose the content.

Bartolome et al. [21] created a multimodal guide for people with VIs to touch a tactile representation of artwork and give voice commands to hear audio descriptions. Ahmetovic et al. [2] developed MusA, an augmented reality application for people with low vision to frame museum artwork with their smartphone and play a description in "chapters" with visual highlights. Ahmetovic et al. [1] created a touchscreen exploration of artwork to hear attributes or a hierarchical description based on their finger's location. Our research expands upon these works by creating a scalable approach to gather accessible artwork descriptions. Laypeople can participate via mobile device, reducing expert work and cost to implement in a museum space.

3 Artworks Used in Both Approaches

We chose eight two-dimensional artworks from a public domain collection from the University of Iowa Museum of Art ranging in medium, date of origin, and region of origin. Ranging across four centuries, three continents, and multiple complex intersections of movements, styles, and subjects, these works reflect a comprehensive selection of artworks. Further, the artworks did not include violence, nudity, or sexually explicit content. We present each artwork and caption information in[1].

4 Baseline Approach Descriptions

We created four baseline tasks inspired by prior HCI research [4, 15, 17, 18] that engaged patrons in the museum space to see whether these already result in accessible descriptions. We deployed these tasks to MTurk. Then, we had four docents rate the responses using the ABS Accessibility Guidelines. Because the docent ratings were low, we waited to engage with people with VIs until we could curate better artwork descriptions. We iterated on the design of our questions so the ABS Approach would not be redundant with the Baseline approach. Below we present the task designs, method for curating the artwork descriptions, and docent ratings.

[1] Our project had 10 artworks, but removed two due to errors in the survey of people with VIs.

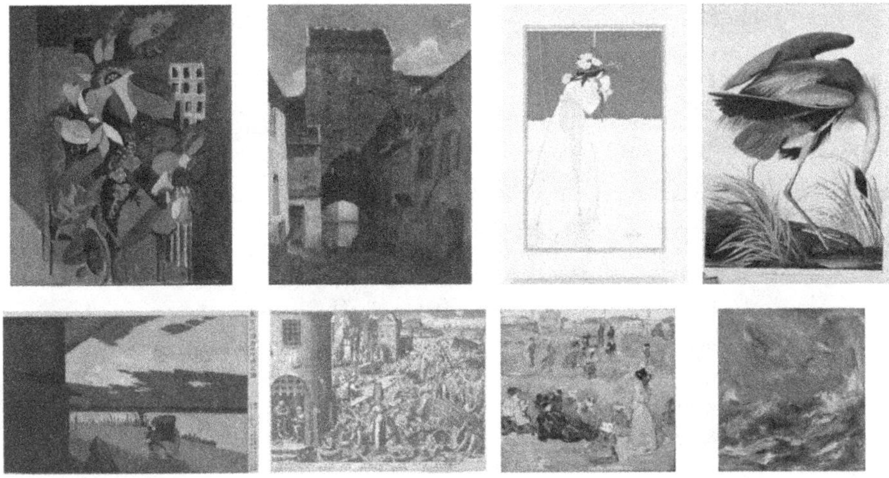

Fig. 1. Our eight selected artworks and captions from top left to bottom right: 1) Agnes Weinrich, Still Life (Sun Flowers), 1921–1926, Oil on canvas, Gift of Henry W. Starker 1973.185; 2) George Henry Yewell, Courtyard and Water Gate, Moret, France, 1856–1861, Oil on canvas, 12 ½ × 9 in. (31.75 × 22.86 cm), Gift of Oscar Coast 1927.21; 3) Aubrey Vincent Beardsley, Isolde, from The Studio, VI, 1896, Chrom-lithograph, 11 1/8 × 7 ½ in. (28.26 × 19.05 cm), Gift of Kenneth J. Oberembt 1983.59; 4) Robert Havell, Great Blue Heron (Ardea Herodias. Male) (after a drawing by J. J. Audubon), 1834, Engraving and aquatint, 38 × 25 ¼ in. (96.52 × 64.14 cm), Estate of Ann U. Morse 2007.56; 5) Kobayashi Kiyochika, Tokyo! Ryogoku Hyappongu Akatsuki No Zu (Dawn by the Hundred Pilings at Ryogoku in Tokyo), July 1879, Woodblock, 9 3/8 × 13 5/8 in. (23.81 × 34.61 cm), Gift of Owen and Leone Elliott 1968.212; 6) Pieter Bruegel, Spes (Hope), plate 2 from The Seven Theological and Cardinal Virtues, published by Hieronymous Cock, c. 1559, Engraving on paper, 8 7/8 × 11 ½ in. (22.54 × 29.21 cm), Museum purchase 1976.16; 7) Maurice Brazil Prendergast, Springtime, 1896–1897, Watercolor and pencil on paper, 9 ½ × 10 ¼ in. (24.13 × 26.04 cm), Gift of Frank Eyerly 1963.1; 8) Lil Picard, Waves, 1957, Oil on Canvas, 36 ¼ × 32 in. (92.08 × 81.28 cm), Lil Picard Collection 2012.209.

4.1 Task Designs

We created four baseline tasks (Baseline Approach) inspired by prior works [4, 15, 17, 18] (Table 1), though the prior works contain more than only crowdsourced descriptions; said works were deployed in physical spaces with in-person interactions. To ensure our layperson contributions resembled the prior works, we replicated each prior work's task in both content and mode of input (i.e., mobile[2], computer[3]). We created four smaller tasks because we wanted to simulate a person's ability to visit artwork for varying time. People could make a single contribution, or if they chose to engage with artwork for a longer period, they could do multiple tasks. We intentionally did not use the ABS

[2] The research in which we based our BL_Emotions task had participants write emotions on paper while they moved around the museum [4]. Therefore, we chose mobile device.

[3] The research in which we based our BL_Story task had people author stories on stationary touchscreens in the exhibition [15]. While the touchscreen dimensions are not mentioned, Fig. 1 in the article shows they are larger than mobile devices. Thus, we chose computer.

Table 1. Baseline (BL) Approach tasks by name, device, and content.

Name	Content
BL_Words/Phrases [18]	"Write words or short phrases reflecting on the work of art displayed above. You may write as many as desired (separate using commas)"
BL_Tags [17]	"Select tags that you feel apply to the artwork above (black, blue, children, circle, clouds, diamond, green, orange, oval, people, pink, play, rain, rectangle, red, snow, square, triangle, white, yellow, none apply); Provide other tags that you feel apply to the artwork by typing them in the box below (separate tags with commas)"
BL_Emotions [4]	"Select the emotions you feel in response to the artwork above (choose all that apply). (anger, disgust, fear, happy, sad, surprise, indifferent, other with a text box)"
BL_Story [15]	"Compose a story about this artwork"

Guidelines in the Baseline (BL) Approach because we wanted to assess the potential of prior HCI approaches to soliciting accessible descriptions.

4.2 Method for Curating Artwork Descriptions

To collect artwork descriptions, we created Qualtrics surveys that had an artwork image, caption information (see Fig. 1's caption), and a task. In Baseline Approach, each artwork had four tasks, so we had 32 surveys. We collected survey responses through Amazon's Mechanical Turk [5], with five MTurkers completing each survey (for redundancy). We informed MTurkers that these descriptions were for people with VIs but did not ask them to follow the ABS Guidelines. We compensated MTurkers for BL_Words/Phrases: \$0.60/task, BL_Tags: \$0.75/task, BL_Emotions: \$0.50/task, and BL_Story: \$1.25/task. Based on average completion times (below), the average hourly rates would amount to \$16.06, \$22.31, \$23.08, and \$19.74, respectively.

We collected these 160 artwork descriptions from 132 MTurkers (demographic information in Table 2). The MTurkers completed a mean of 1.21 tasks, with 112 MTurkers completing 1 task, 14 MTurkers completing 2 tasks, 4 MTurkers completing 3 tasks, and 2 MTurkers completing 4 tasks. We did not filter for colorblindness because museum-goers with colorblindness could provide artwork descriptions. The mean(SD) duration in seconds for MTurkers was BL_Words/Phrases = 134.5(157), BL_Tags = 121(96.5), BL_Emotions = 78(50.3), and BL_Story = 228(350.5).

Table 2. Demographic information for each task in Baseline Approach. All demographics are uncertain due to the anonymity on MTurk. Native/bilingual (NB): *"has complete fluency in the language, including breadth of vocabulary and idiom, colloquialisms, and pertinent cultural references."* Full professional (FP): *"makes only quite rare and minor errors of* pronunciation *and grammar"* and *"can handle informal interpreting of the language."* Professional Working (PW): *"has a general vocabulary which is broad enough that he or she rarely has to search for a word."* Limited Working (LW): *"can usually handle elementary constructions quite accurately but does not have thorough or confident control of the grammar."*

Attribute	BL_Words/ Phrases	BL_Tags	BL_Emotions	BL_Story
Age range (Mean)	21–62 (34.29)	20–63 (32.53)	22–60 (33.26)	22–62 (33.82)
# M/F/Non-binary	24/15/0	25/14/0	29/10/1	26/13/0
# Self-reported Artists	1	3	6	3
Years artist experience range (Mean)	10	13–31 (19.67)	10–26 (14.33)	5–23 (16)
# Museum employees	0	1 with 4 years	0	0
# with colorblindness	0	1 red green	0	1 no answer
Language proficiency	39 NB	39 NB, 1 FP	36 NB, 2 FP, 1 PW, 1 LW	35 NB, 3 FP, 1 LW

4.3 Docent Ratings

To measure the extent to which MTurkers' descriptions follow the ABS Accessibility Guidelines, we recruited four docents (all females, ages 30–62) to rate them. We recruited docents from the Association of Academic Art Museums and Galleries professional listserv and a university museum compiled volunteer docent list. We screened for visual disorders. No docents considered themselves artists. Their experience ranged from 6–28 years with three as educators. We did not assess docent experience with ABS Guidelines, but the docents were qualified to learn and apply the guidelines during our study. First, the target audience of the ABS Accessibility Guidelines includes museum staff [8], so the information is approachable to docents. Second, each guideline is brief, with one paragraph of description followed by an example.

To avoid participant fatigue, we had each docent rate half the descriptions. To mitigate ordering effects, we randomly ordered the artwork descriptions and had the docents begin from different starting points, such that two docents responded to each description. We collected 2240 ratings.

We met each docent via Zoom Video Communications because we wanted to facilitate them visiting the ABS Accessibility Guidelines before starting the survey. After verbal consent, the docent read the ABS Accessibility Guidelines [6]. Then, they rated each artwork description using each of the seven ABS Accessibility Guidelines on a

5-point Likert scale from *"Strongly Disagree"* to *"Strongly Agree."* We encouraged the docent to take breaks during the survey. Not including training time, docents completed the ratings in 00:42:29, 00:29:47, 01:08:01, and 00:50:25. Due to the synchronous format and length of the sessions, we compensated each docent $20. The docents rated the Baseline Approach artwork descriptions low; only 0.27% of the ratings were at least a 4, where 5 is the best possible score.

5 ABS Approach Descriptions

5.1 Task Designs

Since the artwork descriptions from the Baseline Approach were inaccessible, our team designed four tasks to better fulfill the ABS Accessibility Guidelines (ABS Approach, Table 3).

Table 3. ABS Approach names, targeted guidelines, and content. All tasks were mobile. ABS_Reenact applied to artworks 3 & 5–7.

Name: Guidelines	Content
ABS_General: 2,3	We wanted the three most prominent elements, so the below text was repeated three times "Pick the most prominent element of this painting. Tap the image to place dots around the element to outline or bound it. Use at least 3 dots (you can use up to 10 dots). Once a dot is placed, it can be moved by touching and dragging it. Do not hold your finger down on the image; just tap the image Write a description of this element"
ABS_Literal: 3, 6, 7	We wanted the three most prominent elements, so the below text was repeated three times "Pick the most prominent element of this painting. Tap the image to place dots around the element to outline or bound it. Use at least 3 dots (you can use up to 10 dots). Once a dot is placed, it can be moved by touching and dragging it. Do not hold your finger down on the image; just tap the image Write a literal description of this element that does not include emotion or opinion"
ABS_Reenact: 3, 11	"Pick one subject or aspect of this painting. Tap the image to place dots around the subject or aspect to outline or bound it. Use at least 3 dots (you can use up to 10 dots). Once a dot is placed, it can be moved by touching and dragging it. Do not hold your finger down on the image; just tap the image Imagine that you have to describe the selected aspect of the painting using a human pose or movement. Write step-by-step instructions for how another person would hold that pose or perform the movement. You can be as general or specific as you would like"
ABS_Senses: 9, 10	"Imagine you are in the painting above. How would you describe your surroundings using touch, taste, smell, or sound?"

The authors who are HCI researchers were a graduate student and advisor in Accessibility. The advisor worked directly with people with VIs for 8 years with prior experience in artwork accessibility. The associate curator was an art professional with 8 years of experience in public institutions, including art museums. Their research includes access to the arts. When we designed ABS Approach, we had 3 considerations.

1. We experienced a design tension between the collaborators. The authors who identify as HCI researchers wanted creative responses from laypeople. However, the associate curator's concern was that laypeople are more likely to give inaccurate information. Thus, in the ABS_Literal task, we told the MTurkers to exclude emotion and opinion. In the ABS_Senses task, we allowed creative responses.
2. The research team studied the few MTurker contributions that scored at least 4/5 compared to the contributions with lower scores. Clear language, facts, and inclusion of absolute or relative positions of elements resulted in more accessible descriptions - in line with ABS Accessibility guidelines. Therefore, in all tasks except for ABS_Senses, MTurkers could draw outline around the elements they discussed, which we converted to descriptions that included position (Sect. 4.2).
3. We determined that unambiguous language in our task prompts could help MTurkers better answer the questions. Therefore, we included *"subject," "aspect,"* and *"element"* so that MTurkers could respond depending on the targeted guideline and level of abstractness of an artwork. The ABS_General and ABS_Literal tasks use "elements" because we wanted MTurkers to select objects and regions, regardless of whether they are literal or abstract. ABS_Reenact uses *"subject"* or *"aspect"* to cue selections that have a human form.

5.2 Curating Artwork Descriptions

We followed the same procedures as the Baseline Approach. ABS_Reenact applied to half the paintings that had a person or people. Therefore, instead of 32 surveys as in Baseline Approach, we had 28. ABS_General, ABS_Literal, and ABS_Reenact tasks involved tapping to place dots around the element, subject, or aspect of the artwork that a person would describe. To implement these dots, we used the Qualtrics Heatmap question. We collected the coordinates of each dot to programmatically convert it to region descriptions, which we used to augment the artwork descriptions.

We compensated workers $1.25 for each ABS Approach task to be the same as BL_Story. Based on average completion times (below), the average hourly rates would amount to $19.27, $17.54, $15.52, and $28.57. With 28 surveys for ABS approaches, we collected 140 artwork descriptions from 98 MTurkers (demographic information in Table 4). MTurkers completed a mean of 1.43 tasks with 73 MTurkers completing 1 task, 14 did 2 tasks, 8 did 3 tasks, and 1 did 4, 5, and 6 tasks, respectively. There was an overlap of 11 MTurkers between the two approaches. The mean(SD) duration in seconds for MTurkers was ABS_General = 233.5(140.5), ABS_Literal = 256.5(224.5), ABS_Reenact = 290(276.8), and ABS_Senses = 157.5(133).

Table 4. Demographic information for each task in ABS Approach. ABS_Reenact only applied to half the paintings, and therefore has approximately half the workers.

Demographic	ABS_General	ABS_Literal	ABS_Reenact	ABS_Senses
Age range (Mean)	19–58 (31.6)	19–55 (31.7)	22–50 (31.16)	19–62 (31.86)
M/F/Non-binary	26/12/0	25/12/0	12/7/0	24/13/0
# Self-reported artists	7	3	2	1
Years artist experience range (Mean)	4–35 (19)	8–30 (16)	0.25 & 10	20
# Museum employees	0	0	1 w/6 months	0
Language proficiency	37 NB	35 NB, 1 PW	17 NB, 2 FP	36 NB, 1 FP

5.3 Docent Ratings and Comparison to Baseline Approach

We had the same four docents, compensated $20 again, rate the new descriptions. We chose a within-subjects study to compare within each docent. While expectation bias is possible, a 3-month break between docents rating the first descriptions (May-Jun. 2019) and second descriptions (Aug.-Sep. 2019) made it unlikely that they would recall their prior ratings. ABS_Reenact applied to only half the artworks, so there were 1960 ratings. Not including training time, docents completed the ratings in 00:43:34, 00:37:56, 01:49:35, and 01:04:42. The docents rated 6.61% of ABS Approach artwork descriptions 4/5 or above.

Table 5. The guideline number and statistical tests for Task-Artwork and Task. **All statistics have $p < 0.001$**, where p values multiplied by 28

Guideline #	2	3	6	7	9	10	11
Task-Artwork	167.07	190.6	154.05	184.11	199.77	113.71	184.45
Task	140.09	171.16	121.93	161.3	182.78	85.967	168.8

Comparing between all tasks, we note the Task-Artwork interaction had a statistically significant effect on docent ratings, but Artwork did not have a statistically significant effect. Therefore, we focus on Task, which influenced docent ratings for all guidelines (Table 5). ABS Approach outperformed Baseline Approach in terms of the ABS Accessibility Guidelines. The Appendix has tables showing pairwise differences. We describe three high-level findings below. The descriptions highlighted in the findings were chosen based on the highest ratings from docents.

Responses to ABS_General and ABS_Literal are More Well-rounded than Other Tasks. Docents rated artwork descriptions from the ABS_General and ABS_Literal tasks higher than other tasks. ABS_General and ABS_Literal were rated higher than all tasks for three guidelines (#2, #3, and #7). ABS_General and ABS_Literal were rated higher than Baseline tasks for three guidelines (#9, #10, and #11). Finally, ABS_Literal was rated higher than all tasks and ABS_General was rated higher than Baseline tasks for one guideline (#6). For instance, in artwork 7, both docents rated the following MTurker's response a 4/5. The MTurker described the relative locations and orientations of the people and elements:

"On the bottom left: There is a woman lying down in a grass field wearing a black dress. Her left arm is tucked underneath her body, propping her up [off] the ground. She is also wearing a black hat that has a white ribbon wrapped around it. You cannot see her face because she is looking towards a city, so you are seeing [t]he back [of] her.

On the right side: There is a woman wearing a white dress with polka dots in the same field as the other woman. She is standing instead of lying down. The dress has long sleeves, and her hair is styled. She is also wearing a small black hat with a red ribbon. Her hair is a light brown color, and her skin is fair.

On the bottom center: In between these two women is a little girl that is seated. She is wearing a red-orange dress with a white hat on. You cannot see her face because she is turned away from you."

Docents rated this Artwork 7's description 1/5 and 2/5, which was briefer:

"On the right side: The lady standing on the left side of the foreground.

On the bottom center: The crowd in the background.

On the bottom: Lady sitting on the right side of the foreground".

ABS_Senses Strong in Refer to Other Senses as Analogues for Vision and Addressed Other Guidelines. Docents rated the artwork descriptions from the ABS_Senses task as higher than all other tasks. For example, Artwork 6 had two contributions rated by both docents as 4/5: *"I can smell and taste the salty ocean air all around me. I feel my feet rest firm[ly] on the hard[-]stone ground of the platform I am standing on. I hear the tumultuous waves crashing towards me and the chaos of men on wooden boats that seem to be capsizing. I can feel the gritty stone walls of the tower beside me. I smell the stink and hear the groans of prisoners, laborers, and beggars around me."*

> *"It smells of human sweat and dirt mixed with rusted metals. You can taste the salty sea air as your feet stomp along on the pier. The sound of the waves crashing does little to mask the hustle and bustle of the town nearby."*

Further, ABS_Senses responses were rated higher than all Baseline Approach tasks for two guidelines (#7 and #11). ABS_Senses responses were rated higher than BL_Words/Phrases for two guidelines (#6 and #10).

ABS_Reenact Best Fulfills the Encourage Understanding through Reenactment guideline. ABS_Reenact performed better than all other tasks, because it enabled clear descriptions when people were in the artwork. Both docents rated this description as a 5/5 for artwork 3: *"Bend your back forward slightly. Then bring your hands to your face as if you're holding a bowl of warm soup to take a sip."* The docents rated this Artwork 3's description 1/5 and 2/5: *"The wine curtain that serves as a wall. The human lays close to the curtain and remains standing."*

6 People with Visual Impairments' Ratings of and Comments on Artwork Descriptions

Once we had artwork descriptions that better met the ABS Accessibility Guidelines, we gathered ratings and justifications on the understandability of artwork descriptions from people with VIs. We conducted an unsupervised Qualtrics survey because they were answering questions based on their opinion; no initial review of guidelines was needed like with the docents. 31 people with VIs (9 males, 22 females) ages 19–68 mean(SD) = 40.2(15) filled out the survey, labeled P1-P31. Five were artists (from 2–45 years of experience), 23 were not artists, and three did not specify. No one considered themselves a museum employee. Thirteen participants were totally blind from birth, and another eight were totally blind from 10–56 years mean(SD) = 23(14.31). Two were legally blind from birth, and another four were legally blind from 1.5–29 years mean(SD) = 13.63(11.73). Two had low vision since birth, and another had low vision for ten years. Finally, one had a degenerative condition since childhood and cannot discern details. We wanted to pay people with VIs at the same rate of docents., so we used $5 Amazon gift cards, predicting the surveys would take < 30 min.

First, our survey listed the purpose of the study, which was "*... to determine if written descriptions of artwork provided by sighted people are useful to people who have a visual impairment.*" After agreeing to the study, people with VIs rated descriptions for the 8 artworks. Our survey had two pages per artwork. These pages were presented in a random order to offset the learning effect; specifically, half the artworks (i.e., 3–6) showed the collection of Baseline Approach descriptions first, and half the artworks (i.e., 1–2, 7–8) showed the ABS Approach descriptions first. For completeness, we wanted people with VIs to evaluate all descriptions, so they were shown regardless of redundant or contrary content. The survey did not mention that different pages pertained to different approaches. On each page, we presented the artwork, its metadata, and the collection of descriptions from either the Baseline Approach or ABS Approach[4]. The metadata included artist, title, year, medium, dimensions, and how the artwork was acquired; refer to Fig. 1's caption for this information. We asked: "*On a scale of 1 to 5, where 1 is Strongly Disagree to 5 is Strongly Agree, rate how much you agree with the following statement: I am able to understand most elements or objects of this artwork from the provided descriptions.*"

Then, we asked them to "*Write one sentence explaining the rating that you selected in the previous question.*" Participants completed the survey at a minimum of 00:07:44, a maximum of 1 day + 19:16:41, and a median of 01:19:41, which may (we cannot know) have included interruptions.

To assess the difference between people with VI ratings for Baseline versus ABS approaches while controlling for differences in artwork and participant demographics, we used a Linear Mixed Model. Artwork and approach were repeated variables. Artwork, approach, and artwork * approach were fixed effects, while participant, age, gender, and level of vision were random effects. Whether the person was an artist was considered a redundant covariate and therefore not included in the analysis. We found that approach

[4] The ABS Approach Artwork 3 had the descriptions but was missing the relative positions for ABS_General and ABS_Literal descriptions.

influenced *"I am able to understand most elements or objects of this artwork..."* ($F(1) = 82.8$, $p < .001$). They gave higher ratings when responding to descriptions from ABS Approach (mean(SD) = 3.98(1.05)) than descriptions generated from the Baseline Approach (mean(SD) = 3.25(1.24)).

We conducted a qualitative analysis of the optional justifications that people gave. We coded based on the ABS Accessibility Guidelines, the tasks themselves, or other comments from people with VIs. We iterated on our codebook for the first 60% of responses, and two researchers finished coding the remaining 40% responses by dividing in half.

6.1 Qualitative Findings for Baseline Approach

Overall, people with VIs had more negative (155) than positive (132) comments about the Baseline Approach descriptions. There were multiple reasons for criticism. First, 46 comments related to the descriptions not being vivid. P19 commented on artwork 4's description: *"... I have no idea on what the bird is doing or what the scenery looks like outside the fact that there seems to be some kind of lake involved."* The description for artwork 2 had P2 asking follow-up questions: *"How tall is the building; from what angle do we view it? No people?"* Second, participants spoke to flaws relating to the General Overview guideline, with participants not understanding what was occurring in the artwork (n = 38). For instance, with artwork 4, P13 stated that they needed *"... more physical descriptions about what is actually happening."* Third, P13 noted contradictions between the descriptions: *"The [descriptions] varied so much that it was hard to tell what was actually going on."*

BL_Story Best of Baseline Approach. Out of the 132 positive comments about Baseline Approach's descriptions, 51 were related to the BL_Story descriptions. For instance, P22 reflected on an MTurker's story for artwork 1: *"I loved the point of view of the person who said it was a painting of vibrant flowers against a dull background, reflecting the title, Still Life. I could distinguish the contradiction between the vibrant life of the flowers and the dullness beyond."*

Participants appreciated the descriptions from BL_Story for reasons including that *"the stories make the artwork come alive"* (P16, artwork 2) or that P24 *"was able to experience this picture through the stories"* (artwork 6). Second, the artwork descriptions had specific words to add more details: *"I love the description of the meadow and grass, and I can picture a warm spring breeze as families play at the park; I also love how one individual used the descriptive word energizing to describe the weather"* (P22, artwork 7).

Participants did have negative comments about BL_Story descriptions (n = 18). Stories lacked specific information about the layout and location of objects or figures in the artworks. P23 said, *"I need more description about what is happening in each part of the painting and in the painting as a whole not composed in a story."*

Other Baseline Tasks Less Useful. Participants made only 11 positive statements about BL_Words/Phrases. While words or phrases were helpful: *"Strong words like peaceful and waves bring me back to laying down by the ocean."* (P29, artwork 8), overall, the short descriptions left participants confused, with 14 negative comments. Participants felt that the words or phrases *"are not very descriptive"* (P17).

Participants had mixed feelings toward descriptions from BL_Tags, with 17 positive comments and 16 negative comments. Participants noted that the tags were understandable. For instance, in artwork 6, P7 said: *"I found the tags useful, at least in terms of getting a sense of what colors and what some of the elements it might have been."* But participants said that the tags were not helpful. More specifically, P17 felt that the *"tags are very short and generic."* Further P9 *"found the tags confusing because there was a lot of repetition."*

Participants had the fewest positive comments about emotions (n = 7). People could see the benefit; for instance, on artwork 5, P16 said *"Basic elements are understandable, the emotions and stories are helpful in relating what I would see and feel."* However, participants gave 19 negative comments, due to being confusing. MTurkers could choose multiple emotions, and this led to contradictions: *"Emotional responses are confusing with repeated tags and things like "angry, happy, sad" all on the same line, followed by happy elsewhere. What is this supposed to mean?"* (P9, artwork 5).

6.2 Qualitative Findings for ABS Approach

The positive comments given about descriptions from the ABS Approach (n = 179) outweighed the negative (n = 86). Unlike Baseline Approach, people had more positive comments related to the general overview (#2), orient (#3), and vivid (#7) guidelines. For instance, P7 made a comment related to general overview: *"The descriptions of elements, the pose descriptions[,] and the sensory evocations all work together to give me a really good sense of what this is an image of and what emotions it evokes."*

There were 29 comments related to the helpfulness of orienting the reader with specific layout descriptions. For instance, P17 said artwork 5's descriptions *"did a good job with describing the positioning of the different elements of the painting."* Further, P13 spoke to how the layout helped them visualize the artwork: *"The details at the beginning were very helpful, as I was able to understand the layout of the painting, which helped me visualize how a sighted person would see it."*

Speaking to the vivid guideline, P22 stated that they were *"... able to identify each individual part of the painting and imagine the sun reflecting off the water, with room for imagination too"* (artwork 2). With that said, the vivid guideline had the most negative comments (n = 29). The descriptions also could leave participants asking follow-up questions. For example, while P13 knew aspects of artwork 2, they also asked questions: *"I understand that there is an arch and a window that might be rundown, along with a building that may be older, but don't know if there is grass, gravel on the ground, if there is a staircase leading down, or if the window is part of the archway."*

ABS_General. Participants spoke positively about the descriptions from ABS_General (n = 26) far more than negatively (n = 2). P17 stated that for artwork 6 the descriptions *"did a good job of indicating the most prominent features of the painting."* Further, for artwork 5, P7 said: *"The description of elements was incredibly useful, giving me concrete items to imagine and the sentence descriptions were also really useful here* *emphasizing the importance of the sun in the image."* The two negative comments were saying descriptions from ABS_General were less useful than the ABS_Literal or ABS_Reenact descriptions.

ABS_Literal. Participants gave the most positive comments about ABS_Literal descriptions (n = 40) as opposed to only 3 negative comments. The three negative comments wished for more details about the artwork or mentioned the ABS_Reenact descriptions being more useful. P2 said the descriptions from ABS_Literal was *"essential for [artwork 7]."* Regarding artwork 3, P2 said the *"the literal descriptions were the most comprehensive."* More specifically, for artwork 3, P17 said: *"I liked the description of what the woman was wearing and how she was positioned."*

ABS_Reenact. Participants appreciated the descriptions of reenacting a depicted figure, with 14 positive comments versus 3 negative comments (this task pertained to only half the artworks). Participants understood the value of the reenactments when people were in the artwork, with P2 saying about artwork 6, *"The reenactments were most helpful since the woman is the focal part of the picture."* P19 commented that they *"...really like the description of the painting the highlighted the woman, her features, and her pose against the background."* P7 further appreciated the instructions to hold the posture themselves: *"Very clear descriptions of the figures in this piece of artwork, and the instructions as to how to simulate those figures postures were really useful in helping me understand what was presented."*

The negative comments were about artworks 3 and 5, where people with VIs felt the descriptions from ABS_Reenact were unhelpful. P24 said, *"I thought that the [reenact] descriptions detracted from the description, in this case,"* and P16 said, *"the descriptions of body movements are interesting, but not as helpful as actual story scenarios."*

ABS_Senses. Participants wrote 21 positive and 5 negative comments about senses. P24 spoke to artwork 1 *"coming alive": "The statements about the senses made the description more alive, and they agreed with each other enough so that I felt as though I understood the basic meaning of the picture."* P22 wrote about artwork 4: *"I loved how some of the descriptions discussed using one's sense of hearing to enjoy the scenery in the artwork, and I was easily able to follow what was happening."*

Three of the negative comments were saying the ABS_Senses descriptions as unnecessary: *"Description of senses experienced is not necessary, [a] blind viewer can fill in with their own emotional interpretation"* (P17).

6.3 Suggestions for Improvement

Participants gave 53 suggestions for improvement across both approaches. 26 these comments asked specific follow-up questions about the artworks with another 16 saying that they wanted more details. P18 suggested a better ordering of the descriptions: *"Why can the description not be[]given in an order that is more comprehensive instead of skipping all over and back again? Too confusing! But the thoughts about the setting are fantastic!"*.

7 Discussion

7.1 Limitations

While we took careful measures to curate layperson artwork descriptions informed by Art Beyond Sight guidelines, we acknowledge that our study has limitations. It is possible that the level of pay had an influence on MTurker responses, but we opted to be fair in terms of predicted time per task; we did not want to financially disadvantage workers who got longer, more time-consuming tasks. There were MTurkers who completed multiple tasks within one approach, and multiple tasks across approaches. Our goal was to understand the potential quality of descriptions in a museum setting; it is within the realm of possibility that patrons give multiple descriptions for multiple artworks, along with others who make a chance encounter supplying only one description. While people could get better with practice, we did not give feedback on the content of their descriptions.

We chose an online study to assess the potential of this approach before deploying to a museum. These results may differ from in-person settings: MTurkers were paid (patrons would not be paid), MTurkers may be less invested in the artwork than patrons, and MTurkers experienced artworks online rather than in person. These differences may have unequally affected the tasks with respect to docent and people with VIs. For example, MTurkers may have composed better stories because they had to think more about the artwork than labeling with tags. Future work is needed to confirm these findings for in-person.

We did not ask for docent familiarity with the ABS Guidelines. However, these guidelines are meant for docents, we gave the docents time at the beginning of the session to familiarize with the guidelines, and we reminded them of the definition in every question. While docent ratings were low overall, Baseline Approach had only 0.27% of scores of at least 4/5 (5 being the best), while ABS Approach had 6.61%.

We did not ask for people with VIs' familiarity with the artworks, but these artworks were randomly selected to represent a wide range of ages and styles. While our goal was to assess their understanding of the artwork based on MTurker contributions, we did not give a ground truth description. Despite this limitation, we still observed a statistically significant difference in their responses. Finally, we did not capture how long survey sessions took for people with VIs because we did not account for breaks.

7.2 Lessons Learned and Resulting Guidelines

Our goal was to develop and evaluate a novel approach for laypeople to generate accessible descriptions for people with VIs. Quantitatively, descriptions generated by ABS Approach were rated more highly and received more positive than negative comments, while the reverse was true for Baseline Approach. Qualitatively, with Baseline Approach, descriptions had insufficient details, while ABS_General and ABS_Literal led to helpful information about layout and orientation. The ABS_Literal descriptions make the artworks more vivid, which was not achieved by Baseline Approach. ABS_Reenact gave a new dimension that otherwise may have been missed by contributors, encouraging the descriptions to be more specific, particularly about human figures in the artwork. Finally, a positive aspect that arose in both approaches was that BL_Story and ABS_Senses made the artwork come alive.

While MTurkers spent longer completing ABS Approach than Baseline Approach tasks, people with VIs gave them higher scores in terms of understandability and docents rated those descriptions as higher per the ABS Accessibility Guidelines. We confirm a tradeoff in time needed to complete each task versus the accessibility of the description. One hypothesis is that MTurkers did not have to supply as comprehensive of responses to tasks from the Baseline Approach (except for BL_Story). Further, taking creative liberty is not acceptable for audiences looking for facts. This was a tension while we designed the ABS Approach tasks – the museum is a trustworthy institution that does not want to risk losing patron trust due to inaccurate descriptions [27]. People with VIs said the BL_Story was *"silly"* or the BL_Words/Phrases and BL_Tags were *"not useful." We recommend that artwork descriptions are curated via tasks grounded in ABS Guidelines. Further, it is important to disclose to patrons that descriptions were collected from other museumgoers.* We raise further questions: how do we allow patrons to answer questions if they are quickly passing an artwork? How do we allow creativity while gathering factual descriptions?

Further, we uncovered that two of our ABS Approach tasks better fulfilled ABS Guidelines than the Baseline Approach; two were more focused on a singular guideline. This coverage is beneficial, because it allowed MTurkers to focus on one concept at a time, and combining the statements together helped with understandability. Therefore, *we recommend multiple tasks, where some approach the artwork from a high level and other tasks approach from a low level.*

7.3 Future Work: Moving from Online to Museum

While these results show the potential of improving artwork descriptions for people with VIs in museums, there are opportunities for future research. First, there is an opportunity to vote on the best descriptions via a collaboration between patrons and museum curators. There could be incentives for patrons including virtual awards for popular descriptions (much like incentives for Google Local Guides for Google Maps [31]). Finally, a curator and/or accessibility expert could do a final proofread of the best voted descriptions before they become publicly available. This reduces the level of effort for a museum employee from creation to vetting.

Second, our designed tasks are virtual, so one should deploy them in a museum. Patrons might notice different details—the size, texture, and finer aspects of the material reality of art objects; these are hard to convey through digital images. While we used the term "painting" for the online tasks, which is different from "artwork," MTurkers and docents had the same experience: viewing a 2D image. Our caption information described the medium and method, but people did not experience it personally. Further, people in museums are in a formal setting, answering questions about artwork physically in front of them, so they are less anonymous. These factors can influence the statements we receive.

Third, a risk is that deploying this technology to the museum could make accessibility an afterthought. Curators should collaborate with patrons to make the descriptions. Crowdsourcing works toward another goal of museums: teaching audiences to look deeply at art (e.g., [22]). By creating experiences that guide novice visitors through the process of visually analyzing artworks, we achieve this pedagogical goal and aid people with VIs. Our work shows that scaffolding this task is difficult but possible. Future research must learn about the types of descriptions gathered in the museum and measure their accessibility compared to descriptions gathered online.

Fourth, we should explore how to present statements from patrons. For our survey of people with VIs, we included all MTurker artwork descriptions from each approach. However, we found that user contributions differed in content and quality and people with VIs did not always prefer the statement ordering. There are opportunities to explore how to effectively present these statements. User interfaces could present statements in order from most to least prominent elements, regions in clock notation, or moving from general descriptions to detailed descriptions about specific elements.

Finally, there are open questions for the experience of interacting with the written descriptions. A system could play statements through bone-conduction headphones serially or based on user choice. We could present descriptions via a proxemic interface where the user hears more detailed descriptions as they move toward or spend longer with the artwork [24]. This interaction could be physical (via user position) or phone-based using VoiceOver selections.

8 Conclusion

We designed and implemented tasks to help laypeople compose more accessible descriptions of artwork than prior HCI research. Through our framework of using the ABS Accessibility Guidelines and our multidisciplinary team, we were able to curate descriptions from MTurkers that 31 people with VIs and 4 docents rated higher than the descriptions from Baseline tasks. Integrating poses, senses, and orientation with the descriptions of elements allowed people to visualize the artwork and brought them to life. We hope our work will help researchers interested in accessible art exploration and who want to curate artwork descriptions at a larger scale from laypeople.

Appendix

General Overview: Subject, Form, and Color

Table 6. * = p < 1.79e–3, ** = p < 1e–6, where significance is alpha/28

Versus	Tags	Emotions	Story	General	Literal	Reenact	Senses
Words/Phrases	694.5	1000*	595.5	195.5**	142**	417	570.5
Tags		1100*	701.5	250.5**	199**	468	691
Emotions			400**	80**	20**	320	360**
Story				318*	267.5**	517	794.5
General					783.5	713**	1318.5**
Literal						740**	1371**
Reenact							272

Orient the Viewer with Directions

Table 7. * = p < 1.79e–3, ** = p < 1e–6, where significance is alpha/28

Versus	Tags	Emotions	Story	General	Literal	Reenact	Senses
Words/Phrases	800	800	700	120**	140**	300*	480*
Tags		800	700	120**	140**	300*	480*
Emotions			700	120**	140**	300*	480*
Story				223.5**	259**	360	611
General					901.5	697.5*	1334**
Literal						677.5*	1281**
Reenact							337.5

Use Specific Words

Table 8. * = p < 1.79e–3, ** = p < 1e–6, where significance is alpha/28

Versus	Tags	Emotions	Story	General	Literal	Reenact	Senses
Words/Phrases	740	1000*	835	300**	165**	300	440*
Tags		1040*	890.5	364*	237.5**	334	519.5
Emotions			640	180**	80**	220*	280**
Story				285**	159.5**	288	419*
General					705.5	562	1016
Literal						628*	1163.5*
Reenact							331.5

Provide Vivid Details

Table 9. * = p < 1.79e–3, ** = p < 1e–6, where significance is alpha/28

Versus	Tags	Emotions	Story	General	Literal	Reenact	Senses
Words/Phrases	820.5	840	761	198**	158**	342	447*
Tags		820	739.5	171.5**	130.5**	330	417.5*
Emotions			720	160**	120**	320	400**
Story				221**	179**	362	482*
General					764.5	674*	1153*
Literal						698*	1211*
Reenact							270

Refer to Other Senses as Analogues for Vision

Table 10. * = p < 1.79e–3, ** = p < 1e–6, where significance is alpha/28

Versus	Tags	Emotions	Story	General	Literal	Reenact	Senses
Words/Phrases	819	840	760	377**	358**	339	29**
Tags		820	742	370.5**	352**	331.5	36.5**
Emotions			720	340**	320**	320	20**
Story				414*	396*	358	38**
General					794	544	155**
Literal						551	140**
Reenact							40**

Explain Intangible Concepts with Analogies

Table 11. * = $p < 1.79e{-3}$, ** = $p < 1e{-6}$, where significance is alpha/28

Versus	Tags	Emotions	Story	General	Literal	Reenact	Senses
Words/Phrases	800	820	718.5	379.5**	338.5**	309.5	477.5*
Tags		820	718.5	379.5**	338.5**	309.5	477.5*
Emotions			700	360**	320**	300*	460*
Story				491*	449*	353.5	573
General					738.5	511	852.5
Literal						534	904.5
Reenact							328.5

Encourage Understanding Through Reenactment

Table 12. * = $p < 1.79e{-3}$, ** = $p < 1e{-6}$, where significance is alpha/28

Versus	Tags	Emotions	Story	General	Literal	Reenact	Senses
Words/Phrases	800	800	800	360**	320**	0**	500*
Tags		800	800	360**	320**	0**	500*
Emotions			800	360**	320**	0**	500*
Story				360**	320**	0**	500*
General					772	46.5**	947.5
Literal						48**	980
Reenact							770**

References

1. Dragan Ahmetovic: Touch Screen Exploration of Visual Artwork for Blind People (2021). http://dragan.ahmetovic.it/pdf/ahmetovic2021touch.pdf
2. Ahmetovic, D., Bernareggi, C., Keller, K., Mascetti, S.: MusA: artwork accessibility through augmented reality for people with low vision. In: Proceedings of the 18th International Web for All Conference (W4A 2021), pp. 1–9 (2021). https://doi.org/10.1145/3430263.3452441
3. Ahmetovic, D., Gleason, C., Ruan, C., Kitani, K., Takagi, H., Asakawa, C.: NavCog: a navigational cognitive assistant for the blind. In: Proceedings of the 18th International Conference on Human-Computer Interaction with Mobile Devices and Services (MobileHCI 2016), pp. 90–99 (2016). https://doi.org/10.1145/2935334.2935361

4. Alelis, G., Bobrowicz, A., Ang, C.S.: Exhibiting emotion: capturing visitors' emotional responses to museum artefacts. In: Marcus, A. (ed.) DUXU 2013. LNCS, vol. 8014, pp. 429–438. Springer, Heidelberg (2013). https://doi.org/10.1007/978-3-642-39238-2_47
5. Amazon Mechanical Turk: Amazon Mechanical Turk. https://www.mturk.com/. Accessed 24 Feb 2018
6. Art Beyond Sight. AEB's Guidelines for Verbal Description. http://www.artbeyondsight.org/handbook/acs-guidelines.shtml. Accessed 16 Feb 2019
7. Art Beyond Sight: About Art Education for the Blind. http://www.artbeyondsight.org/sidebar/aboutaeb.shtml. Accessed 3 Aug 2020
8. Art Beyond Sight: How Were These Tools Developed? Theory and Research. Retrieved August 3, 2020 from http://www.artbeyondsight.org/handbook/acs-toolsdeveloped.shtml. Accessed 3 Aug 2020
9. Asakawa, S., Guerreiro, J., Ahmetovic, D., Kitani, K.M., Asakawa, C.: The present and future of museum accessibility for people with visual impairments. In: Proceedings of the 20th International ACM SIGACCESS Conference on Computers and Accessibility - ASSETS 2018, pp. 382–384 (2018). https://doi.org/10.1145/3234695.3240997
10. Audio Description Project: American Council of the Blind. 2020. Museums Which Offer Audio Description. http://www.acb.org/adp/museums.html. Accessed 16 Apr 2020
11. Bahram, S., Lavatelli, A.C.: Using Coyote to Describe the World – MW18: Museums and the Web 2018. https://mw18.mwconf.org/paper/using-coyote-to-describe-the-world/. Accessed 16 Mar 2019
12. Bigham, J.P., et al.: VizWiz: nearly real-time answers to visual questions. In: Proceedings of the 23rd Annual ACM Symposium on User Interface Software and Technology (UIST 2010), pp. 333–342 (2010). https://doi.org/10.1145/1866029.1866080
13. Bigham, J.P., Jayant, C., Miller, A., White, B., Yeh, T.: VizWiz::LocateIt - enabling blind people to locate objects in their environment. In: 2010 IEEE Computer Society Conference on Computer Vision and Pattern Recognition - Workshops, pp. 65–72 (2010). https://doi.org/10.1109/CVPRW.2010.5543821
14. Burton, M.A., Brady, E., Brewer, R., Neylan, C., Bigham, J.P., Hurst, A.: Crowdsourcing subjective fashion advice using VizWiz: challenges and opportunities. In: Proceedings of the 14th international ACM SIGACCESS conference on Computers and accessibility (ASSETS 2012), pp.135–142. https://doi.org/10.1145/2384916.2384941
15. Clarke, R., Vines, J., Wright, P., Bartindale, T., Shearer, J., McCarthy, J., Olivier, P.: MyRun: balancing design for reflection, recounting and openness in a museum-based participatory platform. In: Proceedings of the 2015 British HCI Conference (British HCI 2015), pp. 212–221 (2015). https://doi.org/10.1145/2783446.2783569
16. Cock, M., Bretton, M., Fineman, A., France, R., Madge, C., Sharpe, M.: State of museum access 2018: does your museum website welcome and inform disabled visitors? VocalEyes (2018). https://vocaleyes.co.uk/state-of-museum-access-2018/. Accessed 8 Oct 2019
17. Cosley, D., et al.: A tag in the hand: supporting semantic, social, and spatial navigation in museums. In: Proceedings of the SIGCHI Conference on Human Factors in Computing Systems (CHI 2009), pp. 1953–1962 (2009). https://doi.org/10.1145/1518701.1518999
18. Cosley, D., et al.: ArtLinks: fostering social awareness and reflection in museums. In: Proceedings of the SIGCHI Conference on Human Factors in Computing Systems (CHI 2008), pp. 403–412 (2008). https://doi.org/10.1145/1357054.1357121
19. Coyote. Coyote. *Coyote*. https://coyote.pics/. Accessed 7 Feb 2019
20. Hoonlor, A., Ayudhya, S.P.N., Harnmetta, S., Kitpanon, S., Khlaprasit, K.: UCap: a crowdsourcing application for the visually impaired and blind persons on android smartphone. In: 2015 International Computer Science and Engineering Conference (ICSEC), pp. 1–6 (2015). https://doi.org/10.1109/ICSEC.2015.7401406

21. Bartolome, J.I., Quero, L.C., Kim, S., Um, M.-Y., Cho, J.: Exploring art with a voice controlled multimodal guide for blind people. In: Proceedings of the Thirteenth International Conference on Tangible, Embedded, and Embodied Interaction (TEI 2019), pp. 383–390 (2019). https://doi.org/10.1145/3294109.3300994
22. Lin, V.C.-W.: Slow looking: the art and practice of learning through observation. J. Museum Educ. **44**(2), 218–222 (2019). https://doi.org/10.1080/10598650.2019.1576012
23. MoMA: Accessibility|MoMA. The Museum of Modern Art. https://www.moma.org/visit/accessibility/#individuals-who-are-blind-or-have-low-vision. Accessed 4 Aug 2020
24. Rector, K., Salmon, K., Thornton, D., Joshi, N., Morris, M.R.: Eyes-free art: exploring proxemic audio interfaces for blind and low vision art engagement. Proc. ACM. Interact. Mob. Wearable Ubiquit. Technol. **1**(3), 1–21 (2017)
25. Reich, C., Lindgren-Streicher, A., Beyer, M., Levent, N., Pursley, J., Mesiti, L.A.: Speaking Out on Art and Museums: A Study on the Needs and Preferences of Adults who Are Blind or Have Low Vision. Museum of Science, Boston and Art Beyond Sight (2011)
26. Sato, D., Oh, U., Naito, K., Takagi, H., Kitani, K., Asakawa, C.: NavCog3: an evaluation of a smartphone-based blind indoor navigation assistant with semantic features in a large-scale environment. In: Proceedings of the 19th International ACM SIGACCESS Conference on Computers and Accessibility (ASSETS 2017), pp. 270–279 (2017). https://doi.org/10.1145/3132525.3132535
27. Schweibenz, W.: Museums and Web 2.0: some thoughts about authority, communication, participation and trust. In: Styliaras, G., Koukopoulos, D., Lazarinis, F. (eds.) Handbook of Research on Technologies and Cultural Heritage: Applications and Environments, pp. 1–15. IGI Global (2011). https://doi.org/10.4018/978-1-60960-044-0.ch001
28. Smithsonian American Art Museum: Verbal Description Tours. Smithsonian American Art Museum. https://americanart.si.edu/events/verbal-description-tours. Accessed 16 Mar 2019
29. Smithsonian American Art Museum. Calendar. Smithsonian American Art Museum. https://americanart.si.edu/calendar. Accessed 7 Feb 2019
30. The Andy Warhol Museum: Accessibility. The Andy Warhol Museum. https://www.warhol.org/accessibility-accommodations/. Accessed 8 Oct 2019
31. Local Guides. https://maps.google.com/localguides/home. Accessed 14 Sep 2020

Promoting Social Inclusion Around Cultural Heritage Through Collaborative Digital Storytelling

Vanessa Cesário(✉), Albert Acedo, Nuno Nunes, and Valentina Nisi

ITI/LARSyS, IST University of Lisbon, Lisbon, Portugal
vanessa.cesario@iti.larsys.pt

Abstract. We present a case study to understand how migrant communities embrace and connect with their host city's cultural heritage. To achieve this, we deployed a study with ten adult migrants (first- and second-generation Lisbon dwellers) articulated into two stages: (i) a five-day photo-challenge involving storytelling elucidated by pictures and short textual descriptions, followed by (ii) a four-hour audio recorded co-creation workshop, in which participants explored the material they had captured and co-created stories around specific sites, linking them to their memories. This method enabled the participants to express their opinions and experiences on social, cultural, and historical matters. By exploring their connections with the places they inhabit through their own, personal narratives and sharing these with their peers, the participants activated a discussion process exploring the role of storytellers. This case study focuses on the lessons learned and the limitations of the practical work carried out.

Keywords: Cultural heritage · Immigrants · Urban challenges · Digital storytelling · Co-creation · Qualitative analysis

1 Introduction

Cultural heritage institutions are described as places that materialize and visualize knowledge [1]. Their goals are to collect, preserve and share that knowledge with the public. These institutions are slowly but surely moving away from being collections of exhibits, to become dynamic centres where people can engage and empower their knowledge by discovering and challenging themselves [2, 3]; visitors are turning from passive to active participants [4, 5]. Storytelling has been known to be an effective way to convey ideas and beliefs; museums and cultural heritage institutions not only tell us stories but also build those stories through the meaning-making process in which the visitors engage. This fact allows museums' audience to indulge in narratives that aid the construction of meaningful memories as well as providing the fulfilment of a complete experience.

This research was conducted under the European-funded project MEMEX promoting social inclusion by developing collaborative storytelling tools related to cultural heritage. MEMEX will deploy three distinct pilots to analyse different expectations from fragile

M. Wölfel et al. (Eds.): ArtsIT 2021, LNICST 422, pp. 248–260, 2022.
https://doi.org/10.1007/978-3-030-95531-1_17

migrant communities in three regions of the European territory – in Lisbon, Barcelona and Paris. The project aims to give the communities a voice through an advanced but easy-to-use ICT tool on a smartphone for non-expert users to create their own stories. Capturing and mobilizing migrants' intangible heritage while hosted by foreign cultures is deemed essential to foster their social integration [6]. A problem arises when migrants' stories and attitudes towards the hosting country's heritage are misrepresented, ignored, or reinterpreted by our governmental systems. This paper aims to answer the following question: How do migrant communities connect with their host city's heritage, and how do they talk about it? To answer this research question and to inform the above ICT tool, we developed a preliminary case study with first- and second-generation communities of migrants in Lisbon, Portugal. This case study was important for the MEMEX team as it laid the groundwork to start understanding how migrants might capture and discuss Cultural Heritage in connection with their daily life in their hosting city, how they might relate to it, and how they might convey these experiences as stories, before MEMEX team starts creating the ICT tool.

2 Digital Storytelling Research Methods with Migrants

This section describes related work on digital storytelling research methods with migrant participants that inspired us.

Photo-elicitation and short textual descriptions were the means used to prompt the participants to connect their stories and memories to the hosting city and culture. England [7] explored how the photo-elicitation methods helped young immigrant women in Halifax, Canada, narrate the spaces they encountered in their day-to-day lives. Tinkler [8] outlines that photo-elicitation is "*a tool for social and historical research*" and that photographs can be "*deliberately 'inserted' into interviews to prompt discussion, reflection and recollection*". In his study, England [7] proposes that photo-elicitation could express the participants' social positions and experiences by exploring their relationships with the urban spaces in which they dwelt. Yoon and Park [9] conducted four in-depth interviews with female Korean immigrants within the United States to understand their acculturation experiences. The same authors characterize acculturation as *"the process of reconstructing one's identity by negotiating between cultures in a foreign location"* [9]. They propose that identity can be shaped through a person crafting narratives about their experiences and that this process can give rise to a *"concept of temporal unity of the narrative identity"* [9]. With this statement, the authors suggest that in telling such stories, a narrator can create a sense of their identity by bringing together the past and the present, while contemplating their future. Gil-Glazer [10] used the dual methods of photo-monologue and photo-dialogue in a workshop with Arab and Jewish students in Israel to discuss family photo-albums and memories associations that they conveyed. This method inspired discussions about belonging to specific places and being uprooted from them and the experiences of people migrating to new locations and subsequently finding a sense of belonging. In their conclusion, the author outlines that the workshop highlighted the need for young people to engage in further discourses around the experiences of being uprooted, migrating, and belonging to a place and sharing historical knowledge contained within family histories. Moreover, Bødker and Iversen [11] argued that shared "where-to" and "why" artefacts are essential to the successful design of interactive systems. Co-creation is an act of collective creativity, conducted by a group of people [12]. It encourages the development of collaborative knowledge from individuals, through the articulation of their creativity. While a designer-researcher mediates the process and provides tools to activate the process, participants ideate, conceptualize, and develop the final concept or output [13]. Although the co-creation process needs to be established through a focus group [14, 15], the method is usually determinant [10].

3 Participatory Design and Co-creation

This section describes the importance of participatory approaches in including dialogues and interaction between participants to establish an active and effective co-curation.

Participatory Design (PD) has developed from its Scandinavian origins (see [17–20]). PD incorporates several methods and theories, while the core philosophy is to include the final users as active participants in the technology design process [20–25]. Taxén [25] pointed out that PD is a strategic approach to producing user-oriented information technologies. Cesário and colleagues found that co-creation sessions can gave participants a "voice" and engage them enthusiastically in the design process [22, 23, 27]. It allows the creation of collaborative knowledge from individuals, articulating their creativity. While it exists a designer/researcher who mediates the process and provides tools for activate the process, participants ideate, conceptualize and develop the final concept or output [13]. Although the co-creation process needs to be set up based on the focus group within the process [14, 15], the method used in the process is determinant [16]. Mutibwa and colleagues [28] found that the creation of a face-to-face dialogue and interaction helped to establish an effective co-curation. In turn, the comparison of presential and remote collaborative experiments has been studied to understand the different stakeholders' experience (i.e., creators, collaborator and viewers) [29].

4 Case Study: Promoting Social Inclusion

To engage and better understand local migrant communities' cultural heritage [30] to inform the MEMEX project, we deployed a first case study in Lisbon, adopting a method of storytelling elucidated by pictures and co-creation, customized for the migrants' time and spatial constraints. The experiment was conducted in collaboration with one local Non-Governmental Institution (NGO) – *Instituto Marquês de Valle Flôr* and a privately-owned company dedicated to the relationship between museums, audiences and communities – *Mapa das Ideias*. The NGO advertised the study through posters in their premises which caught the attention of potential participants. Those potential participants were informed about the nature and purpose of the study. Finally, ten young adults (first- and second-generation migrants with Brazilian, Cape Verdean and Mozambican roots) between 25–38 years old were willing to participate in this study articulated into two stages: (i) execute a five-day photo challenge in Lisbon, and (ii) attend a co-creation workshop in which they should explore each other's photographs and co-create stories around the featuring sites, memories and experiences. The workshop was audio-recorded, and the researchers took notes. Next, a qualitative analysis on the transcripts was conducted.

Participants were asked to use their own smartphones to take photographs and a consent form about the aims of the study and explaining the protection and privacy treatment of the data was also delivered, explained and signed by all participants. Furthermore, we offered a €25 Gift Card to compensate each participant for their time dedicated to the activities described above.

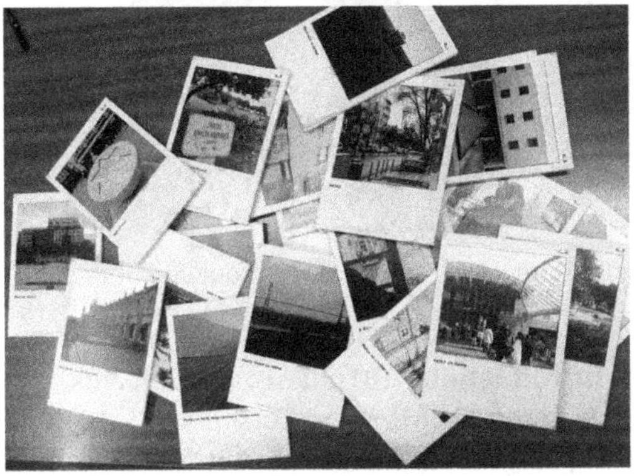

Fig. 1. Examples of the photos taken during the five-day photo-challenge.

4.1 Photo-challenge

For the first stage, participants were asked to take five/six photographs of sites in Lisbon (buildings, public spaces, heritage objects) that they could relate to their past and family history over a five-day period. The participants were asked to provide a textual description per each photograph. This text contained the image's title and a short outline of a memory or story accompanying the photo. Participants sent the photos and their descriptions to a contact person at the collaborating NGO, before being forwarded to the researchers with details of authorship removed. Photographs were edited to deny identification of people and vehicles by means of blurring faces and car plates (Fig. 1). The dataset was anonymized, and each participant was coded with one letter, in alphabetical order.

4.2 Co-creation Workshop

One week later, the participants attended a four-hour co-creation workshop facilitated by the first author and an employee of the privately-owned company. The workshop took place in the NGO's premises. Each participant was given an envelope with a randomly chosen set of another participant's photographs, including the titles, but not the story descriptions. The participants were then split into two groups to discuss and co-create stories around the photographs assigned to them. They were asked initially to work on their own, before co-creating as a group. Care was taken to avoid placing participants in the group where their own photographs were discussed. Upon the participants' consent, the sessions were recorded through audio. The schedule of the workshop is outlined below.

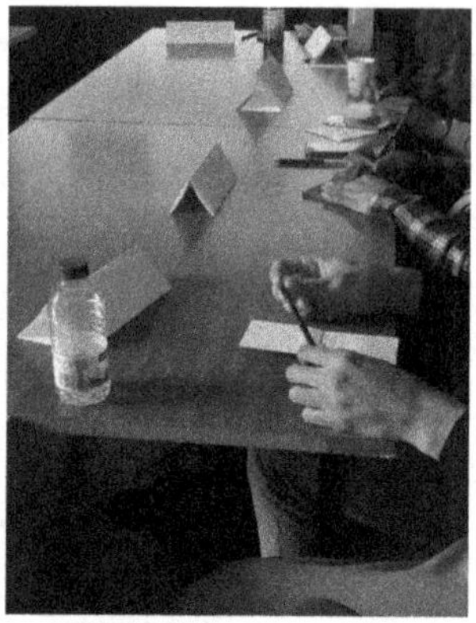

Fig. 2. Participants taking notes during the introduction of the co-creation workshop

Welcoming and introductions (01h05m). Participants were welcomed in the premises and introduced to the MEMEX project aims and study goals (Fig. 2). Inspired by England's approach [7], participants and facilitators took time to introduce themselves and get to know each other. A facilitator presented the structure of the co-creation workshop, and the Informed Consent form was signed. At the end of this stage, each participant was given an envelope containing another participant's set of photographs (Fig. 3), as well as a pencil, and a small notebook.

Storytelling dynamics (01h40m). Participants were split into two five-person groups occupying two adjacent rooms. One group was alone, while the second group, due to logistical reasons, had to share the room with a collaborator from the NGO, who was doing regular work, and always kept her headphones on. Participants from each group were given 10 min to open their envelopes and individually create a story around those photographs – the pictures were numbered to know the order they were shot. Afterwards, participants had one hour to share and discuss their stories and photographs with the rest of the group. Part of the task was to co-create three new stories as a group, based on three sets of the given photographs. The final co-created stories were written-up on a flipchart. During this process, participants talked about how they related to each other's photographs, and how they perceived their host city. The timing was kept by the two facilitators, who waited outside the rooms.

Coffee-break (30min). During this time, participants broke out of their specific group's formations, although some kept discussing the topics that emerged during the session.

Plenary session (45min). Finally, during the plenary session, each group selected a spokesperson to present the co-created stories. Afterward, a facilitator asked the original author of the photographs to comment on the narrative and explain the rationale behind it.

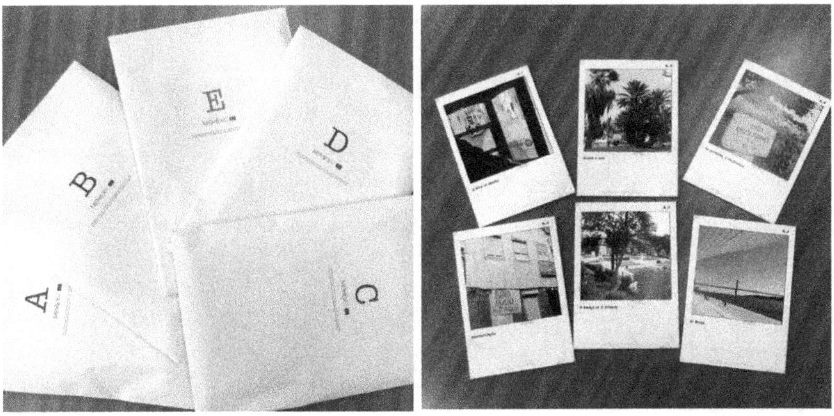

Fig. 3. Envelopes containing the individual set of photos; and the numbered set of photos from participant A.

5 Results

This section presents the analysis of the recordings from the workshop and the notes gathered during the plenary session.

5.1 Thematic Analysis

The audio recordings of the session were transcribed in Portuguese and English. The researchers used thematic analysis to organise and describe the data, identifying, examining, and reporting patterns within the studied transcripts [31]. The analysis was performed through NVivo 12 software by the first author, and then discussed with the others. Firstly, the researcher became familiarized with the transcripts via multiple readings and defined codes. Codes across the whole set were then collated into broader themes and given exact names and definitions to capture the essence of each one. While codes identify significant phenomena in the data, themes are interpretations of the codes and the data. Two overarching themes were identified from the analysis: 'Workshop dynamics' containing four codes, and 'Memories' containing six codes. In the scope of this article, we will focus only on the latter.

The theme 'Memories' comprises six codes in total (Table 1) described in detail below.

(i) Daily lives: routine & transport. Participants talked about their everyday lives as a trajectory through a repetitive routine where they wake up early, use public transport, go to study at the university, go to work, and finally return home. Various forms of public transport in Lisbon (tram, subway, boat, and train) came up in their conversations and storytelling, while no one mentioned private means of transportation. Someone noted that a certain tram, serving the *Bica* area, has become a tourist attraction, hence too expensive for them to use, so they prefer to walk this route instead. Participants noted trams and subways that serve touristic areas are often very crowded. Public transport in general is also often late or out of service. Some participants use the boat to cross the river,

Table 1. Map of codes identified under the theme 'Memories' along with examples of the transcripts assigned to those codes.

Code	Transcripts
Daily lives: routine & transport	**C:** *At the end of the day, these are all routine. We are all made of routines*
Sites of interest	**K:** *(…) Have you been to MAAT* [museum]*? I never got in there. Is it worth it?* **C:** *Yes, I think it is, for those who love art and so*
Relationships with family, friends, and music	**B:** *Ah, this is my godmother's house! Who took a picture of my godmother's house? This is so cool!*
Immigrants' challenges	**H:** *Varina's life was very complicated. Luís's mother's life too, but what really mattered to her was […] she could count on the support of her friends, equally immigrants*
Gentrification & solitude	**K:** *These really lovely pictures […] I love the fact of representing gentrification, which is a reality here in Lisbon* **B:** *The fact that you're with a bunch of people on the transports, but you're alone. At least, I speak for myself. I make this journey always by myself… (…) So here, we could make a connection…* **D:** *Of a lonely journey*
Cultural Heritage from their country of origin	**G:** *And I started, from the assumption of this path… I thought about the persistence of these characters, from our past, what it took them to make their fight possible. (…) To conclude… bearing all these monuments, we can drive our lives to a good port if we have enough persistence in our dreams*

travelling from one side of the city to the other. The ferry was praised because it offered an important, restful moment of contemplation in their day. Contemplative moments and opportunities for relaxation were also considered valuable moments in the routines of the participants. In between going to study and work, participants also stumbled upon urban parks and gardens, where they spent time with friends and recharging their batteries.

(ii) Sites of interest. Participants recognized the photographed various sites and used them to organize their stories. Participants identified the university, as a place of personal growth where they study; museums, specifically the *Museum of Art, Architecture and Technology* as a place for art lovers and beautiful building; heritage sites such as *Mosteiro dos Jerónimos* were highlighted by defining the local residents as lucky, as they can enjoy the sight of these places and (specifically for the *Mosteiro*) attend mass there; and family dwellings. In particular, one participant recognized her godmother's house in a photograph and recalled memories related to that building. Finally, participants identified specific areas of Lisbon such as *Baixa, Martim Moniz,* and *Rua Augusta* as places of great diversity and multiethnicity of people, where commerce and tourism flourishes. They also spoke about the *Tejo* riverbanks where they relax listening to the soothing sound of its waves, and *Rossio*, in which streets it is traditional to celebrate New Year's Eve.

(iii) Relationships with family, friends, and music. When co-creating the stories, participants addressed family, friends and romantic relationships: the subjects of these stories ranged from a child's memories of his Mozambican mother and Portuguese father, a goddaughter remembering her godmother, to the blossoming love story between a boy and a girl at the university. Regarding music, some stories revolved around immigrant friends playing the drums together, in reference to the Cape Verdean tradition of the female drum playing, or a song from a famous Portuguese singer (*Rui Veloso*). Additionally, participants mentioned hearing from their parents that when they immigrated, the city of Lisbon was very different: less developed and less gentrified; there were not so many tourists, big malls or shopping centres. One participant also recognized a photo featuring her old house in Lisbon, sharing how the building is different from when they lived there.

(iv) Immigrants' challenges. Participants narrated about the difficulties of arriving in a foreign country. They highlighted the hardships of not having familiar support and an established network of people to overcome their daily lives challenges. They underlined how guidance and support from other immigrants is essential in helping people integrate into a new society.

(v) Gentrification and Solitude. Participants expressed how journeying through public spaces can be lonely, even if encountering many people along the way. They also underlined how it is not easy to integrate in a new culture. At the same time, they highlighted the value of solitude as these times can be used for reflecting, contemplating, and recharging.

(vi) Cultural Heritage from their country of origin. Participants often recalled their cultural heritage from their country of origin and expressed interest in its history from an autochthone perspective. A Cape Verdean participant focused on the African tribal drumming as an emotional expression of energy. By looking at photographs of monuments celebrating the Portuguese discoveries, one participant talked about the symbolism of the Age of Discovery connected this with the idea of freedom and adventure that setting off for the unknown might bring about.

To summarise, participants addressed memories of their daily lives in Lisbon, in various ways. From their daily experiences and knowledge of the urban area, from gentrification to solitude and how this affected their lives. Participants expressed themselves through memories regarding family, friends, and love, and highlighted a strong relationship with music. When organizing their stories for the exercise, they talked about specific places in Lisbon, which included Universities, museums, cultural heritage and family homes, and specific urban areas. The difficulties they encountered as immigrants in Portugal were also raised frequently in their stories, highlighting how the help of other immigrants was essential for their integration into their new society. Portuguese and African histories were mentioned and valued.

5.2 Notes from the Plenary Session

The plenary session at the end of the workshop highlighted how symbolic interactions can open up opportunities for meaning-making out of co-created stories. Such processes can help develop understanding about how participants relate to their hosting culture as well as each other's cultural backgrounds and heritage, as the following examples illustrate:

Personal meaning and value were found in assets curated by others. One participant identified with someone else's co-created story, around her photograph: *"Yes, it's kind of my daily routine, but well… I don't stay in college till late night.* [laughs] *I just shot it when I had some availability, but yeah it's my routine!"*. Individuals found validation in the recontextualization of their photos by others; more than one author thanked the group for the stories they developed around his or her photographs, one of them saying "*I really liked the story because it's interesting to see how you saw what I shot. That's not the story I had in mind. I didn't have a specific story, though, I just wanted to connect the places that tell me something. And I was happy to see your interpretation of that.*" One went as far as to thank them for their effort in making meaning out of a disconnected collection of unrelated photos: *"I'm very happy […] I think it's spectacular. Thank you."*

Individual narratives and co-developed stories can sometimes coincide. The author of the photos received one of the co-created stories as a similar narrative as the one imagined during photo collection process: *"It's all about it! There's one picture that says 'I won't Move Out' on a wall, which is this one. And then I was inspired to write a poem about gentrification."*

The creation of fictional characters through empathy and imagination can be the starting point of a co-created story. One participant proposed to compose a story from the point-of-view of a young second-generation migrant boy asking his mother questions about life as an immigrant, facing a new city and a new culture. The group accepted this imaginative perspective as a legitimate starting point for a collective narrative: one of them pointing out *"I really like this story* [perspective]*."*

Storytelling is often an entirely subjective task. Two participants happened to photograph the same site, focusing on different facets of the place, effectively telling different stories from different perspectives about the same material space. One of them stated: *"I also took a picture here, she took in landscape mode, and I captured only a female statue that it is this one here* [pointing to the photograph]… *But then, look, both of us, in the same place, I mean, I just focused on her statue…"*.

6 Concluding Remarks and Lessons Learned

An individual's act of sharing information in a collective activity implies self-expression and personal reasoning. The process of storytelling entails a reflection about what information to deliver and in which format. Furthermore, when this process takes place around images, even if captured by another person, the storytelling process takes the form of recognition and reflection starting from another person's experiences. These findings open avenues for a co-creation process that stimulates creativity through abstracting a common collaborative discourse. Specifically, for immigrant communities, the participatory and co-creation process mediated by images presents a valuable and practicable framework to contrast and develop connections between old and new places related to a person's heritage. Such approaches also allow us to consider a migrant's relationship with their new urban environments, as well as offering a view into how the process of co-creation between different cultures can develop. This case study focused on understanding how ten young Lisbon dwellers (first- and second-generation migrants) connect with their host city's heritage and highlights their attitudes towards their hosting country's heritage that is usually ignored or reinterpreted by our governmental systems. Below we specifically reflect on the lessons learned of the method, illuminating how institutions and researchers could appropriate it to engage migrant communities in sharing their stories and appreciation of cultural heritage.

Localisations of the photographs. Out of privacy concerns, participants were asked not to annotate their pictures with the GPS coordinates of the location where they were shot. However, having access to the photographs without knowing their location prompted exciting discussions amongst the participants about the sites and their neighbouring areas. These conversations also acted as an icebreaker, fostering introductions and new connections among the participants. Something that we feared could have been a limitation of the methodology ended up working as an advantage.

Timelines and sequence of photographs. The photo-challenge offered the participants freedom to take five/six photos in any location as a sequence over five consequent days. The window of time between photographs allowed the participants to reflect and eventually plan how to capture the desired places. However, as no photograph time stamp was required, we do not know if the participants stuck to these rules. The conversations captured during the recorded sessions revealed that most participants took their time to think about the photographs and places they wanted to capture. Some expressly displaced themselves to capture specific places. These conversations highlight how participants reflected and took their time to execute the task. This level of care is encouraging and might suggest that the participants found the exercise engaging. Nevertheless, the very personal, almost diaristic style of the narratives highlighted a lack of plotting or characterization, which are often considered critical to a storytelling activity. Future studies could reconsider the structure of the task, perhaps starting from the writing of a narrative first, before illustrating this.

The co-creation activity. Different participants took photographs from the same location, denoting an interest or relationship to essential urban sites and connections. It is important to note that when shooting the photos, participants were not asked to construct an overall narrative and connect the descriptions/memories of each picture to the next one. However, in the workshop, participants were required to co-create a story following the sequence of the author's photographs. Participants were encouraged to imagine a tale following a sequence shot by someone else and wondering about the site's location where the photo was taken. As a result, participants embraced each other's views of the city and came together in a collective effort to create meaning out of a sequence of images, consciously, or subconsciously trusting the original author's sequence. The collaborative effort, the overall respect between the participants, the creativity that emerged from the workshop, as well as the sense of gratitude of the pictures' owners to the storytellers, generated respectful and genuine atmosphere of interest in each other's experience. The workshop thus demonstrated that co-creation can be a successful exercise to generate inclusive meaning for migrants.

7 Limitations

The workshop innovative methodology raised some issues regarding its limitations. One of the two groups had to share the space with the NGOs staff. Although the staff had their headphones on, the fact of having other people in the room not taking part in the activity process might have disturbed the participants. This concern was not evidenced in the transcripts of the workshop, though this might be for reason that they feared being overheard.

Acknowledgements. The authors would like to acknowledge researcher Dan Brackenbury, Ivo Oosterbeek and Ilídio Louro from *Mapa das Ideias*, and Mónica Silva from *Instituto Marquês de Valle Flôr* for their timely support during the development and deployment of the case study. This research was supported by MEMEX (MEmories and Experiences for inclusive digital storytelling) project funded by the European Union's Horizon 2020 research and innovation programme under grant agreement No 870743; and the ARDITI's postdoctoral scholarship M1420–09-5369-FSE-000002.

References

1. Fyfe, G.: Sociology and the social aspects of museums. In: Macdonald, S. (ed.) A Companion to Museums Studies, pp. 33–49. Blackwell Publishing, UK (2006)
2. Falk, J.H., Dierking, L.D.: Learning from Museums: Visitor Experiences and the Making of Meaning. AltaMira Press (2000)
3. Hawkey, R.: Learning with Digital Technologies in Museums, Science Centres and Galleries. NESTA Futurelab Research (2004)
4. Simon, N.: The Participatory Museum. http://www.participatorymuseum.org/. Accessed 24 Sep 2016
5. Mancini, F., Carreras, C.: Techno-society at the service of memory institutions Web 20 in museums. Catalan. J. Commun. Cult. Stud. **2**, 59–76 (2010). https://doi.org/10.1386/cjcs.2.1.59_1
6. Nisi, V., Oakley, I., Boer, M.P.: Locative narratives as experience: a new perspective on location aware multimedia stories. In: Proceedings of the 5th International Conference on Digital Arts, pp. 59–64. International Association for Computer Arts (2010)
7. England, S.: Picturing Halifax: young immigrant women and the social construction of urban space. J. Undergrad. Ethnogr. **8**, 3–21 (2018). https://doi.org/10.15273/jue.v8i1.8620
8. Tinkler, P.: Using Photographs in Social and Historical Research. SAGE Publications Ltd, London (2014). https://doi.org/10.4135/9781446288016
9. Yoon, G., Park, A.M.: Narrative Identity Negotiation between cultures: storytelling by Korean immigrant career women. Asian J. Women's Stud. **18**, 68–97 (2012). https://doi.org/10.1080/12259276.2012.11666132
10. Gil-Glazer, Y.: Photo-monologues and photo-dialogues from the family album: Arab and Jewish students talk about belonging, uprooting and migration. J. Peace Educ. **16**, 175–194 (2019). https://doi.org/10.1080/17400201.2019.1587744
11. Bødker, S., Iversen, O.S.: Staging a professional participatory design practice: moving PD beyond the initial fascination of user involvement. In: Proceedings of the Second Nordic Conference on Human-computer Interaction, pp. 11–18. ACM, New York, NY, USA (2002). https://doi.org/10.1145/572020.572023

12. Zwass, V.: Co-creation: toward a taxonomy and an integrated research perspective. Int. J. Electron. Commer. **15**, 11–48 (2010). https://doi.org/10.2753/JEC1086-4415150101
13. Nielsen, L.: Personas in co-creation and co-design. In: Proceedings of the 11th Human-Computer Interaction Research Symposium, pp. 38–40 (2011)
14. Frauenberger, C., Good, J., Keay-Bright, W.: Designing technology for children with special needs: bridging perspectives through participatory design. CoDesign **7**, 1–28 (2011). https://doi.org/10.1080/15710882.2011.587013
15. Theng, Y.L., et al.: Children as design partners and testers for a children's digital library. In: Borbinha, J., Baker, T. (eds.) Research and Advanced Technology for Digital Libraries. Lecture Notes in Computer Science, vol. 1923, pp. 249–258. Springer, Heidelberg (2000). https://doi.org/10.1007/3-540-45268-0_23
16. Mazzone, E., Read, J., Beale, R.: Understanding children's contributions during informant design. In: Proceedings of the 22nd British HCI Group Annual Conference on People and Computers: Culture, Creativity, Interaction, Vol. 2, pp. 61–64. BCS Learning & Development Ltd., Swindon, UK (2008)
17. Bødker, S.: Third-wave HCI 10 Years Later—participation and sharing. Interactions. **22**, 24–31 (2015). https://doi.org/10.1145/2804405
18. Brandt, E., Binder, T., Sanders, E.: Tools and techniques: ways to engage telling, making and enacting. In: Routledge International Handbook of Participatory Design (2012)
19. Halskov, K., Hansen, N.B.: The diversity of participatory design research practice at PDC 2002–2012. Int. J. Hum Comput Stud. **74**, 81–92 (2015). https://doi.org/10.1016/j.ijhcs.2014.09.003
20. Simonsen, J., Robertson, T.: Routledge International Handbook of Participatory Design. Routledge (2012)
21. Muller, M.: A participatory poster of participatory methods. In: CHI 2001 Extended Abstracts on Human Factors in Computing Systems, pp. 99–100. ACM, New York, NY, USA (2001). https://doi.org/10.1145/634067.634128
22. Cesário, V., Coelho, A., Nisi, V.: Co-designing gaming experiences for museums with teenagers. In: Brooks, A.L., Brooks, E., Sylla, C. (eds.) Interactivity, Game Creation, Design, Learning, and Innovation. Lecture Notes of the Institute for Computer Sciences, Social Informatics and Telecommunications Engineering, vol. 265, pp. 38–47. Springer, Cham (2019). https://doi.org/10.1007/978-3-030-06134-0_5
23. Cesário, V., Matos, S., Radeta, M., Nisi, V.: Designing interactive technologies for interpretive exhibitions: enabling teen participation through user-driven innovation. In: Bernhaupt, R., Dalvi, G., Joshi, A., Balkrishan, D.K., O'Neill, J., Winckler, M. (eds.) Human-Computer Interaction – INTERACT 2017. Lecture Notes in Computer Science, vol. 10513, pp. 232–241. Springer, Cham (2017). https://doi.org/10.1007/978-3-319-67744-6_16
24. Cesário, V., Coelho, A., Nisi, V.: An unlikely seamless combination - future curators designing museum experiences towards the desires of actual teenagers. In: Proceedings of the 1st International Conference on Design and Digital Communication, pp. 101–109. IPCA - Instituto Politécnico do Cávado e do Ave, Barcelos (2017)
25. Cesário, V., Coelho, A., Nisi, V.: Cultural heritage professionals developing digital experiences targeted at teenagers in museum settings: lessons learned. In: 32nd British Human Computer Interaction Conference, pp. 1–12 (2018). https://doi.org/10.14236/ewic/HCI2018.58
26. Taxén, G.: Introducing participatory design in museums. In: Proceedings of the Eighth Conference on Participatory Design: Artful Integration: Interweaving Media, Materials and Practices, Vol. 1, pp. 204–213. ACM, New York, NY, USA (2004). https://doi.org/10.1145/1011870.1011894

27. Cesário, V., Coelho, A., Nisi, V.: Word association: engagement of teenagers in a co-design process. In: Lamas, D., Loizides, F., Nacke, L., Petrie, H., Winckler, M., Zaphiris, P. (eds.) Human-Computer Interaction – INTERACT 2019. Lecture Notes in Computer Science, vol. 11749, pp. 693–697. Springer, Cham (2019). https://doi.org/10.1007/978-3-030-29390-1_65
28. Mutibwa, D.H., Hess, A., Jackson, T.: Strokes of serendipity: Community co-curation and engagement with digital heritage. Convergence (2018). https://doi.org/10.1177/135485651 8772030
29. Bhimani, J., Nakakura, T., Almahr, A., Sato, M., Sugiura, K., Ohta, N.: Vox populi: enabling community-based narratives through collaboration and content creation. In: Proceedings of the 11th European Conference on Interactive TV and Video. pp. 31–40. Association for Computing Machinery, Como, Italy (2013). https://doi.org/10.1145/2465958.2465976
30. Mohr, F., Zehle, S., Schmitz, M.: From co-curation to co-creation: users as collective authors of archive-based cultural heritage narratives. In: Rouse, R., Koenitz, H., Haahr, M. (eds.) Interactive Storytelling. Lecture Notes in Computer Science, vol. 11318, pp. 613–620. Springer, Cham (2018). https://doi.org/10.1007/978-3-030-04028-4_71
31. Braun, V., Clarke, V.: Using thematic analysis in psychology. Qual. Res. Psychol. **3**, 77–101 (2006). https://doi.org/10.1191/1478088706qp063oa

Resonant Webs: An International Online Collaborative Arts Performance for Individuals with and without a Disability

Jonathan Duckworth[1(✉)], Shigenori Mochizuki[2], Ross Eldridge[1], and James Hullick[3]

[1] School of Design, RMIT University, Melbourne, Australia
{jonathan.duckworth,ross.eldridge}@rmit.edu.au
[2] College of Image Arts and Sciences, Ritsumeikan University, Kyoto, Japan
mochiz@im.ritsumei.ac.jp
[3] Jolt Sonic and Visual Arts Inc., Melbourne, Australia
james.hullick@joltarts.org

Abstract. In this paper we discuss the creation of an international online sound arts performance developed in collaboration between artists in Australia and Japan. The COVID-19 global pandemic has made community activities related to the arts more challenging due to restrictions on personal mobility and international travel. We report on a hybrid workshop approach that combines online and face-to-face interaction, which has enabled sound artists with intellectual disability to participate in activities with international outreach. Specifically, we describe the design and technological implementation of a live performance and an interactive audio-visual artwork that translates audio data to visual effect from an Australian ensemble of sound art musicians with intellectual disability, and a professional Japanese Noh singer and performer. Based on the artists initial reflections in developing the performance we discuss opportunities and advantages of online digital spaces for international collaboration and co-creation in community arts contexts that may be utilised in the future.

Keywords: Interactive art · Disability · Sonic arts · Participation · Interaction design

1 Introduction

The significant impact of COVID-19 on the arts, cultural and creative industries are among the most adversely affected industry sectors due to measures to control the spread of the virus such as local government social distancing requirements and closure of physical venues, prohibiting not only public indoor performances but also rehearsals [1]. For many in the skilled, resource-intensive, and highly collaborative performing arts and music sector has seen most activities postponed or cancelled. According to Deloitte Access Economics, in Australia the pandemic resulted in an estimated AU$6

M. Wölfel et al. (Eds.): ArtsIT 2021, LNICST 422, pp. 261–274, 2022.
https://doi.org/10.1007/978-3-030-95531-1_18

billion forecast loss in revenue between April and June 2020 for the arts sector [2]. The Australia Council for the Arts found that only 47% of businesses in the arts and recreational services sector were trading in the week commencing March 30, 2020, with 94% of arts and recreational industry adversely affected by government restrictions arising from COVID-19, as compared to 90% of businesses as a whole [3]. The situation was similarly dire in Japan, with the Government Agency for Cultural Affairs reporting that 80% of cultural events were postponed with 60% cancelled indefinitely [4].

In response to the crises, individuals, and arts organisations with resources to do so have adapted existing materials to the newfound restrictions, luring wider audiences via digitized archives, tours of virtual exhibition spaces, and streaming performances for what would otherwise be localised public events [5]. However, the rapid shift toward digital service delivery has been unevenly distributed across cultural institutions, artist collectives and individuals. The provision of digital services assumes the availability of digital connectivity, access to devices, data, necessary software, and hardware platforms along with the ability, staffing, skills, and resources to access those platforms [6]. A lack of funding for Artists and those working in the community arts industry has made access to appropriate digital resources challenging and the long-term outlook for the sector remains precarious. Furthermore, those with a disability have been identified as being at greater health risk of COVID-19 in Australia [7], which requires organisations to provide additional levels of support and care to ensure their safety in public settings.

The rights of access to the creative arts and the opportunity to live 'an ordinary life' is a statutory requirement of many agencies that serve to protect and foster participation of marginalized groups [8]. And yet, those who need specialised support and who wish to participate in such activities are often excluded by a lack of availability, accessibility and/or the capacity of creative arts organisations to accommodate their needs [9].

We have been working toward creating opportunities for individuals with and without disability to jointly collaborate in the arts using interactive digital technologies through various workshops, performances, and exhibitions [10]. Our prior research discussed how social aspects of group interaction combined with the affordances of digital technology may be exploited to enhance the participation of people with a disability in co-creative, artistic activity [10]. We define participation as an approach that may lead to improved person-related constructs such as heightened sense of self-efficacy, preferences, belonging to a group, and the development of specific competencies that can be carried forward [11]. Indeed, several other examples of inclusive technology design in the arts has been shown to further enhance the opportunities and developmental needs for people with a disability and act as a catalyst that extends the invitation to participate in cultural activities and expand individuals' preferences [12–14].

Our community arts partners provide excellent examples of successful implementations of online technology that facilitate collaboration and creativity during COVID-19. Slow Label is a non-profit organisation in Japan that generates opportunities for forms of co-creation that transcend national and disciplinary boundaries through the arts with a specific focus on disadvantaged and diverse communities in developing stage performances since 2014. Slow Label has produced and developed several successful initiatives and performances for diverse audiences including Slow Circus project, a circus school and workshop program that utilizes the circus arts to support disadvantaged and disenfranchised youth [15]. In 2019, Slow Label developed a social circus program which assists people with a disability to participate in society through practicing and learning circus skills. Their social circus program has conducted numerous workshops and circus schools which resulted in the first Social Circus performance held in Japan [16]. Due to the impact of COVID-19, Slow Label's circus program has hosted online workshops using online video streaming services, where participants can practice moving their bodies and performing while watching videos of the instructors and other participants [17].

Similarly, Jolt Sonic and Visual Arts (JOLT), a non-profit arts organization based in Melbourne, Australia provides specialist training in the arts for people with intellectual disability and disadvantaged communities since 2008. JOLT is an inclusive sonic arts organisation that creates in-house sonic works, whilst also supporting and presenting the works of other auditory creators. Sonic arts access has become central to JOLT's identity having supported and mentored The Amplified Elephants – a sound art ensemble with intellectual disabilities [18]. JOLT has developed an online workshop program since the beginning of the pandemic to facilitate collaborative learning and rehearsals for sound art performances with sound engineers and other auditory creators.

These examples embrace the idea of inclusivity and foster participation that provide an environment for everyone to contribute when they are afforded opportunities for their involvement whether online in virtual space or face-to-face. However, the feasibility of digital technology and hybrid online activities for individuals with a disability during the disruptions caused by COVID-19 are little understood. We report on the development and technical implementation of a sound art performance developed through a hybrid workshop program that combines online interactions between Australia and Japan. We reflect upon the experiences of the artists participating in the workshop program and performance which offers some preliminary insights on how individuals with disability were able to collaborate with international artists and to connect with others in the development and presentation of a performance mediated through digital live streaming technology during the pandemic.

2 The Resonant Webs Live Performance

The Resonant Webs project is an arts and cultural collaboration between RMIT University and Ritsumeikan University to develop a proof-of-concept live online public sound and visual performance aimed at enhancing the participation of artists and individuals with a disability in the arts industry in Australia and Japan. This project builds upon common interests between the researchers who explore inclusive design and rehabilitation technologies for populations with motor and cognitive impairment [19]. This relationship was further augmented to include two community arts organisations, JOLT and Slow Label, with whom the researchers had forged long term collaborations to develop audio-visual technology for several live performances.

2.1 Project Design and Scope

A twelve-week program was designed to facilitate the online collaboration between the partner organisations and the artists between January-March 2021 in preparation for the live performance. With the disruption of COVID-19 remote learning in tertiary institutions and online collaboration became the new normal. In line with government restrictions and health and safety advice in Japan and Australia, Slow Label and JOLT had suspended all face-to-face activities and transitioned to online delivery of their respective workshop programs and events for professional artists and musicians with disability. The Resonant Webs project provided an opportunity to leverage our existing experience and skills working with online video platforms to develop the performance and rehearse remotely. During COVID-19 lockdowns earlier in 2020 The Amplified Elephants had already successfully transitioned to working online using the Zoom™ video conferencing platform prior to the commencement of this project.

The workshop structure was delivered via a blended mode of online sessions with the artists in Japan and face-to-face in Melbourne when restrictions permitted. The workshops were divided into three consecutive four-week activities. The first stage focused on the development and ideation of the performance; the second stage consisted of experimentation and co-creation with sound making technologies; and the final stage on the composition and performance rehearsals.

The scope of the project was to develop a proof-of-concept performance using online technologies to facilitate participation across international borders. Our role was to facilitate and implement the use of technology for the artists and to reflect upon the development of the performance. Our goal was to gain some initial insights into how the technologies supported the creative and collaborative process. We didn't wish to intervene in the existing creative processes of the artists or attempt to establish formal scientific evidence through empirical observations and qualitative analysis. Rather we focused on the artists existing relationships with technology to develop the performance motivated by our desire to facilitate increased participation outcomes for the artists. Our enquiry is framed by introducing the technical aspects of the performance as well as sharing the initial responses of the JOLT artists through the lens of our conceptual model known as the family of Participation Related Constructs (fPRC) which blends current theory on participation, interaction design and community art [10, 11]. After the workshops and performance JOLT provided written reflections based on their observations of the overall experience which we elaborate upon in the outcome section.

Throughout the twelve-week program JOLT mentored and managed The Amplified Elephant musicians who participated in the development of the performance. Their participation is supported through an arts program funded by the Australian National Disability Insurance Scheme (NDIS). The NDIS also made it possible for The Amplified Elephants to purchase a variety of computers and electronic instruments which they used for the performance. Participation into the program is voluntary and recruitment, training and individual consent is coordinated by JOLT management in communication with the participating individual and parent or legal guardian. The Amplified Elephants are supported through the rehearsal process by support workers and JOLT's Health Officer – a dedicated fully registered nurse qualified for mental health triage. Both support workers and the Health Officer are trained by JOLT for the specific care requirements of each participant. Ethics approval was received from RMIT to obtain consent from the artists to use the publicly available outputs (e.g., performance and symposium) for publication and public dissemination.

2.2 Performance Ideation and Conceptualisation

The online workshops provided a space to facilitate the ideation, design and development of the performance and experimentation with sound making instruments and technologies. Common interests expressed among the group related to digital media art making practices, inclusive design approaches, building and sharing knowledge in sound art making skills, and interest in finding new ways and strategies for reaching audiences via online media.

In the initial conceptualization stages of the project, we gravitated toward developing a contemporary interpretation of *Hagoromo*, a traditional Japanese story told through Noh theatre performance. Noh theatre is a form of musical dance drama that crystallised in the 14th century in Japan. In traditional Noh theatre, the stage is a symbolic and minimal space whereby the stage set is usually adorned with hand drawn or painted pine or bamboo trees on the background wall. The main actor called *Shite* plays in the center of the stage, and the Noh orchestra called *Hayashi* and the chorus called *Jiutai* are placed at the side of the main stage [20]. The plot of Hagoromo is as follows: a fisherman finds the Hagoromo, the magical feather-mantel of *Tennin*, an aerial spirit or celestial dancer, hanging upon a tree bough. The celestial dancer demands its return. The fisherman argues with her, and finally promises to return it, if she will teach him her dance or part of it. She accepts the offer, performs the dance to the fisherman and then ascends into the celestial realm with the feathered mantel [21].

We expressed the collaboration as an elision of cultures between Japan and Australia via Noh theatre, contemporary sonic art, and technology as workshop themes to explore in developing the performance. To initiate the cultural exchange, Slow Label invited vocalist and singer Ryoko Aoki to perform Hagoromo as a form of traditional Japanese prayer for the well-being of all people during COVID-19. JOLT invited The Amplified Elephants to create an audio-visual performance as an interpretation of the Noh orchestra to accompany the Noh chanting. With the easing of some COVID-19 restrictions intermittent face-to-face assembly to workshop and rehearse was possible in each country. The rehearsals between the countries were combined online using video conferencing technology between Spiral Hall, Minato-ku, Tokyo, and Kindred Studios in Melbourne, Australia. The basic concept for the event was to hold simultaneous broadcast of the performance in each venue via live streaming and distributed online to the public. The online distribution was made available via YouTube Live as an event in the SLOW MOVEMENT Showcase and Forum on March 28, 2021 [22].

2.3 Technical Implementation

Representing the collaboration between the Japanese and Australian performers who were connected online but physically separated was an important issue. For the performance we adapted an existing digital media artwork developed by the authors called *Disruptive Critters*, to visualize and connect the musical performances of the

Japanese and Australian performers during the online streamed performance. Disruptive Critters is an audiovisual interface originally designed to augment live vocalized sound art performances [23]. The interface consists of a 42-inch multi-touch tabletop display; a graphical menu of six sound generating entities, or *Critters*, at either end of the display that users can select. The six Critter types (or strains) were conceived as an ecology of evolving sonic entities. The six strains are called (i) Pixel, (ii) Line, (iii) Spin, (iv) Flip, (v) Shape, and (vi) Cubic (see Fig. 1).

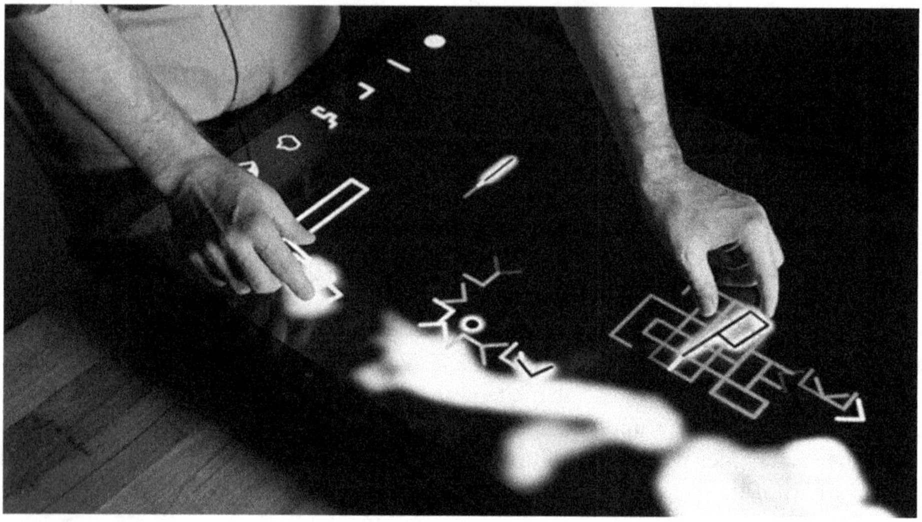

Fig. 1. A user interacts with the Disruptive Critters interface. The six critter types can be selected from the graphical menu at the edge of the screen near the user.

Each strain of the Critter has its own sound world and gestural repertoire that increases in complexity as it evolves from one form to the other in a linear fashion. The critters evolve in graphic and sonic complexity as it transitions from the first strain (e.g., Pixel) through to the final manifestation (e.g., Cubic) over a period of time. For example, the pixel, which is visually represented by a dot, will transition and stretch into a line triggering more complex sounds over time. The rate of transition may occur forwards or backwards at different speed depending on the Critters behaviour within the virtual environment, other Critters, and the performer. The movement of the Critters uses a ballistic physics collision model, which propels them around the virtual environment. Users can drag and place multiple critters into the scene using finger touch gestures. Each computer-generated critter outputs unique vocalized sound sample produced from a database of 456 pre-recorded abstract utterances that resemble human-like emotions. Once selected and placed, the critters become autonomous co-performers moving around the screen seemingly striving to communicate in unpredictable ways with the performers and each other alike. The Disruptive Critters interface was used by The Amplified Elephants during the performance.

In developing a hybrid version of Disruptive Critters for the performance we selected the 'Flip' critter as a central motif and virtual avatar to represent the Japanese and

Australian performers. Avatars are used to visually represent the performers in virtual space rendered and composited as an overlay onto the live video stream (see Fig. 2).

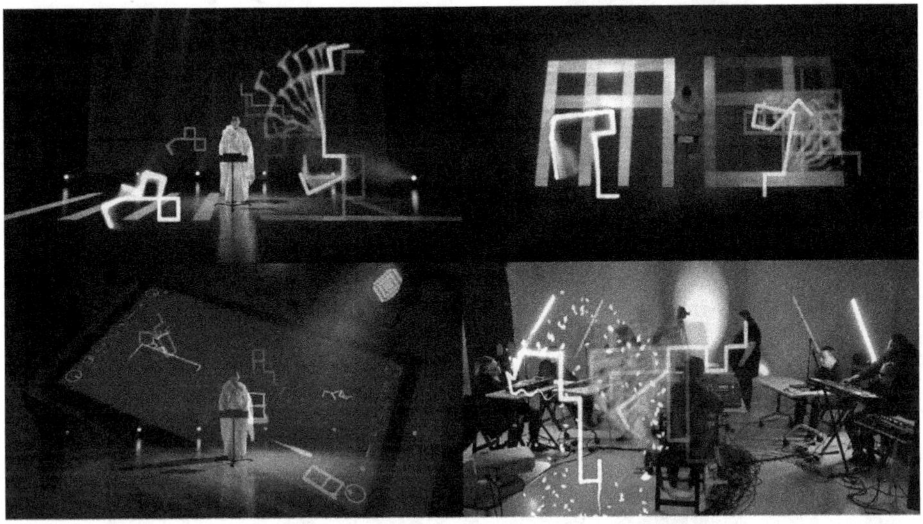

Fig. 2. Examples of the composited overlay of the virtual avatars on the live video stream.

The 'Flip' critter avatar is visually represented by a vertical graphical line divided into twelve equal segments. Segments can rotate by pivoting at the connecting joints. Joints rotate in increments of 90 degrees but are forbidden from flipping back upon the previous segment (180-degree angle). When the audio input signal amplitude exceeds a given threshold a random segment will be rotated 90 degrees either clockwise or counterclockwise. While the audio input remains above the threshold, a random segment will be flipped at a rapid interval. In this way a continuous loud amplitude will cause the critter to rapidly change shape, whilst a momentary sound will create small movements (see Fig. 3).

The heavenly maiden and the fisherman in the Hagoromo story are each represented by a Critter. The movement of each Critter is linked to the audio input of the singing voice of Japanese performer, Ryoko Aoki and the sounds generated by The Amplified Elephant performers in Australia. The 'Flip' Critter representing the Japanese performers was configured to rotate its segments more slowly and with a larger interval between each segment rotation. This Critter had additional visual effects applied throughout the performance: motion blur, ribbon trail and feather particles. Both Critters had a smoke-like fluid simulation effect and a waving cloth simulation applied at various points during the performance. In addition, the audio values of the overall performance were used to trigger and activate stage lighting patterns at Spiral Hall (see Fig. 4).

For the performance we used YAMAHA SyncRoom™ to monitor the audio from each performance venue. Several broadcast 1080p resolution video cameras were setup at Spiral Hall and Kindred Studios to capture the performance from multiple viewpoints. The video from Australia was transmitted to Japan using LiveU™ suite of broadcasting

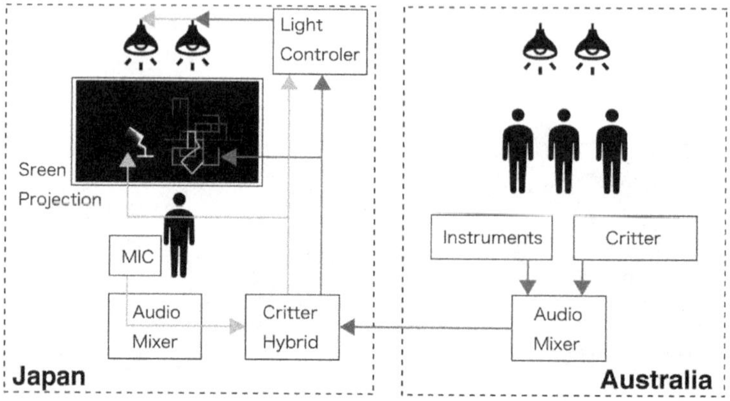

Fig. 3. Audio input and output diagram of the Disruptive Critter hybrid version.

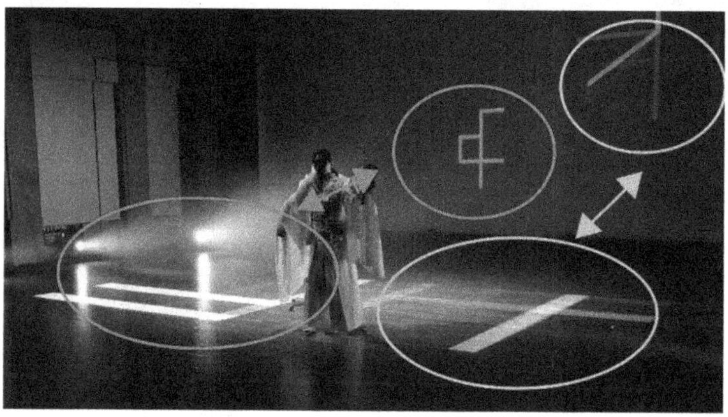

Fig. 4. The stage lighting effects, and movement of the rear wall projected critter avatars are triggered by the corresponding audio input.

technology which can transmit video with low latency and high quality (see Fig. 5). The live video stream from Australia was mixed in Japan with live video footage from Spiral Hall before being transmitted for broadcasting (see Fig. 6).

3 Outcomes and Reflections of the Artists

The online performance of Hagoromo was successfully broadcast on YouTube Live on March 28, 2021, as part of the event SLOW MOVEMENT Showcase and Forum vol.5 and was positively received by the audience as evidenced by their responses in the live chat [22]. The number of pre-registrations was 141 (93 in Japan and 48 in Australia), and the number of viewings during the performance was 204 (186 in Japan and 48 in Australia). However, the exact number is likely higher as many single registrants included the participants extended family members watching the event. The archive of the performance reached 554 views as of July 28, 2021 (354 in Japan, 200 in Australia).

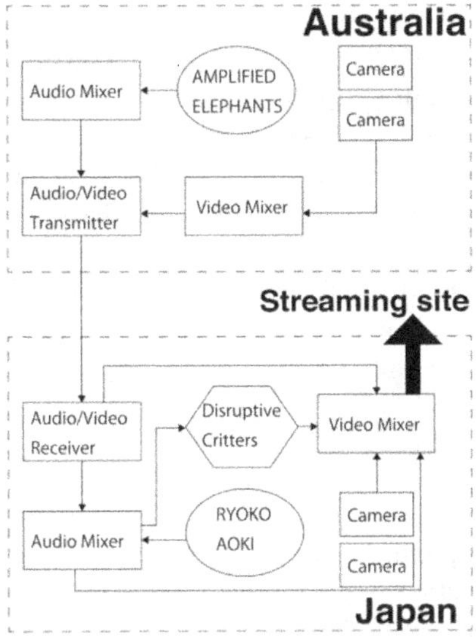

Fig. 5. Wiring diagram of the audio video inputs and live streaming output.

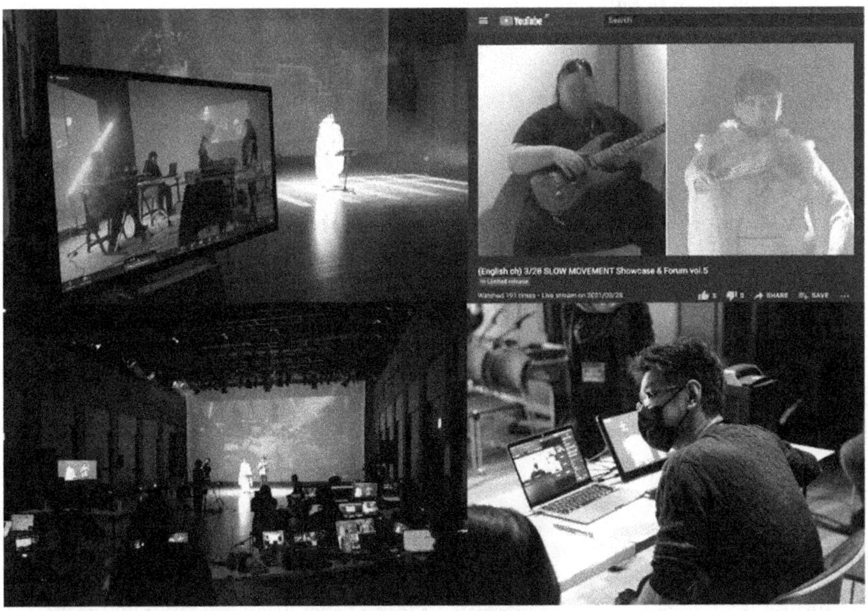

Fig. 6. Photographs of the Hagoromo stage and online streaming video (top right) of the SLOW MOVEMENT Showcase & Forum vol.5. (Image courtesy of Slow Label)

After the performance the JOLT organisation provided written informal reflections based on their observations of the workshops and rehearsals, as well as the perspectives of the artists who discussed their experience during the public symposium that was held after the live performance. The written reflections were prompted by four themes derived from our conceptual model (fPRC) which takes an integrated approach to understanding the role of interactive technology in disability we first presented at the International Conference of Arts and Technology, Interactivity and Game Creation (ArtsIT) hosted in Aalborg, Denmark 2019 [10]. The four themes provide an initial appreciation of (a) the individual's perspectives on interactive digital media; (b) the flexibility of the online technology to enable participation; (c) how the online and face-to-face workshops were designed to afford opportunities for people with a disability to feel included during COVID-19, and (d) the ways in which social-cultural forms of participation can promote a sense of agency for the individual.

3.1 Perspectives on Interactive Digital Media

The Amplified Elephants are an ensemble who regularly use technology in their performances. Standard equipment includes synthesizers, microphones, mixing desks, guitars, computers, and tablets. Since 2007 the ensemble has engaged with new technology including amplified orchestras, robotic sound machines [24], and interactive tabletop audio visual technology originally design for movement rehabilitation [25]. The ensembles engagement with the process of developing and performing with custom designed technology began in 2014 through JOLT's ongoing partnership with RMIT University. In every custom technology project, including the development of Hagoromo, The Amplified Elephants are active drivers of the design and creation of that technology, and user testing is undertaken in arriving at the electronic instrumentation that they can use effectively and creatively. In this context, technology design that accounts for user experience alongside co-creation ensures smart, effective, feasible, and meaningful solutions that incorporate the individual perspectives. JOLT observed that given the opportunity to express their preferences and perspectives on technology use provides a sense of self-efficacy and the development of their skills that can be carried forward by the individual. JOLT reported that the level of engagement along with the ensembles long history of exploring new technologies as part of their arts practice also enabled them to rapidly transition to online modes of collaboration.

3.2 Participation Through Online Technology

In Melbourne, State Government COVID-19 strict lockdown orders were delivered swiftly and hard over a three-month period in 2020 with only five reasons to leave home (food shopping, essential work, medical treatment, getting tested, and exercise). The ensemble had to transition rapidly to working online with very little time given to adapt. The Amplified Elephants were faced with significant challenges of having to learn new software tools that many had not used previously. Through the families of the ensemble members JOLT was able to organise support staff to visit the houses of the artists to assist them set up a computer or tablet at home and provide software training so they could join the workshop video conference calls. By the time The Amplified Elephants joined the Resonant Webs project in 2021, online activity had become a ubiquitous part of their everyday life. Indeed, JOLT reported that through the individual's level of resilience, persistence, and hard work, using online technology eventually became second nature.

The Amplified Elephants were able to access the workshops from wherever they were and those that were unable to regularly attend the previous face-to-face rehearsals due to mobility and health issues were able to access the sessions from home. This suggests that with the appropriate level of support, patience and perseverance, online activities do offer new contexts and flexibility for participation for those with a disability outside of traditional settings such as physical workshop and rehearsal spaces.

3.3 Inclusion Through Online and Face-To-Face Collaboration

Whilst JOLT found that The Amplified Elephants could adapt to online workshops, they were having to think harder to maintain their level of engagement. A full program day (6h) of 'in person' participation was truncated to two 2-h sessions when online to reduce fatigue. Ensemble members' families became the support workers for the online sessions which meant that the whole community was attending the workshops. Much effort goes into administrating and organizing the ensemble, who due to their disability, require very long lead times for rehearsals and fixed schedules and dates. The Japanese Slow Label team arranged a fixed date and the Spiral Hall venue for the performance early in the collaboration and issued schedules and timetables in discussion with the artists and provided the technical specifications for the broadcast. JOLT reported that having a fixed and highly organised schedule assisted with reducing the level of anxiety for The Amplified Elephants.

In the last four weeks of the program, the Victorian Government had successfully suppressed the virus and eased restrictions. This enabled the ensemble to work together face-to-face on Hagoromo. In Melbourne, JOLT booked the Kindred Film Studio, a space the artists were familiar with from previous rehearsals and workshop activities, and large enough to accommodate capacity limitations on people as mandated by the Government. For live streaming purposes film or photographic studios can be more suitable and flexible than theatre and concert halls offering affordable options that are resourced with studio lighting and neutral film backdrops. Extra face-to-face rehearsals were scheduled in the studio when restrictions permitted and Zoom™ was used as a rehearsal tool between Japan and Australia in the weeks leading up to live stream performance. As reported by JOLT, the environment and context were an important factor in regulating the participation and sense of inclusion for The Amplified Elephants. Their familiarity of working in the studio space environment within the context of a highly organised schedule assisted in reducing anxiety and focusing their engagement on the creative process.

3.4 Social and Cultural Participation

Culturally, a feature of common ground was that both the Japanese and Australian teams liked to talk and were highly communicative during their online interactions. The online conference calls provided a space to conduct conversations both casual and work related around the project. This was crucial for many reasons – boosting morale, fostering friendship through laughter and sorrow, and facilitating creative ideation. This dialogue and exchange provided the artists with opportunities to share resources, recordings, and videos to understand Noh theatre and Japanese culture more broadly. Through

Ryoko Aoki's personal insights, The Amplified Elephants were able to understand Noh theatre and develop their own sound world that would complement her vocals. During a workshop the artists might offer suggestions to the group by playing an instrument or making a drawing of a stage layout as well as sharing YouTube videos as a way of pollinating ideas and creativity. Through this process the ensemble chose clearly to be influenced by Noh culture whilst maintaining their own identity through creating, sharing, and accepting sounds to use, and incorporating sounds of other members in the group into their sonic repertoire. Through Hagoromo the cross-cultural collaboration was expressed as a balance between sounds that were Noh and sounds that were The Amplified Elephants' auditory electronica. The workshop program was designed to support individual experiences that enable the participants to exercise control and choice through social interaction. Over time, similar approaches have been shown to lead to an enhanced awareness of one's strength, self-identity, and future opportunities for development [26].

4 Conclusion

In reflecting upon the experience of participants in the development of a proof-of-concept performance of Hagoromo we have begun to understand viable processes that can sustain the artists in using technology to create virtual performances. These processes consider the individuals' perspectives on interactive technology, and the flexibility of online media that enables the artists to participate and feel included in community activities. Providing virtual online space for intercultural dialogue and performance to occur can promote a sense of agency for the individual and the development of competencies that can be carried forward. It is clear through our preliminary reflections and the success of the Hagoromo performance that online solutions and interactive arts technology can enable participation specific to individual needs which may improve inclusion and sense of belonging via community arts practice.

We consider our contribution a work in progress toward a longer-term research investigation to understand participation in the context of interaction design and community art to foster inclusion and contribute to positive change in personal (and collective) well-being. However, we see several challenges ahead such as the economic viability of the technology for underfunded community arts organizations, the accessibility and skillsets required to operate the technology, along with the persistent latency issues of online media and virtual technology connected across distance. To realize collaborative performances in remote locations, technologies that reduce the latency of video data sharing are necessary.

In the future, we will develop technologies and performance designs that allow the audience to participate in the performance. To bring about a new audience experience that is different from viewing a performance on-site, for example, changes in the images and sounds generated can be brought about by the operation of the viewer's mobile device to expand the participation of the audience.

Mental health issues caused by self-isolation and COVID-19 restrictions has had a significant impact on the lives of people with disability and those involved in the arts sector that support them. As shown in this project, hybrid online collaborations are opening new possibilities for international artistic expression for artists with a disability. In the event the pandemic continues to restrict peoples travel and mobility, hybrid face-to-face and online performances will continue to be an important option for community art activities.

Acknowledgements. Resonant Webs is supported by grant funding from the Australia Japan Foundation of the Department of Foreign Affairs and Trade; Toyota Foundation D19-ST-0015 (Interactive Arts and Disability: Creative Rehabilitation and Activity for Individuals with a Disability), and JSPS 17K00740. The authors wish to thank Slow Label, Ryoko Aoki and Minato City: Cultural Program for their support.

References

1. Flew, T., Kirkwood, K.: The impact of COVID-19 on cultural tourism: art, culture and communication in four regional sites of Queensland, Australia. Med. Int. Australia **178**, 16–20 (2021)
2. Deloitte. https://www2.deloitte.com/au/en/pages/media-releases/articles/covid-19-australias-60bn-income-pain-290420.html
3. Australian Council for the Arts: Select Committe on COVID-19 inquiry into the Australian Government's response to the COVID-19 pandemic. Australia Council for the Arts (2020)
4. Agency for Cultural Affairs, Government of Japan. www.bunka.go.jp/koho_hodo_oshirase/hodohappyo/92738101.html
5. Rae, P.: How Will the Arts Recover from COVID-19. University of Melbourne, Melbourne (2020)
6. Halcombe, J.: COVID-19, digital inclusion, and the Australian cultural sector: A research snapshot. Digital Ethnography Research Centre (2021)
7. Australian Government. https://www.health.gov.au/news/health-alerts/novel-coronavirus-2019-ncov-health-alert/advice-for-people-at-risk-of-coronavirus-covid-19/coronavirus-covid-19-advice-for-people-with-disability
8. Reddihough, D.S., Meehan, E., Stott, N.S., Delacy, M.J., Group, A.C.P.R.: The national disability insurance scheme: a time for real change in Australia. Dev. Med. Child Neurol. **58**, 66–70 (2016)
9. Dunphy, K., Kuppers, P.: Picture This: Increasing the cultural participation of people with a disability in Victoria. State Government of Victoria, Office for Disability, Department of Planning and Community Development (2008)
10. Duckworth, J., Hullick, J., Mochizuki, S., Pink, S., Imms, C., Wilson, P.H.: Interactive arts and disability: a conceptual model toward understanding participation. In: Brooks, A., Brooks, E.I.B. (eds.) ArtsIT/DLI -2019. LNICSSITE, vol. 328, pp. 524–538. Springer, Cham (2020). https://doi.org/10.1007/978-3-030-53294-9_38
11. Imms, C., Granlund, M., Wilson, P.H., Steenbergen, B., Rosenbaum, P.L., Gordon, A.M.: Participation, both a means and an end: a conceptual analysis of processes and outcomes in childhood disability. Dev. Med. Child Neurol. **59**, 16–25 (2017)
12. Challis, B.P.: Assistive synchronised music improvisation. In: De Michelis, G., Tisato, F., Bene, A., Bernini, D. (eds.) ArtsIT 2013. LNICSSITE, vol. 116, pp. 49–56. Springer, Heidelberg (2013). https://doi.org/10.1007/978-3-642-37982-6_7
13. Gehlhaar, R., Rodrigues, P.M., Girao, L.M., Penha, R.: Instruments for everyone: designing new means of musical expression for disabled creators. In: Brooks, A.L., Brahman, S., Jain, L.C. (eds.) Technologies of Inclusive Well-being, pp. 167–196. Springer Berlin Heidelberg, Berlin, Heidelberg (2014). https://doi.org/10.1007/978-3-642-45432-5_9

14. Brooks, A.L., Boland, C.: Electrorganic technology for inclusive well-being in music therapy. In: Brooks, A.L., Brahman, S., Kapralos, B., Nakajima, A., Tyerman, J., Jain, L.C. (eds.) Recent Advances in Technologies for Inclusive Well-Being. ISRL, vol. 196, pp. 373–390. Springer, Cham (2021). https://doi.org/10.1007/978-3-030-59608-8_20
15. SLOWLABEL. https://circus.slowlabel.info/en/
16. Igarashi, T.: Social Circus Stage Spectacular in Tokyo Sees Impaired Performers Wowing Audiences. The Mainichi Newspapers, Japan (2021)
17. SLOWLABEL. www.slowlabel.info/4068/
18. Hullick, J.: The rise of the amplified elephants. Int. J. Commun. Music **6**, 219–233 (2013)
19. Rogers, J.M., et al.: Co-located (multi-user) virtual rehabilitation of acquired brain injury: feasibility of the resonance system for upper-limb training. Virtual Reality **25**, 719–730 (2021)
20. Konparu, K.: The Noh Theater: Principles and Perspectives. Floating World (2005)
21. Fenollosa, E., Pound, E.: The Noh Theatre of Japan: With Complete Texts of 15 Classic Plays. Dover Publications, Incorporated (2004)
22. SLOWLABEL. www.youtube.com/watch?v=bogvkdovOuM
23. Jolt Sonic & Visual Arts. https://www.joltarts.org/projects/disruptive-critters
24. Hullick, J.: Prosthetic abilities: conceptualizing sound machines for amplified elephants. Leonardo **49**, 148–155 (2016)
25. Duckworth, J., et al.: Resonance: an interactive tabletop artwork for co-located group rehabilitation and play. In: Antona, M., Stephanidis, C. (eds.) Universal Access in Human-Computer Interaction. Access to Learning, Health and Well-Being. Lecture Notes in Computer Science, vol. 9177, pp. 420–431. Springer, Cham (2015). https://doi.org/10.1007/978-3-319-20684-4_41
26. King, G., et al.: Residential immersive life skills programs for youth with physical disabilities: a pilot study of program opportunities, intervention strategies, and youth experiences. Res. Dev. Disabil. **55**, 242–255 (2016)

Facilitating Mixed Reality Public Participation for Modern Construction Projects: Guiding Project Planners with a Configurator

Lena T. Schramm[1(✉)], Anuja Hariharan[1], Tobias Götz[2], Jonas Fegert[3], and Andreas P. Schmidt[4]

[1] CAS Software AG, 76131 Karlsruhe, Germany
{lena.schramm,anuja.hariharan}@cas.de
[2] Justus-Liebig-Universität Gießen, 35390 Gießen, Germany
tobias.goetz@wirtschaft.uni-giessen.de
[3] FZI Research Center for Information Technology, 76131 Karlsruhe, Germany
fegert@fzi.de
[4] Karlsruhe University of Applied Sciences, 76133 Karlsruhe, Germany
andreas_peter.schmidt@h-ka.de

Abstract. Digital public participation formats are an emerging and accessible way to involve diverse groups of citizens in construction projects in their local area. Particularly, mixed reality can help project initiators to visualize the planned changes to the city landscape in an easy and understandable way, enabling people to participate in a creative manner. However, this technology is challenging for most project initiators, as it requires an extensive technical and/or domain experience. Besides that, specialized hardware and experienced staff is required. An easy on-boarding process, which introduces mixed reality step-by-step and offers assistance by external service providers could promote both adoption and usage. In this paper, we present the design process and resulting concept of a configurator for a public participation platform, that aims to guide initiators with different levels of technical knowledge. Besides detailing the design and development process of the prototype, we will present the preliminary results of our evaluation. The interview partners provided positive feedback on the usage of our configurator. Moreover, different approaches are necessary for the public and private sector when configuring and purchasing their participation solution. Finally, we highlight areas that are still in need of further work, such as the compliance with the regulations for public institutions and address further promising areas of research.

Keywords: Public participation · Mixed reality · Construction projects · Product configuration

1 Introduction

Eliciting citizen's participation in public projects has remained a challenge. Although construction projects and urban planning directly affect everyday life of many individuals, it is difficult to motivate people to engage with the projects more in-depth [1].

M. Wölfel et al. (Eds.): ArtsIT 2021, LNICST 422, pp. 275–291, 2022.
https://doi.org/10.1007/978-3-030-95531-1_19

Visualizing ideas using augmented reality and virtual reality seems to be a promising approach to arouse interest and provide information, as well as to foster participation in the form of ideas and discussion about a project [2, 3]. City planners and project initiators are often faced with the complex task of delivering different types of information to distinct audiences [4] that need to be made available. Participation requirements may vary across projects, initiated by the same client. For instance, the extent and kind of information to be provided, and whether citizens ought to be involved in a consulting role or rather as a customer might differ. Participation is seen today as a spectrum [5] of different activities that ranges from informing to empowering citizens by placing decisions in their hands. Although several approaches have been highlighted for modular and configurable e-participation architectures [6, 7], there is a dearth of examples that are suited for construction projects employing visualization techniques such as mixed reality (MR). In addition, although several of these platforms can be customized for different projects, there are few examples that function as a configurator as well as a market place for services and providers in case of insufficient competencies by the initiating institutions. In this paper, we present the concept, design and development of a platform configurator. Project initiators can configure their participation process, customize it by choosing relevant participation modules and features. In addition, they can use the platform to interact with as well as offer interaction opportunities to their target population.

Since MR is still regarded as an emerging technology, the willingness to utilize it is an important factor in our research. The current adoption of MR and more specifically virtual reality (VR) devices is still low, as can be illustrated with the Steam Hardware Survey[1]. According to the reported owner numbers among users of this digital video game store, one of the key target groups of this technology, currently ~2,3% own such a device. This is significantly higher than the adoption in the general US population[2,3], where virtual reality still struggles to gain traction [8]. Accordingly, it must be assumed that less technology-savvy users do not yet have any experience with MR systems, therefore it needs to be introduced to them and its benefits must be demonstrated.

Additional to a flexible configuration, project initiators can be supported with external service providers during the configuration process and find support for competencies (such as mixed reality content) that are not readily available. The prototype is developed as part of the research project Take Part[4]. The design process is described in detail, followed by the presentation of the prototype, and an overview of the evaluation by means of qualitative interviews. First results show the need to adopt different on-boarding processes for private and public sector construction projects.

[1] https://store.steampowered.com/hwsurvey (last accessed 2021/06/30).

[2] https://www.emarketer.com/content/us-virtual-augmented-reality-users-2021 (last accessed 2021/10/07).

[3] https://omdia.tech.informa.com/pr/2020-dec/six-and-a-half-million-consumer-vr-headsets-will-be-sold-in-2020 (last accessed 2021/10/07).

[4] https://takepart-projekt.de/take-part/ (last accessed 2021/10/18).

2 E-Participation and the State of Configurators

E-Participation emerged in the 2000s as an interdisciplinary research area in between computer science, information systems, political science and public policy [9, 10]. Our early exploration of the topic took place in the field of various small-scale experiments, local governments and small startups conducted with the involvement of citizens through basic tools, which were often developed only for a specific use case [11]. With the general shift towards a platform economy, the development of e-participation artifacts also shifted towards more elaborate concepts, which were no single purpose tools but suitable to be easily rolled out and modularized by non-experts. The European platforms CitizenLab[5], Adhocracy+[6]and Dialogzentrale[7] are well-known examples based on the described business model and marketed accordingly. These platform services were then adopted by cities like Berlin, Utrecht, Seattle and London. The content is maintained by employees of the respective city and the software is either operated by the cities or the platform providers themselves. In addition, although several marketplace examples exist for shared cloud computing providers, there is a dearth of configurable platforms that are tied to an appstore and a marketplace-like concept for different types of service and content providers [12, 13]. The technological options, therefore, heavily depend on the platform providers and an extension of the existing modules with new functions is not intended.

With the research project Take Part, we would like to propose an e-participation platform that offers more flexibility for the initiators of public participation processes through its configurability. Therefore, we are suggesting the use of a configurator with integrated marketplace for e-participation, where initiators can both select participation modules as well as commission third parties with technical and domain expertise. In our case studies, we examined this marketplace model with particular attention to the integration of MR in e-participation. The basic assumption was that the technical capabilities of these initiators would neither allow the creation of high-quality 3D models, nor the creation of separate MR applications. To allow them to keep up with major shifts in the technological market, like the mainstream use of augmented reality (AR) on smartphones, we propose a collaborative marketplace [14] where external service providers can be hired by the respective initiator itself. Subsequently, those external providers can then create content in the specified format, which is then seamlessly integrated in the existing e-participation ecosystem. Depending on the digital maturity of the project initiator, if 3D models in required formats are available from the design phase, these can be directly uploaded on the system and be prepared for consumption in virtual or augmented reality. In case new models have to be created, suitable service providers can be suggested by the system, which the project initiator can choose from and collaborate with, to create their design alternatives for public participation.

We see an advantage in shrinking the existing interdependence between the cities and the individual companies, by arguing that project initiators could then more easily start a participation process with a small selection of modules and later move on to more

[5] https://www.citizenlab.co/de (last accessed 2021/06/30).

[6] https://adhocracy.plus/ (last accessed 2021/06/30).

[7] https://www.zebralog.de/node/275 (last accessed 2021/06/30).

complex forms of participation (like MR visualizations). Thereby, they would not be committed from the beginning on towards a specific kind of participation process and could respond in a more agile way to the citizens' wishes. A collaborative marketplace could therefore simplify the implementation of e-participation. This flexible and simple on-boarding process could be attractive for project initiators, that first want to get to know the platform and get comfortable with all the possibilities offered. Furthermore, due to the enormous price differences that can arise from the use of MR visualizations, comparing different agencies and configurations at the time of purchase can help local government agencies to be more cost efficient. The design and development process of the configurator with a collaborative marketplace for e-participation will be detailed in the following section.

3 Design Guidelines for the Platform Configurator

In order to develop a configurator for the participation platform Take Part, an analysis of successful, existing configurators was carried out. Patterns applicable to the configuration problem at hand were identified. For this purpose, the technique reverse wireframing [15] was applied. In reverse wireframing, a system, such as a website or an app, is reduced to its elementary structures and elements and all aesthetic designs, images, and content is omitted. Analysis of the desired design is hence possible without visual noise and distractions, since the focus is only on the structure of the system and its elements. With the technique, also used in re-engineering of existing systems and structure, considerable time can be saved in the development of a new system. The tool "Balsamiq" was used to create the wireframes (see Fig. 1).

3.1 Usability Guidelines

To achieve the goal of guided configuration, the configurator should have a good usability [16, 17] and therefore be intuitive and easy to use. To this goal, we derived a number of principles from existing configurators, the criteria laid out by the "Konfigurator-Verzeichnis"[8] and considered similar existing research [18]. The following central design principles resulted from analysis as well as the evaluation of the wireframes:

Navigation. First, the user should be aware at any time in which step of the process they are situated in, what has already been configured, which options are still available and which attributes of the product are being changed at the moment. The various configuration options should be clearly grouped and, if necessary, divided into steps.

Support. Second, to improve support and guidance during the configuration process, descriptive texts in the form of tooltips or info pop-ups can be utilized to guide and support the user. In each step of the configuration, an info button is available. An info pop-up can be opened, to see a description of the current step. However, reading these texts should not be a prerequisite for easy and correct use of the system. The system should assist the user in observing restrictions. These should either be considered automatically; matching

[8] https://www.konfigurator-verzeichnis.de/ (last accessed 2021/06/30).

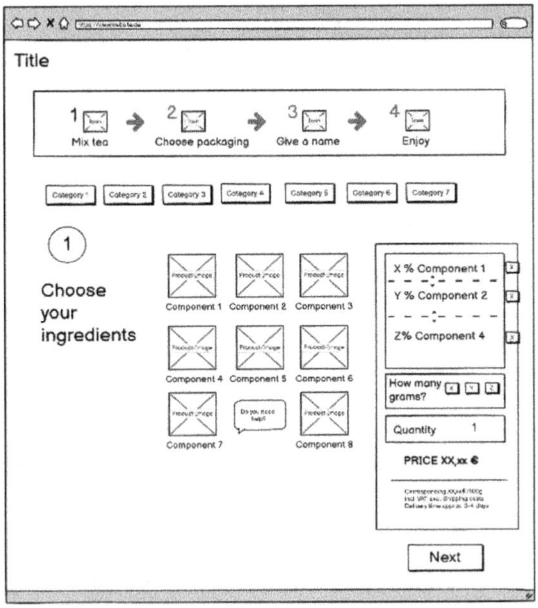

Fig. 1. Mockups using reverse wireframing for the pilot study.

components should only be shown so the user can choose, or a warning message should emerge showing an incompatibility.

Look & Feel. Third, to promote usability, intuitive operating concepts such as drag-and-drop functionalities can be used when appropriate. Another commonly used concept is card design [19–21], where the focus is on the product image and a headline. Further, important information such as the current total price of the configuration, the individual components and important technical characteristics should always be available. In the best case, there should be a list or an information sheet on which the information about the current configuration is displayed.

Short Loading Times. Fourth, short loading and waiting times can have a positive impact on the user experience (UX) in addition to usability. To achieve a short waiting time, data transfer should be efficient. In the configurator, this can be achieved by loading only new page content and keeping the rest of the layout constant. This concept is implemented, for example, in a one-page design, in Progressive Web Apps [22], or a single page application [23].

3.2 Design Dimensions

Decision Support Systems are already extensively researched and sophisticated models exist to represent the use of these systems [24–28]. These, however, are beyond the scope of this research and a simplified model was created instead. This model was based, in addition to the usability findings from the evaluation, on the following design dimensions.

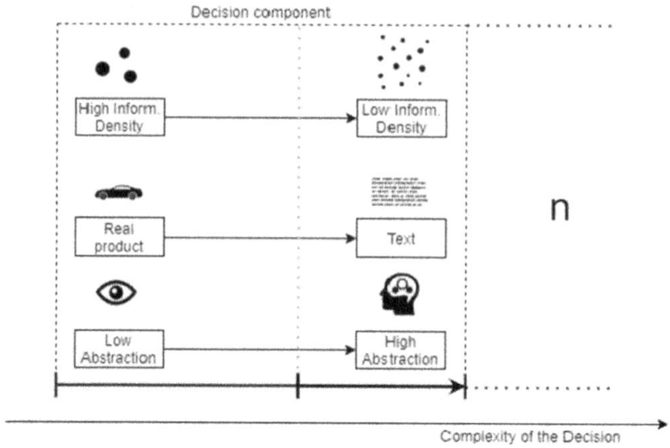

Fig. 2. Design dimensions for user decision complexity.

As illustrated by Fig. 2, a decision consists of one or more decision components that influence it. For example, choosing a set of modules contains multiple decisions about the need of specific single modules. The complexity of a single component is determined by the following dimensions which we applied in our design process.

Information Density: Low - High. The first dimension concerns the presentation of information. It can be decided that the user of the configurator should be provided with as much information as possible on a topic. The more information is offered, the better is the awareness for the topic. However, with more information, cognitive fatigue increases, as the user has to repeatedly decide whether the provided information is relevant to him or not. Hockey refers to this process as "management of control" [29], the decision to do the right thing, which is a major cause of cognitive fatigue. Therefore, the amount of information must be appropriately balanced. In the configuration process, there should always be enough information about a component, an element, a decision step, and the product. However, the user must not be inundated with too much text, whereas product photos are helpful, as they can be easily understood. The analysis revealed that many configurators interact with information tools to provide customers with access to further information if required. This enables non-expert users with a greater need for information to use the configurator better, whereas experts can ignore this functionality.

Product Representation: Textual Description - Real (End) Product.
Another dimension lies in the product representation. The product can be presented to the user in detail. The most accurate presentation is the actual product. For example, in the presentation of a "virtual reality" module, the product can be visualized and presented to the user with VR glasses. This presentation is both time and cost intensive, since technical aids may be required, a demonstration must be possible, or the product might not even be fully developed before a customer places an order. The other extreme is a preview image or textual description of the module. A balance must be found between insufficient product description and marketing cost. The closer the representation is to the real product, the higher the associated cost. For some modules, a demonstration does

not require much effort because it is an available or easily represented software product. This dimension is expected to correlate inversely with information density and abstraction, as an accurate demonstration is more dense in comparison to a highly abstract description of the product.

Decision-making Process: Simple (Binary) - Complex (Non-binary). The problem of making the right decision can be facilitated or encouraged by the complexity of the decision process. The user could be required to make only binary decisions at each step. For example, when selecting suitable components, they could be presented one choice at a time and the user only has to decide whether they are needed or not (yes/no). This, however, lengthens the configuration process by a considerable margin. If several decisions are made in a step, the process shortens but the user has also a higher cognitive load. In summary, the higher the number of decisions in one step and the higher the complexity of each decision, the more complex is the overall decision process. However, in the individualization of a product, the user should also be given the feeling of having a wide range of choices. The number of available options per component should not be too low, but also not too high, as the user could quickly feel overwhelmed [30], whereas a low number may be insufficient for experts. When presented with many configuration options, the user should be supported in the decision-making process by a filter function or automatic preselection. In Fig. 2 this is represented by the number of decision components that are part of a decision.

The dimensions presented were taken into account in the development of designs and the selection of suitable patterns.

4 Platform Configurator Development

After analyzing existing configurators, a process for configuring the Take Part participation platform was developed. The designs and ideas developed were evaluated in a pilot study to determine the most appropriate approaches. The results of the analysis were used to design the process with suitable UI elements and to develop drafts for a prototype. In this section, the different steps and UI decisions made for the configurator that resulted out of the pilot study are described. For the evaluation we chose a combination of a quantitative and qualitative approach. Based on a questionnaire we created polarity profiles and determined the preferred designs. For this we used the "user experience questionnaire" (UEQ[9], short version) and a slightly adapted version of the "system usability scale" (SUS) [31]. Additional to the questionnaire we conducted one-on-one interviews and applied the thinking aloud method [32]. With a few exceptions, the participants were employees of a leading mid-sized software firm and experts in the field of UX and interface design. A detailed summary of the procedure and the results have been published [33]. In the following steps, "user" refers to the project initiator and/or the project coordinator handling the construction process as well as the publicity, marketing and participation experts in charge of the processes.

[9] https://www.ueq-online.org/ (last accessed 2021/06/30).

4.1 Concept

The following steps were derived for the configuration process. At the beginning of the configuration, the user is taken to a start page where the participation platform Take Part and the app are briefly described. A video can be found in which the platform is concisely explained and demonstrated.

Step 1: General Information about the Project. At the beginning of the configuration, the user defines general information about the project needed to advise the user later in the configuration process. This includes, for example, the purpose they are pursuing by providing the participation platform, as well as the geographical range of people they wish to reach. In order to be able to create a basic version of the project page on the platform or to facilitate subsequent consultation, information such as the name of the contact person, the project name, the location of the construction side and the planned project duration, already existing website or brief description are gathered. This information can be used, for example, to determine which citizen groups are notified about the new project on the platform. The range is determined by specifying a radius around the location of the construction project on a map, which is compared to the location of registered citizens. Other definitions of outreach could include specifying a particular city, country, or even targeting a user group, such as a company's employees. If necessary, it must be specified here whether the project is publicly available or should only be visible to a specific audience.

Step 2: Goal of Participation. In the second step, the goal of public participation can be defined using the mentioned Participation Spectrum [5]. It consists of five successive stages in which the citizen's influence on decisions increases progressively, accompanied by promises to citizens, which are communicated implicitly or explicitly. The user selects the desired participation level. These are briefly described and are used to recommend modules in Step 3, *module selection*. In the next step ("module selection"), to give a complete overview, all non-recommended modules are nevertheless present and displayed to the user regardless of their choice.

Step 3: Module Selection. In the module selection step, the project initiator can select the required participation formats that will be available for participating in the project. These are described briefly in the overview to be comparable at a glance, but more detailed information is available as well. A video can be provided for each module, to support the users' understanding. In addition to the attribute-level constraints, there are some inter-module dependencies to consider from a business perspective. For example, the "Surveys" module is only relevant if citizens have previously been informed by the "Information" module or an MR element about the topic on which they are to vote on. However, it is possible that a project initiator may still wish to purchase only one of the modules. The module options should therefore be available and only a recommendation should be given by the configurator.

The modules are thus divided into two lists: recommended modules and other modules. An overview of the modules in the basic package is also provided (Appendix Fig. A). The modules can be filtered by price, interaction options and participation level. Each module is assigned to a participation level. All modules whose assigned level is less than or equal to the level previously selected by the user are displayed as "Recommended". In addition, the user should have the opportunity to get a preview about the available modules and what is offered even before the configuration. This can be provided on a regular website external to the configuration process. Finally, an analysis of the previous configuration indicates the extent of various aspects (information for citizens, feedback collection, interactivity, opportunities for participation). These aspects have to be explored and improved in future research.

Step 4: Additional Functionalities. Once matching modules have been selected, their functionality can be configured. Additional features for each module, such as displaying a video or photo gallery, are presented to the users and they can decide which of them are needed and which remain deactivated.

Step 5: Marketplace for External Service Providers. For a participation process and most modules, certain specific competencies may be required, which the project initiator can fulfill on his own or which an external company can provide in the form of services. For example, the project initiator may already have received a 3D model from an architect and does not need any support in this regard. However, if this is not the case, they must find a provider/a specialized company who can create the required 3D models - compatible for augmented and virtual reality displays. The configurator thus shows the user which skills, content, or even technical equipment they need for the selected modules. The users can then decide whether they provide these themselves or obtain them from a provider.

In the configurator, providers can be suggested from which the project initiator can obtain an offer, or a service can be booked directly during the process (Appendix Fig. B). For this purpose, a partnership can be entered into with providers, or the "Competence Atlas" product from CAS Software AG can be linked via an interface. Similar to the project "farmshops.eu - direct marketer map"[10] of the Open Knowledge Foundation Germany, providers with certain competences can be found via a map. A special focus can thereby lie on local providers, with promotions to support them. The Competence Atlas hence functions as a marketplace, for users to find providers with specific domain expertise or technical competencies in a specific domain.

All available service providers and partner companies are displayed in a list, in case the user needs support. The name of the company and its distance from the project location are displayed. In addition, a short advertising text is available, as well as a link for references, through which the user can further inform himself about the provider. In addition, the location of the providers can be viewed on a map.

Step 6: Summary. The last step of the configuration process is a summary of the selected components (modules, additional functions, service providers) and the purchase. In order to offer the project initiators more flexibility and assurance, the user can send a non-binding appointment request for a consultation. An analysis similar to that during the configuration process is shown at the end of the configuration, which illustrates the selected modules and summarizes the expected participation effect. The various modules (apps) are bundled as a software package and made available through

[10] https://farmshops.eu/ (last accessed 2021/06/30).

the platform. Authorizing other users to assist in managing the project site and publishing content should be possible by default. In addition, a dashboard should be available for project initiators to view a summary of participation results.

There are several approaches to designing the configurator and the individual steps in the configuration process. Mockups were created for each step and the process flow as a whole. Since there are no technical dependencies between the individual modules and the definition of the exact contents is not considered within the scope of this research, a configuration in the direction of a "pick-to-order" configurator is possible. However, since the modules have to be configured with respect to the activated additional functions and added providers, the complexity of the configuration problem is more like an "assemble-to-order" problem. There are simple dependencies that have to be taken into account and the functions are available as prefabricated modules. The entire configuration process can be iterated several times by the user by adding new modules and new content in an agile fashion.

5 Prototype and Evaluation

5.1 Programming the Prototype

The designed configuration process is implemented as a web application in the CRM cloud solution SmartWe[11] [34]. The configurator should enable project initiators to create a new project in the Take Part app for making it available to citizens (horizontal scalability). The project data and activated modules and additional functions are transmitted to SmartWe via a REST API[12] as a JSON object. In the prototype, the Java library "smartdesign" is used to access the Web API of SmartWe. The Take Part app retrieves all the data records located in the database. The newly created project can thus be seen directly in the app, and further content can be added and maintained. After configuring a project, if the construction initiator wishes to book additional modules, the configuration process shown can be run through again.

Each page and some of the UI elements, such as the shopping cart or analysis are organized as components. Communication between the components is carried out with the help of a service. The "Module", "Additional function" and "External provider" features are products that can be purchased by the user during configuration. Additional functions and modules contain a list of jobs: for example, the "Live chat" function which can be activated in the "Discussion" module might require a moderator. The resulting job of moderating the chat can be done by the project initiator himself, or by an external service provider. Each job contains a list of service providers that can be considered for the task, considering inputs such as latitude and longitude, and range (in km) of service providers. Results are aggregated from external databases of third-party providers in Germany and depicted as hotspots and routes on the Yellow map[13] solution. After a project has been successfully created on the platform, the project initiator can manage

[11] https://smartwe.de/ (last accessed 2021/06/30).

[12] https://partnerportal.cas.de/SmartDesignSDK/SmartWe/ (last accessed 2021/06/30).

[13] https://www.yellowmap.com/ (last accessed 2021/06/30).

the content on the platform environment. In further development, the project should only be made public at the request of the project initiator after the content of the project has been completed.

5.2 Evaluation

To evaluate the prototype, the target group – project initiators – were identified and interviewed to assess feedback and potential improvements to the app. To this end, twelve semi-structured qualitative interviews were conducted with experts from different construction project contexts to evaluate the platform. Methodically, we followed a research approach suggested by Kaiser [35]. The interviews began with an introduction into the Take Part app and MR technologies to make the interviewees acquainted with MR. This was followed by concrete questions on specific topics concerning the initial and long-term usage of the app (such as desired participation levels by the initiator, use of configurator, relevant modules, interest in MR, and so on). Although, in this chapter preliminary results of those interviews, based on notes created from these interviews are presented, a detailed analysis based on a full transcription of the interviews could give more insights. For the complete analysis of the study, the interviews will be transcribed and a structured content analysis based on Kaiser [35] performed, using the software MAXQDA. In this paper, we present the qualitative interviews' preliminary results.

The interviews recognized, that the developed configuration process is well accepted by project initiators from the private sector and is suitable for this purpose. The participants rated the prototype as easy to use, well-structured and user-friendly. Further, they evaluated the configuration of the platform as intuitive and all steps were comprehensible. The interview partners stated that a filter option in the list of available service providers would be important to them in the provider selection process. In addition, detailed offers for the required services were reported to be missing. With a detailed service description, which was not available in the prototype, the interviewees reported that they would publish their project on the platform via this channel. However, all initiators insisted on a consultation appointment before making a final purchase decision, in which the contractual framework conditions and modules of the platform would be explained in greater detail. They would only waive this condition if a comparatively low investment value was required. Large companies that want to use the platform in the long term prefer an individual purchase agreement. It is therefore recommended that different price models be made available for SMEs and large companies, and that individual offers will be made possible.

The situation is different for project initiators from the public sector. In this case, cities that want to use the platform for their own projects are bound by the public procurement law that applies in Germany, particularly when commissioning service providers to fulfill its public tasks(Bundesgesetzblatt[14]). Therefore, project initiators from this area cannot select and commission any external service provider as designed but must publish a call for tenders for the required service. The same regulations apply to the platform itself. Take Part's offer therefore must be compared with similar participation platforms before a city can use the platform, unless they are below a certain cost limit. In future

[14] https://www.gesetze-im-internet.de/vgv_2016/ (last accessed 2021/06/30).

development, it must be examined to what extent the public sector can be supported in the tendering of required services.

Regarding the importance and acceptance of MR technologies in public participation processes by project initiators, at least four out of the twelve interviewed initiators found it essential to provide a good "media mix" to citizens, and perceived the introduction of MR elements in public participation processes as an "interesting" element. One initiator reported that for long term usage of public participation, more intelligent interaction methods for users would be necessary, and mixed reality is a promising approach in this regard. More than 50% of the initiators were not convinced of the necessity of MR for a digital participation process, of which two interviewees reported this being potentially owing to ow low levels of experiences with MR technologies. Initiators expressed concerns on acceptance owing to the ability of MR to reach the masses, particularly reaching citizens who are not mobile or techno-affine. The availability of accurate 3D models, achieving a high quality of MR experiences, and maintenance of 3D data in the planning process, were also perceived as hurdles in long-term usage of MR. In summary, the mixed reality aspect was not reported to be the key deciding factor that determined the use of the platform[15]. However, one initiator reported, that he/she believed sufficient marketing and an appealing, suitable presentation of MR content, would pave the way to increase acceptance of MR for public participation processes. Given that citizens had a very positive reaction to the use of MR for visualizing public construction processes, as shown from pilot studies and in final evaluations [36, 37] of the prototype, the move towards MR technologies for public participation could be driven by the increasing acceptance and usage amongst citizens.

6 Next Steps and Outlook

In this paper we developed design guidelines and a simplified dimensional model for designing and developing a platform configurator for public participation. Although the model was only tested specifically in our use case, we achieved good results in the prototype in terms of usability and acceptance. Even if there are additional dimensions to be considered, in our case it was sufficient to only focus on the three main dimensions summarized. In a more detailed analysis the interrelationship between the components and the decision complexity should be investigated further.

[15] Most of the initiators assumed, that they would use first the simpler, more familiar modules (such as providing surveys, information, photos, etc.) and found the networking effect of the platform useful (the ability to find service providers through the marketplace, as well as to connect with citizen pools of projects made publicly available by other initiators).

From the preliminary analysis of the interviews, we developed several insights for the future development of the configurator. The legal framework for citizen participation is major design driver for such platforms. There is considerable difference between the approaches for the public and private sector, both in terms of procurement processes as well as legal requirements for citizen participation. This applies to both using the platform services as well as the marketplace functions. Project initiators from the public sector have more restrictions during the configuration of the platform than those from the private sector. In the future, especially project initiators of the public sector should be able to specify a service description in the configurator, which is then automatically put out to tender. Suppliers can then send bids to the city management. The configuration process and the platform must be checked for conformity with regulations on participation processes applicable in Germany and the EU.

Furthermore, during the development and evaluation of the configurator it was recognized that the use of the levels of participation is not optimally suited. Rather than using a simple linear model for describing participation along several stages, it has become more promising to use a pattern-based approach in which configurations are chosen "by example" and based on successful configurations, which are selected based on similarity. Moreover, in the configuration process, it is important to ensure that the project initiator has thought about the intended participation process in detail in advance in order to avoid unconsidered selection of modules. Hence, a more generic approach of selecting modules based on categories or templates is recommended for specific use cases.

Support and recommendations for the project initiators can be further improved by using data on participation processes that have already taken place. For instance, a knowledge catalog on already completed reference projects can be provided. With this, project initiators can find out about similar projects that used the platform in the participation process and understand which modules were used at what stage of participation in the project. The result of the participation and the acceptance of the modules by the citizens involved can also be described there. A presentation of selected reference projects can also increase trust in the platform. Further, guidance during the process can be improved by providing recommendations for the use of certain modules and functionalities. In future work the effects of modules and functionalities on a participation process and citizens need to be analyzed. After that, an analysis of the participation platform based on the users' configuration and recommendations supported by artificial intelligence, can be implemented. In the later development of the participation platform, the required data to derive recommendations can be drawn from usage analysis during participation processes. For the start, studies on publicly documented participation procedures can serve as the initial data basis.

The availability of external service providers through the marketplace reduces the effort for initiators to develop as well as maintain the participation process on the platform long-term. Our configurator concept introduces initiators to the specialized technologies virtual reality and augmented reality in the context of public participation, which can increase acceptance and use of mixed reality in the long term.

Appendix

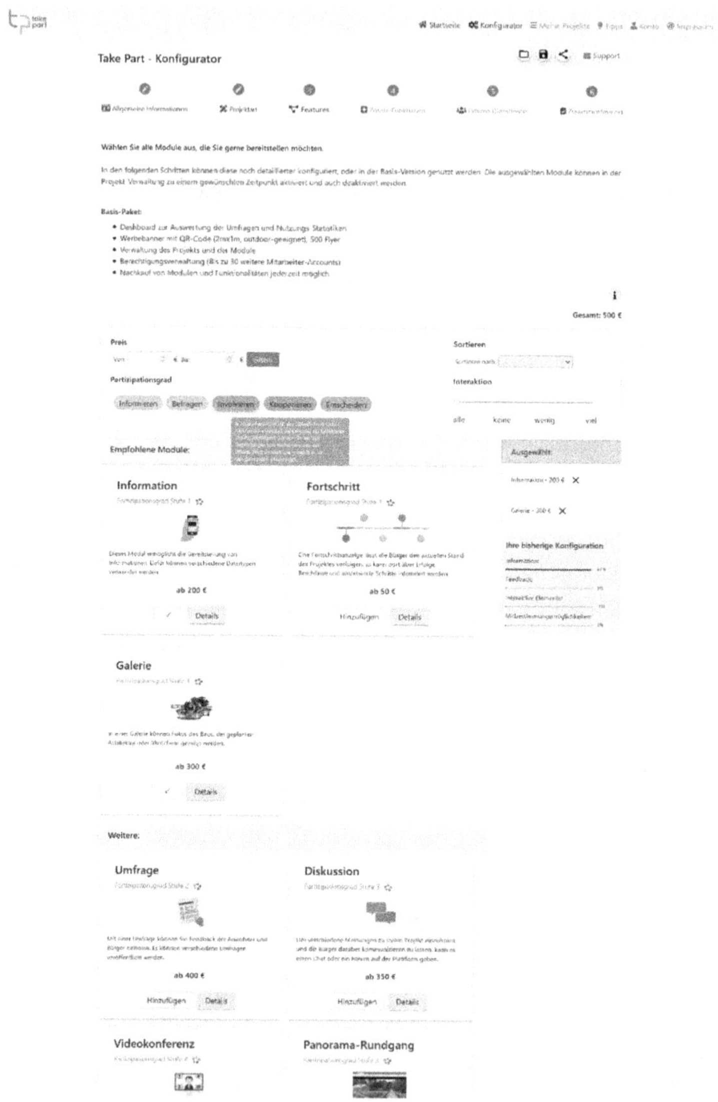

Fig. A. Step 3 (Module selection)[16].

[16] https://github.com/LenaS16/TakePartPaper/blob/b2796b0e68bdf9b7d06744157da64bfe09d850de/Modulauswahl-Screenshot.png (last accessed 2021/10/29).

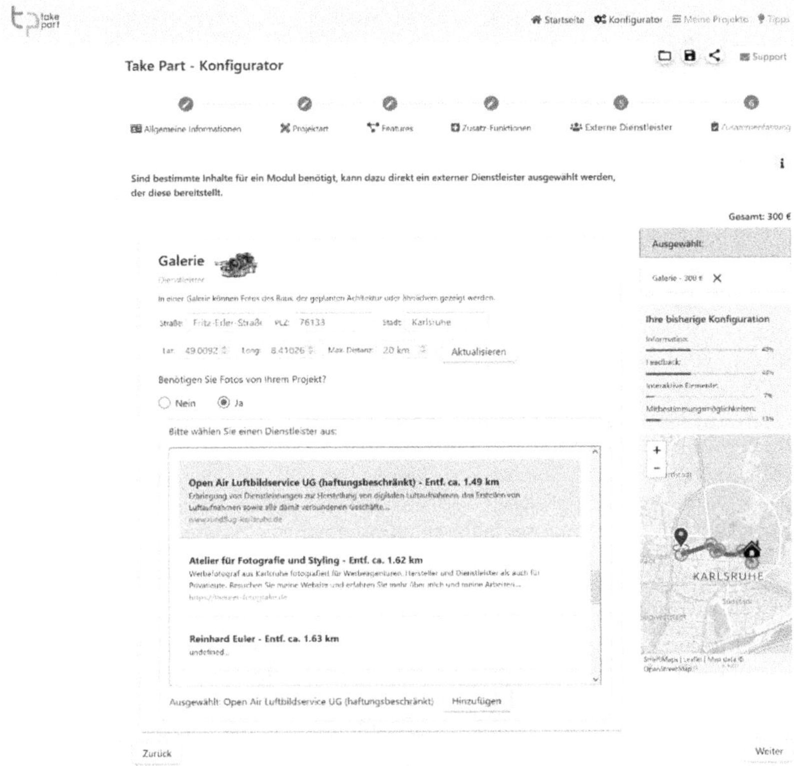

Fig. B. Step 5 (External service providers)[17].

References

1. Zepic, R., Dapp, M., Krcmar, H.: Participatory budgeting without participants: Identifying barriers on accessibility and usage of German participatory budgeting. In: 2017 Conference for E-Democracy and Open Government (CeDEM), pp. 26–35 (2017)
2. Wolf, M., Söbke, H., Wehking, F.: Mixed Reality media-enabled public participation in urban planning. In: Jung, T., tom Dieck, M.C., Rauschnabel, P.A. (eds.) Augmented Reality and Virtual Reality. PI, pp. 125–138. Springer, Cham (2020). https://doi.org/10.1007/978-3-030-37869-1_11
3. Van Leeuwen, J.P., Hermans, K., Jylhä, A., Quanjer, A.J., Nijman, H.: Effectiveness of virtual reality in participatory urban planning: A case study. In: Proceedings of the 4th Media Architecture Biennale Conference, pp. 128–136 (2018)
4. Goudarznia, T., Pietsch, M., Krug, R.: Testing the effectiveness of augmented reality in the public participation process: a case study in the city of bernburg. J. Digit. Landsc. Archit. **2**, 244–251 (2017)
5. International Association for Public Participation: IAP2 Spectrum of Public Participation (2018)

[17] https://github.com/LenaS16/TakePartPaper/blob/b2796b0e68bdf9b7d06744157da64bfe09d850de/Externe%20Dienstleister-Screenshot.png (last accessed 2021/10/29).

6. Alfaro, C., Gomez, J., Lavin, J.M., Molero, J.J.: A configurable architecture for e-participatory budgeting support. JeDEM-eJ. eDemocr. Open Gov. **2**, 39–45 (2010)
7. Cindio, F., Peraboni, C.: Fostering e-participation at the urban level: outcomes from a large field experiment. In: Macintosh, A., Tambouris, E. (eds.) ePart 2009. LNCS, vol. 5694, pp. 112–124. Springer, Heidelberg (2009). https://doi.org/10.1007/978-3-642-03781-8_11
8. Chuah, S.H.-W.: Why and who will adopt extended reality technology? Literature review, synthesis, and future research agenda. Lit. Rev. Synth. Futur. Res. Agenda. (2018)
9. Wirtz, B.W., Daiser, P., Binkowska, B.: E-participation: a strategic framework. Int. J. Public Adm. **41**, 1–12 (2018)
10. Macintosh, A., Coleman, S., Schneeberger, A.: eParticipation: the research gaps. In: Macintosh, A., Tambouris, E. (eds.) ePart 2009. LNCS, vol. 5694, pp. 1–11. Springer, Heidelberg (2009). https://doi.org/10.1007/978-3-642-03781-8_1
11. Nelimarkka, M., et al.: Comparing Three Online Civic Engagement Platforms using the Spectrum of Public Participation (2014)
12. Zissis, D., Lekkas, D.: Securing e-Government and e-Voting with an open cloud computing architecture. Gov. Inf. Q. **28**, 239–251 (2011)
13. Christina, K., Tsarchopoulos, P., Simitopoulos, D., ASI, A.G.Q.: Deliverable 5.1. 1 Body of Knowledge about the Migration of Public Services into the Cloud (2015)
14. Lönn, C.-M., Uppström, E.: Core aspects for value co-creation in public sector. In: Twenty-first Americas Conference on Information Systems. Association for Information Systems, Puerto Rico (2015)
15. Chen, J., et al.: Wireframe-based UI design search through image autoencoder. ACM Trans. Softw. Eng. Methodol. **29**, 1–31 (2020)
16. Bevan, N., Kirakowski, J., Maissel, J.: What is usability. In: Proceedings of the 4th International Conference on HCI (1991)
17. Nielsen, J.: What Is Usability? In: User Experience Re-Mastered, pp. 3–22. Elsevier (2010). https://doi.org/10.1016/B978-0-12-375114-0.00004-9
18. Abbasi, E.K., Hubaux, A., Acher, M., Boucher, Q., Heymans, P.: The anatomy of a sales configurator: an empirical study of 111 cases. In: Salinesi, C., Norrie, M.C., Pastor, Ó. (eds.) CAiSE 2013. LNCS, vol. 7908, pp. 162–177. Springer, Heidelberg (2013). https://doi.org/10.1007/978-3-642-38709-8_11
19. Lee, Y.-J.: Card-Based User Interface on Smart-Phone. J. Digit. Converg. **15**, 555–561 (2017)
20. Rodrigues, J.M.F., et al.: Adaptive card design UI implementation for an augmented reality museum application. In: Antona, M., Stephanidis, C. (eds.) UAHCI 2017. LNCS, vol. 10277, pp. 433–443. Springer, Cham (2017). https://doi.org/10.1007/978-3-319-58706-6_35
21. Roy, R., Warren, J.P.: Card-based design tools: a review and analysis of 155 card decks for designers and designing. Des. Stud. **63**, 125–154 (2019)
22. Tandel, S., Jamadar, A.: Impact of progressive web apps on web app development. Int. J. Innov. Res. Sci. Eng. Technol. **7**, 9439–9444 (2018)
23. Gavrilă, V., Băjenaru, L., Dobre, C.: Modern single page application architecture: a case study. Stud. Informatics Control. **28**, 231–238 (2019)
24. Marakas, G.M.: Decision Support Systems in the 21st Century, vol. 134. Prentice Hall, Upper Saddle River, NJ (2003)
25. Bonczek, R.H., Holsapple, C.W., Whinston, A.B.: Foundations of Decision Support Systems. Academic Press (2014)
26. Pfeiffer, J., Benbasat, I., Rothlauf, F.: Minimally restrictive decision support systems. In: Thirty Fifth International Conference on Information and Systems (2014)
27. Pfeiffer, J., Scholz, M.: A low-effort recommendation system with high accuracy. Bus. Inf. Syst. Eng. **5**, 397–408 (2013)
28. Wang, W., Benbasat, I.: Interactive decision aids for consumer decision making in e-commerce: the influence of perceived strategy restrictiveness. MIS Q. **33**, 293–320 (2009)

29. Robert, G., Hockey, J.: A motivational control theory of cognitive fatigue. In: Ackerman, P.L. (ed.) Cognitive Fatigue: Multidisciplinary Perspectives on Current Research and Future Applications., pp. 167–187. American Psychological Association, Washington (2011). https://doi.org/10.1037/12343-008
30. Gourville, J.T., Soman, D.: Overchoice and assortment type: when and why variety backfires. Mark. Sci. **24**, 382–395 (2005)
31. Brooke, J.: Others: SUS-A quick and dirty usability scale. Usability Eval. Ind. **189**, 4–7 (1996)
32. Olson, G.M., Duffy, S.A., Mack, R.L.: Thinking-out-loud as a method for studying real-time comprehension processes. In: Kieras, D.E., Just, M.A. (eds.) New Methods in Reading Comprehension Research, pp. 253–286. Routledge (2018). https://doi.org/10.4324/9780429505379-11
33. Schramm, L.T.: Gestaltung eines geführten Konfigurationsprozesses einer Bürgerpartizipations-Plattform für Bauprojekte (2021). https://ilin.eu/wp-content/uploads/2021/06/Bachelorthesis-Lena-Schramm.pdf
34. Bordeleau, F., Sillitti, A., Meirelles, P., Lenarduzzi, V.: Open Source Systems. Springer (2019). https://doi.org/10.1007/978-3-030-20883-7
35. Kaiser, R.: Qualitative Experteninterviews: Konzeptionelle Grundlagen und praktische Durchführung. Springer-Verlag (2014)
36. Fegert, J., et al.: Take Part Prototype: Creating New Ways of Participation Through Augmented and Virtual Reality. In: 29th Workshop an Information Technologies and Systems. WITS, Munich (2019)
37. Fegert, J., et al.: Ich sehe was, was du auch siehst. Über die Möglichkeiten von Augmented und Virtual Reality für die digitale Beteiligung von Bürger: innen in der Bau-und Stadtplanung. HMD Prax. der Wirtschaftsinformatik. 1–16 (2021)

Artificial Intelligence in Art and Culture

AI in Art: Simulating the Human Painting Process

Alexander Leiser and Tim Schlippe(✉)

IU International University of Applied Sciences, Bad Honnef, Germany
tim.schlippe@iu.org

Abstract. While AI is being used more and more to generate images, the generation usually does not resemble a human painting process. However, for applications in the field of art, it is useful to simulate the human painting process—e.g. in relation to location, order, shape, color and contours of the areas being painted in each step. Such applications are for example when a robot paints a picture or a program teaches humans to paint. Consequently, in this paper we evaluate and compare different approaches to simulate the human painting process. Additionally, we present our solution for this task which is based on a combination of filters and semantic segmentation. In our survey, this approach was rated as better and more realistic than the most realistic approach for this task so far which is a reinforcement learning approach: In all surveyed categories—location, order, shape, color and contours of the areas being painted in each step—always a significant majority of the participants prefers our approach to simulate the human painting process. When we displayed two time-lapse videos with the painting process of Edvard Munch's The Scream in parallel, even 79% found our generated process more realistic than the reinforcement learning-based process.

Keywords: AI art · AI in Art · Painting · Semantic Segmentation · Semantic Labeling · Machine Learning

1 Introduction

Painting is one of the most basic and oldest forms of art. Early findings of simple rock paintings are dating back almost 40,000 years [1]. While the general outcome of this art form is usually depicting a person or an object on a medium like paper or canvas, painters have developed different painting styles, techniques and methods over the past centuries to achieve this goal [2]. Since painting itself is a very visual form of art, the appeal of paintings can, at least in part, be conveyed through modern technologies. This also allows for simulation of the painting process as well as automated interaction with a brush, the main tool in this art form. Thus painting found its way into modern technologies such as AI and robotics. There are several examples of applications in this field:

M. Wölfel et al. (Eds.): ArtsIT 2021, LNICST 422, pp. 295–308, 2022.
https://doi.org/10.1007/978-3-030-95531-1_20

A team of scientists and engineers from IBM Japan, the University of Tokyo, and Yamaha Motors equipped an industrial robot with a camera and a paintbrush to explore the realms of creativity in machines and AI [3]. Other teams such as *AI_NORN*[1] and *cloudpainter*[2] also experiment with AI art, more specifically painting, to explore the outcomes when machines, which are capable of handling a paintbrush, are combined with modern AI technologies. The company Nvidia has a dedicated AI Art Gallery on their homepage[3] to show and support art projects which are generated or supported by AI. However, most of the works do not simulate a realistic human-like painting process or their simulation still has shortcomings. Consequently, the focus of this paper is on the simulation of the human painting process.

In the next section, we describe the painting process in more detail. In Sect. 3, we present the latest approaches of other artists and researchers. Section 4 characterizes our approach to simulate the human painting process. Section 5 describes our survey and the feedback on the approaches. We conclude our work in Sect. 6 and suggest further steps.

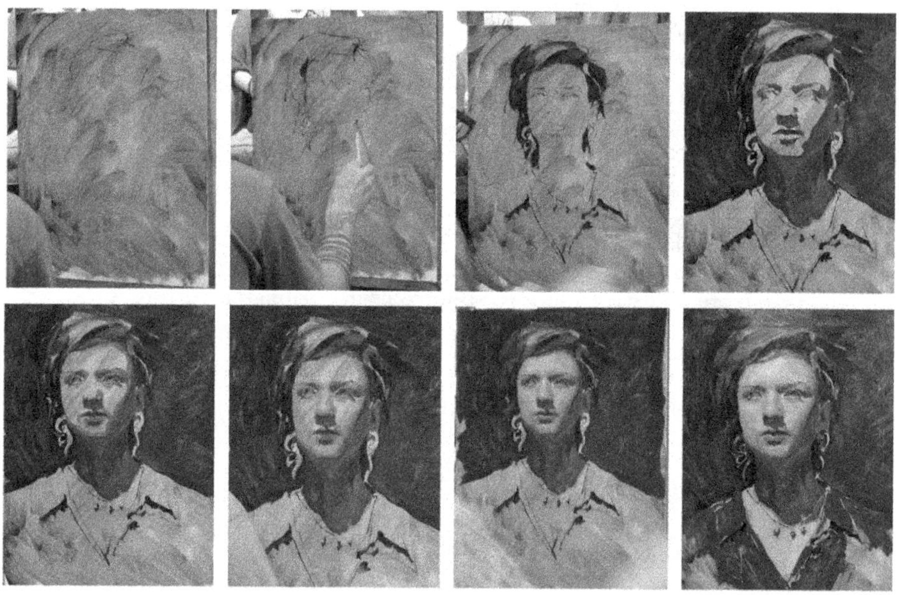

Fig. 1. Human painting process with a layering strategy [4].

[1] https://ainorn.art
[2] https://www.cloudpainter.com
[3] https://www.nvidia.com/en-us/deep-learning-ai/ai-art-gallery

2 The Human Painting Process

The painting process of humans usually follows a specific pattern from coarse to fine details. While there are many different techniques in painting which largely depend on the paint and medium used [5,6], many artists use the common layering strategy to paint objects with a background [7]. This is a widely used process especially for realistic paintings like portraits or still-life images. Our work will focus on the common layering strategy described in [7]. As shown in Fig. 1, in this strategy one layer is painted over another, until the final picture is created. The layering process starts with a uniform background and then gradually adds details in size decreasing regions until all the desired details are contained in the painting.

With the simulation of such a layering process all kinds of images could be created—from very abstract images to very realistic detailed images. The focus of this paper is on the creation of images which contain one or more objects and have a background in relation to the aspects location, order, shape, color and contours of the areas being painted in each step.

3 Related Work

Several studies and practical machine learning based approaches to replicate or generate images and paintings and to simulate the process of painting itself have been conducted in the past. While some methods capture the general picture and thus the outcome of the painting process very well, they often fail to reproduce the process itself in a human-like manner. Examples are [8–11].

[12] use a generic algorithm which resembles images well with the help of certain constraints. However, the painting process is not human-like since it uses semi-transparent polygons to approximate the original painting as close as possible. [6] uses a differential renderer in combination with a generative adversarial network to generate human-like brushstrokes. While the resulting brushstrokes and pictures have a human-made appeal, the process of generating them is not in the way most humans would paint a picture since the intermediate images show random brush strokes which do not seem to help achieve the end result of the picture. An example are arbitrary purple and yellow brushstrokes in the middle of the image, where in the final image is a snow-covered mountain that contains white and blue tones.

To the best of our knowledge, there is no database which contains enough images of intermediate steps of the human painting process to train generative adversarial networks for more human-like simulations. Google's "Quick, Draw!" [13] database contains 50 million drawings across 345 different categories to train neural networks for the human doodling process. However, the images are mostly in the form of simple doodles. Since the data set only contains the vectorial information of doodles with their start points, end points and tractorial information, the images have a limited resolution and are in only one color, the dimensionality is too low for the simulation of a full painting process. For example, [14] used the collected data of sheep doodles to generate 10,000 sheep published in the book "Dreaming of Electric Sheep". The lack of training images with intermediate steps of the painting process is one reason why we did not choose generative adversarial networks or reinforcement learning approaches but the leaner approach, described in Sect. 4.

[15] trained a convolutional neural network with 117 collected, 4-minute long time-lapse videos of real and digital paintings, to synthesize the time-lapse video of new paintings. While the algorithm outputs decent time-lapse videos, it is not suitable for the simulation of the human painting process: As visualized in Fig. 2, the transitions between the different painting stages are blurry and do not resemble the single painting process steps. Furthermore, in a step different colors simultaneously appear in different regions of the image.

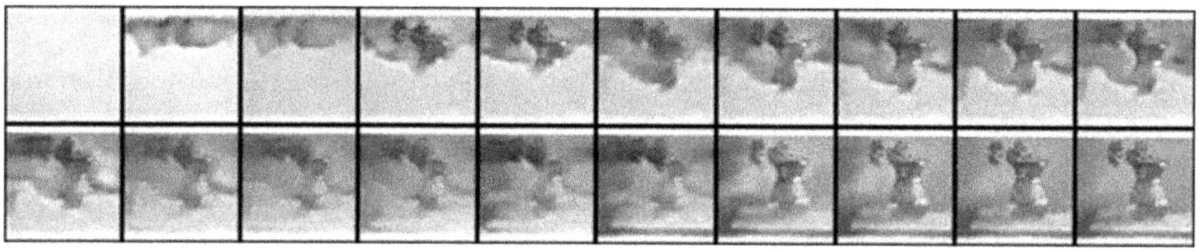

Fig. 2. Intermediate screenshots from synthesised time-lapse videos based on [15].

[16] and [17] apply reinforcement learning for the sequential decision-making. In reinforcement learning usually an agent is programmed to interact with an environment and improves its interactions with the given feedback coming from this environment [18]. Whereas [16] also focuses rather on the final result than on the actual process of generating the image, [17]'s approach tries to imitate the human painting process. [17] is close to the idea of [6] but captures the underlying picture sooner in a more human-like manor. The order of the brush strokes is not always human-like or intuitive. Nonetheless [17] serves as a good baseline since the results resemble the human painting process the closest of all described methods. Consequently, this approach was also evaluated in our survey for comparison with our own method.

4 Simulating the Human Painting Process Using Filters and Semantic Segmentation

While other methods focus on the resulting image or use not-so-human-like shapes or colors in the step-by-step expansion of the image during the painting process, our goal was to develop an algorithm that represents the human painting process as realistically as possible. Particularly with regard to the location, order, shape, color and contours of the areas being painted in each step, the generated painting process should be coherent. Additionally, it was important to us that we did not need huge amounts of image or video material of the painting process for training our computer vision models as is often the case with deep learning algorithms. With these conditions in mind, we developed an algorithm which is modular and thus flexible, which is lean, can be set up quickly, and emulates the layering process of a painter. Our simulation of the human painting process consists of the following components and steps, which are also visualized in Figs. 3, 4 and 5:

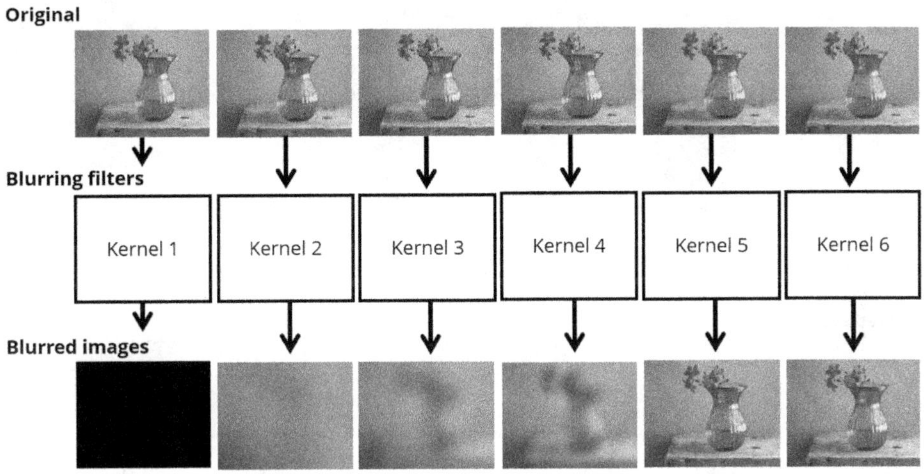

Fig. 3. Simulating the human painting process: Blurring filters.

1. *Blurring filters*: With the goal of coloring large areas first, the image is first blurred with various filters. As demonstrated in Fig. 3, the goal is to dilute the edges and colors in the image to different degrees so that the segmentation algorithm in step 2 outputs a different number of segments based on the details to be detected in the image. For our experiments we used 5 Gaussian filters with different kernel sizes to generate 5 blurred images. The implementation was done with OpenCV [19].
2. *Semantic segmentation*: We apply a semantic segmentation algorithm to the images blurred with different degrees to obtain smaller and smaller areas to be painted. As visualized in Fig. 4, the retrieved segments are given the color which has the most occurrence in that segment in the original image (*coloring*). For our experiments, we applied the unsupervised convolutional neural network based semantic segmentation described in [20] which minimizes similarity loss and spatial continuity loss to each blurred image.
3. *Stepwise adding colored areas*: Our goal is to add painted areas step by step. To avoid reapplying colors already applied in the painting process in the same place, we remove the areas that have the same color as the image on the left as shown in Fig. 5. Individual images are then created from the different color areas, with each new image corresponding to a step in the painting process,

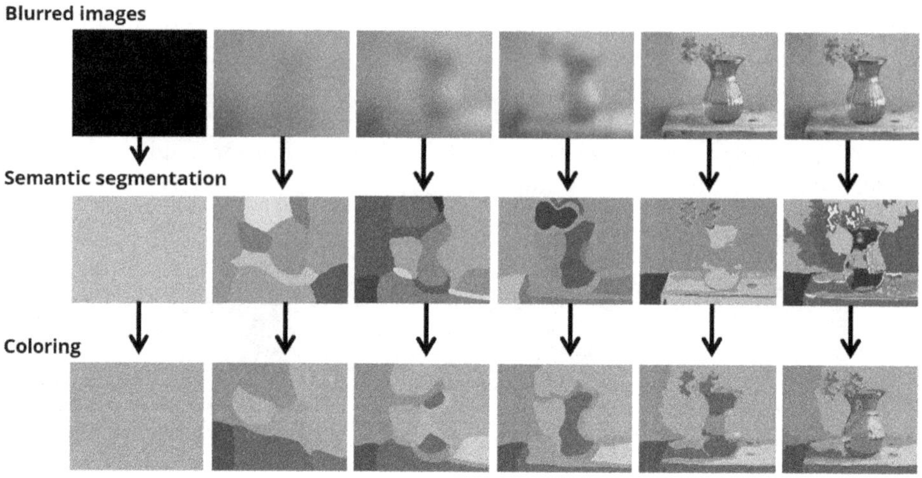

Fig. 4. Simulating the human painting process: Semantic segmentation.

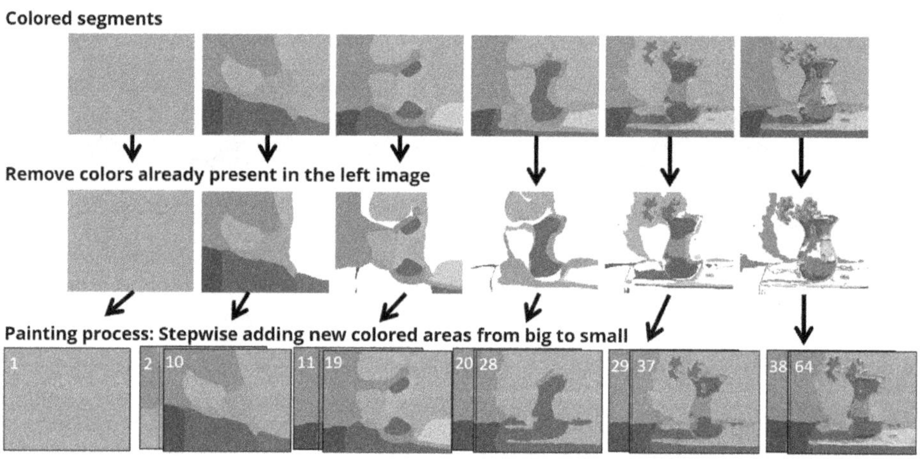

Fig. 5. Simulating the human painting process: Stepwise adding colored areas.

such as adding only one color to an area. The individual images are sorted in such a way that the images with the large color areas come first before the images with the smaller color areas.

Our approach requires no pre-training. It also eliminates the need for a pre-trained neural renderer to make the training process possible in a suitable amount of time as used by [20]. Due to the slim implementation, the algorithm can entirely be used locally or in cloud services such as Google Colab [21]. Single components like the *blurring* step and the *semantic segmentation* step can be replaced or extended. Images are generated in each intermediate step and can be extracted for further or other use. This makes our presented approach really versatile and accessible for future improvements. For example, in our experiments we used 5 Gaussian filters with different kernel sizes to generate 5 blurred images. But the number and the strength of the blurring effect could also be calculated based on the number of motives or the level of detail in the image. As illustrated in Fig. 6, our method paints in regions to slowly fill the canvas similar to the layering painting technique introduced in Sect. 2 instead of making semi-transparent brush strokes in seemingly random areas and combine them to the target image.

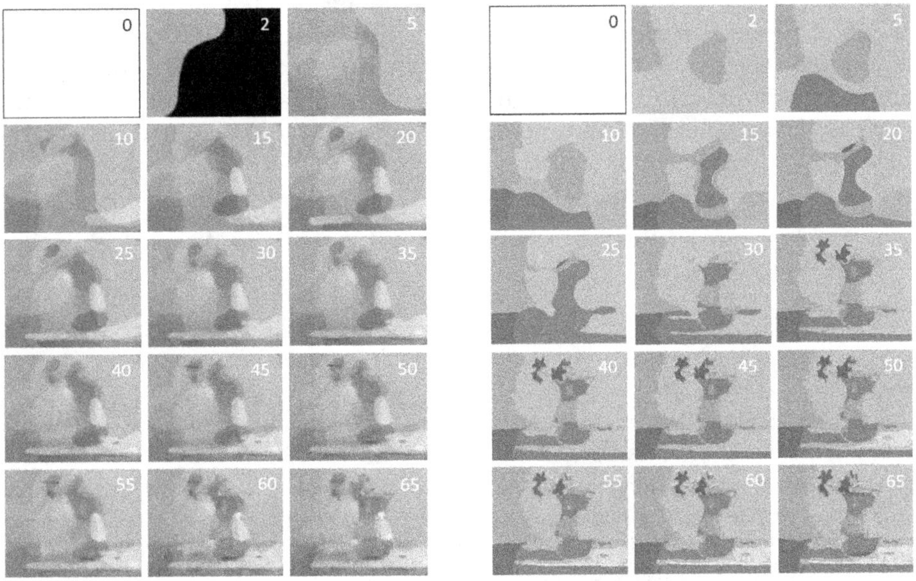

(a) [17]'s reinforcement learning approach.

(b) Our approach.

Fig. 6. [17]'s reinforcement learning approach vs. our approach.

5 Experiments and Results

5.1 Experimental Setup

We evaluated our approach in comparison to the most realistic approach [17] (*reinforcement learning*) for simulating the human painting process in a survey. The study examined the location, order, shape, color and contours of the areas being painted in the intermediate steps. The progress of the painting processes was demonstrated as a time-lapse video and an image sequence for each of 3 pictures—a vase, a lemon, and Edvard Munch's The Scream. Of course, we did not tell the participant what painting processes they were shown. For the picture with the lemon, we also conducted a Wizard of Oz experiment: In addition to the two AI-based processes, we asked questions about the painting process that a human had actually performed, without telling the participants. For the pictures of the vase and the lemon, the participants were always shown only one simulation per page in the questionnaire, so that they could rate one approach without the influence of another approach. However, in order to also evaluate the direct comparison, for Edvard Munch's The Scream we displayed two time-lapse videos in parallel with the reinforcement learning approach and our approach. The participants evaluated most questions with a score. The score range follows the rules of a forced choice Likert scale, which ranges from (1) *strongly disagree* to (5) *strongly agree.* 24 people (14 female, 10 male) filled out our questionnaire. The participants of our user study were randomly selected volunteers between 19 and 71 years old who participated free of charge. The participants' painting routine varies from once a week to once a year or even never. Most people indicated that they are interested in art, but there are also some who are not interested in art. We appreciate these distributions as it was important to us to get feedback from different people.

5.2 Evaluation in Relation to the Location of the Areas Being Painted

We asked the participants in our questionnaire how human-like they find the painting processes in relation to the location of the areas being painted. The goal was to find out whether the painting progress always happens in the right place. Figure 7 illustrates the feedback on the location for the vase and the lemon. While *reinforcement learning* was rated on average with 3.00 for *vase* and *lemon*, *our implementation* was rated better with 3.46 (*vase*) and 3.38 (*lemon*) on average. Thus, *our implementation* with regard to the location is rated 15% better for the vase and 13% for the lemon than *reinforcement learning.* The Wizard of Oz painting process *human painting* wins with an average of 3.83.

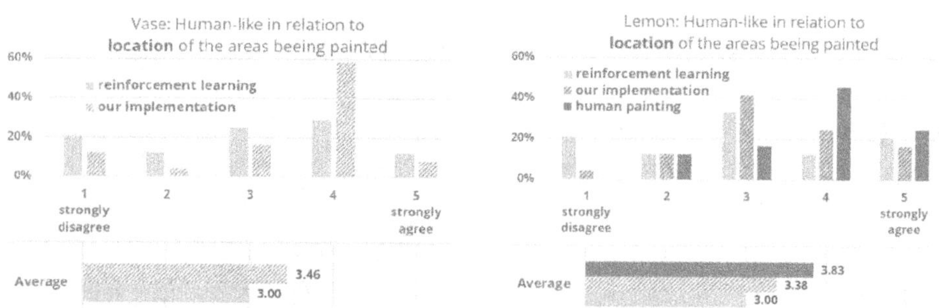

(a) Painting a vase. (b) Painting a lemon.

Fig. 7. Feedback on the location of the areas being painted.

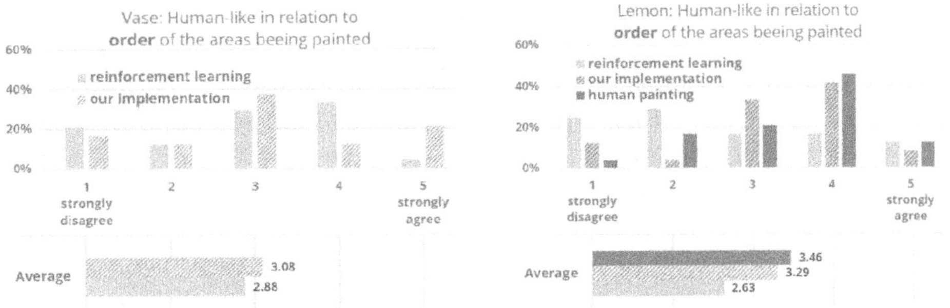

(a) Painting a vase. (b) Painting a lemon.

Fig. 8. Feedback on the order of the areas being painted.

5.3 Evaluation in Relation to the Order of the Areas Being Painted

Figure 8 illustrates our evaluation in relation to the order of the areas being painted. Overall, this evaluation is a bit worse than the evaluation of the location. But here, too, our method is rated on average between *reinforcement learning* and *human painting*: While *reinforcement learning* was rated with averages of 2.88 (*vase*) and 2.63 (*lemon*), *our implementation* was rated better with 3.08 (*vase*) and 3.29 (*lemon*) on average. This means that *our implementation* with regard to the order is rated 7% better for the vase and 25% for the lemon than *reinforcement learning*. *Human painting* outperforms the other approaches with an average of 3.46.

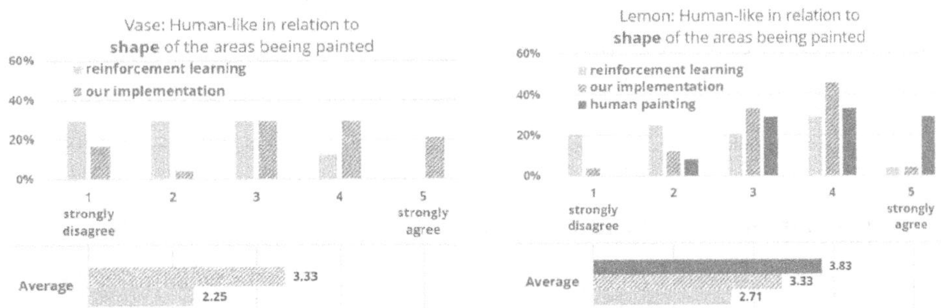

(a) Painting a vase. (b) Painting a lemon.

Fig. 9. Feedback on the shape of the areas being painted.

5.4 Evaluation in Relation to the Shape of the Areas Being Painted

In the category *shape* we are significantly better than *reinforcment learning*: As shown in Fig. 9, the question if the shape of the areas being painted is human-like was rated only with an average score of 2.25 (*vase*) and 2.71 (*lemon*). *Our implementation* was rated better with 3.33 on average for *vase* and *lemon*. Comparing the scores shows the significance of the shape: *Our implementation* with regard to the shape is rated 48% better for the vase and 23% for the lemon than *reinforcement learning*. *Human painting* again performs best, in this category with an average of 3.83.

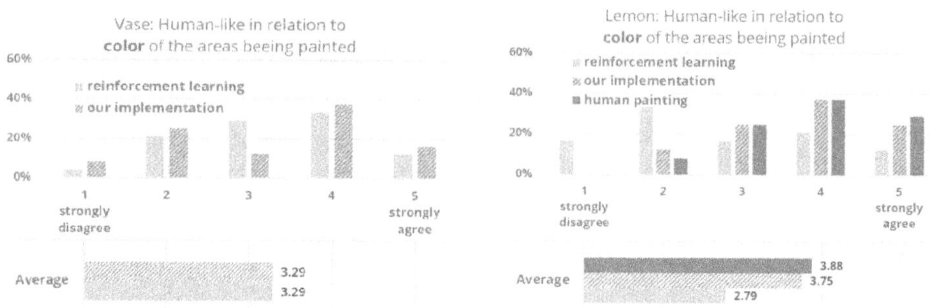

(a) Painting a vase. (b) Painting a lemon.

Fig. 10. Feedback on the color of the areas being painted.

5.5 Evaluation in Relation to the Color of the Areas Being Painted

Then we asked the participants in our questionnaire how human-like they find the painting processes in relation to the color of the areas being painted. The results are demonstrated in Fig. 10: While for *vase* both approaches *reinforcement learning* and *our implementation* were rated 3.29 on average, *reinforcement learning* performed with 2.79 and *our implementation* with 3.75 on average for *lemon*. Thus, *our implementation* with regard to the color is rated equal for the vase but for the lemon 34% better than *reinforcement learning*. *Human painting* outperforms the other approaches again, this time with an average of 3.88.

5.6 Evaluation in Relation to How and When Edges Are Painted

The final aspect which we evaluated was how human-like the painting process is in relation to how and when edges are painted. As illustrated in Fig. 11, the trends are as in the other aspects: For the *reinforcement learning* the question was rated with an average score of 2.54 (*vase*) and 2.88 (*lemon*). *Our implementation* was rated significantly better with 3.33 (*vase*) and 3.21 (*lemon*) on average. This means that *our implementation* with regard to the edges is rated 31% better for the vase and 12% for the lemon than *reinforcement learning*. *Human painting* again performs best, in this category with an average of 3.63.

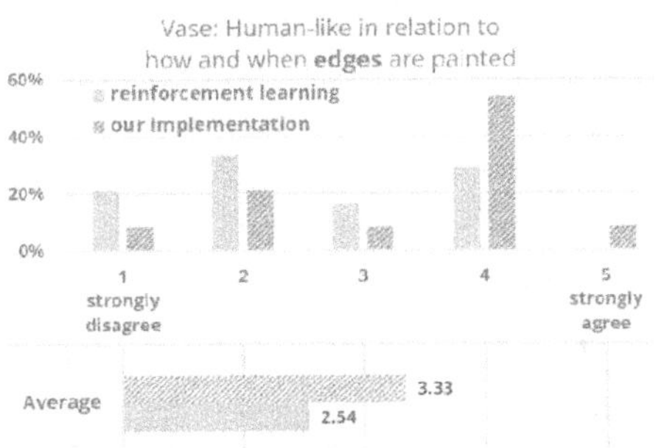

(a) Painting a vase.

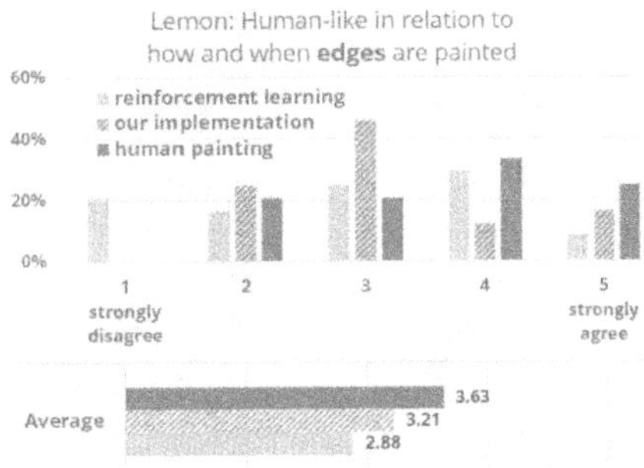

(b) Painting a lemon.

Fig. 11. Feedback on how and when edges are painted.

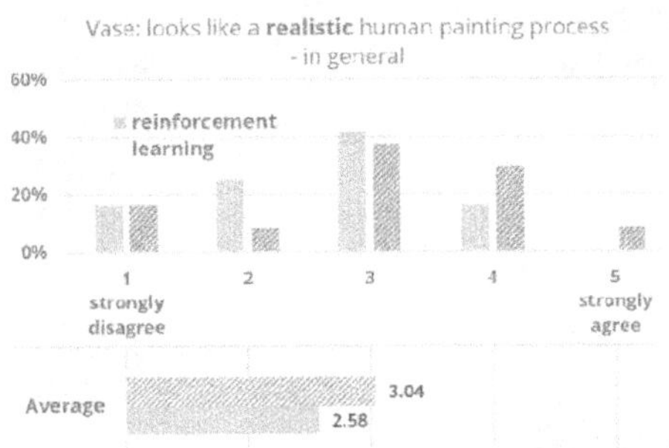

(a) Painting a vase.

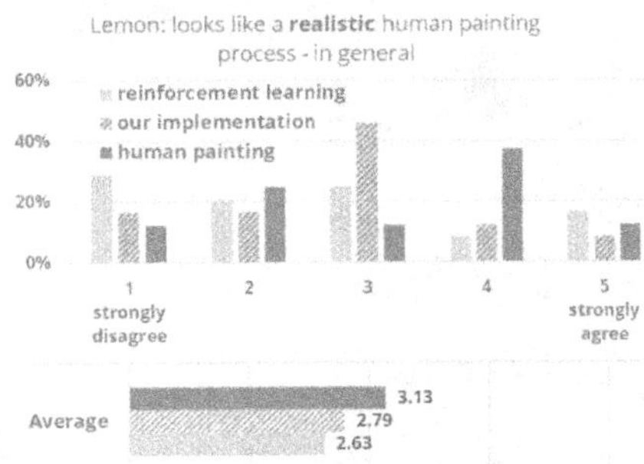

(b) Painting a lemon.

Fig. 12. Feedback on how realistic the human process is in general.

5.7 Evaluation of the General Realistic Look

In the previous questions we asked about the individual aspects in order to analyze the strengths and weaknesses of the procedures. But we also wanted to use the questionnaire to find out how human-like the participants find the procedures in general. As visualized in Fig. 12, for the *reinforcement learning* the question was rated with an average score of 2.58 (*vase*) and 2.63 (*lemon*), whereas *our implementation* was rated with 3.04 (*vase*) and 2.79 (*lemon*) on average. This shows that the general impression of *our implementation* is by 18% and 6% better. *Human painting* achieves an average of 3.13.

In order to also evaluate the direct comparison, for Edvard Munch's The Scream we displayed two time-lapse videos in parallel with the *reinforcement learning* approach and *our implementation.* Figure 13 indicates that here, too, participants find our implementation significantly more human-like in each single aspect and also in general. Even 79% find *our implementation* more realistic than *reinforcement learning.*

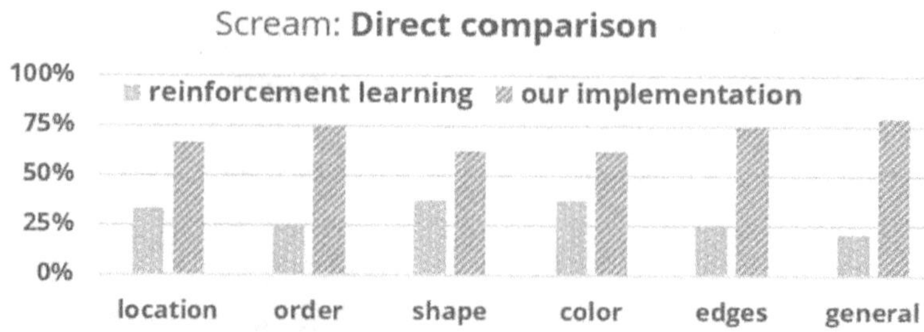

Fig. 13. Direct comparison of location, order, shape, color, edges and in general for the paining process of Edvard Munch's The Scream.

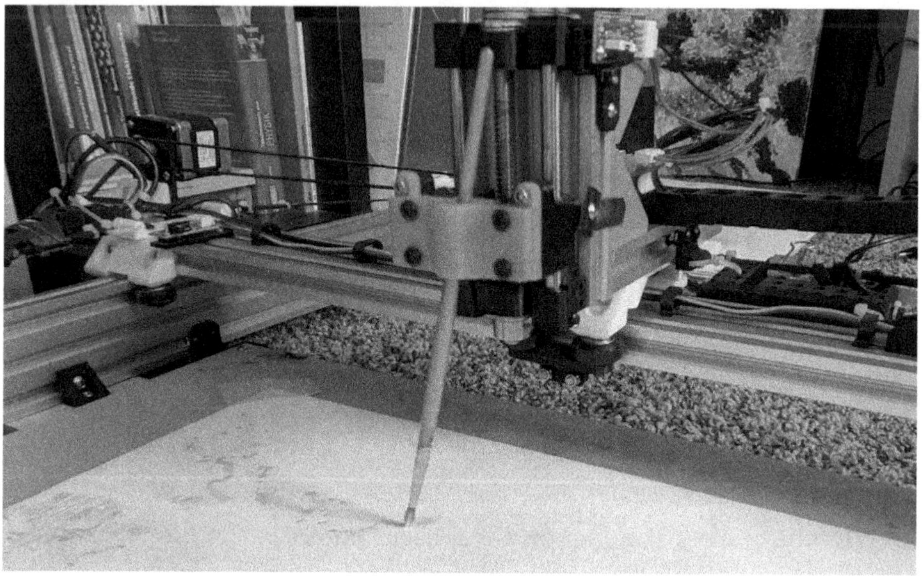

Fig. 14. Painting robot.

6 Conclusion and Future Work

In this paper we have evaluated and compared different approaches to simulate the human painting process. Additionally, we presented our solution for this task which is based on a combination of filters and semantic segmentation. In our survey, this approach was rated as better and more realistic than the most realistic approach for this task so far which is a reinforcement learning approach. When we look at the individual aspects evaluated, we perform best on the shape of the areas being painted. Our advantage over other methods is that the use of filters and semantic segmentation does not approximate shapes, which would be necessary in other deep learning approaches to limit the parameter space [12,17].

Future work may include the combination of our implementation with a robot which possesses the equipment to paint or software which teaches how to paint. Figure 14 demonstrates our painting robot Paintbot[4]. The painting robot is similar to a CNC machine and built in a standard 3-axis design, focusing on a large and flat work area to paint canvases. The machine has 60 different acrylic paints directly accessible. It also has a sponge, water reservoir and a cloth for cleaning the brush between color changes. The idea is to use our algorithm to generate the intermediate images of the painting process which will serve as input of the G-code generator software to control the robot and paint the image in a human-like manor.

Our approach is suitable for paintings that clearly have coarse to fine levels (e.g., those with a clear separation of background and foreground). Future studies may explore how to treat other types of paintings, e.g., those with local fine structure and no global arrangement, or—the other way around—with global structure but no local details. To refine our algorithm we plan to add edge detection to draw outlines after painting the background as demonstrated in the second step of the human painting process in Fig. 1. Having achieved that the image is divided into separate painting areas/layers, we could further divide these areas into individual brush strokes. When combining our algorithm with our painting robot, this would be done implicitly by the painting robot, but not simulated beforehand in our algorithm. Whereas the goal of our study was to evaluate the feedback of people with different interest and experience in art, the perception of people with more experience compared to those with less experience can be investigated in future studies.

References

1. Aubert, M., et al.: Pleistocene Cave Art from Sulawesi. Indonesia Nat. **514**, 223–227 (2014)
2. Driscoll, S.: Painting. Salem Press Encyclopedia (2019)
3. earthryse: An AI-based Robot that Creates Fine Art Paintings (2021). https://earthryse.prowly.com/130623-an-ai-based-robot-that-creates-fine-art-paintings. Accessed 24 May 2021
4. 3KICKS fine art studio: Sean Cheetham's Demo in Advanced Portraiture Class (4/25/11) (2011). http://3kicks.blogspot.com/2011/05/sean-cheethams-demo-in-advanced.html. Accessed 24 May 2021
5. Durani, B.: Acrylic Painting Techniques: A Series of Nature Themed Acrylic Paintings. Ph.D. thesis, Yeshiva College, Yeshiva University (2020)
6. Nakano, R.: Neural Painters: A Learned Differentiable Constraint for Generating Brushstroke Paintings. ArXiv abs/1904.08410 (2019)
7. Reyner, N.: How to paint with layers - in acrylic and oil (2017). https://nancyreyner.com/2017/12/25/what-is-layering-for-painting/. Accessed 26 May 2021

[4] https://vialps.com

8. Singh, J., Zheng, L.: Combining Semantic Guidance and Deep Reinforcement Learning for Generating Human Level Paintings. In: Proceedings of the IEEE/CVF Conference on Computer Vision and Pattern Recognition (CVPR), pp. 16387–16396, June 2021
9. Kotovenko, D., Wright, M., Heimbrecht, A., Ommer, B.: Rethinking Style Transfer: From Pixels to Parameterized Brushstrokes. In: Proceedings of the IEEE/CVF Conference on Computer Vision and Pattern Recognition (CVPR), pp. 12196–12205, June 2021
10. Zou, Z., Shi, T., Qiu, S., Yuan, Y., Shi, Z.: Stylized Neural Painting. In: Proceedings of the IEEE/CVF Conference on Computer Vision and Pattern Recognition (CVPR), pp. 15689–15698, June 2021
11. Liu, S., et al.: Paint Transformer: Feed Forward Neural Painting with Stroke Prediction. CoRR abs/2108.03798 (2021). https://arxiv.org/abs/2108.03798
12. Johansson, R.: Genetic Programming: Evolution of Mona Lisa (2008). https://rogerjohansson.blog/2008/12/07/genetic-programming-evolution-of-mona-lisa. Accessed 24 May 2021
13. Google Creative Lab: The Quick, Draw! Dataset. https://github.com/googlecreativelab/quickdraw-dataset (2017), accessed: 2021-05-24
14. Diaz-Aviles, E.: Dreaming of Electric Sheep (2018). https://medium.com/libreai/dreaming-of-electric-sheep-d1aca32545dc. Accessed 24 May 2021
15. Zhao, A., Balakrishnan, G., Lewis, K.M., Durand, F., Guttag, J., Dalca, A.V.: Painting Many Pasts: Synthesizing Time Lapse Videos of Paintings. In: 2020 IEEE/CVF Conference on Computer Vision and Pattern Recognition (CVPR), pp. 8432–8442 (2020)
16. Ganin, Y., Kulkarni, T., Babuschkin, I., Eslami, S.M.A., Vinyals, O.: Synthesizing Programs for Images using Reinforced Adversarial Learning. In: Dy, J.G., Krause, A. (eds.) Proceedings of the 35th International Conference on Machine Learning, ICML 2018, Stockholmsmässan, Stockholm, Sweden, July 10–15, 2018, Proceedings of Machine Learning Research, vol. 80, pp. 1652–1661. PMLR (2018)
17. Huang, Z., Zhou, S., Heng, W.: Learning to Paint with Model-based Deep Reinforcement Learning. In: 2019 IEEE/CVF International Conference on Computer Vision (ICCV), pp. 8708–8717 (2019)
18. François-Lavet, V., Henderson, P., Islam, R., Bellemare, M.G., Pineau, J.: An Introduction to Deep Reinforcement Learning. Found. Trends Mach. Learn. **11**(3–4), 219–354 (2018)
19. Culjak, I., Abram, D., Pribanic, T., Dzapo, H., Cifrek, M.: A Brief Introduction to OpenCV. In: 2012 Proceedings of the 35th International Convention MIPRO, pp. 1725–1730 (2012)
20. Kim, W., Kanezaki, A., Tanaka, M.: Unsupervised Learning of Image Segmentation based on Differentiable Feature Clustering. IEEE Trans. Image Process. **29**, 8055–8068 (2020). https://doi.org/10.1109/TIP.2020.3011269
21. Pessoa, T., Medeiros, R., Nepomuceno, T., Bian, G.B., Albuquerque, V., Filho, P.P.: Performance Analysis of Google Colaboratory as a Tool for Accelerating Deep Learning Applications, p. 1. IEEE Access (2018)

Unusual Transformation: A Deep Learning Approach to Create Art

Mai Cong Hung[1(✉)], Mai Xuan Trang[2], Ryohei Nakatsu[3], and Naoko Tosa[3]

[1] Osaka University, Osaka, Japan
[2] Phenikaa University, Hanoi, Vietnam
trang.maixuan@phenikaa-uni.edu.vn
[3] Kyoto University, Kyoto, Japan
ryohei.nakatsu@design.kyoto-u.ac.jp, tosa.naoko.5c@kyoto-u.ac.jp

Abstract. In this research, we proposed the concept of "unusual transformation," which realizes transformation between two very different image sets as an extension of CycleGAN. CycleGAN is a new deep-learning-based AI technology that can realize transformation between two image sets. Although conventional CycleGAN researchers have tried transformation between two similar image sets, we applied CycleGAN to the transformation of two very different image sets such as between portraits photos and Ikebana or Shan-Shui paintings. Then to obtain a better result, we improved CycleGAN by adding a new loss function and developed "UTGAN (Unusual Transformation GAN)." We found that by using UTGAN, portrait photos and animal photos are transformed into Ikabana-like and Shan-Shui-like images. Then we carried out an analysis of the obtained result and made a hypothesis that the unusual transformation works well because both Ikebana and Shan-Shui are fundamental and abstracted expressions of nature. Also, we carried out various considerations to justify the hypothesis.

Keywords: GANs · CycleGAN · Image transformation · Ikebana · Shan-Shui · UTGAN

1 Introduction

In recent years, the rapid development of AI and Deep Learning raises the questions about the impact of these advance technologies in the way we create and study art. In the analyzing topic, machine learning techniques were used in the artwork clustering task [1] and the art evaluation task [2]. However, for the fundamental question such as whether AI can create artworks or not, an answer has not yet been obtained.

Style transfer is widely considered as one basic approach of AI toward such a direction. One might use generative models in Deep Learning to transform normal photos or sketches into images that have similar visual effects to artworks with a specific style.

M. C. Hung and M. X. Trang—Contributed equally to this work as first authors.

M. Wölfel et al. (Eds.): ArtsIT 2021, LNICST 422, pp. 309–320, 2022.
https://doi.org/10.1007/978-3-030-95531-1_21

Recently, the appearance of GANs (Generative Adversarial Networks [3]) has made a breakthrough regarding the topic of style transfer. In the training of GANs, a generator network G learns to generate new data while a discriminator network D tries to identify the generated data whether it is real or fake. In game theory terms, this training process can be interpreted as a minimax game. With this interesting mechanism, the training proces of GANs networks can converge provided a relatively small number of training data.

Based on the minimax game of generator and discriminator in the basic configuration of GANs, a large number of variations has been developed by modifying the network structure and the objective loss function. CycleGAN [4] is an elegant variation of GANs which study the mutual transformation between two sets of photos. CycleGAN is effective for art style transfer because of the unpaired training mechanism. It realizes set-to-set level transformation to learn the distribution of the target sets, or art styles.

Classic examples of CycleGAN and other style transfer techniques were developed by achieving the transformation between two sets of data of relatively similar size, themes, or categories. On the other hand, in this paper, we propose the idea of "Unusual Transformation," which achieves a mutual transformation between two image sets with different sizes and themes. In our previous research [5], we gave several examples of portraits and animal photos transformed into Ikebana (Japanese flower arrangement) via CycleGAN. At the same time, however, as there were problems of under transformation and over transformation, we found it necessary to improve CycleGAN [6].

By combining these previous research results, in this paper, we propose "Unusual Transformation" by explaining its concept and also by giving various examples. We also discuss the underlying connection of this concept to other art-related topics.

2 Generative Models in Deep Learning

In the last decade, GANs (Generative Adversarial Networks) [3] have become one of the most essential topics in Deep Learning. The generative model in GANs provides impressive performance on art style transfer even with a small number of training data. The architecture of GANs could be described as in Fig. 1 with the basic configuration of two networks, a generator network (G) and a discriminator network (D). The training of GANs is based on a minimax mechanism in the sense that the generator G learns to generate fake data from random noise while the discriminator D tries to classify the generated data into categories of "real" or "fake." In other words, the training on G tries to maximize the probability of the generated data to lie on the targeted distribution and the training process of D tries to minimize it.

Among the variations of GANs, CycleGAN [4] is an effective approach to set-to-set level learning to study the mutual transformation between two sets of photos. The architecture of CycleGAN consists of two generators and two discriminators as can be seen in Fig. 2.

To perform the mutual transformation of two image sets A and B, the training of CycleGAN learns two mappings $G_{AB} : A \to B$ and $G_{BA} : B \to A$ given the training samples: $\{a_i\}_{i=1}^{N} \in A$ and $\{b_j\}_{j=1}^{M} \in B$ with the data distributions $a \sim p_A(a)$ and

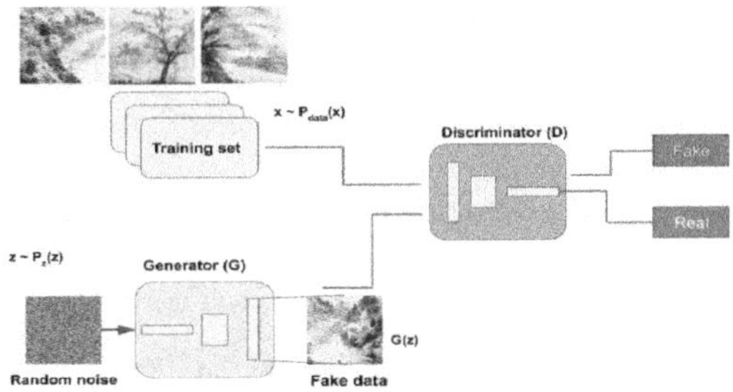

Fig. 1. The basic configuration of GANs

$b \sim p_B(b)$. The respected discriminators are D_A and D_B aim to distinguish between real photos and generated fake photos.

To emphasize the mutual transformation, the objective loss function of CycleGAN includes two components: adversarial losses for matching the generated images to the target set, and cycle consistency loss for preventing the mappings G_{AB} and G_{BA} from contradicting each other.

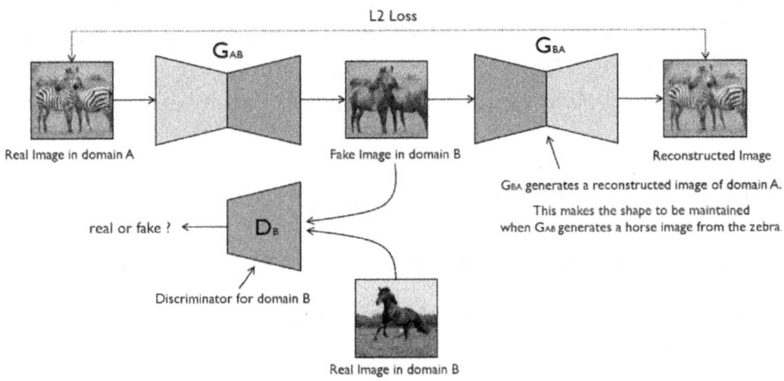

Fig. 2. Basic configuration of CycleGAN ([4])

Adversarial Loss: The adversarial loss applies to both generators.

- For $G_{AB} : A \rightarrow B$ and the respected discriminator D_B:

$$\begin{aligned} \mathcal{L}_{GAN}(G_{AB}, D_B, A, B) \\ &= \mathbb{E}_{b \sim p_B(b)}\left[\log D_B(b)\right] \\ &+ \mathbb{E}_{a \sim p_A(a)}[\log(1 - D_B(G_{AB}(a)))] \end{aligned} \tag{1}$$

- For $G_{BA} : B \rightarrow A$ and the respected discriminator D_A:

$$\begin{aligned} \mathcal{L}_{GAN}(G_{BA}, D_A, B, A) \\ &= \mathbb{E}_{a \sim p_A(a)}\left[\log D_A(a)\right] \\ &+ \mathbb{E}_{b \sim p_B(b)}[\log(1 - D_A(G_{BA}(b)))] \end{aligned} \tag{2}$$

Cycle Consistency Loss: For each image a from domain A, the generated image after applying two transformations G_{AB} and G_{BA} should be similar to a: $a \rightarrow G_{AB}(a) \rightarrow G_{BA}(G_{AB}(a)) \approx a$. We call it forward cycle consistency. We also have backward cycle consistency in the reverse direction: $b \rightarrow G_{BA}(b) \rightarrow G_{AB}(G_{BA}(b)) \approx b$. The cycle consistency loss is a combination of both forward and backward cycle consistency losses:

$$\begin{aligned}\mathcal{L}_{cyc}(G_{AB}, G_{BA}) = \mathbb{E}_{a \sim p_A(a)}\big[\|\mathrm{G_{BA}}(G_{AB}(a)) - a\|_1\big] \\ + \mathbb{E}_{b \sim p_B(b)}\big[\|\mathrm{G_{AB}}(G_{BA}(b)) - b\|_1\big]\end{aligned} \tag{3}$$

The total objective loss function of CycleGAN consists of the adversarial losses and the cycle consistency loss:

$$\begin{aligned}\mathcal{L}(G_{AB}, G_{BA}, D_A, D_B) = \mathcal{L}_{GAN}(G_{AB}, D_B, A, B) \\ + \mathcal{L}_{GAN}(G_{BA}, D_A, B, A) \\ + \lambda\mathcal{L}_{cyc}(G_{AB}, G_{BA})\end{aligned} \tag{4}$$

where constant λ is the weight of the cycle consistency loss.

We note that the generative models in CycleGAN learn the set-to-set level of transformation while the original GANs learn to generate data to fit in a target set. In the task of art style transfer, CycleGAN could learn the mutual conversion between normal photos and art styles as well as more general transformation between two sets of data.

3 Unusual Transformation

3.1 Concept of Unusual Transformation

In classic examples of CycleGAN in [4], the generative models were used to make a mutual transformation between horse and zebra images, landscape photos and Monet paintings, etc. This means that the transformation was made between images of relatively similar size, theme, and category. Because of such similarities between two image sets, obtained results are interesting but not impressive enough. For example, in the case of the transformation from a landscape to a Monet-like image, the obtained image only looks like a Monet-like image and not more than that. This means that, at this stage, AI does not have the capability of art creation.

Here, we should understand that creation can be achieved based on the connection of different things. As has always been indicated, ideas and inventions often come from the connection of two different things [7].

A good example is Surrealism. In artworks of Surrealism such as Dali's artworks, we find that things that never co-exist in our real-world appear together such as the co-existent of day and night scenes, co-existent of a real-world, and a dream world, etc. These artworks inspire our imagination and therefore have been highly evaluated. If two different things could be connected by AI, it may be possible that AI can create art. Although CycleGAN has the capability of connecting two different things, so far what it can achieve is the transformation between two similar image sets.

Based on this, we propose the idea of "Unusual Transformation," a high-abstracted transformation where transformation is achieved between relatively different domains of objects such as macro and micro-size worlds of plants and animals. This concept of unusual transformation is believed to be a key idea to create new art. For example, portraits or animal photos could be unusually transformed into Ikebana-like or Shan-Shui-like artistic images.

Although the definition of unusual transformation is very simple, it contains several fundamentally difficult problems. One is that the unusual transformation is a naturally difficult task with a low rate of successful transfer. To increase the success rate of conversion, we think the original CycleGAN is not enough and we have to develop new style transfer technology by improving CycleGAN.

Another is that if we want to achieve a good transformation, the target data set B should have a conceptual meaning. If we use concrete images such as Monet's drawings as the target image set B, the obtained image only looks like a Monet-like image and not more than that. What kind of image set should be used as B is a crucial issue. Ikebana would be a good example as a target data set because of its minimality and flexibility. Shan-Shui paintings would be another good painting tool as well because in Shan-Shui natural elements such as rocks, streams, mountains would be put in flexible positions. This will be further discussed in Chapter 4.

3.2 Ikebana and Shan-Shui as Target Data Sets

(1) Ikebana

Ikebana (Japanese flower arrangement) is one of the most important art forms in Japanese culture. Ikebana is the art of flower arrangement where the flowers are given life under the conceptual arrangements of Ikebana artists [8].

Ikebana has a deep root in the Japanese philosophy of art under the strong influence of Zen Buddhism. The tradition of arranging flowers on Buddha was brought to Japan from China in the Heian period (794–1185) by Zen Buddhist monks. Then Ikebana grew to be an important art form along with the development of Zen.

Ikebana has a long history of development and has continued to be a great source of inspiration in modern art. For instance, Naoko Tosa, created a video artwork named "Sound of Ikebana" by applying fluid dynamics to the art creation process [9, 10].

The "minimality" and the "flexibility" are the two important properties of Ikebana which support our idea of the unusual transformation. Under the influence of Zen philosophy, "emptiness" plays an essential role in the art of Ikebana. The emptiness appearing in Ikebana artworks is believed to provide meaning and be harmonic to the whole scene. We consider the minimality of Ikebana as the appearance of the emptiness in an Ikebana artwork. At the same time, we call Ikebana flexible as the materials (flowers, leaves, branches, and so on) can be placed in various shapes and arrangements.

(2) Shan-Shui Painting

Shan-Shui refers to a style of traditional Chinese painting that involves or depicts scenery or natural landscapes, using a brush and ink rather than more conventional paints [11]. Mountains, rivers, and waterfalls are common subjects of Shan-Shui paintings.

Shan-Shui painting first began to develop in the 5th century in China, in the Liu Song dynasty. It was later characterized by a group of landscape painters such as Zhang

Zeduan, most of them already famous, who produced large-scale landscape paintings. These landscape paintings usually centered on mountains. Mountains had long been seen as sacred places in China, which were viewed as the homes of immortals and thus, close to the heavens. Philosophical interest in nature, or mystical connotations of naturalism, could also have contributed to the rise of landscape painting. The art of Shan-Shui, like many other styles of Chinese painting, has a strong reference to Taoism/Daoism imagery and motifs, as symbolisms of Taoism strongly influenced "Chinese landscape painting". Some authors have suggested that Daoist stress how minor the human presence is in the vastness of the cosmos, or Neo-Confucian interest in the patterns or principles that underlie all phenomena, natural and social lead to the highly structuralized nature of Shan-Shui.

Shan-Shui painting was first introduced to Japan from China along with Zen as an ink painting during the Kamakura period (1185–1333). At first, many paintings expressed Zen thought, but gradually the form of ink painting changed and Shan-Shui paintings began to be drawn. In the latter half of the 15th century, the famous Shan-Shui painter Sesshu (1420–1506) appeared and completed the Japanese Shan-Shui painting.

4 Preliminary Experiment and Development of UTGAN

4.1 Preliminary Experiment: Transformation to Ikebana

In this section, we introduce some experimental examples of unusual transformation via CycleGAN. We performed the unusual transformation via CycleGAN with the sets A1, A2, and the target set B as follows:

- Dataset A1: Portrait photos in Flickr
- Dataset A2: Kaggle Animal-10 dataset. (https://www.kaggle.com/alessiocorrado99/animals10)
- Dataset B: Ikebana photos in Google Image Search

Figure 3 shows the result of the transformation by CycleGAN. As can be seen from the results, portraits and horse photos turned into Ikebana images while keeping the original shape. This "unusual transformation" concept would inspire a new method to create art via Deep Learning. However, there are some limitations. In some cases of photos with complex background, the experiments failed to transform them into abstract Ikebana. Some photos were over-transformed so that we could not recognize the original shape. We consider the reason is that the structure of the CycleGAN was not designed to learn specific high abstract representation such as the unusual transformation in this experiment.

4.2 Improvement of CycleGAN

In the classic examples, CycleGAN worked well on mutual transform two sets of similar photos in terms of size, themes, or categories. However, as shown in Fig. 3, our experiments showed the over-transformation limitation of CycleGAN in unusual transformation with Ikebana. To improve the performance, we introduce the combination of

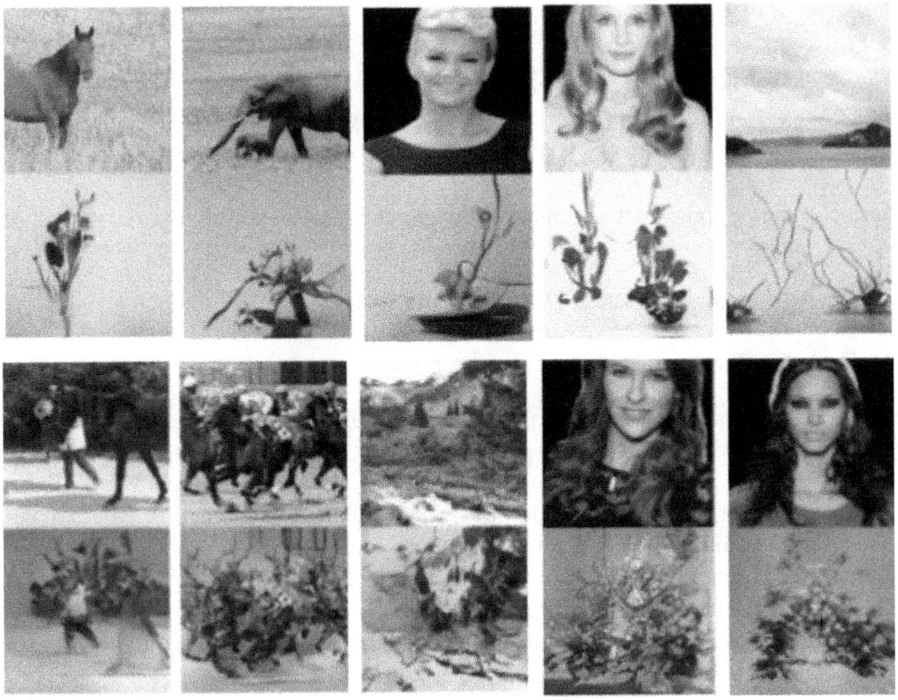

Fig. 3. Results of transformation into Ikebana including several failed transformations.

CycleGAN with several computer vision techniques in an improved modification called "UTGAN." To preserve the original shape of the main objects to a certain level, we apply object recognition techniques to detect the main object, and then an edge detection is used to emphasize the main shape. In the last step, we include an object edge-promoting loss to enforce the model to emphasize the original shapes of the main objects under the transformation. The calculation of the object edge-promoting loss is described in the next paragraph.

From the training photos set A, we generate a set of edge photos $E = \{e_i\}_{i=1}^{N}$ by removing clear edges of the main object in $\{a_i\}_{i=1}^{N}$. For each photo $a_i \in A$, we operate the following steps:

(1) To recognize objects in the photo by using a pre-trained object detector (Mobile_Net_SSD),
(2) To detect edge pixels of objects by the Canny edge detector [12],
(3) To dilate the edge regions, and
(4) To apply Gaussian smoothing in the dilated edge regions.

In our proposed UTGAN, the training target of the discriminator D_A is to maximize the probability of assigning the correct label to $G_{BA}(b)$, the real photos without clear edges of the photos' main objects (i.e., $e_j \in E$) and the real photos (i.e., $a_i \in A$).

Finally, we add object edge-promoting loss to the adversarial loss:

$$\begin{aligned}\mathcal{L}_{GAN}(G_{BA}, D_A, B, A) = \mathbb{E}_{a\sim p_A(a)}\left[\log D_A(a)\right] \\ + \mathbb{E}_{b\sim p_B(b)}[\log(1 - D_A(G_{BA}(b)))] \\ + \gamma\mathbb{E}_{e\sim p_E(e)}[\log(1 - D_A(G_{BA}(e)))]\end{aligned} \tag{5}$$

where constant γ controls the weight of the object edge-promoting loss.

5 Examples of Unusual Transformation

5.1 Transformation to Ikebana

We performed the unusual transformation via UTGAN with the style sets A1, A2 and the set B as follows:

- Dataset A1: Portrait photos in Flickr (https://github.com/NVlabs/ffhq-dataset)
- Dataset A2: Animal photos from Google Image Search
- Dataset B: Ikebana photos in Google Image Search

Figure 4 shows several results of A1 to B transformation and Fig. 5 shows several results of A2 to B transformation.

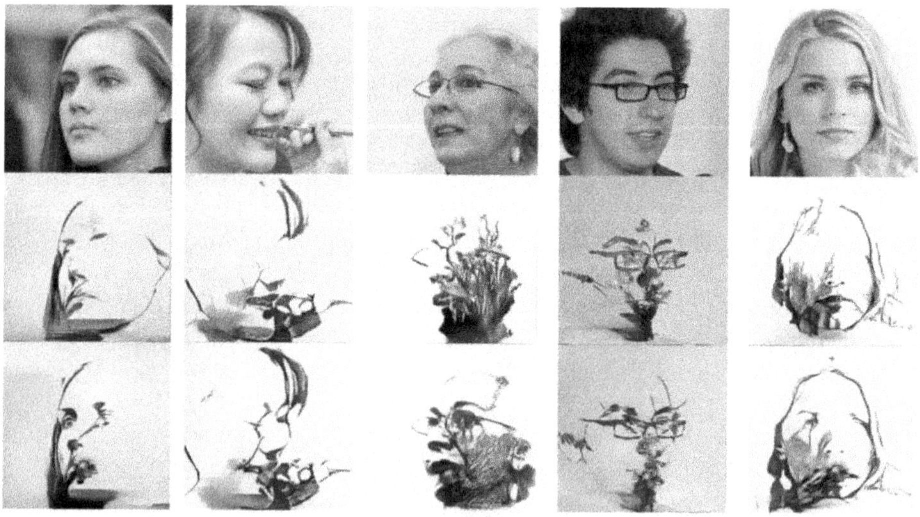

Fig. 4. Experiment result A1-B: the first row is the original photos, the second row is the result by CycleGAN, the last row is the results by UTGAN

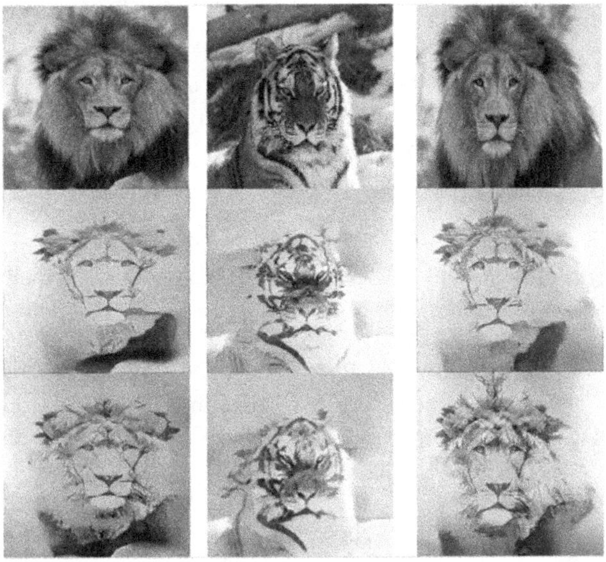

Fig. 5. Experiment result A2-B: the first row is the original photo, the second row is the results by CycleGAN, the last row is the results by UTGAN

5.2 Transformation to Shan-Shui Paintings

We applied the unusual transformation to transfer photos of human faces in Flickr to Chinese Shan-Shui paintings via UTGAN. Below is the data set we used.

- Dataset A: Portrait photos in Flickr
- Dataset B: Shan-Shui artworks in Google Image Search

Some of the obtained results are shown in Fig. 6. Portraits turned into Shan-Shui-like images while one can still recognize the original shape of human faces.

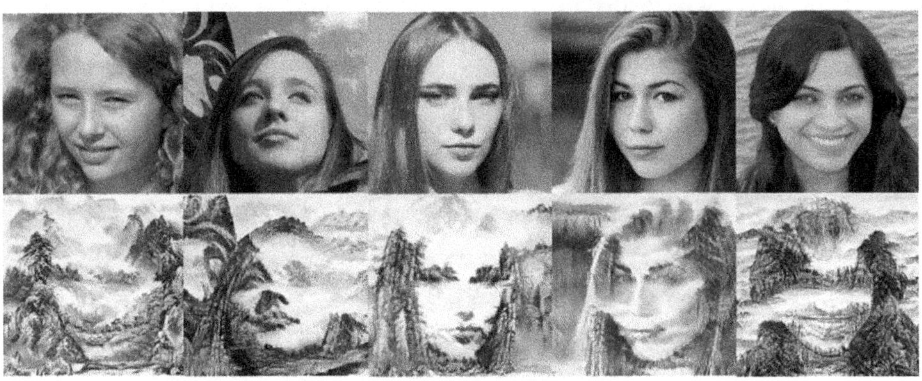

Fig. 6. Experimental results of portrait to Shan-Shui transformation.

6 Discussion

In this chapter, we will propose hypotheses regarding art by considering the functions of CycleGAN and its improved UTGAN. We will also discuss the possibility of clarifying the essence of art by using UTGAN.

In this paper, beyond the scope of transformations so far achieved by CycleGAN, we have attempted unusual transformations by carrying out transformations between image sets that seem to have no similarity at all. We tried to convert between image sets of animals and portraits and image sets of Ikebana photos and Shan-Shui paintings, which are completely different in appearance. As a result, the portraits or animal photos were converted into Ikebana-like images and Shan-Shui-like images while retaining the characteristics of the original image. Rather, obtained images may be unprecedented Ikebana images or unprecedented Shan-Shui paintings. In other words, our unusual transformation has produced paintings that have never been seen before. What does this mean? We think that the following hypotheses can be made.

- Hypothesis 1: Portraits and animal photos are successfully converted into Ikebana and Shan-Shui images because both portraits and animals are natural objects.
- Hypothesis 2: The conversion into Ikebana and Shan-Shui was successful because Ikebana and Shan-Shui paintings contain the essentials of natural objects.

There is a famous Aristotle words that "art imitates nature [13]". As expressed in these words, art represented by paintings used to express nature. The so-called realism paintings are typical examples. For the impressionism that was born after realism represented by the artworks of Monet, Cezanne, etc., artworks created by the Impressionism artists are abstract in the sense that they did not draw nature as it is but drew the impressions of the artists. However, although they have drawn the impression they received, what they tried to draw is clearly understood and not very abstract. After that, however, paintings with a higher degree of abstraction such as Cubism and Surrealism appeared, and it continued to the present extremely high degree of abstraction. This is the brief history of Western art.

Based on CycleGAN's idea to carry out transformation between two image sets, what the Western paintings have tried to express can be shown in Fig. 7. In other words, there is a conversion from the actual landscape to the landscape paintings. (The process of converting a landscape painting into a landscape photograph doesn't make much sense for our discussion, so it's enough to consider only the transformation function G here.)

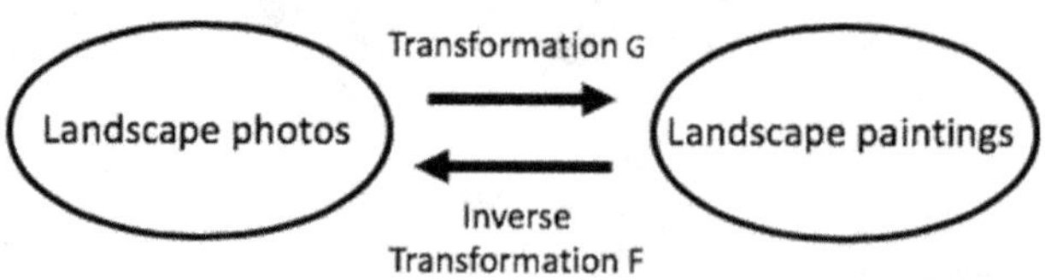

Fig. 7. Relationship between landscape photos and landscape paintings.

To make it even more abstract, Fig. 7 can be expressed as Fig. 8. In other words, art extracts the essential things from natural objects and phenomena.

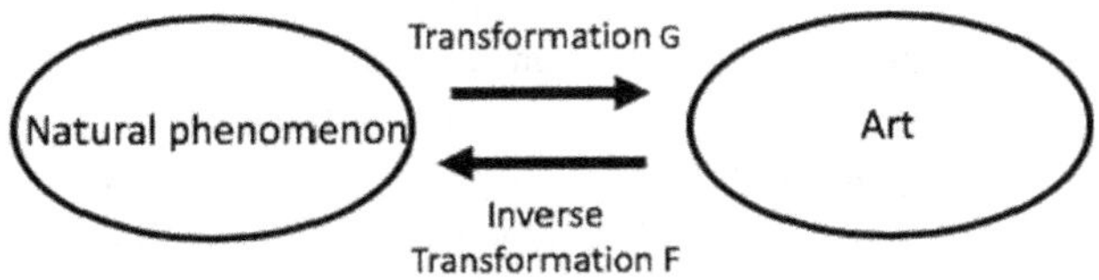

Fig. 8. Relationship between natural phenomenon and art.

If we think in this way, we may find that hypotheses 1 and 2 mentioned above are correct. Furthermore, when examining the characteristics of Ikebana and Shan-Shui paintings, they have the following characteristics and are appropriate for expressing the essence of natural objects and phenomena.

(1) Minimality

Ikebana and Shan-Shui paintings try to remove unnecessary things from natural objects and phenomena and express them with the minimum expression. For example, Ikebana tries to express the scenery of nature with a very small number of flowers and vegetation. Also, it can be said that Shan-Shui paintings express nature by decomposing what constitutes nature into the minimum basic elements (mountains, rocks, water streams, etc.) and by reconstructing them.

(2) Flexibility

As mentioned above, both Ikebana and Shan-Shui paintings are trying to reconstruct nature by breaking down what construct nature into the minimum elements and reconstructing them. In addition, when reconstructing nature, the individual elements have flexibility in their placement. For example, in the case of Ikebana, the arrangement of a small number of flowers and vegetation greatly differs depending on each artist. In other words, the degree of freedom of arrangement itself may lead to the diversity of Ikebana. Also, in the case of Shan-Shui paintings, the individual components such as rocks and water streams can be freely placed in the painting.

7 Conclusion

CycleGAN, which is one of the variations of GANs, enables mutual conversion between datasets without the need for a one-to-one correspondence of data. For example, it is possible to convert landscape photographs into Monet-like images. However, this means that AI merely produces a Monet-like image. At this stage, AI is not yet capable of creating art. The main reason for this is that the style transfer in previous studies only involves conversions between similar datasets, such as between horses and zebras, between landscape photographs and Monet's landscape paintings, etc.

Art and inventions have been a creation based on the connection of different things. Based on this basic principle, this paper proposes the transformation between different types of datasets called "Unusual Transformation." Then, as an example, we tried to convert portraits and animal photographs into Ikebana using CycleGAN. However, it has been shown that under transformation and over transformation often occur. To solve this problem, we proposed UTGAN, in which a new element is added to the loss function, to give CycleGAN a new function to keep the original structure of portraits or animal photos. It was shown that by applying UTGAN, portraits and animal photos can be successfully converted into Ikebana and Shan-Shui.

Based on these results, we considered why portraits and animal photos can be converted into Ikebana and Shan-Shui images. As a result, it became clear that these even seemingly different types of image sets are connected at the root. In other words, since human faces and animals are natural objects, and Ikebana and Shan-Shui paintings are the essences of nature, conversion is successful when there is such a relationship between two image sets. Extending this further, we may be able to approach art more deeply from the science and technology aspect.

References

1. Gultepe, E., Conturo, T.E., Makrehchi, M.: Predicting and grouping digitized paintings by style using unsupervised feature learning. J. Cult. Herit. **31**, 13–23 (2018)
2. Mai, C.H., Nakatsu, R., Tosa, N., Kusumi, T., Koyamada, K.: Learning of art style using AI and its evaluation based on psychological experiments. In: Nunes, N.J., Ma, L., Wang, M., Correia, N., Pan, Z. (eds.) ICEC 2020. LNCS, vol. 12523, pp. 308–316. Springer, Cham (2020). https://doi.org/10.1007/978-3-030-65736-9_28
3. Creswell, A., et al.: Generative adversarial networks: an overview. IEEE Sig. Process. Mag. **35**(1), 53–65 (2018)
4. Zhu, J., Park, T., Isola, P., Efros, A.A.: Unpaired image-to-image translation using cycle-consistent adversarial networks. In: 2017 IEEE International Conference on Computer Vision (ICCV), pp. 2242–2251 (2017)
5. Mai, C.H., Nakatsu, R., Tosa, N.: Developing Japanese Ikebana as a digital painting tool via AI. In: Nunes, N.J., Ma, L., Wang, M., Correia, N., Pan, Z. (eds.) ICEC 2020. LNCS, vol. 12523, pp. 297–307. Springer, Cham (2020). https://doi.org/10.1007/978-3-030-65736-9_27
6. Hung, M.C., Trang, M.X., Tosa, N., Nakatsu, R.: IkebanaGAN: new GANs technique for digital Ikebana art. In: Rauterberg, M. (ed.) HCII 2021. LNCS, vol. 12794, pp. 88–99. Springer, Cham (2021). https://doi.org/10.1007/978-3-030-77411-0_7
7. Jewkes, J., Sawers, D., Stillerman, R.: The Sources of Invention. W. W. Norton & Company (1971)
8. Luu, A., Matsuba, I.: Ikebana Unbound: A Modern Approach to the Ancient Japanese Art of Flower Arrangement, Artisan (2020)
9. Tosa, N., Nakatsu, R., Yunian, P.: Creation of media art utilizing fluid dynamics. In: 2017 International Conference on Culture and Computing, pp.129–135, 10–12 September 2017
10. Pang, Y., Zhao, L., Nakatsu, R., Tosa, N.: A study of variable control of Sound Vibration Form (SVF) for media art creation. In: 2017 International Conference on Culture and Computing, pp.136–142, 10–12 September 2017
11. Law, S.S.-M.: Being in traditional Chinese landscape painting. J. Intercult. Stud. **32**(4), 369–382 (2011)
12. Canny, J.: A computational approach to edge detection. IEEE Trans. Pattern Anal. Mach. Intell. **PAMI-8**(6), 679–698 (1986)
13. Aristotle: The Art of Rhetoric. Oxford University Press (2018)

Synthography – An Invitation to Reconsider the Rapidly Changing Toolkit of Digital Image Creation as a New Genre Beyond Photography

Elke Reinhuber(✉)

SCM School of Creative Media, City University of Hong Kong, Kowloon Tong, Hong Kong
elke.r@cityu.edu.hk

Abstract. With the comprehensive application of *Artificial Intelligence* into the creation and post production of images, it seems questionable if the resulting visualisations can still be considered 'photographs' in a classical sense – drawing with light. Automation has been part of the popular strain of photography since its inception, but even the amateurs with only basic knowledge of the craft could understand themselves as author of their images. We state a legitimation crisis for the current usage of the term. This paper is an invitation to consider Synthography as a term for a new genre for image production based on AI, observing the current occurrence and implementation in consumer cameras and post-production.

Keywords: Artificial Intelligence (AI) · Future of creative photography · Synthetic photography

*"Taking pictures with a cellphone is perhaps the most pervasive digital light activity in the world today, contributing to the vast space of digital pictures. Picture-*taking *is a straightforward 2D sampling of the real world. The pixels are stored in picture files, and the pictures represented by them are displayed with various technologies on many different devices. But displays don't know where the pixels come from."* Alvy Ray Smith [1]

1 Technical Advancements for a Creative Medium

Unlike any other visual media with a century old history, photography is rapidly changing, along with the technical advancements. In its beginning, photography was recognised as a craft which demanded technical and scientific knowledge but also an aesthetic sense. The skilful image composition, the accurate illumination and the particular palette, let alone the chemical and optical process behind the image, deserved elaborate proficiency and year-long training and practice. Nowadays, billions of images are captured every day without any consideration towards exposure, musings on the effect of focal length or focus, aperture, shutter speed or ISO. On top of that, even more images are captured by machines – not necessarily for the human eye, but to be read again by machines, well beyond our visible spectrum; my reflection on this has been addressed in

M. Wölfel et al. (Eds.): ArtsIT 2021, LNICST 422, pp. 321–331, 2022.
https://doi.org/10.1007/978-3-030-95531-1_22

the research on *Phasmagraphy* [2]. Editing software facilitates the improvement not only of exposures but as well of flaws in the motive itself. Thanks to this amount of images, Artificial intelligence (AI) enabled editing software to improve impressively – however, I dare to question if this form of image production may still be called "photography", in its etymological description, based on the Greek terms which are commonly translated as 'painting or drawing with light'[1].

With my background as a photographer, being professionally trained using large format cameras and analogue processes in my practice while appreciating the effortlessness of digital sensors and accelerated post-production, I keep pondering on the development of the medium in the days in which every one – human, animal or machine – is able to take correctly exposed and focused images, even optimised, fully automated. With the ease in shooting and the increasing quality, we have already observed a change in the attitude, in particular a desire to 'over'-beautify or aestheticise the captured reality. Therefore I propose with this paper that the image creation by 'intelligent' apparatuses might pave the path for a new creative medium beyond the classic understanding of photography[2]: *Synthography.* With this term, the methodology of synthetic[3] production relates to AI but also encompasses images rendered by 3D software while the process of 'drawing' is still included in the second part of the term, linked with the 'O' as a remainder of *phōtós.*

1.1 The Autopilot

Perhaps it helps to remember that in aviation, systems have been established since the 1930s that can intervene in the control of aircrafts in a variety of ways as technology has advanced. First it was for airborne stability, then for possible changes in altitude, subsequently to follow the plotted course and finally to control the speed, so meanwhile, from take-off to landing, the entire process has been completely automated [3].

While these devices have been established at sea for 100 years, the beginnings of autonomy in road traffic are only very gradually becoming widespread. There are a myriad of parameters to process and decisions have to be made very quickly – similar to the processes between the photographer's eye and finger.

Piloting a planetary craft through the infinite reaches of space will be a rather monotonous activity, unless there is a flotilla of UFOs waiting in the shadow of a moon, just as navigating on the vast oceans, in the skies above or on hundreds of kilometres long straight and grey highways. But surely photography should be anything but boring – so why automate this activity?

---

[1] φωτός (phōtós) is the genitive of φῶς (phōs), light and γραφή (graphé), drawing.

[2] Although other terms like computational photography have been used, I argue that synthography for the detailed subset is more appropriate.

[3] σύνθεσις (súnthesis), in its original meaning putting together, construct, compound. From 1874 onwards used in reference to products or materials made artificially and from 1934 established as a noun for 'synthetic material'. https://www.etymonline.com/word/synthetic.

1.2 The Autonomous Camera

Not long ago, I shared my observation how the role of the photographer shifted more and more to the automatism of the cameras [4] while today the software behind these basic programmed settings has taken over. It could be argued that a good camera depends today less on the size of its sensor or the optical quality of the lens, but on the processor and the AI faculties behind these features.

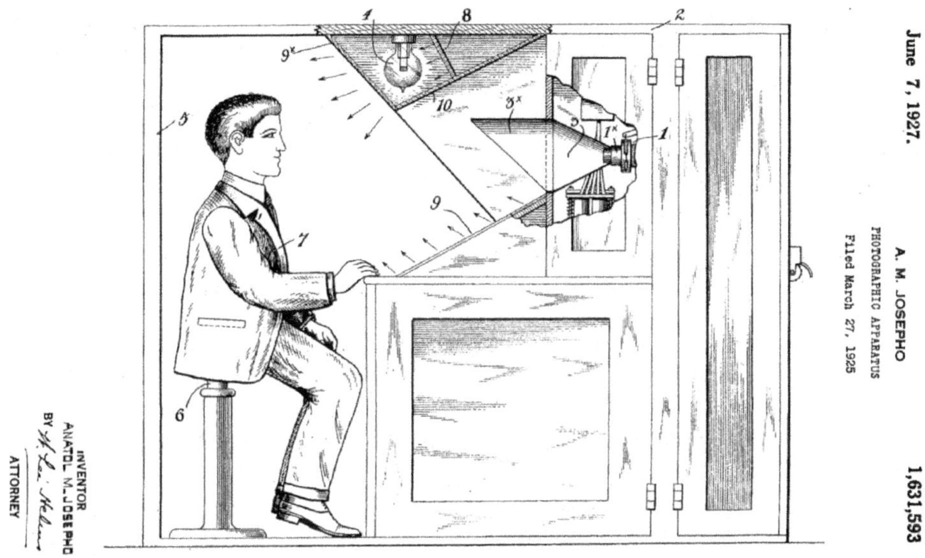

Fig. 1. Patent US1631593A of the *Photographic Apparatus* in 1925, later named Photomaton.

The concept of an automated photographer is not a fancy idea or a futuristic invention but a very reasonable notion, merging the possibilities of image-capturing and -recognition. Although Kodak's George Eastman promised the ease of photography with their legendary slogan "You Press the Button, We Do the Rest" after patenting the roll-film camera [5] in 1888, they could not guarantee that each image would be correctly exposed to the full satisfaction of their clients, photography seldomly takes place in a controlled environment as in the following example: Non-intelligent fully automated photography-machines were already established with the Photomaton in 1925 [6], a coin operated imaging machine which delivered eight in-situ developed portraits within eight minutes (see Fig. 1). This static setup including diffuse lighting, fixed settings and exposure enabled acceptable results and its analogue and digital descendants are still in use today.

Jumping almost a century ahead on the timeline, with a much shorter lifespan, the Google Clips camera was announced with exciting prospects: a small and inconspicuous camera which could be placed anywhere and capture all the important moments in one's life. Equipped with rudimentary AI capabilities, the camera was supposed to know who, what and when to capture and even recognise the quality of images, by applying the rule of thirds and detecting motion blur. The little personal surveillance device was considered 'freaky' by many, but nothing compared to the omnipresent surveillance cameras which are surrounding us today. One well-recognised example are camera traps for wildlife photography. The images captured and recorded by night or day provide insight into their often endangered lives and habitat, analysed and evaluated through AI.

Most of the resulting images nonetheless are unlikely to ever be seen, some will be deleted or simply lost, become unreadable after the next update, or they will disappear in an ocean of data without being missed. The essence of digital photography is itself transient, since these photos exist only as long as you look at them, they are generated by the imaging software instantly just to dissolve again as bits in the stream of data then, and they manifest themselves only for a moment.

With the actual image being gone, the authenticity of the creator becomes arguable, especially so if it is an AI [7]. Images 'inspired by ...' – let us say Rembrandt or Van Gogh – are frequently created, however the development of an independent artistic and aesthetic language will become harder to achieve.

1.3 The Indecisive Moment

The 'decisive moment,' as postulated by Henri Cartier-Bresson, serves as a catchphrase for ambitious photographers to describe their craft, finding exactly the right adjustments and timing for each picture. Photography is for him "the simultaneous recognition, in a fraction of a second, of the significance of an event as well as of a precise organization of forms which give that event its proper expression." [8, p. 102].

This acknowledgement has all been disrupted through recent photographic technology, described in the following features, which can be divided into AI support during the capturing process, in post-production but also in the image creation itself.

The technical history of photography shows plenty of inventions to simplify the act of image-taking by automating certain stages in the process. The approach remained always the same, streamlining the technique to free the person behind the lens from any obstacles, with shutter priority, aperture priority, program mode or autofocus. Today's techniques allow even retrospective decisions. While the tools became more powerful, photographers were always informed by the technical limitations of their machinery, "the possibilities offered by the apparatus" [9], until now.

Different from the hardware technique with its rigid rules provided by engineers, machine learning (ML) algorithms are capable of adjusting and adapting from the large amount of training images. Neural networks use the context of a scene in a split second to determine the section of the image to focus on and to adjust the exposure accordingly.

High Dynamic Range Imaging (HDR/Smart HDR). This is made possible by intentionally over- and underexposing the same picture, weighing the different light values into an image and allowing the recovery of unseen details in bright and dark areas, making it previously difficult to represent moving subjects in this way. Smart cameras enhance the ability of HDR by recognising specific components in the image and manipulating the brightness gradients accordingly. To include dark shadows or to overexpose clouds in a picture becomes virtually impossible, since the routines evaluate every image along the secret formula and process with a generic lighting profile.

This trend provided us especially with homogeneous landscape and architecture images, with uniform lighting and a tendency to be perceived as flat, monotonous and kitsch – albeit every detail well visible. Images captured as RAW supported the adjustment of highlights and shadows to a certain degree through their wider dynamic range, however not to the extent of HDR, and particularly not Smart HDR, featured by many modern day mobile phone cameras. Through computational photography techniques, the best image is composed out of an image series with a range of exposures while touching the shutter.

Pre-Capture. Also known as Pro Capture – eliminates shutter lag and reaction time by recording a series of images while the shutter is only half pressed. If released, no images will be saved, until fully pressed, then the significant moment gets preserved as a still image.

Automatic Shutter Release. For instance, camera traps for wildlife capture images with AI evaluation [10]. Usually, the classification of the generated footage is processed after the fact, utilising large datasets of similar captured animal sightings and ML. The necessary operations of filtering unwanted events from triggering the shutter is provided in arrays of cameras connected to Raspberry Pi microcomputers already in the wild, generating a much more valuable output [11].

Shutter Delay. 'Intelligent' cameras can delay the release of the shutter until the presumed subject is in focus – or even more: over a decade ago, Sony introduced the smile detection algorithm in certain cameras to the effect that all portraits were made with happy faces. However, the intensities of the desired smiles could be adjusted by the photographer in the pre-sets [12]. Today, this feature is not limited to human faces anymore. AI powered content detection such as animal-, in particular bird-detection, supports focussing on the eyes.

Low Light/Night Sight Mode. Available light gets amplified through a series of long exposures which are stitched together, supported by a machine learning algorithm and countermeasures are calculated against involuntary movements while the shutter is open, either with electronic compensation or by optical means on the sensor, the lens or both.

AI-Powered Stabilisation. Beyond in-camera stabilisation, gimbals, with adapted speeds according to focus length are available on the market, stabilising the image even further, mainly for filming. Scene analysis allows locking selected objects, while panning, zooming or following objects.

In-camera Focus Stacking. Through this feature, it became possible to focus retrospectively. The lens is moved in small increments to achieve the maximum depth of field and only the sharpest segments are actually recorded. For smart phones with multiple lenses, the second camera creates a depth map of the captured situation and helps to define the focal plane. In the well-established portrait mode of Apple's iPhone, people are recognised in the image in real time and the desired depth of field can be retrospectively adjusted.

Computational Improvements. Many applications of AI during the capturing process are already well established while others are still in their early days. The implementation of AI in smartphone cameras enhanced their capabilities significantly, which were limited due to their small size, and made them into highly advanced devices. Deficiencies of the size constraints, such as the minuscule sensors and the less elaborate lenses are compensated by computational corrections, eliminating the disadvantages towards 'real' cameras. For mirrorless and DSLR cameras, some applications exist, but here, the emphasis is on the workflow outside of the camera. Crowdfunded startup projects like Alice Camera or Arsenal are exploring the application of AI with attachments to smartphones or cameras. While the software Skylum Luminar AI led the way into AI supported editing, Adobe is improving a wide range of tools based on the experience with machine learning.

Neural Processing Units. NPUs provide the necessary computing power to allow AI processing on board, which is used for tasks like semantic image segmentation, the recognition of elements for the application of specific settings. Saliency mapping is applied to weigh the calculated results according to the centre of interest [13].

Postproduction. Although the decisive moment appears now to have moved into post production, to the selection of the image with the best composition and significance – yet the AI supports the tedious work of sorting the images and tagging the captured results; even the choice from a burst or the correct crop, auto-tilted in the right direction, is machine-provided. Developments such as plenoptic cameras, also known as light field photography, enable the photographer to decide retrospectively on focus and the depth-of-field. Analogously, postponing the perfect framing, while shooting a 360° image in high resolution, one can subsequently choose any aspired angle. The Insta360 One records movies or stills as a full sphere and allows to frame the final image according to simple markers, put into the software viewer with the claim 'Shoot First, Point Later' [14]. Since the framing of the shot constitutes the essential idea of a compelling image similar to the decisive moment, the prospect of finding another perspective retroactively seems propitious and sombre at the same time. Not only because of an excess pixel-resolution and with the extreme wide-angle lenses of omnidirectional cameras, the retrospective framing became easily possible. In the case of Insta360's auto frame possibility, the software suggests central motives and compositions, according to well established principles [15]. This technique comes also handy today in classrooms for hybrid online teaching and is implemented in Adobe Sensei to facilitate cropping video for multiple devices. In the current iPad, a similar feature is included: thanks to a wide-angle camera, a section containing mainly the face of the speaking person will be presented during online meetings, no matter if there is movement involved. The cinematic mode in the current iPhone applies the same technology, allowing the automatic rendering of focus ramps in live video.

The following properties are part of the current AI supported toolkit:

- Image recognition and classification
- Retrospective focus
- Retrospective framing/content aware cropping

- Retrospective depth of field
- Adaptation of different lens characteristics
- Exposure correction
- Retouching of skin while respecting textures and details
- Content aware background fill
- Enlarging (predicting and adding extra pixels)
- Sharpen and blurring (camera shake, movement blur, lens quality)

All the above mentioned features support and facilitate the creation of the image, although they sometimes overshadow the artistic intention of the photographer and need to be turned off or adjusted, if this possibility exists.

Content Creation Through AI. Research in the field of artificial intelligence has meanwhile progressed to the point where findings about individual abilities acquired through machine learning can be tested using the tools of experimental psychology. Knowledge about optical phenomena as the law of completion [16], an idea from Gestalt psychology, can be verified with the experimental set-ups from IQ tests.

As soon as a generative adversarial network (GAN) is trained to create an image which resembles a photograph, I propose to better describe it with the term synthograph.

Phillip Wang attempted to raise awareness and interest for this rapidly improving technology with the viral success of his website, stating in its URL address that 'This Person Does Not Exist' [17]. Several websites now display cats, horses, automobiles, beaches, food and other once favourite snapshot subjects by random while others allow sophisticated fine-tuning. For instance generated.photos [18] advertises 'unique, worry-free model photos' which can almost convincingly be created by adjusting gender, age, hair- and skin tone, mood and further details (see Fig. 2).

The special force of the neural network named DALL·E [19] is the promise to create images from verbal descriptions of objects and their possible attributes [20]. Resulting in imaginative and surreal items, they still appear believable and could support a designer's inspiration (see Fig. 3).

While the generated creations are becoming better and more refined, accordingly, at the same time attempts are being made to reveal images through AI which do not originate from the real world or are intensively manipulated. Unless flaws are obvious, most frequently in the eye area or at the hair or facial contour, the distinction is almost impossible, once the photorealistic synthograph has been generated.

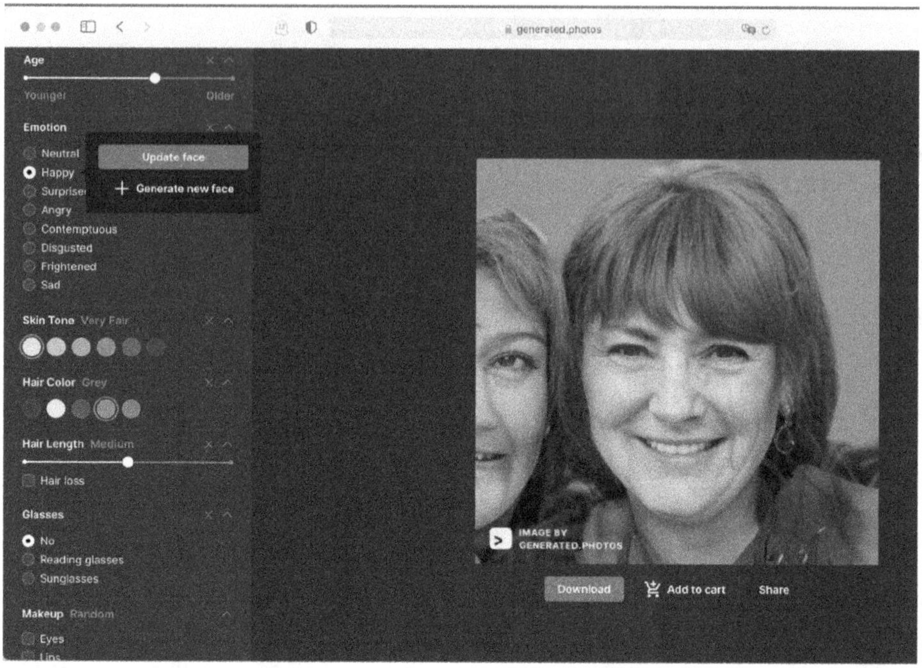

Fig. 2. AI generated portrait according to pre-sets on generated.photos, with mishap (left)

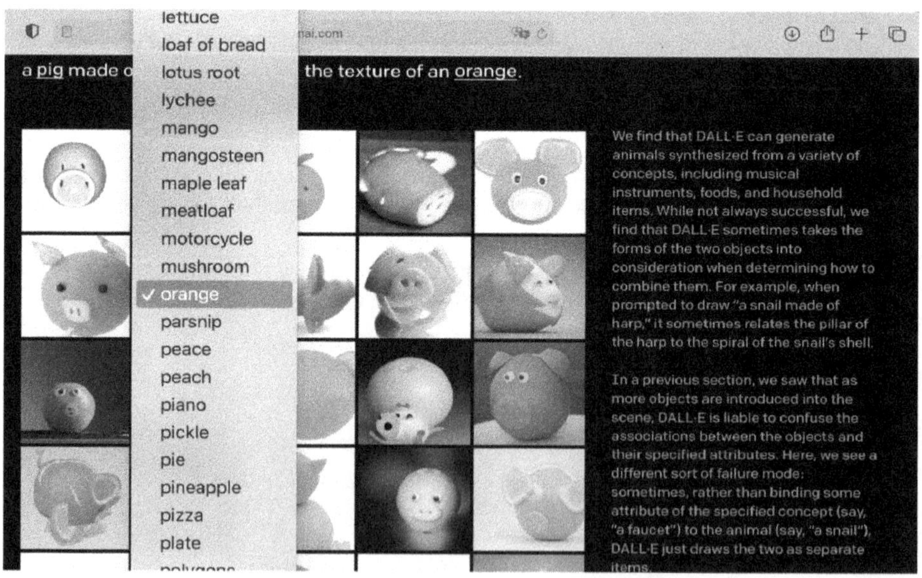

Fig. 3. Synthographs of 'a pig with the texture of an orange' as suggested by DALL·E

2 Autonomous Photographers

Based on the observations of the state-of-the-art, we can only imagine what will be the next technical achievement to facilitate and automate photography, considering all the industrial advances in image recognition and generation.

Certain aspects of the photographic profession will disappear, since repetitive chores are superseded by different means, as in other industries. More than three quarters of the products in IKEA's catalogue are already photorealistic images rendered by CG-artists. Before they manipulated vectors and shaders on their computers, the 3D-illustrators were trained in product photography, to emulate this visual vernacular [21]. Similarly, fashion photography can skip the lenses and shutters, while the imagery is produced on graphical engines [22]. Soon, even no photo models will be necessary to wear the fabrics, since AI generated avatars can provide a fresh face for every look.

This kind of image creation shifts the role of the individual author to a group of people, distributing tasks among many professionals with diverse but circumscribed assignments. Many other images are created without any author or anyone waiting for the decisive moment.

Surrounded by surveillance cameras, the individual photographic apparatus might soon become superfluous, at least for selfies and other concepts to record the proof of an individual's happiness at a certain location or event.

The public spaces around us, cities and crowded places all over the world, are pervasively furnished with surveillance cameras which act as autonomous photographers, framing and recognising faces, following people's movements, and filling databases. Since these devices point in every direction to catch perpetual glimpses of us, we could demand to capture us on our holidays and deliver the images right to our email account, associated with our facial recognition profile[4]. With adjustments for stylistic elements such as basic rules for composition and colour, these postcards from the omnipresent observer could console us in our loss of independence and privacy.

Based on the wide range of existing and analysed images, there will be plenty of results applying a familiar style, the colour palette and lighting of famous artworks are already frequently applied examples [23]. But could an independent practice be generated out of these pre-sets, other than reiterating the already known? Waiting rooms or hotel walls around the world provide plenty of examples. For young artists, it would no longer be the challenge to develop a personal style but rather find sophisticated algorithms and explore idiosyncratic combinations.

Currently, the more interesting artistic positions are the ones which critically examine the development of artificially generated images. The neutral and revealing observations of operational images and surveillance in the work of Harun Farocki [24] could be regarded as a foundation for the investigations of Trevor Paglen. He excites our curiosity in combining diverse aesthetically attractive images with intense backstories. For

[4] "[A]lthough Facebook will delete data on more than a billion faces, the company will retain DeepFace, the AI model trained with that data. [...] The deep learning model was created in 2014 with 4 million images from 4,000 people, the largest dataset of people's faces to date." Kari Johnson, *Facebook Drops Facial Recognition to Tag People in Photos* https://www.wired.com/story/facebook-drops-facial-recognition-tag-people-photos/.

instance the accompanying text of a series of portraits of ten ordinary people entitled 'It Began as a Military Experiment' (2017) reveals them as military employees which are part of a database of thousands of portraits for Face Recognition Technology (FERET), developed by the US Department of Defense in the 1990s. On close inspection, facial features are defined through small white letters and rectangles, ready for automatic identification.

The ubiquity of cameras at any time of day in every corner of the world results unsurprisingly in hardly anything happening unnoticed. But not only the arbitrary activities of anyone will be recorded, so will our surroundings be documented for future generations. In times of unrest and war, these documents can come handy – when the dust settles, an architectural site which lies in ruins could be reconstructed only with the aggregate of the many existing photographs. This restoration would not necessarily depend on a professional photogrammetric assessment. The mass of images from all angles could suffice such as in the reconstruction of Palmyra [25].

3 Diverging into New Disciplines

With these observations, I anticipate a diversion into a new form of art, connected to image creation and improvement through artificial intelligence. The classic photography, already known as slow or late photography [26] with all its flaws and peculiarities will always have a right of existence. Just by looking at the performative aspect of arranging a composition through the optical lens, the fascination for the precision engineering of the apparatus itself, the pleasure of the chemical process which almost like a miracle brings the latent motive into being and the pleasure – but also sometimes the disappointment – which the final result inherits. AI supported image creation will always appear different as long as we do not recognise the originator as a sentient being. Therefore, to observe and treat 'synthesised' images in an independent way is of paramount importance. The paper is an invitation for further discussion, in particular where to draw the boundaries between AI enhanced digital photography, AI supported postproduction, CGI created images – and the obvious application for synthography – images fully generated by AI.

It is comparable to the autonomous driver on the highway, the support by automatic transmission and cruise control, versus the classic convertible car on the country road – all may take us to the desired destination – the same is valid for synthography, AI supported- and 'classic' photography.

References

1. Smith, A.R., Warburton, N. (ed.): Pixel: a biography (2021). https://aeon.co/essays/a-biography-of-the-pixel-the-elementary-particle-of-pictures
2. Reinhuber, E.: Phasmagraphy: A potential future for artistic imaging (2017), Technoetic Arts. https://doi.org/10.1386/tear.15.3.261_1
3. USA FAA: Air Traffic Technology (2021). https://www.faa.gov/air_traffic/technology/. Accessed 04 Nov 2021
4. Reinhuber, E.: Are photographers superfluous? The autonomous camera. In: Allen, R. (ed.) Art Machines: Proceedings of the International Symposium on Computational Media Art, pp. 101–103 (2019)

5. Eastman, G.: Patent for camera with roll film (1888). https://patents.google.com/patent/US388850A/en. Accessed 04 Nov 2021
6. Josepho, A.M.: Patent for photographic apparatus (1925). https://patents.google.com/patent/US1631593A/en. Accessed 04 Nov 2021
7. Wölfel, M.: Artificial Intelligence Assisted Creation – Fostering Inspiration & Raising Moral Issues (2020). https://doi.org/10.13140/RG.2.2.16957.41445
8. Cartier-Bresson, H.: The Decisive Moment. In: Images à la sauvette. New York: Simon and Schuster (1952)
9. Flusser, V.: Towards a theory of techno-imagination. Philos. Photogr. **2**(2), 195–201 (2011). https://doi.org/10.1386/pop.2.2.195_7
10. Schindler, F., Steinhage, V.: Identification of animals and recognition of their actions in wildlife videos using deep learning techniques. Ecol. Inform. **61**, 101215 (2021). https://doi.org/10.1016/j.ecoinf.2021.101215
11. Dawes, R.: Using AI to Monitor Wildlife Cameras at Springwatch. BBC Research & Development blog (2020). https://www.bbc.co.uk/rd/blog/2020-06-springwatch-artificial-intelligence-remote-camera. Accessed 04 Nov 2021
12. Huang, Y., Fuh, C.: Face Detection and Smile Detection. National Taiwan University, Department of Computer Science and Information Engineering (2009). https://www.csie.ntu.edu.tw/~fuh/personal/FaceDetectionandSmileDetection.pdf
13. Apple Developer Documentation: Sample Code: Highlighting Areas of Interest in an Image Using Saliency (2019). https://developer.apple.com/documentation/vision/highlighting_areas_of_interest_in_an_image_using_saliency. Accessed 04 Nov 2021
14. Nicholls, W.: Insta360 ONE: A 4K 360 Camera That Lets You 'Shoot First, Point Later'. Berkeley: PetaPixel (2017). https://petapixel.com/2017/08/28/insta360-one-4k-360-camera-lets-shoot-first-point-later/. Accessed 04 Nov 2021
15. Rivard, W., Feder, A., Kindle, B.: Image sensor apparatus, method and computer program product for simultaneously capturing multiple images (2014). https://patents.google.com/patent/EP3216211B1/en. Accessed 04 Nov 2021
16. Kim, B., et al.: Neural Networks Trained on Natural Scenes Exhibit Gestalt Closure (2020). arXiv:1903.01069 [cs.LG]
17. Lucidrains [Wang, P.]: This Person Does Not Exist. https://thispersondoesnotexist.com. Accessed 04 Nov 2021
18. Generated Media, Inc.: Generated Photos. https://generated.photos. Accessed 04 Nov 2021
19. OpenAi: DALL•E. https://openai.com/blog/dall-e/. Accessed 04 Nov 2021
20. Ramesh, A., Pavlov, M., Goh, G., Gray, S., Voss, C., Radford, A., Chen, M., Sutskever, I.: Zero-Shot Text-to-Image Generation (2021). arXiv:abs/2102.12092
21. Shaw, M.: See How IKEA 3D Models the Rooms in Their Catalogs. https://web.archive.org/web/20210920151627/https://architizer.com/blog/practice/details/see-how-ikea-3d-models-the-rooms-in-their-catalogs/. Accessed 04 Nov 2021
22. Adobe: 3D Visualisation for Fashion (2020). https://substance3d.adobe.com/magazine/3d-visualization-for-fashion/. Accessed 04 Nov 2021
23. Manovich, L.: AI Aesthetics. Strelka Press, London (2018)
24. Elsaesser, T.: Simulation and the labour of invisibility: Harun Farocki's life manuals. Animation **12**(3), 214–229 (2017). https://doi.org/10.1177/1746847717740095
25. Williams, T.: Syria – the hurt and the rebuilding. Conserv. Manag. Archaeol. Sites **17**(4), 299–301 (2015)
26. Campany, D.: Safety in numbness: some remarks on the problems of "Late Photography". In: Green, D. (ed.) Where is the Photograph?, Photoworks/Photoforum (2003). https://davidcampany.com/safety-in-numbness/. Accessed 04 Nov 2021

SOUND OF(F): Contextual Storytelling Using Machine Learning Representations of Sound and Music

Zeynep Erol[1], Zhiyuan Zhang[2], Eray Özgünay[1], and L. C. Ray[2](✉)

[1] Hong Kong University of Science and Technology, Kowloon, Hong Kong
zeynep.erol@metu.edu.tr, eouzgunay@connect.ust.hk
[2] City University of Hong Kong, Kowloon, Hong Kong
zzhang452-c@my.cityu.edu.hk, LC@raylc.org

Abstract. In dreams, one's life experiences are jumbled together, so that characters can represent multiple people in your life and sounds can run together without sequential order. To show one's memories in a dream in a more contextual way, we represent environments and sounds using machine learning approaches that take into account the totality of a complex dataset. The immersive environment uses machine learning to computationally cluster sounds in thematic scenes to allow audiences to grasp the dimensions of the complexity in a dream-like scenario. We applied the t-SNE algorithm to collections of music and voice sequences to explore the way interactions in immersive space can be used to convert temporal sound data into spatial interactions. We designed both 2D and 3D interactions, as well as headspace vs. controller interactions in two case studies, one on segmenting a single work of music and one on a collection of sound fragments, applying it to a Virtual Reality (VR) artwork about replaying memories in a dream. We found that audiences can enrich their experience of the story without necessarily gaining an understanding of the artwork through the machine-learning generated soundscapes. This provides a method for experiencing the temporal sound sequences in an environment spatially using nonlinear exploration in VR.

Keywords: Spatial audio · Virtual Reality Art · Machine learning · t-SNE · Sound Visualization · Nonlinear Listnening

1 Introduction

Our dreams are full of unexplored data, from memories that we seem to have forgotten about to sounds that we hear incompletely but feel completely at home with. Dreams are expressions of our collective memories, much like the way machine learning represents large data sets using a memory-network-based model. To explore the way we represent music and sound in our dreams, we applied machine learning to spatially cluster both whole works of music and sound recordings, converting temporal representations to spatial interactions, These interactions allow audiences to grasp a long piece of music

M. Wölfel et al. (Eds.): ArtsIT 2021, LNICST 422, pp. 332–345, 2022.
https://doi.org/10.1007/978-3-030-95531-1_23

without having to listen to it sequentially, but rather explore fragments of the piece put in related locations by the similarity of their sound features. We first designed these spatial interactions of sound in two related case studies, then uses these prototype findings to create a Virtual Reality (VR) work that narrates a particular looping memory in dreams. Evaluation with exhibition audiences revealed that the sound interactions enriched the experience of the story without affecting the understanding of particular scenes, showing that machine learning-based spatial interactions of sound and music can foster a new way to perceive diverse and potentially incongruent sources of data as one may find in a dream-like state (Fig. 1).

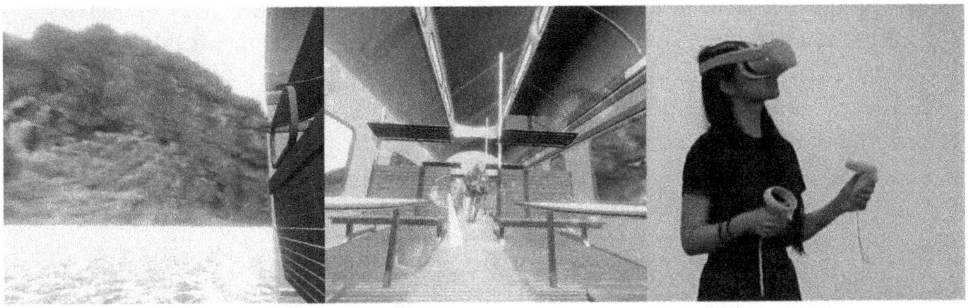

Fig. 1. (**Left**) The natural landscape outside of the train generated from a 360 photo dataset by machine learning algorithm StyleGANS2. (**Middle**) The interior of the train with translucent bubbles being pointed at by the controller as the source of sound clustered by t-SNE. (Right) An audience member experiencing Sound Of(f) using the VR headset and controllers.

Current approaches to understanding music and sound include song-content-based visualization using graphs [19] and reduced dimensionality visualization using musical features [9], part of a general attempt to understand music by visualizing the computation of features. In addition, visualizations of music databases allow intuitive exploration of music catalogs. Music and sounds, however, are part of an environment that evokes nuances that are divorced from visualization based only on computed dimensions. To evoke the feeling of one's home city, for example, we need to consider the spatial relationships between the sounds collected as well as the interactions with these sounds in space [3]. To better grasp the experience of long works of music and large sets of sound recordings, we used the t-SNE algorithm [16] to cluster related sounds for interaction in 3D space in a VR environment.

In the testing phase, we prototyped both segmentation of a *single* work of music to convert a temporal experience to a spatial one, as well as clustering of *different* sound recordings in a single environment to grasp the scene soundscape. We designed spatial interactions like gestural and pointing methods for experiencing the sounds in space, providing an immersive method to experience sounds in the context of the environment they originate from. In the creative phase, these design principles allowed us to create a VR artwork exhibited at a local gallery that attempts to blend different clustered sound and music fragments with particular scenes that convey the mood and nuance of "saying goodbye," "hope," "longing," "misunderstanding," and "silence," conveying the way rich sources of information blend together in a dream.

2 Background

Previous attempts at understanding complex audio data must deal with a large amount of information under consideration, and have included metrics that make the retrieval process more efficient [5]. These approaches rely on efficient classification schemes that resonate with human perception [14] but require a user-centered design perspective to implement. Machine learning has been applied to high dimensional audio classification using features of the sound [23], but these computational approaches do not always produce the phenomenological separations in human sound classification [8]. Similarly, environmental sounds have also been classified using convolutional neural networks [22]. Recent approaches have included using human biometrics data like EEG to automatically and computationally classify the experience of the sound itself rather than its physical properties [25].

One way to overcome the divide between the classification of the sound's features and classification of its experience involves using immersive techniques to allow human interaction with the sound's computational classification. The immersive experience of data has been applied to domains such as data analysis workflows [6], visualization relationship amongst scientific paper corpora [10], musical catalog visualization [7], cultural analyses of musical patterns [9], and previewing audio samples using dimension reduction techniques [4]. While these works have shown the promise of using immersive techniques like VR to help users experience complex audio data, they have yet to focus on the diverse set of gestural and spatial interactions that are possible.

Previous artworks like Blortasia have explored the effect of soundscapes on the unreal state of a virtual environment [17] but uses abstract shapes and colors to represent the abstract world instead of a reality-based transfiguration in dreams.

3 Technology Validation

To test our machine learning technology, we first collected 117 sounds of subway street musicians in New York City to form a sound collection that can be grouped according to musical features by machine learning. We then use t-SNE to populate these sounds in a 2D sphere around the audience. In addition, we also used a single 16:45-long performance of Gershwin's *Rhapsody in Blue* to use for the application of segmenting of a single musical work. For this work, we put sounds in 3D space using t-SNE also clustered by similarity. This allows us to test both the clustering of different sound samples to understand an environment, and the segmentation of a single musical work to transfer temporal experience to spatial experience.

3.1 Audio Processing

For segmenting the single piano piece *Rhapsody in Blue*, we used an onset detection algorithm [2] embedded in a MIR python package (librosa [18]). This algorithm divides the piece into chunks by detecting their onsets, i.e. beginning of the transient parts. For the collection of sound recordings in the New York Subway stations, we do not break them into chunks because we directly use the recordings for clustering, but instead, we take segments of 10 s from each sample for subsequent analysis.

Next, we generate the feature vectors of the sounds to capture the parameters of the recordings and segments. One way to capture these features is to obtain the Mel-frequency cepstrum coefficients and their time derivatives of the sounds, which are used in speech and music information retrieval (MIR) and processing [13, 15, 20]. We get these coefficients, 13 for each recording, and their first and second-time derivatives, called first and second delta features, using librosa [18]. Then we concatenate them to get the feature vectors of each recording; in total, we have a vector with a length of 39 for each recording.

3.2 Dimensionality Reduction

Using the 39-dimensional vectors for each recording, we transform them into 2D or 3D spaces for human visualization. We use the scikit-learn's tool [21] dimension reduction method t-SNE [16] for this purpose. t-SNE is a machine learning technique that assigns every point located in a higher dimension onto a location in the two or three-dimensional space by considering pairwise similarities between these points. t-SNE essentially clusters the audio data while maintaining the local structure of the data by transforming the similarities between vectors into the models of joint probability distributions and minimizing the Kullback–Leibler (KL) divergence [12] (a measure of how much a probability distribution is divergent from another) between them. Compared with linear dimensionality reduction methods like PCA, which focus on separating the low-dimensional representation, t-SNE puts similar neighbors close together in low-dimensions, making it easier to cluster sounds to different genres.

In detail, one problem with clustering algorithms involves the "crowding problem", in which the points that are far away in the high-dimensional space are all packed together in the low-dimensional space because there is less space to work with in low-dimensions [16]. A direct consequence of the crowding problem is that the separated clusters in the high-dimensional space are not clearly divided in the low-dimensional space. t-SNE solves the crowding problem by replacing the joint Gaussian model distribution being minimized with a Student-T distribution. Other non-linear methods like Isomap [1] and LLE [24] are more suitable for unfolding a single continuous low-dimensional manifold, but are not as desirable as t-SNE in this case due to the multiple manifold structure of audio data (Fig. 2).

After applying t-SNE to the extracted MFCC and its time derivatives of the recordings, we obtained 2D and 3D point clouds of the sound data in a dispersed form. Some of the points are extremely close to each other while others are far away. In the 2D output case, they can be transformed into a regular raster cloud without losing their neighborhoods. We used Raster Fairy [11] to assign this diverse point cloud into a circular point cloud while preserving the similarities and dissimilarities between points. Essentially, the Raster Fairy encoding gives us an alternative 2D embedding that can be transformed to a 3D embedding in VR by putting each point in the skybox environment. This would provide a regularized set of points for use in 3D for an evenly distributed encoding.

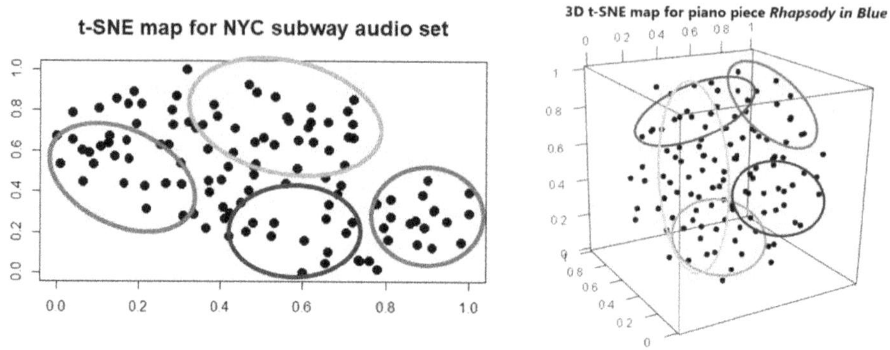

Fig. 2. (**Left**) The 2D point cloud for New York Subway street recordings after applying t-SNE. The red ellipse indicates a cluster of percussive sounds, the green ellipse includes vocals, the burgundy ellipse is string sounds, and the purple ellipse includes brass sounds. (**Right**) The 3D point cloud for a piano performance of *Rhapsody in Blue* after applying t-SNE to 8.5-s segments. Yellow ellipsoid includes fast sounds, green ellipsoid includes mellow sounds, burgundy ellipsoid includes monotonic sounds, purple ellipsoid includes rich sounds and blue ellipsoid includes brisk sounds in the piano segments. (Color figure online)

3.3 Virtual Reality Configuration

Using the coordinates obtained in the t-SNE and the Raster Fairy reduction process (in the 2D case), we place sound sources onto 3D locations in the Unity (2019.4.9f1 URP) development environment. Audio sources are either placed on a sphere (2D) corresponding to a 360 photo mapping in the New York subway case study or in 3D locations corresponding to a 3D t-SNE encoding in the piano performance case study. An application was built to add environmental context as detailed in the design section, then built for Oculus Quest 2 VR headset for subsequent prototyping. Additional augmented content was added as 360 photos or navigable assets in Unity.

To see how a spatial view of audio collections and segments can facilitate user experience, we prototyped two VR designs as case studies for the final artistic output. In one case, we let users experience environmental sounds of the NY subway system by putting sounds clustered by t-SNE around a sphere. In the other case, we put fragments segmented from a single work of music and grouped by t-SNE in 3D space (Fig. 3).

3.4 Testing Study: Sounds of Street Performers in the New York City Subway

Environmental sounds are a strong determinant of the way we experience space. To test an immersive platform for audiences to experience the sonic environment of a soundscape, we recorded 117 clips of street music in the different NYC subway stations and applied the t-SNE/Raster Fairy strategies previously described.

We found that representation of the location-specific properties of the sound requires a spatial distribution of the t-SNE returned samples in a 2D sphere around the user. In such a format, using the reticle of the looking direction of the user provides optimal ability to discern and select different fragments of sound. The user selects the desired spheres that represent the recordings by directly looking at them. We calculate using

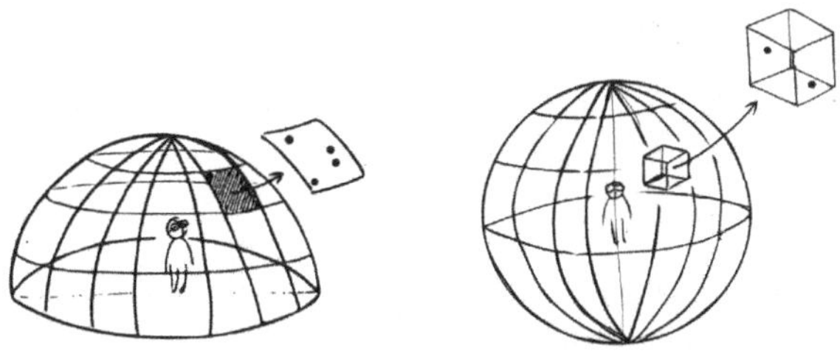

Fig. 3. Different embeddings of audio sources in 2D (left) and 3D (right) spatial projections. Points represent the sounds in the collection. Half-sphere surface 2D mapping is used for the NYC subway music case study (**Left**). The number of sources per unit area is 0.75 given a radius of 5 units. Full-sphere 3D mapping is used for the single piano performance case study (**Right**). The number of sources per unit volume is 0.22 when the radius is 5 units.

ray tracing which of the audio source spheres are selected/highlighted. After that, the user finds herself in the virtual panorama (360 photos) of the station where the street performer is recorded. Moreover, on the bottom half of the sphere, an NYC Subway map in the equirectangular form that surrounds the user appears. The station that the recording took place is highlighted on this NYC Subway map conveying the impression of traveling between different subway stations.

In the case of the Raster Fairy constructed 2D grid, we found that the interaction was not as telling of the subjective distances between the sounds, since Raster Fairy imposes equal distances between the sources. In VR therefore it would be better to perform clustering that optimizes the ability to tell between similarities in the features of the sound using the original t-SNE encoding (Fig. 4).

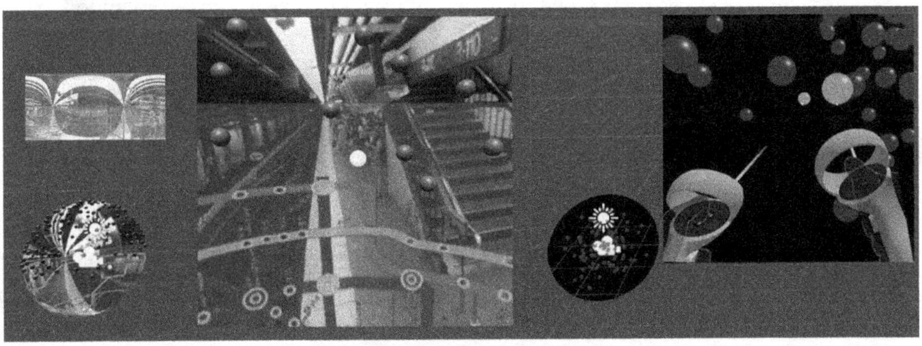

Fig. 4. (**Left**) VR design for Case Study 1: the spheres located on the upper surface of the outer sphere (2D) are representing the recordings of the street performers, and the selected spheres turn yellow. A transparent NYC Subway map is shown in 3D and the station where the performer is recorded is highlighted on the Subway map, along with the background of the 360 images of this station. The audio source spheres are equidistance away at the radius of the 360 photos of the subway station where the sound originated. (Right) VR design for Case Study 2: the floating spheres represent the 8.5 s segments of the piano performance of Rhapsody in Blue as audio sources, with the selected spheres highlighted with light blue in 3D. The sizes of the spheres indicate how far they are away in 3D space. (Color figure online)

3.5 Testing Study: The Structure of a Piano Performance

Longer works of music require sustained attention over the course of the performance on the part of the audience, whose mental state may vary at the beginning and end of the piece. To allow audiences to physically play with and listen to each part of a complex work of music in an interactive, self-directed, nonlinear manner, we divided a 16:45 long work of professional performance of Gershwin's *Rhapsody in Blue* into 117 segments (8.5-s fragments) and produced a 3D embedding of the data using t-SNE. Using a controller, it is possible to let the user interact with the sonic environment in a creative way. We used a 3D encoding here after we found that this allows for the most natural gestures for exploring the structure of a single piece interactively since in this case, the piece does not rely on contextual information, but rather on the ability to finely navigate the particular structure of the piece.

Moreover, in this case, we found that using the joystick controllers to identify and play the fragments best allowed for fine-tuned nonlinear navigation over the different fragments of work. In particular the 3D sense of space allowed users to play better with sounds structurally and provide an immersive experience outside the screen-like sphere encoding in the previous test study. To enhance the sense of place, the following design issues are taken into consideration: 1. two controllers can be used together and selected spheres are highlighted, 2. multiple sounds can be heard at the same time by using triggers in the controllers simultaneously, 3. navigating in 3D space leads to different views of the sources and different ways to use the controllers.

4 Design of SOUND OF(F)

We learned from our previous test studies that placing sounds in 3D using a controller-based interaction best facilitates the music exploration process. Hence we decided to use the 3D t-SNE encoding for our subsequent sound placement.

Sound Of(f) is a narrative VR experience that uses machine learning clustered soundscapes and 360 video landscapes to build an environment around different themes in a dream-like exploration on a metaphorical train. The inspiration for the work comes from a dream where someone close to the dreamer exited the train at her exit without ever saying goodbye. The dreamer then wakes up to find that he has no recollection of the exact person involved but rather a continuous fusing of one person's face with another person's personality, the product of a dream that, like machine learning models, fuse together personalities in creating characters and landscape. The character is intended to have within a single person the characteristics of multiple people in our lives, just as the landscape transitions between many places we know.

The experience takes around five minutes and loops repeatedly. The audience begins inside a train where she can move around using the joystick in the controller but cannot exit the train entirely. The train is going nowhere and everywhere. Noises in the form of music and sounds selected for each theme and clustered by t-SNE are found in the environment. They may be sounds that we don't want to hear waking us up, or fake news that we don't want to pay attention to, or misinformation designed to trigger our behaviors, but each set of sound revolves around a theme. The audience can listen to the spatial audio presented as translucent spheres in the scene grouped by t-SNE so that similar-feature audio clips are close to each other in space. The sounds are played back (while the sphere colors change) when you hover the mouse over individual clips. Audiences can feel like they are inside sound sequences, with the ability to explore them spatially rather than passively listen to them temporally.

The sounds for each thematic scene consist of sets of both music and sound recording fragments. The themes, presented in order, are: *Goodbye*, *Hope*, *Longing*, *Misunderstanding*, and finally *Silence*, each with its own associated character animation of a different way for the character to leave the train. For the *Goodbye* theme, we used Hiroshi Sato (佐藤博)'s *Say Goodbye* (セイグッバイ) and the goodbye final scene from the movie *Casablanca* where the main characters go their separate ways at when Ilsa boards the plane. Both are nostalgic and have different approaches to say goodbye: one is hopeful and the other one is dramatic and sad. For the *Hope* theme, we used a sound recording of a song sung by local Hong Kong people and Martin Luther King's "I have a dream" speech, which is inviting the dreamer to hope for future. For the *Longing* scene, we used the song *Apo Mesa Pethamenos* and the dialog from the car scene in the movie *Before Sunset*, both of which are about longing for someone, focusing after a breakup. For the *Misunderstanding* theme, we used Animals' song *Don't let me be Misunderstood* and the movie *The Switch* that puts the audience inside a fight. The songwriter shouts out about how everybody doesn't understand him and in the movie we hear one couple's dialog consistently fail to understand each other. In this scene, we also change the perspective so that the audience is looking at the character and sound bubbles from above, narrating the misunderstanding idea. For the *Silence* theme, we leave it to similar sounding meditative sounds. This is the literal "sound off" for all noises as well as a goodbye to our character, who now stands outside the train for the first time while the train has stopped moving and the landscape outside has stopped moving. After our intimate character has repeatedly stepped off the train without saying goodbye, we ourselves finally say goodbye ourselves, in order to turn off the sound (Figs. 5, 6, 7, 8 and 9).

In terms of the context, the train is running on a moving ocean. The interior decoration of the train is nostalgic and given bloom effects to emphasize the dreamy scene. As the train moves, the view outside of the windows will transit between many places seamlessly. The landscape skybox is a looping 360 video generated by state space traversal through a StyleGAN2 machine learning model trained using 478 total 360 photos taken by the authors at local landscape locations. As the audience walk in the train, turn around, look out of the windows, and explore the spatial sounds grouped by t-SNE using the controller, they find the ever-changing character on the train that acts on her own to leave the train. Everytime she leaves, the scene transitions to the next segment. The previous character that stepped off the train without saying goodbye is replaced by the same character at a different position. The audience is also relocated to a different location of the train, while the sound bubbles are updated. To see the VR experience, see this link: https://youtu.be/yMyR5DKjGA0.

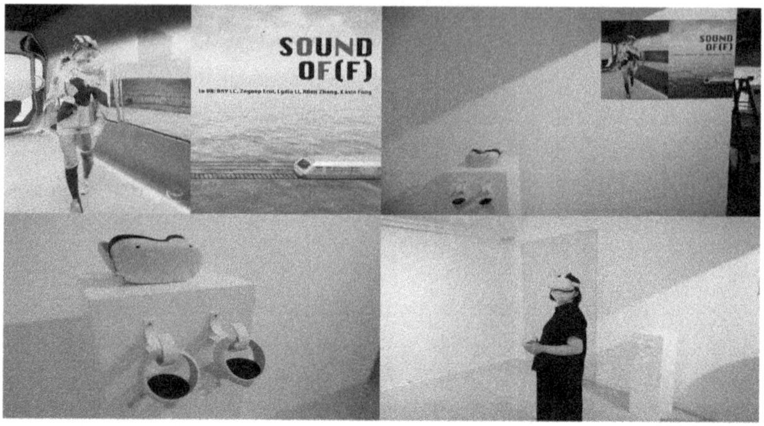

Fig. 5. Installation of the VR artwork. (Upper Left) Poster for the exhibition. (Upper Right) Location of the poster and headset relative to the wall. (Lower Left) Positioning of the headset on top of a plinth with cable entering the box, and two controllers on hinge brackets. (Lower Right) An audience member interacting with the work in the layout designed for the show.

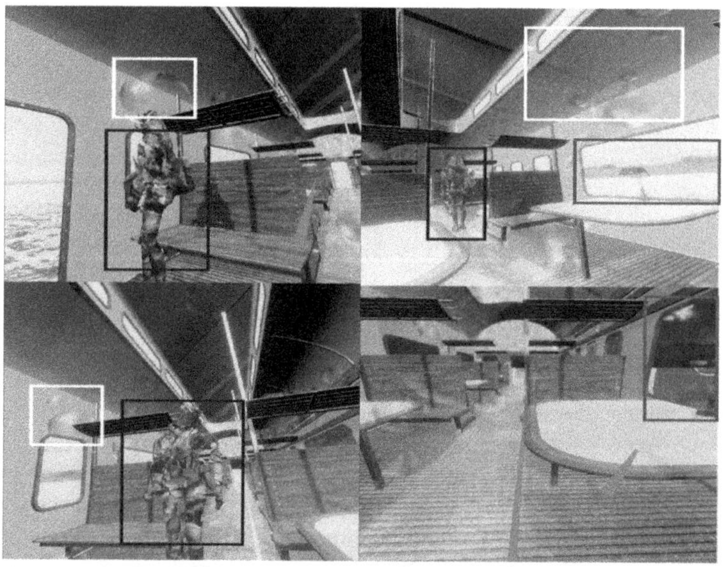

Fig. 6. Interior view of the train. The black rectangle contains the character. The dark blue rectangle contains the landscape, which is the 3D video generated by the machine learning model. The rectangle circle contains the bubbles in the air, which can be triggered to play t-SNE sounds. (Color figure online)

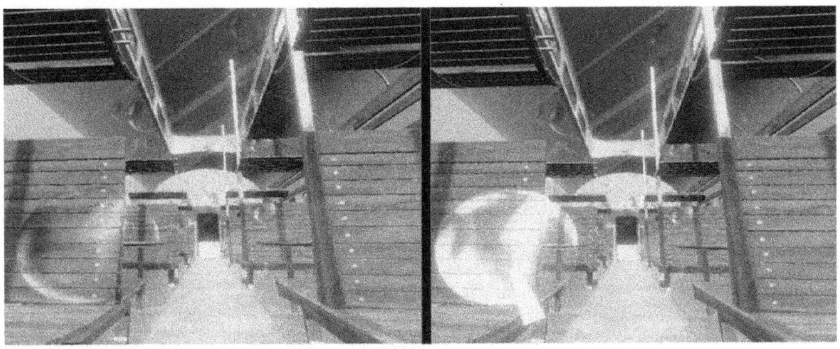

Fig. 7. Interaction with sound by pointing: audience can hear and explore the spatial audio (**Left**) grouped by machine learning using the red laser pointing at the bubble (**Right**).

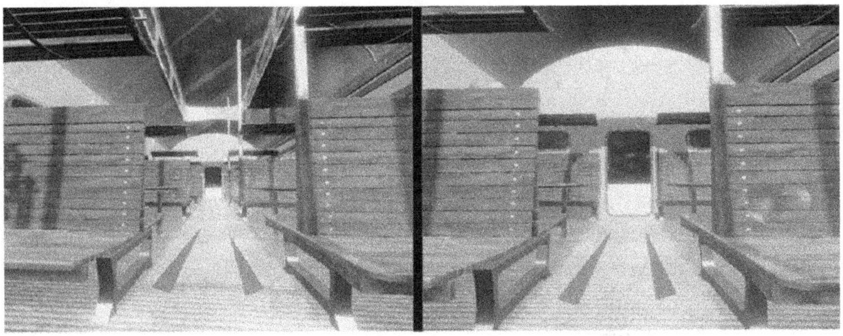

Fig. 8. Interaction with the environment by joystick: audience can walk and explore the inside of the train. (**Left**) Before walking movement. (**Right**) After walking movement.

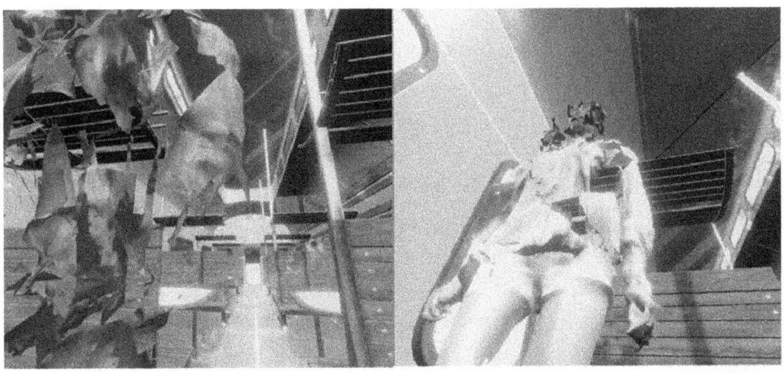

Fig. 9. An intimate character contains the characteristics of multiple people in our lives. During the journey, the character is changing repeatedly and walking off the train. In the last scene our intimate character is seen outside the train and walking off in a new direction without saying goodbye. (**Left**) Close-up of the character to see the shader working. (**Right**) Standing just under the character in an early scene.

5 Evaluation

5.1 Methodology

We surveyed 26 people at the opening of the exhibition immediately after they experienced all five scenes of the art work (14 female). The sample included 18–25 year-olds (13), 26–35 year-olds (5), 36–50 year-olds (5), and over 50 year-olds (3). The experimenters alerted the visitors that the controller can only be used for navigation and pointing and warned about possible dizziness. Visitors were allowed to immediately stop playing when experiencing strong dizziness. If the visitor completed the game, they were asked to participate in the survey immediately after removing the headset. The 12-question survey was given on a tablet, and took approximately five minutes, including 6 likert scale ranking questions (1–7) and 6 open-ended questions.

5.2 Quantitative Findings

Likert scale questions revealed that participants experienced the environment as a dream-like experience rather than a realistic one, and that the clustering of the sounds appear to be well-organized. Interestingly, the sound organization appear to contribute to the experience of the story but not to the understanding of what the scenes mean." This may be due to the difficulty of inferring the theme for each scene from the audio information alone, or to the lack of other identifying information that reveals the theme (goodbye, misunderstanding, etc.) We were also encouraged by the ability of the sounds to capture the mood of the scene, suggesting the effectiveness of spatial interactions to form a milieu of environments rather than sequential sounds like playing music from beginning to end (Fig. 10).

5.3 Qualitative Findings

Most audiences found their experience to be dream-like, as intended by the intervention. Almost half of the participants (11 of 25) chose "a dream-scape" as the answer to "Which do you think is the topic of this VR experience?" For example, one participant described the story as "*dreamy meta good-bye to self*," while another wrote that "*you are on a train in a dreamy situation and there is a perturbed shape of a woman who moves around; everything feels half a dream, half real.*" The dream-like experience may come from the evolving landscape, the moving and stationary train, the eerie soundscape and postprocessing, the change in perspective in one scene, *etc.*, but one unexpected source of the dreaminess is the ever-present character. One audience member noticed that when they get close to the character, she does not interact with them, which gives an eerie feeling of an unreal dream space. Interestingly, participants tended to be attached to the intimate figure. They had a connection with the character and made comments like "*I have liked that person*," going so far as to attempt to follow her out of the train in some scenes. One person said "*the user (me) wants to speak to the character really badly but she leaves the train and cannot be reached!*".

The interactions of different participants reflected their experience and personality. For example, some visitors moved their hands often and heard sound fragments only

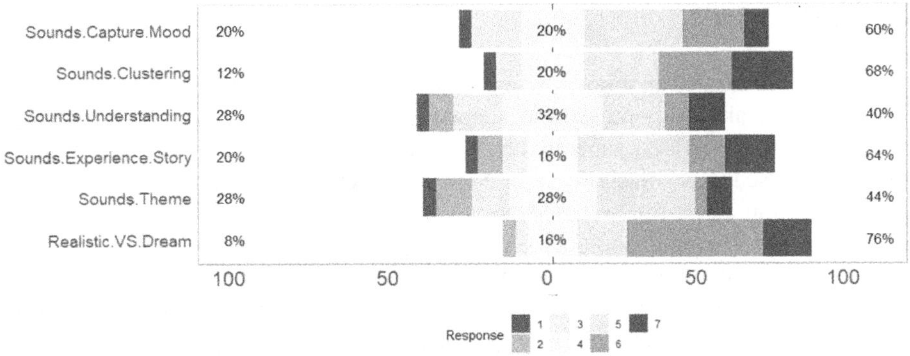

Fig. 10. Quantitative audience evaluation following playthrough of the entire set of 5 scenes. Sounds Capture Mood rated 1–7 for "How well do the sounds of each scene capture the mood of the particular scene?" Sounds Clustering rated 1–7 for "How well are the sounds in each scene clustered into related close-by fragments?" Sounds Understanding rated 1–7 for "How strongly do the sound fragments facilitate your understanding of what is happening in the scene?" Sounds Experience rated 1–7 for "How well do the audio fragments contribute to your experience of the story in VR?" Story Theme rated 1–7 for " Based on playing through each scene, how much have you grasped the theme of the audio in each scene?" Realistic VS. Dream rated 1–7 for "How much does the environment evoke a realistic vs an abstract, dream-like state?"

briefly, while others were more deliberate, hearing the entire sound one bubble at a time patiently before moving on. Some stayed in one place for the duration of the experience while others tried to go outside first and found themselves being stuck inside. Younger visitors appear to adapt themselves easily, having the most fun and interaction throughout the exhibit session. Older visitors take time to get adjusted, and tend to tire and get dizzy quickly. A few did not finish all five scenes, and were left with partial knowledge based on their incomplete experience, holding different opinions about the scenes they did see. However, some older participants found the scene relaxing as they slowly went through the sounds, especially the silence scene.

While participants understood a difference in the sounds and music used in each scene, they often didn't perceive the theme being portrayed. They variously described the sounds heard as "*political scene*," to "*noisy restaurants*," and "*orchestral music in old movies*." One of the most common descriptions, however, involves comparing the experience to a radio. One audience member described the experience as "*this is like searching for a channel on the radio.. clustering the sounds, and trying to find the correct one*." The idea of the radio strongly reflects the idea of spatial navigation of sound, in that people can turn a dial spatially and explore nearby channels to hear fragments of sonic experience instead of temporally listening to the entire piece. The way participants gravitated towards this type of interaction may reflect the need to turn temporal sonic events into spatial movement events for a global view of the soundscape, instead of listening to an entire sonic experience beginning to end.

6 Conclusions

In this work, we created an immersive environment for storytelling using spatial interactions for canonically temporal audiom, shaped by machine learning clustering technique and 360° panoramic video generated by the machine learning. First, we prototyped a musical soundscape of the subways in New York using a spherical embedding of the sound collection and a reticle-based pointing system. This interaction puts the audio sources in a sphere around 360 photos to provide context to the machine learning representation. Next, we prototyped a contrasting case where a single musical work is broken down into segments that are then interactable in 3D space. Here the separate expressive parts of the music are selected and played using controllers to better allow nonlinear exploration of the single musical work in VR.

Using the findings in these prototypes, we then created an artwork that applied the t-SNE strategy of clustering sounds for spatial interaction into a narrative context, exploring a way of interacting with audio data spatially. We have used the interactable 3D space and combined the two case studies' approach: using single audio and multiple sounds, since both approaches have their own outcomes which help telling a story in the VR environment. However we chose to use controller-based operation for its more precise control over sound selection. We further supported the work with GAN-generated 360 video landscapes. The sounds are key elements for storytelling, with a set of five different themes inside the dream-like setting with a unique design of the character that represents the multiplicity of intimate people in our dreams.

Audience evaluation further showed how the experience of the story can be enhanced by the spatial sound interactions while the understanding of the scenes may not be affected. It also showed how spatial interactions of sound may already be present in a simplified form in the case of the radio, and points out the general complementarity of spatial and temporal interactions for sound. By using machine learning to pre-categorize our audio data, we envision a future where single glances and fast spatial exploration in 3D are utilized to convey the essence of entire musical works. It thus allows us to experience the story and its thematic elements as sonic spaces of different sound recordings or fragments of a long piece of music using an augmented form of intuitive understanding in space, in short, the "sound of" an environment.

Supplemental Materials
To see the interactions in VR during gameplay, see: https://youtu.be/yMyR5DKjGA0.

References

1. Balasubramanian, M.: The isomap algorithm and topological stability. Science **295**(5552), 7a–77 (2002)
2. Böck, S., Krebs, F., Schedl, M.: Evaluating the Online Capabilities of Onset Detection Methods
3. Born, G.: Music, Sound and Space: Transformations of Public and Private Experience. Cambridge University Press, Cambridge (2013)
4. Carr, C.J., Zukowski, Z.: Curating Generative Raw Audio Music with D.O.M.E, Los Angeles, p. 4 (2019)

5. Casey, M., Rhodes, C., Slaney, M.: Analysis of minimum distances in high-dimensional musical spaces. IEEE Trans. Audio Speech Lang. Process. **16**(5), 1015–1028 (2008)
6. Cavallo, M., Dholakia, M., Havlena, M., Ocheltree, K., Podlaseck, M.: Dataspace: a reconfigurable hybrid reality environment for collaborative information analysis. In: 2019 IEEE Conference on Virtual Reality and 3D User Interfaces (VR), pp. 145–153 (2019)
7. Flexer, A.: Improving Visualization of High-Dimensional Music Similarity Spaces. ISMIR (2015)
8. Gemmeke, J.F., Ellis, D.P.W., Freedman, D., et al.: Audio set: an ontology and human-labeled dataset for audio events. In: 2017 IEEE International Conference on Acoustics, Speech and Signal Processing (ICASSP), pp. 776–780 (2017)
9. Gomez, O., Ganguli, K.K., Kuzmenko, L., Guedes, C.: Exploring music collections: an interactive, dimensionality reduction approach to visualizing Songbanks. In: Proceedings of the 25th International Conference on Intelligent User Interfaces Companion, Association for Computing Machinery, pp. 138–139 (2020)
10. Klimenko, S., Charnine, M., Zolotarev, O., Merkureva, N., Khakimova, A.: Semantic approach to visualization of research front of scientific papers using web-based 3D graphic. In: Proceedings of the 23rd International ACM Conference on 3D Web Technology, Association for Computing Machinery, pp. 1–6 (2018)
11. Klingemann, M.: Raster Fairy (2016)
12. Kullback, S., Leibler, R.A.: On information and sufficiency. Ann. Math. Stat. **22**(1), 79–86 (1951)
13. de Leon, F., Martinez, K.: Enhancing timbre model using MFCC and its time derivatives for music similarity estimation, p. 5
14. Li, D., Sethi, I.K., Dimitrova, N., McGee, T.: Classification of general audio data for content-based retrieval. Pattern Recogn. Lett. **22**(5), 533–544 (2001)
15. Logan, B.: Mel frequency Cepstral coefficients for music modeling. In: Proceedings of the 1st International Symposium Music Information Retrieval (2000)
16. van der Maaten, L., Hinton, G.: Visualizing data using t-SNE. J. Mach. Learn. Res. **9**(86), 2579–2605 (2008)
17. Mack, K.: Blortasia: a virtual reality art experience. In: ACM SIGGRAPH 2017 VR Village, Association for Computing Machinery, pp. 1–2 (2017)
18. McFee, B., Raffel, C., Liang, D., et al.: librosa: audio and music signal analysis in Python, pp. 18–24 (2015)
19. Muelder, C., Provan, T., Ma, K.-L.: Content based graph visualization of audio data for music library navigation. In: 2010 IEEE International Symposium on Multimedia, pp. 129–136 (2010)
20. Müller, M.: Information Retrieval for Music and Motion. Springer, Heidelberg (2007). https://doi.org/10.1007/978-3-540-74048-3
21. Pedregosa, F., Varoquaux, G., Gramfort, A., et al.: Scikit-learn: machine learning in Python. Mach. Learn. Python, 6
22. Piczak, K.J.: Environmental sound classification with convolutional neural networks. In: 2015 IEEE 25th International Workshop on Machine Learning for Signal Processing (MLSP), pp. 1–6 (2015)
23. Rong, F.: Audio classification method based on machine learning. In: 2016 International Conference on Intelligent Transportation, Big Data Smart City (ICITBS), pp. 81–84 (2016)
24. Roweis, S.T.: Nonlinear dimensionality reduction by locally linear embedding. Science **290**(5500), 2323–2326 (2000)
25. Yu, Y., Beuret, S., Zeng, D., Oyama, K.: Deep learning of human perception in audio event classification. In: 2018 IEEE International Symposium on Multimedia (ISM), pp. 188–189 (2018)

Questions and Answers: Important Steps to Let AI Chatbots Answer Questions in the Museum

Stefan Schaffer[1(✉)], Aaron Ruß[1], Mino Lee Sasse[1], Louise Schubotz[2], and Oliver Gustke[2]

[1] German Research Center for Artificial Intelligence (DFKI), Berlin, Germany
stefan.schaffer@dfki.de
[2] Linon Medien, Berlin, Germany
chim@linon.de

Abstract. In this paper, we describe our work within the research project "CHIM - Chatbot in the Museum". CHIM is an AI-based chatbot prototype that enables conversational interaction using text and speech input: visitors can ask questions about certain artworks and receive answers in multimodal formats (text, audio, image, video). The application will be tested in the Städel Museum, Frankfurt/Main, Germany. To develop a proper Natural Language Understanding module, we adapted an existing categorization approach, gathered visitor questions, and structured them into twelve distinct content types. The preliminary results suggest that our approach to subdivide the previously overloaded content type *meaning* into further categories was successful, leading to a more balanced distribution of the data. We further describe the Natural Language Processing mechanisms employed here; these follow a multi-tiered approach using techniques like Rasa, BERT, and cosine-similarity to generate answers with different degrees of effort. Future steps are the implementation of dialog management, the refinement of the NLP strategies by integrating additional answers for selected exhibits, and the implementation of the final layout and interaction design. We are planning to test and evaluate the CHIM prototype on site in the Städel Museum in late 2021.

Keywords: Chatbot · Conversational interaction · Digital Museum Guide

1 Introduction

In recent years, artificial intelligence, or AI, has gained increasing attention in museums all around the world. In this context, AI was initially mainly used

We want to thank the Städel Museum Frankfurt for their support. This research is part of the CHIM project of the research initiative "KMU-innovativ: Mensch-Technik-Interaktion", which is funded by the Federal Ministry of Education and Research (BMBF) of the Federal Republic of Germany under funding number 16SV8331.

M. Wölfel et al. (Eds.): ArtsIT 2021, LNICST 422, pp. 346–358, 2022.
https://doi.org/10.1007/978-3-030-95531-1_24

for example for image recognition or database analysis in general, which can be considered as "classical AI domains" [8]. Today, we find a wider range of applications in museums using AI, utilizing neural networks, machine learning, robotics, computer vision, deep learning, or natural language processing. Moreover, there are exhibitions focusing on AI itself.

Our approach in the research project "CHIM - Chatbot in the Museum" is to use AI within a chatbot museum guide application aimed at improving visitors' museum experience. The chatbot AI should be able to answer specific questions about certain artworks and thereby help to eliminate known pain points in knowledge transfer and learning situations in museums: Often, a personal human guide is not available at a given time. Additionally, in the current pandemic situation, tour groups that cluster in front of an artwork are no longer allowed. Digital media guides, which allow for a more individualized experience, normally offer only "one-way-information" such as audio guide texts but cannot reply to specific questions. According to the concept of "free-choice learning" [10], the kind of learning that occurs in museums fundamentally differs from the type of learning that happens in schools. Whereas in schools, one is forced to learn content that is not self-selected, in the museum, you can choose to learn about objects and artworks that interest you. This kind of learning is described in a "contextual learning model" [10]. One key factor to improve free choice learning is to keep visitors activated and personally involved with the story. To accomplish this, content delivery has to take into account visitors' motivations, expectations, and their personal leisure values by giving them a maximum of choice and control in how they want to learn about an artwork. Our chatbot allows visitors to ask any questions they have about an artwork. On the one hand, this open question functionality complicates finding an appropriate answer by the AI. On the other hand, it allows us to offer information tailored to the specific users' interest at that particular moment. The chatbot could become a sort of "virtual guide" that is available at any time or place to answer visitors' questions. In contrast to a guided tour, where visitors could shy away from asking "stupid" questions in front of other group members, these social context barriers are usually lower when interacting with a machine. Compared to traditional media guides, a chatbot allows visitors to ask the questions that interest them at that moment. They do not have to choose from predefined content. In this way, we hope to simplify the learning process, boost visitors' attention, and ultimately increase visitors' satisfaction. In addition, by evaluating visitors' questions and interactions with the chatbot, museums will be able to improve their educational offers, since they will learn more about what visitors want to know. To make the CHIM chatbot available to a wider audience, we intend to implement it as a smartphone application. Unlike some previous chatbot applications developed for museums [11], the system is not specifically aimed at attracting younger audiences, but ideally caters to museum visitors of diverse ages and backgrounds.

We developed the current version of CHIM to be used in the Städel Museum, Frankfurt/Main. This has two main reasons. Firstly, the museum has recognized the importance and promoted the use of innovative digital applications in the cultural heritage sector for several years now and has set up a team specifically dedicated to digital aspects of their educational agenda. We are extremely grateful to the museum and its staff for their kind support in the development of CHIM and the hosting of the on-site evaluation of the prototype in late 2021. Secondly, we have access to a large corpus of audio guide texts, written and produced by Linon Medien specifically for the artworks exhibited at the Städel Museum. As we will elaborate in the following sections, in the CHIM project we explore whether we can use these existing texts in order to find answers to visitors' questions.

Regarding previous work, we want to point out theoretical approaches, especially in the field of digital humanities. Some scholars postulate that digitalization and the massive application of AI technologies could lead to new methods in analysis and rating patterns in art history [13]. A content-focused chatbot AI allows us to gain insights into topics such as user-generated content. Furthermore, these insights can provide important impulses for the discussion about the sovereignty over the interpretation of art and cultural heritage.

With respect to relevant technical aspects, well-known chatbot and dialog platforms like Alexa (Amazon), Dialogflow (Google) and others need to be mentioned: they enable intention detection for many fields but are not sufficiently "case sensitive". If one asked Google questions of the kind that we collected and evaluated (see Sect. 3), regarding a specific artwork, one would get internet and Wikipedia hits, but not necessarily a proper answer. However, the number of AI-based conversational guiding systems, specialized in the field of cultural heritage or museums is growing [5]. A wide variety of approaches can be found, starting from systems that provide audio or media guide information via platforms like WhatsApp by typing numbers [2], to more conversational chatbot applications [1].

The goal of CHIM is to develop a learning, multimodal dialog system for knowledge transfer in museums. While working towards the envisioned chatbot, we explored different methods for making the system understand visitor questions and for finding suitable answers, e.g., by extracting the answers from existing audio guide texts. In this paper, we describe the steps we undertook in building a Natural Language Understanding (NLU) model for the classification of visitor questions. Adopting an approach from [6], we identified distinct content types for questions asked about selected artworks from the Städel Museum and developed Natural Language Processing (NLP) strategies for generating answers by using these content types, complemented by additional annotations. One novel contribution of CHIM to the field is that the system allows for user generated questions, rather than relying on pre-scripted dialogues, as other German language museum chatbots currently do [3,4]. Moreover, the advantage of developing our own NLU und NLP models as opposed to relying on for example Dialogflow, is that it enables us to store and process our data in accordance with German data protection laws, a non-trivial aspect of the project.

2 About CHIM

The main objective of CHIM is to develop a chatbot that can answer questions by museum visitors about objects in the museum. CHIM enables conversational interaction based on text and speech. Visitors can ask their questions and receive answers in multimodal formats (text, audio, image, video). In addition, the application will offer customized tours based on the interests and needs of the respective visitors to create a personalized experience.

In the process of developing CHIM, we explore different methods to extract answers from our corpus of existing audio guide texts. On the one hand, we explore how large language models, such as BERT [9], can be used in the museum chatbot context to find answers in unstructured or partially structured data. On the other hand, we explore how established methods for NLU can be efficiently integrated into the process of creating chatbot tours [7].

A crucial step in the creation of the CHIM chatbot is to build an NLU model for the classification of visitor questions. To collect relevant questions from potential museum visitors, we created a website designed specifically for this purpose. Our approach is to first identify the content types of the questions asked by museum visitors. To this end, we categorized the collected questions according to their content type. The question collection itself, the procedure for question categorization and the results of the categorization are described in the following section. In Sect. 4, we outline our planned and partially realized NLP strategies.

In a subsequent step, we will refine the content types by adding annotations for entities and relations. Further, to extract matching answers from the existing corpus of audio guide texts, the texts will also be labelled with content types.

3 Question Collection

3.1 Experimental Procedure

A website was built to gather relevant questions about 14 selected exhibits of the Städel Museum. To find as many contributors to the question collection as possible, a campaign was initiated in cooperation with the Städel Museum via the Städel Blog. In this way, we collected a total of 2182 questions from 203 unique user sessions during the period from December 22, 2020, to March 23, 2021. Each user session corresponds to one participant.

On the home page of the question collection website, we briefly described the procedure and purpose of the collection. The participants were presented a sub-selection of the 14 artworks, one at a time, and their task was to ask one or two questions per artwork. For each interaction, the date on which the interaction occurred, the input form (text input or voice input), as well as the browser used were anonymously stored. As input of the participants, the questions about the objects and optional comments about the application, as well as optional information about age, gender and education level were stored. About 50% of the participants provided demographic information. The average age of these participants was approximately 43 years (min. 17/max. 71). The participants had the following gender distribution: 63% female, 33% male, 2% other, 2% (explicitly) no indication. The educational background was distributed as follows: 80% university, 13% university of applied sciences, 5% high school, 2% other.

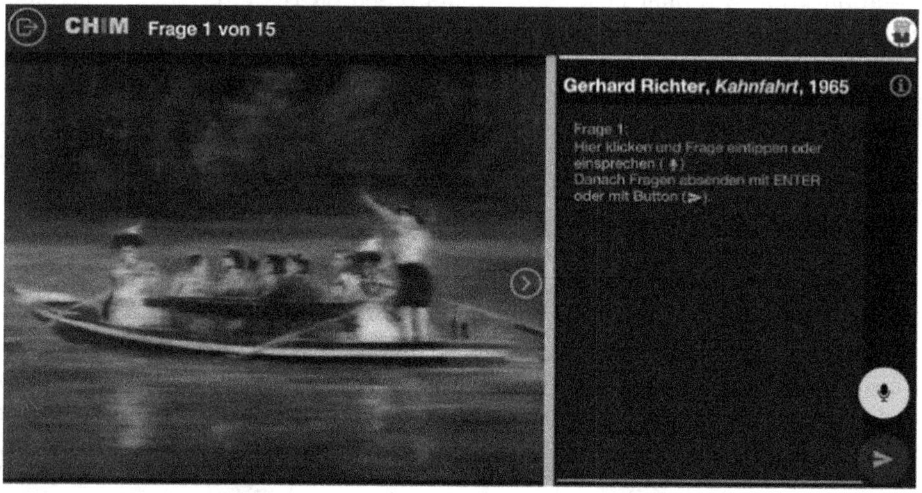

Fig. 1. Question Collection Website. Artwork is displayed in the Städel Museum. Photo: © Gerhard Richter 2020 (0217).

Figure 1 shows the user interface of the question collection website. Each participant was asked to enter a total of 15 questions. On the left side, below the question number, an image of the artwork was displayed for which questions were to be entered. At the top right, basic information like the artist's name, the title of the artwork and the year of creation were shown. Below this was a text field for entering the questions. Questions could be entered either via keyboard or by using the microphone symbol on the bottom right. Speech input was transcribed into text using automatic speech recognition. The recognized text was displayed in the text field. After entering 15 questions, a short questionnaire was displayed for demographic data and for comments about the application.

3.2 Question Categorization

The categorization of the questions is based on an approach of [6]. They categorized visitor questions into content types to explore which types of content voice-based AI conversational systems should attend to in order to meet visitors' expectations in a museum. We intend to utilize this approach to generate an annotated set of visitor questions that can be used to build an NLP model that is able to categorize questions about artworks into such content types. This model will be one of the building blocks in our NLP pipeline. The following 8 categories were used in [6]:

- *fact*: questions related to who is the artist, when the artwork was made, its size, or where it has been exhibited;
- *author*: visitor utterances about the artist's life, which art movement they were part of, or stylistic influences;
- *visual*: questions about colors and materials used, brushing techniques, etc.;
- *style*: questions about the style of the artwork, which school it belonged to and its characteristics, or artworks with style;
- *context*: inquiries about the historical, political, or social context where the artwork was produced;
- *meaning*: questions related to intentions, meanings, or whys, and the stories possibly behind the people and elements depicted in the artwork;
- *play*: utterances of playful engagement with the artwork, questions beyond the scope of the work, such as which soccer team a character roots for;
- *outside*: groups questions related to the conversational guide itself, its technology, or unrecognized utterances.

In their analysis, [6] revealed that far more than the half of the questions were about the *meaning* of the artworks (about 60%), followed by factual questions (17%), and questions about the artist's biography (7%). About 10% of the questions were not understood or were outside the scope of the artwork. The other 4 content types, together, corresponded to under 7% of the questions. Further, it was shown that the distribution of question types did not significantly differ per artwork.

As the content type *meaning* is overused in [6], we refinded this category by adding the following four content types:

- *content*: questions related to what or who is depicted in the artwork, both overall and in detail. Examples: "Is that the baby Jesus on her lap?", "Who are these people?", "Is the dog really sleeping or just pretending?"
- *model*: questions about the original models that were used. This is about the portrayed real person or object that have a real background. Examples: "Did the painter really work with a nursing mother as a model for this picture?", "Is the dog real?", "Surely this is painted from a photograph?"
- *response*: questions related to the response that the artwork triggers in/the effect the artwork has on the viewers, both historically and contemporary. Examples: "How did people back then react to the image of a bare breast?", "I think the picture is stupid, you can hardly see anything?", "What makes the painting so peaceful?"
- *provenance*: questions/information regarding the chronology of the ownership, custody, or location of the artwork. Examples: "How did this painting end up at the Städel Museum?", "Who commissioned the painting?", "Was this painting stolen by the Nazis?"

With the questions collected via our website, we ran a blind manual classification with five annotators. One main annotator created annotations for all questions, while the remaining four annotators annotated about 25% of the data each. Disagreements between the main and the other annotators were resolved jointly. When no consensus could be reached, the annotation of the main annotator was used. In this way, each question received exactly one annotation. In the next subsection, we will give an overview of the preliminary analyses of the annotated data.

3.3 Results and Discussion

Questions on All Artworks. Figure 2 illustrates the distribution of the questions we collected across the twelve content types in percentages. Approximately one third of the questions were about the *content* of the artworks (about 32%), followed by questions about the *meaning* (26.5%). The content types *visual*, *artist*, *model* and *context* all range between 5 and 9%. The content types *play*, *fact*, *style*, and *response* all range between 2 and 4%. Less than 2% of the questions were categorized as *outside* the scope of the artwork or as *provenance*.

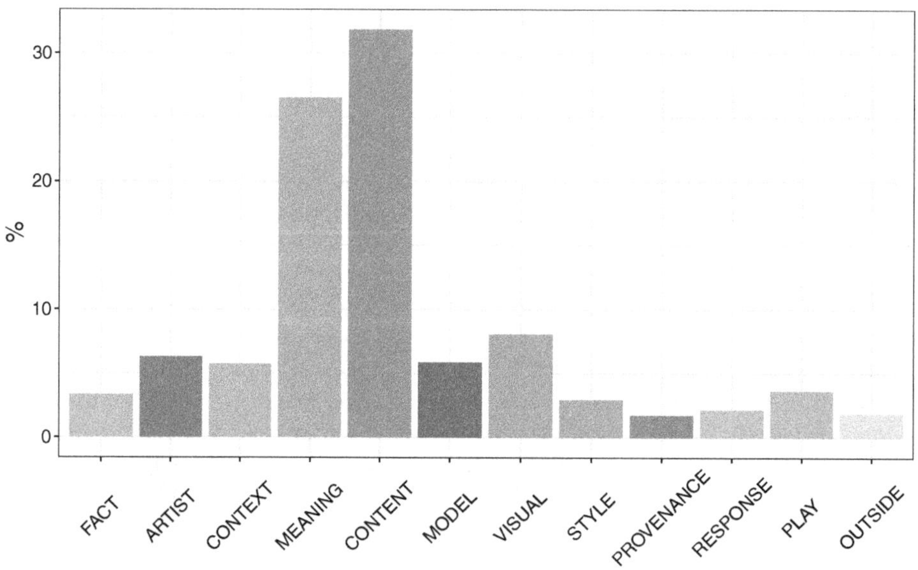

Fig. 2. Questions all artworks: distribution of questions according to content type in percentages. N = 2357.

Compared to [6], the content type *meaning* was considerably reduced from 60 to 26.5%. The largest contribution to this shift was made by the new content type *content*. The new content type *model* was the third most frequency, albeit contributing far less than the new content type *content* to the reduction of *meaning*. Out of the new content types, *response* and *provenance* were used the least. We conclude that adding more content types to split up the category *meaning* was successful in our case, since overall, a more balanced distribution of questions across the different content types was achieved.

The content type *fact* was used considerably less in our dataset than in [6]. This may be in part attributable to the inclusion of the new content type *provenance*, which can be seen as a subtype of *fact*. However, as *provenance* accounts for only 1.75% of the total questions, we also consider another explanation for this difference: The user interface of our question collection site already provides essential information for the category *fact* using text labels, displaying the artist's name, the title of the artwork and its year of creation. We deliberately chose this design, since in the Städel Museum, too, basic information is available on text labels displayed next to the artworks. However, it must be mentioned that as far as we know, the Pinacoteca museum in Brazil (the museum where the [6] application was tested) also displays basic information about the objects. We assume that sometimes this information is not easily visible for those visiting the exhibition. When collecting data in the future, we will consider not displaying such information on the website, so as to more closely mimic the actual situation in the exhibition.

Another clear difference is that the content type *outside* was used much less in our study. This can be explained by the fact that in [6], the category *outside* was used for annotation if the question was not understood by the system or was outside the domain of the artwork. In our study, so far no technical module is used to classify the questions, therefore, corresponding false detection in language understanding cannot occur.

Overall, the questions in our study are distributed more evenly across the content types than in [6]. In particular, our extension of the set by three additional content types may have contributed considerably to shift the distribution of the questions. A more balanced distribution is desirable for the creation of an NLP model: on the one hand, more training data is available for the classes of the model, avoiding biases due to uneven training data distribution. On the other hand, we hope that more clearly separated content types will lead to better precision determining the answers in further processing.

Looking into Data of Single Exhibits. Figure 3 shows a subset of our data. As is clearly visible, the distribution of the content types shows large differences for the individual images. For the object 'Lucca Madonna', the frequency of *context* and *meaning* is almost opposite of that of the overall distribution presented in Fig. 2. This painting from the field of Christian art is full of symbolic objects and imagery. We assume that this is one of the main reasons, why the questions are strongly concentrated on the meaning rather than the content.

Another notable difference can be seen with the object 'Boat Trip'. The content type *visual* is considerably more frequent than in the other distributions. Looking at the actual questions, we found that an above-average number of the questions relate to the technique the artist used to create the artwork.

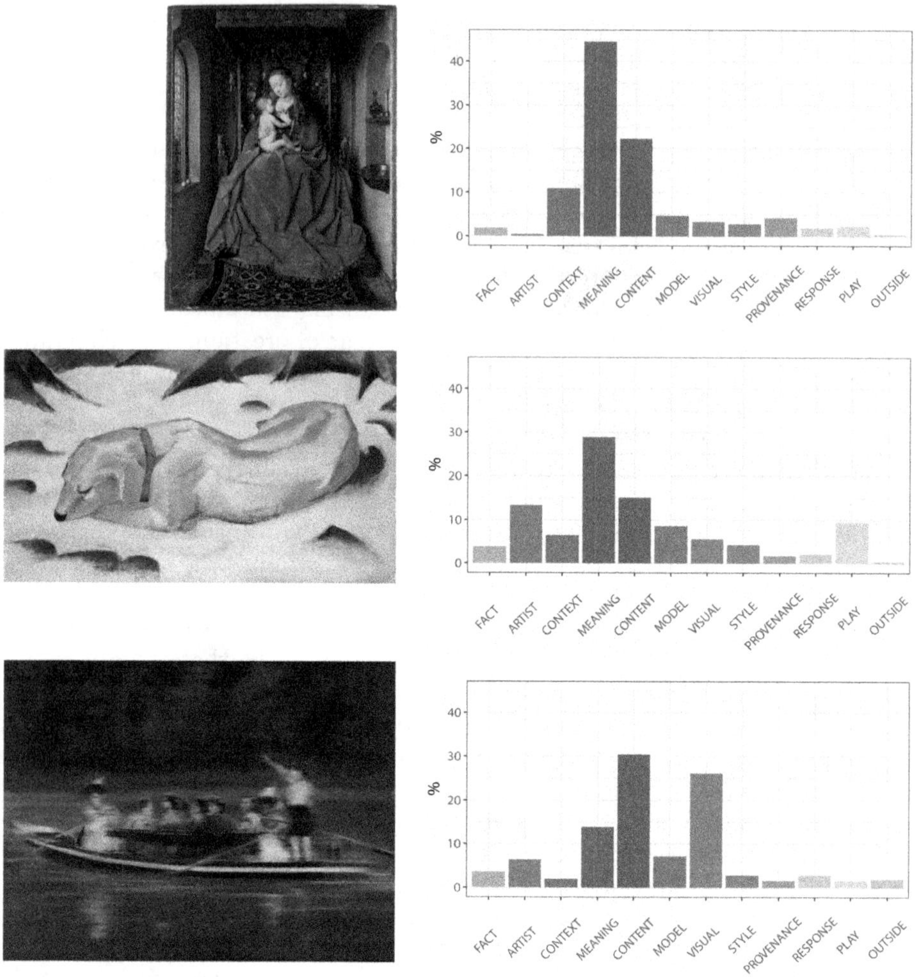

Fig. 3. Objects from top to bottom: Lucca Madonna[a], Boat Trip[b], Dog Lying in the Snow[a]; each with the distribution of questions across content types to the right. Photos: (a) CC BY-SA 4.0 Städel Museum, Frankfurt am Main.; (b) © Gerhard Richter 2020 (0217)

For the object 'Dog Lying in the Snow', increased usage of the content type *artist* can be observed. Again, looking at the actual questions, we found that many participants asked whether this was the artist's own pet dog, or if the artist liked to paint animals in general. Also noteworthy here is the use of the content type *play*. With this object, playful questions such as "does it bite?" were asked more frequently.

These preliminary results suggest a noticeable effect for the individual artworks on the frequency of specific content types in questions. However, this contrasts with the results of [6]. They found no significant correlation between artwork and content type. One possible explanation is that for the participants of our survey, each artwork represents a domain of its own. Across different domain, the content types may differ. However, this is only a preliminary finding. So far, we have not been able to extensively investigate the data of all the works. Going forward in our project we plan to investigate this difference and possible explanations for it.

4 Answering Strategies

The main task of the chatbot is to give a satisfactory answer to users' questions. This can be framed within the classical NLP problem of Question Answering (QA), i.e., based on a question, finding the correct document or excerpt within a document that contains the answer.

For the documents containing the answers, we considered two options: the first is to create dedicated answers specifically designed (= written) for the chatbot, whereas the second is to utilize existing text documents and descriptions for the exhibits in question - in the case of our project, the corpus of existing audio guide texts. When using existing texts, different degrees of enriching the text with metadata are possible (see Table 1), that allow better "machine understanding".

For example, the sentence "His way of painting was radically different from the International Gothic style, which at that time had been prevalent across Europe." could be annotated with metadata "artist: Jan van Eyck" making the artist in this sentence explicit; or annotated with some metadata like "style" as a content type to indicate that this sentence deals with the style of a painting.

While creating a specific answer for each question would be ideal, it is also the costliest option regarding time and effort. In addition, these dedicated answers can only cover those questions, or answers to those questions, that occur in the corpus of collected questions. Topics that are not brought up by these questions will in principle not be answerable by dedicated answers. In this event, the existing audio guide texts can be used as a fallback, since these texts usually are written with the goal to cover a wide variety of informational needs.

4.1 Degree of Enrichment

The effort for utilizing existing text depends on the degree of "enrichment". In our project, we follow a multi-tiered approach where we apply different degrees of enrichment and effort for the answers: for a few selected exhibits, we will create new, dedicated answers as well as highly metadata-enriched texts. For the rest of the exhibits, only question-clusters that crystallized as "frequently asked" by different users during our annotation phase will get dedicated written answers, and only, if these answers do not already exist in the available texts. Furthermore, these remaining exhibits' text descriptions will receive a middle to low degree of effort regarding metadata enrichment. One goal in our project is to find out exactly which degree of effort is minimally necessary or is enough to be able to create a satisfiable user experience.

When designing dedicated answers, it is useful to classify questions as factoid-type and open-ended-type questions: generally, factoid-type questions aim at short, specific answers (e.g., "when was this painted?") while open-ended-type questions require more elaboration (e.g., "what is the meaning of the painting"). Preliminary evaluations have shown that depending on the type of question, different NLP approaches are more - or less - successful at delivering satisfying answers [14,15].

Accordingly, our system uses multiple NLP techniques in stages, depending on how likely they are to deliver a satisfying answer. The mechanisms are designed so as to always return some kind of relevant information; if fallback mechanisms (see Sect. 4.2) are used, the system explicitly states that it may not have found *the* answer, but some information potentially related to the requested answer. In addition, the chatbot will employ further interaction strategies for dealing with errors and failures, which are not discussed in this paper.

Table 1. Used NLP mechanisms and their required vs. optional metadata-enrichments. Abbreviations: Entity (E), Relationship (R), Event (Ev), Content Type (CT).

	Required enrichment	Optional/additional enrichment
Factoid-ER module	E/R/Ev annotation in (set of) questions	
Factoid-BERT module		CT ("intent") annotations (in answer sentences)
Open-ended intent module	CT ("intent") annotations (in answer sentences)	E/R/Ev annotations (in questions & answers)
Open-ended similarity module	"section annotations": annotated answer sentences w.r.t. "same topic"/coherence Text follows	CT ("intent") annotation of answer sentences; E/R/Ev annotations (in questions & answers)

4.2 NLP Techniques

During the first stage, Intent Recognition using the tool Rasa [7] trained on the content type annotations is applied as well as a classification for factoid- or open-ended-type of question. For factoid-type questions, an Entity Relation Extraction mechanism will try to identify the question-target and -topic (e.g., "when was the image painted?": target is image, topic is time-of-creation). If successful, the corresponding factoid-datum is retrieved from a database and a natural language answer is generated. If unsuccessful, a BERT [9] model, pre-trained for QA is utilized for finding a matching answer in the text documents available for that particular exhibit.

For open-ended-type questions, answer candidates are retrieved from the dedicated answers and the annotated audio guide texts, if their annotated content type matches the recognized content type of the user's question with sufficiently high confidence. If the answer candidates comprise a continuous section of text, this longer explanation will be selected as answer. If the answer candidates correspond to multiple, separate sections, we plan to use Entity Extraction to reduce answer-candidates further down.

If the confidence for recognizing the question's intent is not sufficiently high to extract answer candidates based on this feature, a fallback mechanism is used that calculates a cosine-similarity [12] between the question and all answer-sentences, and then selects the encompassing text-section of the sentence with the highest similarity. From this, a chatbot answer is created, stating that no matching document could be found, but maybe the returned text contains some related information.

5 Conclusion and Future Work

In this paper, we report essential steps that we undertook in the development of CHIM, the prototype of a chatbot AI that answers questions of visitors in the museum. We adapted an existing approach for categorizing questions of potential museum visitors according to content type. In our adapted approach, we increased the number of content type categories from eight to eleven and used them to categorize a set of questions about exhibits in the Städel Museum. In the original approach, the distribution heavily skewed in favor of the content type *meaning*. Preliminary results from our extended approach show a more balanced distribution of the questions across the different content types. The annotated questions are used to set up a multi-tiered NLP approach in which we apply different degrees of effort to generate answers. Compared to building a complex NLP model using a multitude of fine-grained intents and entities, categorizing questions into rather rough content types comprises a comparably low degree of metadata enrichment and thus less annotation work for human annotators. To ensure scalability during the production of museum chatbots that make use of existing audio guide content, we suggest making use of highly metadata-enriched texts only for a few selected exhibits. For the rest of the exhibits, we reduce the overall effort for enrichment by applying NLP mechanisms trained with content type annotated questions and a follow-up NLP mechanism like BERT that can be applied to unstructured data. By using this approach, we consider that there are finite resources for creating "chatbot content", balancing the effort of creating new content and making existing content usable by enriching it to varying degrees. Our future work in the project CHIM will include the implementation of dialog management, a further refinement of the NLP strategies, and the integration of additionally created answers for selected exhibits. Further results will be reported after the final implementation and evaluation of the system during a field test at the Städel Museum in late 2021.

References

1. The field museum. https://www.fieldmuseum.org/exhibitions/maximo-titanosaur?chat=open. Accessed 30 July 2021
2. Jüdisches museum berlin. https://www.jmberlin.de/whatsapp-guide-hey-und-herzlich-willkommen. Accessed 30 July 2021
3. Kunsthalle karlsruhe: Art of chit-chatting. https://www.moodfor.art/chit-chatting. Accessed 02 Nov 2021
4. Ping! die museumsapp. https://www.museum4punkt0.de/ergebnis/ping-die-museumsapp-spielerisch-durchs-museum. Accessed 02 Nov 2021
5. Zentrum für kunst und medien. https://zkm.de/de/talk-to-me-chatbots-in-museen. Accessed 30 July 2021
6. Barth, F., Candello, H., Cavalin, P., Pinhanez, C.: Intentions, meanings, and whys: designing content for voice-based conversational museum guides. In: Proceedings of the 2nd Conference on Conversational User Interfaces, pp. 1–8 (2020)
7. Bocklisch, T., Faulkner, J., Pawlowski, N., Nichol, A.: Rasa: open source language understanding and dialogue management. arXiv preprint arXiv:1712.05181 (2017)
8. Ciecko, B.: Examining the impact of artificial intelligence in museums, February 2017
9. Devlin, J., Chang, M., Lee, K., Toutanova, K.: BERT: pre-training of deep bidirectional transformers for language understanding. CoRR abs/1810.04805 (2018). http://arxiv.org/abs/1810.04805
10. Falk, J., Dierking, L.: Learning from museums: visitor experiences and the making of meaning, January 2000
11. Gaia, G., Boiano, S., Borda, A.: Engaging museum visitors with AI: the case of chatbots. In: Giannini, T., Bowen, J.P. (eds.) Museums and Digital Culture. SSCC, pp. 309–329. Springer, Cham (2019). https://doi.org/10.1007/978-3-319-97457-6_15
12. Huang, A.: Similarity measures for text document clustering. In: Proceedings of the Sixth New Zealand Computer Science Research Student Conference (NZCSRSC 2008), Christchurch, New Zealand, vol. 4, pp. 9–56 (2008)
13. Kohle, H.: Digitale Bildwissenschaft. Hülsbusch, Glückstadt (2013). http://nbn-resolving.de/urn/resolver.pl?urn=nbn:de:bvb:19-epub-25747-3
14. Zaman, M.M.U., Schaffer, S., Scheffler, T.: Comparing BERT with an intent based question answering setup for open-ended questions in the museum domain. In: 32. Konferenz Elektronische Sprachsignalverarbeitung. Elektronische Sprachsignalverarbeitung. Elektronische Sprachsignalverarbeitung (ESSV-2021). TUDpress, Dresden (2021)
15. Zaman, M.M.U., Schaffer, S., Scheffler, T.: Factoid and open-ended question answering with BERT in the museum domain. In: Proceedings of the Conference on Digital Curation Technologies. Conference on Digital Curation Technologies (QURATOR-2021). CEUR Workshop Proceedings (2021)

Poetic Automatisms

A Comparison of Surrealist Automatisms and Artificial Intelligence for Creative Expression

Andreas Kratky(✉)

University of Southern California, 3470 McClintock Avenue, Los Angeles, CA 90089, USA
akratky@cinema.usc.edu

Abstract. Inspired by the recent controversy about art created by artificial intelligence (AI) algorithms and its successes in the art market, we are analyzing the use of automatisms as creative processes in visual arts. Without an attempt at exhaustiveness, we focus on two examples that mark two significant moments in art history: We compare surrealist automatisms and the automatisms used in recent AI artworks. Our interest is to understand the nature of the associated automatisms and the intentions and poetics motivating their use. The paper discusses the criteria of selection of which automatisms to analyze and locates them in their art historical context. To facilitate this analysis, we propose a framework to assess the poetic intentions and correlate them with the creative processes developed by the different artist groups. The overarching question motivating this investigation is to understand what has changed in our perception of the creative process and how it became possible for computational art to, today, occupy a seemingly uncontested place in the art market and discourse, after decades of heated controversy about its impossibility words.

Keywords: Artificial intelligence · Surrealism · Neural networks · Poetics

1 Introduction

Until not long ago it seemed more or less impossible that human creativity could be challenged by computers. While computing has entered nearly every aspect of our lives, the domain of artistic expression seemed to be one of the last bastions in which computers would not be able to replace human beings [7]. The controversy whether computers have a place in the arts and can possibly be creative unto themselves goes back to the time when computing just entered the imagination of people beyond the specialist circles in research centers, the military or big corporations. In the 1960s, even before access to any real computers was available to average people, the kind of thinking and potential the machines embodied, inspired several artists to employ computation-like procedures in their work. The question about creativity raised by this new style of work, in particular in these early days, became a heated topic, which mostly revolved around what it meant to create art, what it meant to be creative, and what the role of art should be in society. The reactions ranged from the hopes to found a new aesthetic, to Gustav Metzger's criticism

M. Wölfel et al. (Eds.): ArtsIT 2021, LNICST 422, pp. 359–378, 2022.
https://doi.org/10.1007/978-3-030-95531-1_25

that artists who engage in new computational media will be "eaten up by big business and manipulated by technology," [27], to the alleged sabotage of a computer installed to be part of the exhibition titled "Software," one day before the opening in 1970, so that the computer did not work [34].

Today, computational processes are a normal phenomenon across the entire creative field. Since many years, computational tools have supported the creative work of their users. For example, the painting and photography tool *Photoshop*, originally released in 1990, has become one of the most popular digital image editing tools. It has, for instance, been used by the artist David Hockney to create drawings [18]. In the last five years, a growing number of computational processes employing artificial intelligence (AI) have become available to automate tasks that used to be carried out by human users. Examples are tools using AI to automatically enhance photographic images (adjust, sharpen, resample, replace parts etc.) with synthesized information based on large amounts of images as training data for machine learning. And finally, there are tools that automatically create entire images and – for that matter – poems or sculptures, based on artificial intelligence algorithms. In this latter case it might be a matter of discussion if these automatic creation processes can be referred to as a "tool," since the connotation of the term "tool" is that it is an implement used by a human to do something, suggesting that it is still the human who is the creative force behind the product [44]. For the purpose of this paper, we will not go deeply into this discussion and adopt a rough categorization, according to which we take the first two categories, including software such as *Photoshop*, *ProCreate* or others, as tools comparable to a brush or other implements used to support artists in the process of creating artworks, which do not automatically create works. The last type delivers automated creation processes of partial or complete works. This line dividing automated creation from the support of human creation may not be perfectly sharp, but it allows us to see, in the latter type, that the activity of creating a work is controlled by rules and decisions of automated processes rather than by a human creator. As the process of creation of an artistic product is taken over, to a degree, by a machine that displays a certain amount of autonomy, without the human creator as the sole instance of creative decision making, we see ourselves confronted with the same question that has surfaced in the 1960s, when computers - or the inspiration of computers - entered the creative domains for the first time. In this context we understand "machine" as a complex device that is designed and set in motion to accomplish a certain task or produce a certain product [42]. Compared to the controversy of the 1960s, today, this question arises in a much more moderate form without the passion and radicalism that it had earlier. In comparison, earlier, when artists used computation-informed automatisms without actually employing machines to execute them, this question did not emerge at all.

1.1 AI in the Creative Process

In the last four years, artworks created by AI software have been exhibited in art shows and sold in art auctions. To circumscribe the timeframe of this development, we refer to a few events that mark the appearance and wider reception in traditional art press. In 2018, the Grand Palais in Paris had a show entitled "Artistes & Robots," which presented works of about 40 artists under the notion of "artificial imagination" to trace a development that proceeds from automatisms of a more analogue kind, like Jean Tinguely's machine sculptures and other material incarnations, to an algorithmic form of art creation [34]. Other commercial galleries followed suit to embrace AI-generated art [33].

Also in 2018, the auction house *Christie's* sold for the first time an AI-generated art piece in a major auction [6]. In 2019, the auction house of the *Sotheby's* corporation also began selling AI-generated art pieces [36]. This is not to suggest that something being sold in an art auction is any kind of useful definition, but the fact that some of the largest established art auction houses are including AI-generated art in their business portfolios indicates that there is a growing acceptance, at least among a commercial art-audience. In the press about these two auctions, a new kind of controversy is surfacing, which no longer focuses on the question whether artistic expression is a uniquely human characteristic or whether it could be taken over by computational processes; the new controversy is about the question how autonomous the artificial intelligence systems have to be and how much human interference can be tolerated for the product to be a piece of 'AI-art.' The French artist collective *Obvious*, which was the pioneer in having their works go on auction in *Christie's*, was criticized for using "a straightforward application of an algorithm that has been available since 2015 and their pieces involved a large amount of human intervention – deciding when a portrait was finished and framing it like an Old Master" [28].

With the "admission" of automatically-generated artworks to the art market modifies the traditional notion of originality and the idea that one unique person, the artist genius, has to be unequivocally associated with the creation of an art piece for it to have value, another aspect of art-market value, the idea of verifiable property that confirms ownership, lineage of origin, and the fact that the value associated with a certain artwork indeed belongs to its owner, has become more important. One of the big hurdles of digital art to enter the art market through galleries, beyond festivals and exhibitions, was that digital artifacts could be infinitely copied, and every copy was indistinguishable from any other. This fact made it practically impossible to uphold the idea of one unique original that could warrant value. The introduction of non-fungible tokens (NFT) in the art market was one way to address this problem. The 'double spending problem' of digital artifacts is mitigated – just like in digital currencies - through a cryptographic record in a blockchain that verifies the ownership of a certain item and, thus, makes value attribution possible. The adoption of NFTs in the art market exploded in the last couple years, indicating yet another sea-shift in how artworks and their value is seen [9, 11]. An almost satirical episode in this story is the acquisition and subsequent destruction of a real artwork, once it had been digitized and an NFT had been created for it. A company operating a platform to calculate blockchain transactions purchased an artwork of the English artist Banksy, turned it into an NFT and then destroyed the original, which they considered disposable after its existence as NFT was warranted [29]. We may just interpret this as an artifact in itself, or the display of the art market as a purely commercial endeavor - but it shows that it is worthwhile to analyze the concepts of artistic creation and poetics associated with the automated creation of artworks and how they have shifted.

Different visions about automatic art-creation have circulated for a long time. If we step back from the current debate about artificial intelligence in the arts and look at art history in a slightly broader perspective, artists have had an interest in exploring automatic processes as a means of art creation already long before computation entered the stage. We could take the engine of the *Academy of Lagado*, which was described in Jonathan Swift's 1726 book "Gulliver's Travels" as an early example of a 'computational' device for creative output [37]. But while Swift's engine is a thought experiment, other automatic processes have been actively conceived and used. An outstanding example is the extensive range of different automatisms conceived by the surrealists and can be interpreted as a continuous line of development in which today's AI processes mark the current endpoint. In this paper we will analyze these automatisms and the concepts of artistic creation and poetics inherent to them. Different artist groups or movements associated different intentions with the use of automatisms, and we are investigating how these intentions represent different ways of thinking about the process and purpose of artistic creation.

2 Computation, Art and Creativity: Human vs. Automatic Creation

The overarching question motivating this investigation is to understand what has changed in our perception of the creative process that, first, made it possible for computational art to, now, occupy a seemingly uncontested place in the art market and discourse, after decades of heated controversy about its impossibility and, second, what made it conceivable to discard the actual incarnation of an artwork in favor of its digital record of existence. So far, digital reproduction was simply a form of documenting artworks and it was generally understood that the perceptual experience of the piece is the real, original experience that is intended by the artist for the audience to have. So far, this real, original experience could only be hinted at by a digital representation but not substituted by it. We are familiar with a separation between the intellectual concept and the material incarnation of an art piece from the works of concept art of the 1960s and 70s, but this separation is different. The blockchain records just store information about the creation and successive transactions of the NFT, there is no information about the artistic concept or in fact anything pertaining to the thoughts, aesthetics, or materiality of an artwork. It is possible to correlate an image or description of the piece with the NFT, but this is again in the realm of documentation. What changed in the current relationship to the experience of an artwork that makes this paradigm shift possible, and what are the philosophical and poetic concepts related to this shift? While the scope of this paper is limited, we will present a preliminary framework and methodological considerations to approach these questions.

As we are still at an early stage of the wave of AI-generated artworks and NFTs, there is still a lot of speculation whether this is just a temporary fad or whether this is a sea-change of lasting impact [10, 19]. In particular in this moment, it is important to track and understand the transformations in the perception of the creative process that are responsible for these – maybe transitory – changes. We have a brief moment of capturing them in their raw state of emergence and see the discourse shaping around them.

Systematic research to develop and assess creativity in computational processes has existed since a number of years. These research endeavors tend to not make a claim to generate valuable and appreciable art; the goal is rather to investigate the question whether computers can be creative at all and how to design such automated creative processes. Creativity, in this context, is not limited to artistic creation, but includes scientific problem solving, mathematics, engineering problems and other areas of creative activity. It is a complex of several methods such as pattern matching, idea generalization or contextual thinking, which are applied in numerous contexts besides art creation. The majority of computational creativity research has been more focused on having humans and computers be collaborators in the creative process rather than fully automating it [8, 29]. But while the scientists are more conservative, trying not to venture too far into the realm of art, it is artists themselves and businesspeople in the art market that seem to more readily embrace computational processes as valid and valuable sources of artworks. What is behind this change of mind and how does this relate to the processes of creative expression?

3 Two Moments in Time - Comparing Two Seemingly Radically Disparate Art Phenomena

To approach these questions, we will trace some key positions in the controversy about automatic processes in the arts and formulate an outline of the poetics of automatisms employed by different artists and artist groups. For the purpose of this paper, we will adopt a concept of the term *poetics* that defines it as the theory of form and the "creative principles informing any literary, social or cultural construction" [43]. The focus on the creative principles and the combination of aspects of the literary, social and cultural is suitable for our investigation, since the phenomenon we are approaching is characterized by a complex confluence of these areas. Considering the wide range of automatisms that have been devised by different artists and artist groups over time, traversing the multitude of automatism concepts since Swift's *engine* until today's AI methods, it is clear that it will be difficult to formulate a single concept of poetics that will be able to adequately represent the intentions and principles of all of these approaches. Therefore, we will begin with a consideration to select certain kinds of automatisms to be included in this analysis.

3.1 Selection of Artistic Automatisms

The types, intentions and usage scenarios of automatisms in artistic production is abundant and, for the scope of this paper it will be necessary to select a limited number of significant instances. This choice is by no means representative, but it is intended to support an understanding of the associated motivations and cultural contexts. As stated earlier, the *engine* of the Academy of Lagado is a theoretical concept inspired specifically by the combinatorial structure of language and its inherent linguistic principles of creating new meaning. The principle idea of this engine is simple combinatorial action, in which no logical or other heuristics are employed that would shape the outcome of the machine. In Swift's description, this selection process is done by the students and professors of the Academy. Even though we may consider this idea of an engine a close relative of the step reckoner conceived by Swift's contemporary, Gottfried Wilhelm

Leibniz, which is often seen as the first computer, we will keep the field of automatisms narrower and start significantly later in the history of automatism imagination.

An automatism that is much closer to the algorithmic nature of today's computers is Emmett William's poem entitled "IBM." It is a poem based on a principle he referred to as a game or "do-it-yourself poem," which he devised in 1956, without actually having access to a computer. The basic algorithm goes like this:

And now, back to the very beginning. Here are the rules of the game, vintage 1956:

1. Choose 26 words by chance operations – or however you please»
2. Substitute these 26 words for the 26 letters of the alphabet, to form an alphabet-of-words.
3. Choose a word or phrase (a word or phrase not included in the alphabet of words) to serve as the title of the poem.
4. For the letters in the title word or phrase substitute the corresponding words from the alphabet-of-words. This operation generates line one of the poem.
5. Repeat the process described in step 4 with the results of step 4.
6. Repeat the process with the results of 5.
7. Et cetera" [39].

In recursive application, the algorithm delivers a complex field of word and sentence transformations. At the time of its creation, computers were still very rare and expensive devices that were not available for artists to do creative experiments with them [17]. Only later, in 1966, Williams had the opportunity to use a computer to carry out the algorithm and named the result "IBM," in an "understandable tribute to the muse's assistant" [38]. What Williams specifically appreciated in the process of using a computer for this purpose was the "indefatigability of the computer," which allowed him to introduce several other dimensions of transformation that would have been hard to carry out manually. The purpose behind those was to "relieve monotony, and to thicken the plot" [39].

Williams' process is a good example for a certain type of use of computational processes, in which the machine is used for its ability to process large amounts or complex transformations according to different combinatorial systems, while the artist controls the process by adjusting the parameters and functional aspects of the transformations. In this case, the artist controlled which transformation rules get combined and the input into the system. As Williams describes, the input was established through chance operations that "reflect the bewilderment of an expatriate returning to the United States after an absence of 17 years" and adds that he "might have cheated" in the process of generating the seed-word lists [39].

In the "IBM" poem of Emmett Williams we can see a turning point where creative principles that are quasi-computational and inspired by the idea of computing, but that were not actually executed on a computer, slowly give way to principles developed with an opportunity to use computers to carry out the procedures. From his own description of the process we understand that the automatic process was intended to yield textual material with qualities such as a plot, and associative likeness (e.g. the "bewilderment of an expatriate"). Care was taken to shape the transformations in such a way that they deliver enough complexity for readers to imagine a plot in the lines of the poem and to provide associative hooks to guide their interpretation. We can see two main strategies in the construction of the automatism: one is the aim to provide a remainder of meaningful structure by using sufficiently suggestive start-words, and the second is to have sufficiently complex transformations that are neither too simple to be immediately transparent nor too unstructured to make the results appear to be completely random and meaningless.

A group that never used computers but devised various forms of algorithmic procedures and automatisms is the French *Oulipo* group, the "Ouvroir de littérature potentielle," the workshop for potential literature. This group was founded on November 24, 1960, and became a heterogeneous assembly of writers, mathematicians and scientists, whose motivations to use automatisms varied somewhat between the different members. One of the members, Georges Perec, who became famous for his approach to constraint-based writing techniques, considered himself as "a writer, but of a rather unusual kind – one with no imagination to speak of" [1]. For Perec, the use of automatisms was a way of filling in the absence of imagination with more reliable tools; in his case, automatisms were a tool to support his own writing process, which means that he adhered to a collaborative relationship between human and automatic creation.

The *Oulipo* group was founded in 1960 by poet and novelist Raymond Queneau and the chemical engineer, mathematician and poet François Le Lionnais. Queneau's use of automatisms was directed at leveraging the potentiality of literary texts. When he was 21 years old, he encountered the surrealist movement and participated avidly in their activities. We can assume that his interest in automatisms formed during his work with the surrealists and then, extended by his interest in mathematics, lead to the practices of the *Oulipo*. Following the 'success' of the *Oulipo*, several other workshops emerged, dedicated to a variety of forms of expression, such as painting (*Oupeinpo*), music (*Oumupo*), composition and others, employing similar algorithmic methods. For our purposes, though, we instead turn toward the surrealist automatisms, which are conceptual precursors to the work of *Oulipo* and in distinction to which the Oulipists define their own creative practice.

3.2 Psychology and Concepts of the Brain

The surrealists employed an elaborate apparatus of automatisms in service of artistic creation. The movement formed around the French artist André Breton, who, with his publication of the Surrealist Manifesto in 1924, gave the official starting impulse of the movement. It evolved from the Dada movement and we can trace numerous connections between the art practices and ideas. Nevertheless, in the case of surrealism the concept of using automatic processes, their purpose and practice was a new concept. The automatisms were intended to appeal to the unconscious and free the imaginative energy from the restrictions normal life imposes on it. The beginning of the movement saw the opening of the Bureau of Surrealist Research, created to "gather all the information possible related to forms that might express the unconscious activity of the mind" [12]. This focus on research and a methodical investigation of the subconscious as an artistic practice geared to creative expression was a new phenomenon. The direct precursor, Dadaism, was rather a systematic defeat of any systematic approach and instead dedicated to forms of absurdism. Several members of the surrealist movement were active in Dada, several of their practices, such as collage and cut-up writing, are similar, and they share the skepticism of rational logic that both movements saw in direct connection with the traumatic experience of World War I. Nevertheless, the notion of research and the attempt to theorize the practices of the surrealist movement distinguish it rather clearly from its antecedent [2].

Breton encountered the writings of Sigmund Freud in 1917, while he worked in the psychiatric center of a hospital, taking care of soldiers with mental distress from the battles of the war. From this work he described the "astonishing images" that he heard about from his patients [31]. The focus on automatisms as a way to free the subconscious mind and the imagination from the restrictions imposed by a rational and utilitarian society was the result of the combination of Breton's interests in psychoanalysis and in poetry.

The concept to use automatisms to bypass rational control of the mind and set the imagination free, and the importance the surrealist practices had in art history, make the surrealist automatisms a very suitable subject of comparison for this investigation. The contemporary counterpart, artificial intelligence algorithms, in a very similar way, take their origin from a theory about the functional principles of the human brain. It is worthwhile to compare both moments in respect to the history of artistic creation and conceptualization of the function of the human brain. Breton and Freud being contemporaries, Breton responded strongly to the theories of Freud and had a sense of their possible implications for imagination and creative expression. It took him until 1924 to formulate a "research agenda" based on these ideas, but nevertheless, we can say that the embracing of the scientific theories of the subconscious for artistic ends was rather swift. The embrace may have strayed somewhat from the scientific approach, for example, Breton may have erroneously taken Freuds free association technique as equivalent to automatic writing [13]. In the other direction, as we understand from Polizzotti's biography of Breton, Freud was somewhat uninterested in engaging with Breton's ideas about the subconscious and the role his ideas could play in liberating humans from the oppressions of their surrounding society. Even though Freud was the source of inspiration, he did not engage with Breton beyond a limited exchange of conversation. Later though, psychologist and psychoanalyst Jacques Lacan, was deeply inspired by both Freud and Breton. Lacan and Breton were friends, and Lacan even published some texts in the surrealist magazine *Minotaure*. Several of the early texts by Lacan appeared in the *Minotaure*, and in particular in consideration of the surrealist's efforts to create methods to access the unconscious and irrational, the proximity of Lacan's concerns is evident. In the text "Le problème du style et la conception psychiatrique des formes paranoïaques de l'expérience," which was published in the first issue of *Minotaure* in 1933, Lacan analyzes the experience of states of paranoia in respect to their stylistic potential and the potential of symbolic expressions. He considers the phenomena he observes as extremely productive in terms of poetic production and states that they exclude normal ethic and rational consideration in favor of a freedom that he describes as "imaginative creation." Lacan goes on the consider experience of paranoid states as a form of original syntax, stating that the knowledge of this syntax represents an indispensable introduction to understand the symbolic values of art, and specifically of the problems of style [22]. The mutual inspiration between Lacan and Breton evident and both state an indebtedness to each other's thinking.

In comparison, it took much longer for artists to embrace artificial intelligence algorithms as a meaningful tool of artistic expression. Some of the foundational concepts of artificial intelligence, in particular machine learning, i.e. the possibility that machines can learn and improve by themselves, also go back to findings in psychology. The book "The Organization of Behavior" by Donald O. Hebb, published in 1949, described the functioning principles of the so-called Hebbian learning process and which role neurons and the Hebb synapse play in it [16]. Hebb began to investigate learning processes in neuron networks already in 1932, when, in his Master's Thesis, he described the role of neurons in explaining reflexes and inhibitions. He produced multiple papers on this topic until he finished "Organization of Behavior." This work was the basis for Warren McCullough and Walter Pitts to work on a logical calculus, which mathematically formulated the learning behavior described by Hebb. Their 1943 paper "A logical calculus of the ideas immanent in nervous activity" presented the foundational concepts for the McCullough-Pitts artificial neuron, which, in turn, was the basis for any neural networks [25]. First implemented by Frank Rosenblatt in 1957, the perceptron was a first learning algorithm, consisting of a network of multiple artificial neurons. While this work was picked up very quickly in the engineering community, artists took until quite recently to embrace AI as a possible source of creativity – even though, we might assume a conceptual proximity between the fact that the perceptron was geared toward visual perception and image recognition, tasks not unrelated to at least the visual arts. For reasons of conceptual proximity, for this comparison we will focus on recent examples that have been exhibited in recent shows and the topic of discussion in the art press.

These considerations suggest that both the surrealist automatisms and neural networks are useful objects of analysis for the purpose of this study. Without ignoring or prioritizing certain automatisms over others, this selection will serve as opposing poles and endpoints of a spectrum of different poetic concepts in the use of automatisms for creative expression. This focus on AI, though, should not distract from the fact that the field of computational art is of course much larger and comprises a wide range of different approaches to the use of algorithms and creative expression than those building on concepts of Hebbian learning and machine learning processes. For the purpose of this article, we will focus on two moments in time, taking into consideration artworks as well as conceptual texts produced in this period. The first moment comprises the period of surrealism between the first manifesto, published in 1924, which marks the founding moment of the movement. It includes the second manifesto of surrealism, which was published in 1929 and considers examples of the automatic writing work done in this period, up to some of the cadavre exquis works, done collaboratively by Yves Tanguy, Jeannette Tanguy and André Breton in 1938. The second moment looks at a timeframe beginning in 2018 up to now, with the first widely discussed entries of artworks created with artificial intelligence algorithms into the traditional art market and discourse.

3.3 Criteria of Comparison

The historic context, the tools and procedures employed in the creative processes, and the personalities and societal embedding of the artists seem wildly different when comparing the surrealists with current AI-artists. With this difference, what can be criteria of comparison that can be applied in a reasonable way and deliver outcomes that are meaningful? We are interested in particular in two aspects, the poetics and creative intent, and the surrounding discourse of the larger context. Since, in particular for the recent AI-art pieces, art market value has been a central area of discussion, we will include that aspect into the analysis of the larger discursive context. We cannot say that there is anything like a coherent movement of AI art currently, we rather have individual actors who are adopting techniques of creation that are rooted in AI, nevertheless, from some of them, the classic insignia of an art movement, specifically a manifesto, do exist, and we have rather consistently formulated theories about the functioning and the supposed creative principles leveraged with the described automatisms for both the surrealists and AI art.

To determine the poetics and formal qualities of the works produced by these groups of artists we will refer to the theories and manifestoes they have formulated themselves as a way of communicating their intentions and practices. Using the accounts about the creative principles of the automatisms by those who actively use them for creative ends seems to be more meaningful than to refer to any categorizations and stylistic patterns that have been ascribed to those groups by art historians, critics or other uninvolved observers. As Mary Ann Caws is arguing in the introduction to her anthology of surrealist painters and poets, the artist's own self-characterization seems to be one of the most meaningful criteria for such a comparison [5]. Even though we might be able to identify common stylistic elements among the surrealist works, and possibly some for AI-created art pieces, but the variety of AI artworks is such that there is not necessarily a meaningful common trait. Other criteria, such as group membership, are also not useful; Breton, for example, 'expelled' several surrealist artists in the second manifesto, stating they were not surrealists.

4 Surprising Differences and Similarities

4.1 A Collection of Automatisms

Some Surrealist Automatisms

To get a sense of the functioning principles of the different automatisms and to be able to assess and compare them, we will begin with a description of a selection of the main automatisms from both the surrealists and from AI art creations.

The surrealists have devised a rather large set of different procedures of different degrees of automation. Alastair Brotchie identifies 33 different procedures, which he calls surrealist games. He also mentions several "provocations" and more procedures that he does not examine more closely [3]. Brotchie does not categorize all of these procedures as automatisms, but we can find common principles that connect them to the definitions given in the Surrealist Manifesto and other descriptions of their application by several surrealists. The use of the term automatism by the surrealists needs further examination, which will allow us to identify what types or traits of automatisms should be primarily considered as part of this comparison.

In the (first) Manifesto of Surrealism from 1924 Breton describes an experience that prompted him to adopt a certain kind of procedure that later is referred to as automatic writing. One evening, before falling asleep, Breton encountered a phrase that came to him without any apparent relationship to his situation or experience prior to this moment; he described it as "knocking at the window" [2]. Intrigued by its rare quality, he decided to incorporate it into the material of his poetic construction. And once he had done that, a sequence of phrases came to him so fast than he could not even write them down. Breton formalized this process in a section of the manifesto entitled "Secrets of the Magical Surrealist Art," which became the concept of automatic writing.

The procedure is described as follows:

> After you have settled yourself in a place as favorable as possible to the concentration of your mind upon itself, have writing materials brought to you. Put yourself in as passive, or receptive, a state of mind as you can. Forget about your genius, your talents, and the talents of everyone else. Keep reminding yourself that literature is one of the saddest roads that leads to everything. Write quickly, without any preconceived subject, fast enough so that you will not remember what you're writing and be tempted to reread what you have written [2].

The suspension of rational control is the main aspect that is to be achieved by this form of automatism. The formulation of this concept as a creative practice is influenced by several theories, Breton referred to Sigmund Freud's free association technique as a way of uncovering experiences and thoughts that have been relegated to the unconscious or repressed. He is also making a direct reference to Pierre Reverdy's statement that images are a pure creation of the mind, invoking the role of mental activity in creative expression. Another theory that was influential at the time and with which, we can assume, Breton was familiar given his interest in psychology, is Pierre Janet's book on psychological automatisms from 1889. Even though his ideas were published and in circulation, Janet gets mentioned only in the second manifesto of surrealism. Janet proposes a theory of elementary human activities, which are normally ignored in favor of higher forms of activity, such as acts of the will and decision, even though the simple activities are tremendously impactful on our actions and could serve to explain many of the more complex activities of humans. Janet coins the term psychological automatism for these low-level activities [20].

The automatic writing procedure is probably the most well known and most influential procedure of the surrealists. Along with it, automatic drawing, a technique very similar to automatic writing, with the difference that the activity consisted in drawing lines on sheets of paper and making what we might call "doodles." Another well-known automatism is the "exquisite corpse," which is based on the collaborative effort of several (minimum three) artists working together on one creation. The "exquisite corpse" exists as both, a textual exercise as well as an exercise in drawing, painting or collage. The idea is that the first collaborator writes down an article and an adjective, folds the paper such that the next participant cannot see what was written by the first, and then passes it on to the next participant, who contributes a noun, the next a verb, then another article and finally another noun. At the end the sentence is read aloud. The same principle exists with drawing, where the first participant draws a head, the next the body and the last the legs. This automatism is interesting to mention, because not only is there an unintended inspiration that emerges from the not consciously controlled collaboration of multiple participants, it also is a break with the idea that one artist is the sole author of an artwork. The artwork is rather the result of a collaborative process, rather than the conscious creative act of its author. We could refer to this as collaborative authorship, but it becomes quite clear from Breton's descriptions that the automatisms are considered as quite detached from individual or even collaborative authorship – they are more akin to an unknown force that "knocks at the window." The artist serves, so to speak, as a medium that captures what has been presented to it from an unconscious instance. The surrealists did not directly employ this terminology and rather made clear that the processes enabled by automatisms are the "actual function of thought: dictated by thought in the absence of any control exercised by reason, exempt from any aesthetic or moral concerns [2]. While with the earlier versions of surrealist automatisms the artists took great care that no human intervention interfered with the results of the process (not even reread what was written in a session of automatic writing), later many artists turned toward a more collaboratively structured model, in which the results from an automatic process were the beginning of further, conscious creative work by the artists [3].

Artificial Intelligence Automatisms

In AI art, again, we encounter a range of different algorithms that are used in the process of creation. For the purpose of this comparison we will focus on a few principles that are significant for the underlying concept of AI art. The central process the automatisms of AI art are based on machine learning. Realized in several different algorithmic approaches, the foundation of AI arts are learning processes in which a large databases of examples were presented to an algorithm, which made inferences based on the data presented to shape the output similar the input data. The learning process in most of the examples of AI art is to feed examples of existing artworks to the machine learning algorithms and then use the trained algorithms to create new works based on the principles that were learned. For this purpose, a generative model is employed that can generate output that closely resembles the characteristics of the input (training) data. In the case of the artists we mentioned earlier, Klingemann and *Obvious*, a particular kind of generative modeling was used, so called generative adversarial networks, or GANs.

This approach to neural networks has been proposed by Ian Goodfellow et al. in 2014 as a way to transfer the successes made in deep learning models to distinguish complex data characteristics to deep generative models [15]. Significant progress was made in machine learning through improving deep learning models through techniques such as back propagation, a specific way of training a neural network, to learn how to distinguish complex features in data sets presented to the algorithm. This is referred to as discriminative modeling, as the machine learning algorithm differentiates between classes of data and labels them according to the categorizations it learned. Transferring these successes to generative models, which, instead of labeling classes of data, learn how to generate data that closely resemble the input (or training-) data, proved difficult. Generative adversarial networks presented an efficient answer to this problem. They consist of two types of models, a discriminative and a generative model, which compete with each other and, by way of competing, improve each other's performance. The generative model (generator) learns from the training data set what kind of patterns exist in it and then generates output that corresponds to these patterns. Its adversary, the discriminative model (discriminator) learns how to distinguish the output generated by the generator from the original training data. By competing against each other, both parts of the adversarial network improve their modeling. Goodfellow explains the functioning principle with the following metaphor:

> The generative model can be thought of as analogous to a team of counterfeiters, trying to produce fake currency and use it without detection, while the discriminative model is analogous to the police, trying to detect the counterfeit currency. Competition in this game drives both teams to improve their methods until the counterfeits are indistinguishable from the genuine articles [15].

This means that the output of a GAN closely resembles its input data and produces subtle variations close enough to the original to be considered as part of the original domain. While the AI component in this process is based on Goodfellow's et al. algorithm and the various improvements that have been made to it since then, the main action of the artist who employs a GAN automatism to produce AI artworks, consists in choosing the training data and adjusting the parameters controlling the learning process of the GAN. It is clear that the choice of training data will significantly shape the possible outcome of this process. The particular way how the generator iterates through the probability space of its model creates a rather specific kind of distortion that has often been described as resembling paintings of the British artist Francis Bacon. This specific look has been described as the "defining look of contemporary AI art" [41].

AI art is not limited to visual output, even though these examples have received most public attention. To give an example of a language-oriented model we are looking at a recent example called Deep-speare, by Jay Han Lab et al. [24]. Based on training data curated from William Shakespeare's sonnets this AI implementation produces new sonnets in the style of Shakespeare. While the approach to use training data from existing artworks to produce new works that are very similar, the actual algorithms used for text production differ from those used for image production. Deep-speare uses multiple Long short-term memory networks, or LSTMs. In distinction to models like GAN, LSTM are specifically tailored to include time-based context into their learning process. This makes them particularly suitable for language-oriented applications, such as speech modeling and translation, handwriting recognition, analysis of audio and video data etc. In addition to multiple layers of artificial neurons, LSTMs comprise memory cells, which can store time-based information and take temporal context into account in the learning process. Deep-speare uses one LSTM to build a language model, one for a pentameter model, and one for the rhyme model. Shakespeare's sonnets are written in iambic pentameters, i.e. lines of ten syllables, consisting of five pairs of an unstressed syllable followed by a stressed syllable and the pentameter model learns this structure of poetic meter. The rhyme model learns the structure of Shakespeare's sonnets, which consist of 14 lines structured as 3 groups of four lines, the quatrains, and two groups of two lines, the couplets. The rhyme scheme of these possesses several variants, with a typical structure being ABAB CDCD EFEF GG. In the generation procedure the context of preceding lines is taken into context. Since the rhyme structure of the lines is important, Deep-speare generates lines beginning with the last word, which is adjusted so that it fits the rhyme scheme and then the line is generated building backward from the last word.

4.2 Poetic Intentions

The Program of Surrealism

From the analysis of the functional principles of the automatisms we get a sense of what the artists aim to accomplish with their use of automatisms, but to get a real understanding of their intentions it is useful to look at what they state about the poetic intentions themselves to really get a picture.

Drawing from the Surrealist Manifestoes we get the sense that the intention of the surrealists was to shift the human brain into a different state of mind that suspends considerations of utility etc. which normally govern our decision-making processes. The use of automatisms serves two main purposes: it is a way of producing perceivable traces, Breton tends to refer to "images," which, when externalized and perceived, put the human brain into a state of imaginative activity that goes beyond the normal responses; the second purpose is, by way of producing those traces that trigger mental images, they reshape the human perception of the world and the value sets that are brought to it.

About the first purpose, the particular qualities of the perceivable traces – we use this formulation to be inclusive of the range of different creative products of surrealists including text, images, objects, moving images etc. – Breton sates that they are as strange to the artist as they are to anyone else. He goes on to say that:

> ...poetically speaking, what strikes you about them above all is their extreme degree of immediate absurdity, the quality of this absurdity, upon closer scrutiny, being to give way to everything admissible, everything legitimate in the world: the disclosure of a certain number of properties and of facts no less objective, in the final analysis, than the others [2].

The way this inspiration works is likened to a spark that jumps between the different images brought together by an automatism such as automatic writing: "a particular light has sprung, the light of the image, to which we are infinitely sensitive" [2].

The second purpose of the exercise of surrealism is summarized at the end of the first Manifesto, where Breton points out that surrealism is an expression of complete nonconformism, concluding the manifesto with the statement that "Surrealism is the 'invisible ray' which will one day enable us to win out over our opponents" [2].

AI Intentions

The French artist group *Obvious*, consisting of three members, also formulated a manifesto from which we can glean some insights into their ideas and intentions. In the manifesto the members of the group introduce themselves as "limited by their creativity" [14], which might explain the motivation to turn to machine learning automatisms to make art, which, as they state, can empower the natural creativity." Their mission statement says that they "wish to demonstrate that algorithms help us complete our understanding of how we function as humans and push us to outsmart our current level of creativity." With their work they intend to shed light on the emerging tools available and believe "that a new generation of creators will rise, one that will know how to build and manage algorithms that will help in an innovative process." The intentions we read from this text are predominantly educative, to introduce the audience to new emerging tools for creativity and invite them to better understand how humans function. This statement resonates with some of the opinions associated with – in particular the early stages of – artificial intelligence research, that, treating human beings as symbol processors, would allow us to simulate and better understand the procedures of human intelligence [34]. We would assume that, in the case of *Obvious*, the idea is that by simulating creative processes, we might learn something about human creativity, which is also a common position in computational creativity research, a subfield of artificial intelligence research.

The concluding statement of the manifesto section explaining the intentions goes as follows: "This is why *Obvious* focuses on accompanying the emergence of benevolent and harmless ideas, by promoting alternative uses for it, and unveiling its true creative potential." The focus on benevolent and harmless ideas is in strong contrast to the radical statements of the surrealists, which expressed nonconformism and were motivated by an idea of a "war" against the limiting dominant structures of the contemporary society that would eventually have to yield to the forces set free by surrealism. A hint of a similar desire for change may be found also in the *Obvious*-manifesto, where it is stated that expanding creativity can help to "destroy our current mental boundaries."

In contrast to the *Obvious* group, Mario Klingemann does not have a manifesto and states that he rarely writes about his work; nevertheless, a few passages about his interests are available on his website, where he describes himself as an "artist, and a skeptic with a curious mind." His areas of interest, he says, are "manifold and in constant evolution." In a similar way as *Obvious*, he stresses a desire to understand: "If there is one common denominator it's my desire to understand, question and subvert the inner workings of systems of any kind. I also have a deep interest in human perception and aesthetic theory" [22].

In contrast to the visual artists Klingemann and *Obvious*, the makers of Deep-speare do not identify as artists but as scientists; their aim is to investigate computational creativity and how neural models can be employed in this process. Along with the difference in self-identification, the evaluation criteria and methodologies they use to determine the performance of their systems differ significantly. While the first two follow traditional art-context criteria for success, such as participation in exhibitions, critical response and the resale value of their works in art auctions, Lau et al. employ a precise method of assessing specific criteria of their system. A first round of evaluation is done by crowd workers, anonymously recruited online workers who get paid a minimal amount per task ($0.05 in this case), who have to determine whether a sonnet is human made or computer generated. A second round was done with expert judgement, in which a professor of English evaluated the sonnets in respect to their meter, rhyme, readability and emotion. Their findings were that their system is able to produce formal characteristics such as meter and rhyme well but lacks in terms of readability and emotional expression.

5 Discussion

The most significant distinctions between surrealist automatisms and AI automatisms pertain to the creative intent and the sources of inspiration. The surrealists draw from the human unconscious, seeking for experience traces that are hidden or repressed, but nevertheless exist and influence human behavior. They bring these to the level of perceivable formulation by surfacing the traces through automatic processes insulated from rational control and then use them to inspire new forms of thinking to the audience of their works. The central focus thus is human experience; AI art engages – so to speak – with second hand human experience: it draws from a curated set of existing artworks, which, as traditional artworks created by human artists, are highly likely to express human experience, and uses them as input data for the machine learning processes. Through the analysis of those human-made artworks, the AI learns the traces of expressions of human experience as part of the patterns it processes, but not as a targeted expression meant to express a specific experience. Human experience a residue that, in an unspecific form, is contained in the output of the algorithm. Since the curation of the training data is one of the main influences an artist working with AI algorithms has, we can speculate that the use of artworks as training data in the cases we are discussing here, is either a form of self-referential statement about the creative process in the arts, or it is the attempt to "warrant" the art-status of the generated product: since the training data are art-historically sanctioned works, the resulting works should be equally eligible to be sanctioned as art; in respect to the Deep-speare system the choice of training data originating from human creation is in line with common practice in computational creativity research. In this area of research, machines are supposed to learn what human creativity is and, for that purpose, results of human creativity are presented to the learning algorithm. In both cases, whatever the machines learn will contain the inscription of the "secrets of the magical art" as Breton called it.

We can conclude, though, that in the cases where it is known to the audience members that the work was created by AI algorithms, the perceptual and interpretational stance of the audience toward to work is different. In the press about exhibitions of both Klingemann's piece "Memories of Passersby I" and *Obvious*' piece "La Famille de Bellamy," which both employ GANs and generate visual output that has been likened to the paintings of Francis Bacon, we found no comment that perceives them as violent or unsettling, descriptions that are very often attributed to Bacon's paintings. Famously, Bacon came to his style of painting seeking to express the "brutality of fact" [37] and developed forms of painting that could render a form of brutal realism. In particular in work that focuses on benevolence and harmlessness it would be a surprising aesthetic choice to use forms that indeed evoke brutality, violence and upheaval in the audience. This is a clear sign that knowledge about the creator – or creation process for that matter – plays into the interpretation the audience. Knowing that human expression and a direct relationship to what we would refer to as reality is only existing as a decontextualized and indirect form in the artwork, shapes the audience interpretation as potentially harmless.

Even though surrealist artists also employ methods in which they assume more of a status as an externally controlled medium that responds to or channels experiences that are not under their rational control, the connection to human experience and the knowledge about the artist enters the audience interpretation in a different degree. In creativity research this is often expressed as a question of autonomy: A work is considered creative when we can read a degree of autonomy in it. The lack of emotional expression observed in the AI sonnets of Deep-speare is an indicator of a similar phenomenon.

In their use of automatisms, it seems that artists using surrealist automatisms and those using AI-automatisms have the opposite problems: the surrealists try to keep rationally classified human experience away from their works to get to the "raw" content of unconscious elements of human experience; and AI artists are trying to somehow infuse aspects of readable human experience into their creations. Surrealist artists would probably not respond to research in learning, they rather respond to research in finding or encountering. Learned things are what those artists actively tried to subvert. The stated intention of the surrealists is very much about human experience that needs to be liberated, it is not about conscious learning and reproduction, but about the revelation of already inadvertently learned experiences. The radical approach of the surrealists resonates with the strong criticism that was brought up against computer art in the 1960s, when the first works of computer art surfaced, and which we see in Metzger's criticism of the combination of technology and art as an aestheticization of modern warfare and totalitarianism.

Nevertheless, when we wonder how it was possible that computationally generated art could now circulate in the art market without triggering a major critical discourse, we are relegated to two aspects: The first, and we may say the less interesting one, is the connection to novelty, the connotation of high-tech, of which the perception has fundamentally changed in comparison to the associations with the military industrial complex that was present in the 1960 and 70s; predominantly, though, we may conclude that it is also a business speculation aspect that plays a role in this. It is probably not by accident that the artists who engage in AI art do not come from traditional arts training, and some of them even have a business background. The group *Obvious* was also awarded by the business magazine Forbes in their annual prize of "30 under 30", a selection of particularly influential young entrepreneurs [40]. The subsequent explosion of prices and sales of the NFT market strongly indicates that the main motivation of activity in this sector is not artistic expression but financial revenue. Nevertheless, besides engineers and businesspeople who entered the market, also established artists discovered the NFT market as a source of distribution and revenue. The growth of this market was such that the amount of energy that is consumed by the blockchain calculations necessary for NFT trading became a subject of concern and criticism. Websites like "carbon.fyi" allow to calculate the carbon emissions related to specific, given addresses of the blockchain-based digital currency Ethereum [4]. Some artists, like French artist, Joanie Lemercier, began to engage in criticism and activism against the environmental effects of this stepped-up energy consumption. Lemercier, besides participating in protests, also started a project in which he called out the software company Autodesk for its environmental irresponsibility and hypocrisy regarding standards of sustainability and thoroughly documented the exchanges with the company executives [21].

The second, more interesting aspect, for the appeal of AI art may be rooted in what we just discussed: the connection to a trace amount of reality. With generative models AI becomes interesting as it connects to elements of surprise and the gesture of 'bringing up something from the hidden depths of *something* - in this case it is not human experience, but maybe the likeness to paintings like Bacons or old classics. But the connection to other alienation techniques as they were employed by, for example dadaists and surrealists, is noticeable as a remainder of likeness of reality that makes the automated products appealing.

References

1. Bellos: Georges Perec's thinking machines. In: Higgins, H.B., Kahn, D. (eds.) Mainframe Experimentalism: Early Computing and the Foundation of the Digital Arts. University of California Press, Berkeley, California (2012)
2. Breton, A., Breton, A.: Manifestoes of Surrealism. University of Michigan Press, Ann Arbor (1972)
3. Brotchie, A., Gooding, M. (eds.): A book of surrealist games: including the little surrealist dictionary. Shambhala Redstone Editions: Distributed in the United States by Random House, Boston (1995)
4. carbon.fyi: Calculate the CO2 Footprint of an Ethereum Address. https://carbon-fyi-e9mk5li4h-brendanmc6.vercel.app/. Accessed 04 Nov 2021
5. Caws, M.A. (ed.): Surrealist Painters and Poets: An Anthology. The MIT Press, Cambridge (2001)
6. Christie's: Is artificial intelligence set to become art's next medium?|Christie's, https://www.christies.com/features/A-collaboration-between-two-artists-one-human-one-a-machine-9332-1.aspx. Accessed 16 June 2021
7. Colton, S., Wiggins, G.A.: Computational creativity: the final frontier? In: ECAI (2012)
8. Cornell Tech: Cornell Tech - Can Machines Be Creative? https://tech.cornell.edu/news/can-machines-be-creative/. Accessed 16 June 2021
9. Dean, S.: $69 million for digital art? The NFT craze explained. https://www.latimes.com/business/technology/story/2021-03-11/nft-explainer-crypto-trading-collectible. Accessed 16 June 2021
10. Dudley, A.: Fast Trend or Stand-Alone Direction: Is NFT Art Here to Stay? - Art Business News. https://artbusinessnews.com/2021/06/fast-trend-or-stand-alone-direction-is-nft-art-here-to-stay/. Accessed 16 June 2021
11. Duffy, R.: The NFT Market Tripled Last Year, and It's Gaining Even More Momentum in 2021. https://www.morningbrew.com/emerging-tech/stories/2021/02/22/nft-market-tripled-last-year-gaining-even-momentum-2021. Accessed 16 June 2021
12. Durozoi, G.: History of the Surrealist Movement. The University of Chicago Press, Chicago (2009)
13. Esman, A.H.: Psychoanalysis and surrealism: André Breton and Sigmund Freud. J. Am. Psychoanal. Assoc. **59**(1), 173–181 (2011). https://doi.org/10.1177/0003065111403146
14. Fautrel, P., et al.: Obvious: Artificial Intelligence for Art (2020). http://obvious-art.com/wp-content/uploads/2020/04/MANIFESTO-V2.pdf
15. Goodfellow, I.J., et al.: Generative Adversarial Networks. ArXiv14062661 Cs Stat (2014)
16. Hebb, D.O.: The Organization of Behavior: A Neuropsychological Theory. L. Erlbaum Associates, Mahwah (2002)
17. Higgins, H., Kahn, D. (eds.): Mainframe Experimentalism: Early Computing and the Foundations of the Digital Arts. University of California Press, Berkeley (2012)
18. Hockney, D.: Digital : Works|David Hockney. https://www.hockney.com/index.php/works/digital. Accessed 16 June 2021
19. Holland, O.: How NFTs are fueling a digital art boom - CNN Style, https://www.cnn.com/style/article/nft-digital-art-boom/index.html, last accessed 2021/06/16
20. Janet, P.: L'automatisme psychologique: essai de psychologie expérimentale sur les formes inférieures de l'activité humaine. Félix Alcan, Paris (1889)

21. Lemercier, J.: Autodesk and coal mining. http://joanielemercier.com/autodeskearth/#1Page. Accessed 04 Nov 2021
22. Klingemann, M.: About|Quasimondo. https://underdestruction.com/about/. Accessed 01 July 2021
23. Lacan, J.: Le problème du style et la conception psychiatrique des formes paranoïaques de l'expérience. In: Revue Minotaure, Éditions Albert Skira, Paris, n. 2 (1933)
24. Lau, J.H., et al.: Deep-speare: a joint neural model of poetic language, meter and rhyme. ArXiv180703491 Cs. (2018)
25. Macey, D.: Lacan in Contexts. Verso, London (1988)
26. McCulloch, W.S., Pitts, W.: A logical calculus of the ideas immanent in nervous activity. Bull. Math. Biophys. **5**(4), 115–133 (1943). https://doi.org/10.1007/BF02478259
27. Metzger, G.: Automata in History [Part 1], Studio International (1969)
28. Miller, A.: Can machines be more creative than humans? http://www.theguardian.com/technology/2019/mar/04/can-machines-be-more-creative-than-humans. Accessed 16 June 2021
29. Morris, D., Secretan, J.: Computational creativity support: using algorithms and machine learning to help people be more creative. In: Proceedings of the 27th international conference extended abstracts on Human factors in computing systems - CHI EA 2009, p. 4733 ACM Press, Boston (2009). https://doi.org/10.1145/1520340.1520728
30. Patterson, D.: Blockchain company buys and burns Banksy artwork to turn it into a digital original. https://www.cbsnews.com/news/banksy-nft-injective-destroy-art-digital-token/. Accessed 16 June 2021
31. Polizzotti, M.: Revolution of the Mind: The Life of André Breton. Black Widow Press (Massachusetts). Place of Publication not Identified (2010)
32. Rea, N.: AI-Generated Art Just Got Its First Mainstream Gallery Show. See It Here—and Get Ready, https://news.artnet.com/art-world/ai-generated-art-gallery-show-1339445. Accessed 03 June 2021
33. Réunion des musées nationaux – Grand Palais: Artists & Robots. https://www.grandpalais.fr/en/event/artists-robots. Accessed 16 June 2021
34. Shanken, E.A.: In forming software: software, structuralism, dematerialization. In: Higgins, H.B., Kahn, D. (eds.) Mainframe Experimentalism: Early Computing and the Foundations of the Digital Arts, pp. 51–62 University of California Press, Berkeley (2012)
35. Simon, H.A.: Studying human intelligence by creating artificial intelligence: when considered as a physical symbol system, the human brain can be fruitfully studied by computer simulation of its processes. Am. Sci. **69**(3), 300–309 (1981)
36. Sotheby's: Artificial Intelligence and the Art of Mario Klingemann. https://www.sothebys.com/en/articles/artificial-intelligence-and-the-art-of-mario-klingemann. Accessed 16 June 2021
37. Swift, J., et al.: Gulliver's Travels. Oxford University Press, Oxford (2005)
38. Sylvester, D., et al.: The Brutality of Fact: Interviews with Francis Bacon. Thames and Hudson, New York (1988)
39. Williams, E.: A Valentine for Noël: Four Variations on a Scheme. Something Else Press, Barton (1973)
40. Under 30 Europe 2020: Art & Culture. https://www.forbes.com/30-under-30/2020/europe/art-culture/. Accessed 01 July 2021
41. A never-ending stream of AI art goes up for auction - The Verge. https://www.theverge.com/2019/3/5/18251267/ai-art-gans-mario-klingemann-auction-sothebys-technology. Accessed 30 June 2021
42. "machine, n." Oxford English Dictionary Online, Oxford University Press, September 2021. www.oed.com/view/Entry/111850. Accessed 04 Nov 2021

43. "poetics, n." Oxford English Dictionary Online, Oxford University Press, September 2021. www.oed.com/view/Entry/318383. Accessed 04 Nov 2021
44. "tool, n." Oxford English Dictionary Online, Oxford University Press, September 2021. www.oed.com/view/Entry/203258. Accessed 04 Nov 2021

Approaches and Applications

Design Patterns of Health Animation – Scaling Pattern Languages Into a New Domain

Katja Thyra Pedersen[1], Peter Vistisen[2(✉)], Mette Terp Høybye[1,3], and Janni Strøm[1,3]

[1] Research Unit, Elective Surgery Center, Silkeborg Regional Hospital, Silkeborg, Denmark
[2] Department of Communication and Psychology, Aalborg University, Aalborg, Denmark
vistisen@id.aau.dk
[3] Department of Clinical Medicine, Aarhus University, Aarhus, Denmark

Abstract. This paper presents the results of a Danish study on the scaling of the design approach of design pattern languages into the context of citizen-oriented health animation. We propose that the use of design patterns, and the development of an emerging pattern library of health animation patterns, can support the design of more informative and useful animations visualizing health information. We mapped 72 Danish citizen-oriented animation products into 23 design categories, including both form-related and content-related elements. We used the design pattern approach to systematize the state-of-art animations to enable an overview of approaches typically applied in health animation across different institutions, producers, and target audiences. We discuss how design patterns can be appropriated from previous uses in e.g. architecture and digital design into a health communication context, and through a pilot split-test we discuss both the benefits but also the limitations of using the design pattern approach to design new health animations.

Keywords: Health animation · Design patterns · Danish study

1 Introduction

Over the past 15 years, there has been a significant increase in the usage of animated films within the area of municipal, regional, and state communication to the public. This form of animation is categorized by its use outside the context of art and entertainment, also labeled 'functional animation' [1]. In this domain, animation is used to promote facts and reduce complexity of information for audiences of different literacy levels across diverse fields such as e.g. governmental communication, science dissemination, interest group communication, and health communication [2]. Previous studies have indicated that health animations can have a positive impact on citizens with low health literacy and their ability to recall health information [3, 4]. Since animations can use various modalities such as visualizations together with text and sound, it has been suggested that this will decrease the cognitive overload of the recipient of the health information [3]. This also seems to be the driving force behind the creation of various health animations that in many cases target people with low health literacy [4]. A health animation can

M. Wölfel et al. (Eds.): ArtsIT 2021, LNICST 422, pp. 381–397, 2022.
https://doi.org/10.1007/978-3-030-95531-1_26

therefore be described as an animation that delivers visualized health information in a simple manner to citizens/patients with different literacy levels and does this by using different elements within the animation (icons, speech, text pieces, etc.).

When designing health animations, it is therefore important to find applicable principles and standards to ensure both proper communication as well as a certain level of predictability in terms of production cost and expected outcomes. Surprisingly, there is a lack of research documenting which principles and patterns govern these types of design processes. The majority of research focuses on evaluating how well a designed health communication product worked, without consideration of which design decisions are behind the measured effect. This is a well-established problem in the field of design and has been an area of concern since Christopher Alexander [5] and later contributors sought to bring established conventions to design using pattern libraries. A promising point of venture is thus to ask whether the well-established design tradition of building and working with design patterns can be scaled up to also contribute to the field of health animations?

In this paper, we present the results of a Danish study exploring the application of the design pattern languages method in the context of healthcare-focused citizen-oriented animated explainer videos. The aim is to develop a design pattern library and to use this as a basis to critically question health animation products and generate detailed descriptions and justifications for the way these products are organized and developed. Thus, the knowledge generated in this study aims to qualify both the development and the evaluation of health animations. We work methodically based on previous frameworks for creating "design pattern libraries" – a method deriving from the studies in the field of design illustrating how one specific designed solution exemplifies a broader pattern of the treatment of functional, aesthetic, and communicative issues within a given object field [6–8]. In recent decades, this has become an established approach in a wide range of design fields, including industrial design, architecture, and digital design.

Knowledge about combining possibilities of form and content is collected through design pattern libraries; this knowledge not only helps to systemize the state of art within a given field but also strengthens the basis of design solutions over time. Thus, fewer decisions are based on subjective opinions and styles rather than analytical principles and standards [9]. Particularly the use of analytical principles and standards provides crucial support for the spread of an emerging design field such as health animation by gathering norms, standards, principles and best practices across applications and sectors. In the health animation field, such analytical standards could improve the presentation of health information and as a result make the information more comprehensible to the recipient. An inadequate comprehension of health information can lead to more hospitalizations and poorer health for the citizens [10]. Therefore, an effective communication design through expressive mediums such as health animation can be a crucial part of creating better health management and healthcare behavior among citizens.

Today a wealth of available toolsets has made it possible to gather the analytical standards and best practices, which consequently makes it unnecessary to start a design from scratch. These toolsets exist in many forms and formats including design patterns – a solution to a recurring design problem that can be used repeatedly within a specific context [6, 11].

The purpose of design patterns is to identify the best practices in the field by acquiring existing sustainable solutions formed from the knowledge and experience of designers. This way design knowledge is shared among experts along with novice learners, obviating any need to reinvent the wheel and start a design from scratch [12]. As a result, resources can be managed effectively and ultimately reduce the production cost of e.g. animations, which are generally expensive to produce [1]. Thus, design patterns are an established and pervasive methodology among many design disciplines, including visual design fields in which design patterns have emerged based on e.g. gestalt psychology [13] and patterns of orthodox motion translated into the 12 principles of cartoon animation [14]. This paper takes a special interest in extending the use of design patterns within the domain of designing animation – specifically "functional animation" [1].

The aim of mapping patterns in health animation is to generate new knowledge concerning the types of animation and the approaches typically used to convey health information. We propose that the use of design patterns, and the development of a pattern library of health animation patterns, can support the design of more informative and useful health animations for citizens. Furthermore, it will better articulate how and when health animations borrow patterns and principles from traditional art and entertainment-based animation, and when they diverge into their own unique patterns. As in other design disciplines, this is important to assess what meaningful combinations are possible and for what purposes different elements can be combined. Later, we discuss how the design pattern can be appropriated from its previous uses in e.g. architecture and ICT design into a health communication context, and discuss the limitations of using the approach in the context of animation. Through the developed pattern library, form-related decisions in designing health animations can be analytically compared with existing idioms, conventions, and standards rather than being subject to individual styles and artistic opinions alone.

1.1 Design Patterns – A Broad Approach

The concept of architectural design patterns was introduced by Christopher Alexander in 1964 [5] and later described in more detail by Alexander et al. [6]. They presented more than 250 patterns consisting of reusable architectural solutions to common problems within e.g. regional planning and interior design [15]. Furthermore, the book presents Alexander's theory that the use of his patterns, and thereby his "timeless way of building", will ingrain environments with what he considers the "unnamed quality" ("beauty", "alive", "free", etc.). Alexander's theory received immediate attention along with a high level of criticism among peers. A comprehensive description of the critics of Alexander's theory and his associated patterns can be found in Dawes & Ostwald [15]. The criticism is primarily concerned with Alexander's ontological and epistemological positions, such as his view that there is only one right way of building (the timeless way), his definition of science, and his confusion of subjective and objective phenomena. Furthermore, he received criticism of the logic and reasoning behind the development of his patterns (primarily based on his argument that there is only one right way to build), while he was also criticized for his lack of documentation and proper testing of his patterns. A core criticism concerned the concept of design patterns, namely the difficulty of testing patterns in general, especially regarding sizable design pattern libraries – the point being that patterns would be difficult to test independently due to the connection between patterns. Therefore, new potential design patterns should be documented properly and subjected to testing. Considering the criticisms of Alexander's theory along with the concept of design patterns, it seems of absolute importance to create patterns that are not based on a rigorous ideology. Design patterns should instead be viable tools that provide proven solutions and a common terminology formed by knowledge and experience of designers that can be used for inspiration. As such, design patterns are a pragmatic tool for design, to balance ideologies and trends, establishing a solid base of experience for dealing with the ultimate particulars of design.

The idea of implementing design patterns in the Human- Computer Interaction (HCI) community was initially mentioned by Donald Norman and Stephen Draper in 1985 [16]. In recent decades, there have been more than 250 HCI patterns published in books and on online sites [11] – patterns involving problems such as how to create a structure to manage pictures and videos and how to design a search area on websites [17]. To ascertain if the solutions in the design patterns are good and worth of being reused, it is of utmost importance to evaluate and validate the patterns. Previously Elisabeth Bayle et al. [18] suggested differentiating patterns into two groups: *Design Patterns* and *Activity Patterns*. Design patterns are proven solutions (across time and circumstances) to a repetitively occurring problem within a specific context. The approval of the solution/pattern can e.g. be done by empirical verification and an overall agreement on the pattern by users [19]. Activity patterns, on the other hand, describe the solutions as they are and present them in a pattern without evaluating on how or if the pattern is worth being preserved [18].

1.2 Practices When Designing Health Animation Products

Traditional animation design patterns exist, most famously embodied by the 12 animation principles of Disney [14]. These are all based on crafting the illusion of life in artificially created motion, by adhering to principles deduced from the early years of cartoon animation. These original patterns of animation design have also been transferred into more functional design domains, such as interface design [20]. We argue that the design pattern approach might also be applicable within domain-specific areas of communication design. Here factors of both form and content must be considered to ensure a positive user experience of the designed form as well as the comprehensibility of the information presented. One such domain is health communication.

In the domain of health animation, studies have aimed to explore knowledge acquisition through the dissemination of functional animations compared to a more traditional mode of information, such as verbal or written. We find studies within a wide range of specialties targeting for example glaucoma patients [21], patients with diabetes [4], patients receiving opioid analgesics [22], patients with periodontitis or periodontal health issues in general [23, 24] and several others [25–28]. These studies all share an overall focus on the effect of animations regarding perceptions, attitudes, beliefs, or knowledge gain. Thus, the focus is on educational aims, and providing recipients with knowledge and skills to prepare for, prevent, or overcome health challenges. While all studies provide a rigorous description of the sampling and effect evaluation, only few provide any details on the actual design choices made. In the few instances where details are given, the rigor is significantly lower than the other parts of the studies. In Polk et al. [29] the authors provide a technical description on how they translated a 3D model of an infant's cranium into a cartoon flash animation but provide no rationale for the design choice behind this configuration. Another example is Narimatsu et al. [30], which details the digital design of a health e-learning platform and its intended user flow, but only includes one sentence explaining the rationale for the form of its animated contents. We argue that this lack of rigor and transparency in the existing research and state of art is an important design issue to be dealt with – potentially through the development of design pattern libraries.

2 Building and Applying a Pattern Library for Health Animation

In this study, we mapped 72 Danish citizen-oriented functional animation products into 23 design categories including both form- and content-related elements (see Fig. 1). We used the design pattern approach to systematize the state-of-art animations to enable an overview of typically applied approaches in health animation across different institutions, producers, and target audiences. Thus, the outcome of the systematization was the development of a design pattern library. The collection of health animation patterns was restricted to animation products created in and aimed at a Danish context. These animation products were collected from their respective sources (e.g., YouTube, health-related websites or digital applications) – identified both through searching among known animation providers of Danish health animations, through health communication sources in Denmark, and through users of health animations. Afterwards, the collected animations were mapped into 23 design categories according to existing knowledge concerning the form and content of animation products [31–34]. These categories encompass both form-related elements (e.g., visual fidelity, orthodox vs. exaggerated animation, diegetic vs. non-diegetic sounds) and content-related elements (e.g., narrative structure, discourse, and subjective vs. objective communication). Throughout the mapping, some categories were elaborated, and other categories were added because of the pragmatic use of the framework. One example is the sound/narration category, which got elaborated as patterns vaguely started to appear and other elements seemed of importance, e.g., the specific pronouns used by the speaker (you, we, he or she etc.) were included as well as the use of different types of narrators than expected (e.g., visual representation of a third-person narrator). Another example is the category "narrator's gender" which was added later when we discovered through the mapping that the animations generally use a male voiceover. This showed that it was unfeasible to determine all the relevant elements and categories beforehand, and only after watching and mapping the animations was it possible to adjust the categories to fit the pragmatic world.

After categorization, common features, deviations, and variations were identified in order to induce specific elaborated patterns that were then collected and described as patterns for health animation products to support an increased substantiated design process of new health animations. Our aim was to create a framework involving the three important elements in a design pattern: a problem, a context, and a solution [6, 7]. Inspiration was found in the existing frameworks within the Human-Computer Interaction field presented in e.g. [17] and [8]. This led to our version of a design pattern framework consisting of multiple categories, where the emphasis was given to the following categories:

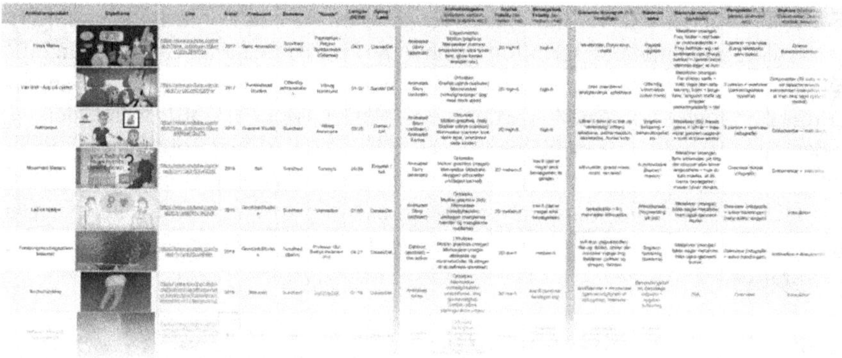

Fig. 1. A snapshot of the mapping of 72 Danish health animations, which forms the basis for inducting specific clusters into design patterns. Each animation is indexed within 23 categories on the horizontal axis. The total mapping can be seen in Appendix 1.

tendency, occurrence, approach, outliers, interpretations, and examples (see Fig. 2). The ultimate objective was to locate potential patterns in the mapping of health animations and turn them into a design pattern by addressing the categories above. In the category *tendency*, we present a broad description of a pattern and its elements that are brought into play. *Occurrence* pinpoints the theme of animations that a pattern encompasses e.g., health animation or public administration. The category *approach* covers the approaches employed and solutions given by previous animations using different elements such as voiceover, perspectives, and icons. *Outliers* is the category the captures the inconsistent data of the concerned pattern. Furthermore, *interpretations* encompass the reasoning for why a specific solution is sought. In *interpretations*, we strive to provide a reasonable argument behind the choice of the adopted approach, though it remains up for debate. Lastly, as the name suggests, the category *examples* include examples of a solution with either a picture, a text or both.

We followed the top-down and bottom-up approaches to discover typically applied solutions in the mapping and turn them into design patterns by addressing the categories in our design pattern framework. Employing the top-down approach allowed us to investigate the data through a general lens, for instance: do animations adopt a specific form while targeting children? Likewise, the bottom-up approach was used to examine whether certain categories, e.g. 2D or 3D animations, contained similarities within the category and therefore form a pattern. In summary, we created a design pattern framework that allowed us to deconstruct and elaborate on the patterns observed in the data from the mapping of 72 health animations. This allowed us to build an emerging design pattern library of the best practices within the field of functional health animations. This design pattern library is constructed as an online accessible database, from general patterns to isolated examples, usable as a reference to support the identification of suitable patterns for a given communication challenge, but also to search for examples of the pattern's use in existing animation practice within health animations. Thus, the database is usable in designing specific health animations, but also serves as a general database for the health sector overall (which could gradually grow in size) comparable

PATTERN NAME	
TENDENCY	
OCCURENCE	
SOLUTIONS/APROACH	
OUTLIERS	
INTERPRETATIONS	
EXAMPLES	

Fig. 2. Our version of the design pattern framework, which include the following categories: tendency, occurrence, solution/approach, outliers, interpretations and examples.

to the design pattern libraries used in other sectors such as software design [35]. The next section will provide an example of the development of a specific pattern from our pattern library. Afterwards a pilot test applying the pattern library in a specific health animation project will be described and analyzed with the purpose of serving as a proof of concept regarding the applicability of the design pattern method within the domain of health animations.

2.1 Inducting a Pattern – An Example

Forming design patterns is predominantly an inductive process of inferring from multiple single instances to a broader general description. During the mapping of the functional animations, some patterns were immediately apparent while others only emerged from details after more careful analysis. This is due to the fact that some patterns were clear formulations of recurring problems, while others arise from sets of 'competing forces' as can be seen from the pattern's scope of applicability. All our identified patterns for health animation can be found in Appendix 2. In the following section we use the pattern "Icons in sickness explanations" to demonstrate how the patterns were inducted from the mapping, while highlighting what the different categories in the pattern contain (see Fig. 3). Later in the text, we discuss two patterns that have gone through a validation process.

"Icons in sickness explanations" is a pattern that deals with the use of similar iconography in animations that explain how a sickness affects the body e.g., bacterial infections or carcinogenic tumors (tendency). Eleven animations embodying the theme "sickness explanation" were located within the mapping (occurrence) and afterwards thoroughly

PATTERN NAME: Icons in Sickness Explanation	
TENDENCY	The animations under the theme: 'sickness explanation' have similar iconography. 'Sickness explanations' refers to the animations where the speaker is explaining how the sickness/problem affects the body e.g. bacterial infections, carcinogenic tumor, overall sickness, etc. As such, the pattern deals with the problem of communication the relation between symptom and cause in relation to health.
OCCURENCE	13 instances in the mapping of Health Animations (link)
SOLUTIONS / APPROACH	There are two types of icons that stand out in this type of animation: 1. The use of the transparent body to visualize the location of the health issue, e.g. in an animation about asthma, users can see the lungs in the chest. 2. The visualization of the sickness process. For example, how cancer cells affect the body.
OUTLIERS	No substantial outliers in this pattern. However a small portion do not use any visual icons, but resort only to audio explanations
INTERPRETATIONS	It can be overwhelming and abstract to learn a topic with complex processes and associated terms, that average users are unaware of. Therefore, the visualization of the processes can make it more concrete and understandable. A transparent body in the animations provides a context to the user and knowing the context is a part of understanding the sickness/problem.
EXAMPLES	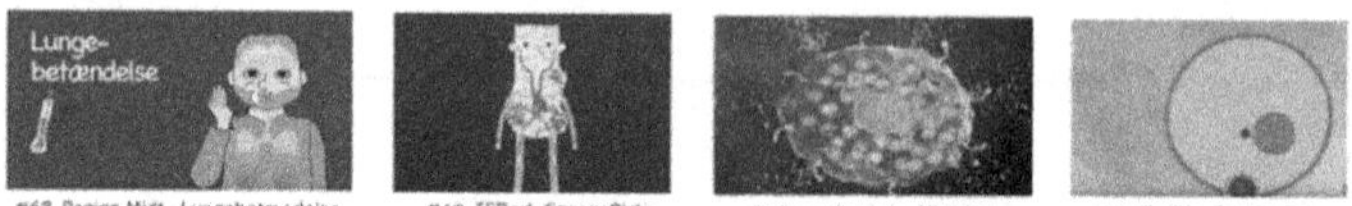 #68 Region Midt - Lungebetændelse #10 TEDed- Cancer Risks #23 sundhed.dk - Allergies #26 TedED - Feeling sick
LINKS	Focus on sickness explanation: [illegible] Focus on cause of treatment (with some sickness explanation) [illegible]

Fig. 3. Example of one of the eight design patterns of health animations emerging from the mapping of the 72 Danish animation products from the health sector. The full overview of patterns can be found in Appendix 2.

examined for similarities. In this process, we discovered that the animations employed two specific types of icons – 1. A transparent body and 2. The visualization of the sickness process (solution/approach). The 'transparent body' icon enables the animations to visualize the location of the health issue and thereby permits the viewers to see the location of a disease origin, e.g. observation of lungs through the chest in an animation about asthma. The "visualization of the sickness process" icon shows the impact of health issues on internal organs along with cells, i.e., how cancer cells evolve inside the colon. The icons above are observed as core elements of the pattern "Icons in sickness explanations". The next step of action was to determine why these specific icons are chosen. We found that these icons are used to contextualize the health issue, making it less abstract and easier to understand (interpretations). The design problem of conveying complex health issues, in this specific context, can thus be approached by applying the two icons. Consequently, this can be considered a design pattern, since it contains the three important elements: a problem (conveying complex health issues), a solution (two icons: the transparent body and the visualization of the sickness process), and a context (sickness explanations).

The next step in the process was to determine if there were any outliers in the established pattern and if so, would we then be able to discover the reason for this anomaly. Three animations were found to deviate by only using icon 1, a transparent body (outlier). However, these animations focus on treatment of the disease and have only a few details regarding explaining the sickness, whereas the rest of the animations main focus is explaining the sickness. The pattern of these animations is therefore the employment of the transparent body as well as the visualization of the sickness process, which together constitute a template of how to create further animations within the theme of "sickness explanation". This inductive process and the visual framework were utilized to create a total of eight patterns in the initial pattern library for health animations. The framework itself is based upon a mix of traditions of representing design patterns from both Alexander's original notations [5] as well as later developments within e.g. HCI. The pattern library is available in Appendix 2. A step not accounted for in this stage of the pattern library is whether the patterns represent activity- or design patterns, or whether the characteristics of health animation patterns merit a new interpretation altogether. Therefore, the next section details insights from a pilot split-test based on the pattern library.

3 Putting Patterns to the Test

The exploration of scaling the design pattern approach to health animation was part of the Danish research project "Animation på Tværs" (Cross-Sector Animation). The project's aim is to explore the development, the effect, and the implementation of animation across sectors in the healthcare system. The project's aim was to increase the acquisition of health information regarding treatment and course of disease among citizens with low health literacy when the course of their disease involves treatment across several healthcare sectors. This project was designed to leverage the insights gained from the emerging pattern library by informing the design of 12 health animation videos for citizens with lower back pain. The animations were developed in collaboration with an established health animation company that already had a fundamental structure of visuals and aesthetics along with a repertoire of animation elements from previous projects. Therefore, a majority of the 23 design categories (graphic fidelity, third-person narrator etc.) was already predetermined and unchangeable. However, in this design process we identified a possibility to critically test the assumptions about some of the form and content decisions along with testing the relevant patterns in order to compare the insights from both parts.

We used the health animation videos in the project as the testbed for a split-test exploring the application of two patterns from the pattern library. The split-test was designed based on the knowledge and the general practice gathered from the mapping of the 72 functional animations and the two (out of 8) inducted patterns. The purpose of the test was to validate our design patterns in terms of the patterns' solutions and interpretations along with potential benefits and limitations regarding using the design pattern method. However, a focus was also laid on validating the form and content decisions in the 12 animations from the Cross-Sector Animation project. We differentiate between design patterns and activity patterns (an existing pattern which not necessarily should be reused), whereby our patterns would be considered activity patterns until they have been properly validated by e.g. users. Therefore, the split-test consisted of a focus group of six citizens watching variations of the same health animation. There were three sections within the split-test: facts vs. emotions, male vs. female voiceover, and with text vs. without text. Each section began with showing an animation and then receiving feedback from the citizens. The same animation would then be shown, but it would contain one tweaked variable, e.g., the first animation would have a male voice-over and the second would have a female voice-over, whereas the rest of the animation would be the same (icons, events, information given etc.). Considering the size of the test, we cannot yet say for sure whether these patterns are recurring enough to be preserved or accepted fully by the users; however, they gave us an insight into which benefits along with problems that might arise for a design pattern library for digital health animations. This way the split-test functioned as a proof of concept and is not an attempt to draw statistical conclusions at this stage.

3.1 Pattern Example 1: Male vs. Female Voiceover

The male vs. female narrator pattern explains the role of narrators in our mapped 72 animations. In the scrutinization process, 55 out of all mapped animations used a third-person view through an omniscient narrator. Further, 89% of the 55 animations employed a male narrator, whereas the remaining 11% used a female narrator. The remaining 17 animations incorporated various other techniques; two animations involved a child's voice, seven used a visual representation of a third-person narrator, two used conversations between animated characters, and the last six did not involve any narrators. The observation shows that male narrators are a dominant choice in functional animations, which makes it a predominant pattern.

To test this pattern, we showed the same animation with the only difference being the gender of the voiceover (see Fig. 4). The animation portrayed a protagonist (a woman with lower back pain) sitting in a chair in her home while being gloomy. Meanwhile the voice-over comments on how people with lower back pain often fear the time of sick leave from their job, since they are afraid, they will get replaced. The speaker then goes on to explain how the job center can help to maintain their relation to the job and employer along with helping with some health courses.

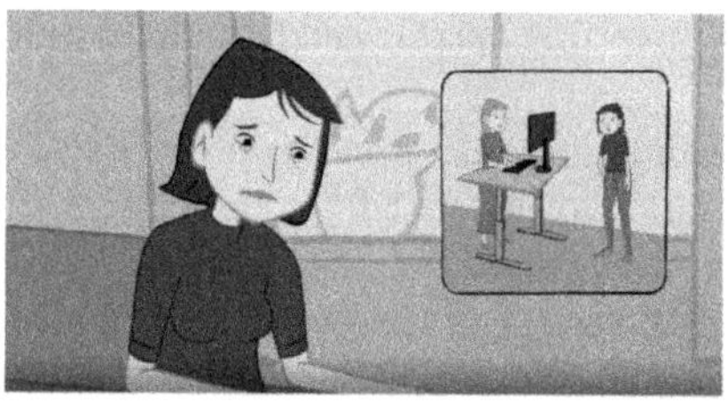

Fig. 4. Still image from the health animation from the split- test, featuring a citizen thinking about her illness with a third-person narrator – in one version a male and in another a female.

In this test, the participants preferred the male voiceover over the female in delivering this message. They described the male with the following words: "I think it is a very pleasant voice" and "He has a good voice" etc. On the other hand, the female voice was described as "... total no-go" and "I just think, it is a bit tiresome" and the participant expressed feeling "a bit uneasy". Overall, the participants perceived the female voice as less pleasant than the male voice. Two of the participants also experienced confusion in understanding the content of the animation as they linked the female voiceover to the main character (a woman). One participant said: "In the beginning, I think, you have doubts about if it is her thoughts or if it is the narrator's (an omniscient narrator)".

While these qualitative remarks are inconclusive about the multitude of different biases there might exist for the interpretations of a voiceover, it does show that if the gender of a voiceover is (wrongfully) associated with the animated character, it can create a potentially unconstructive dissonance for the viewer. To summarize, the male narrator was preferred over the female narrator in the split-test. The possible reasons for this preference could e.g., be a general preference for a male voice, liking and disliking of these specific male and female narrators, or the confusion regarding the female narrator.

A multitude of different biases can therefore be in play, which also can be the case of the remaining 23 design categories of which we mapped the 72 animations into. However, our objective of this test was not to unquestionably define whether the male voice or the female voice would be the right choice in every context. Instead, it was an exploration of the design pattern method's ability to discover specific patterns and test if these patterns could be of value to the users of the health animations and thereby a useful tool within this domain. The result being that this design choice (male vs female) was noticed by the participants in the split-test and did affect their experience with the animated information. Furthermore, the test highlighted the complexity of separating one variable within an animation that consist of various modalities. It indicates the competing forces within a potential pattern and how demanding it can be to sort out. In addition, it shows a potential need for patterns to evolve over time through the testing and evaluation of the patterns.

3.2 Pattern Example 2: No Speaker

The last pattern we wish to present is the pattern we named "no speaker". This pattern consists of animations that do not include a speaker or spoken conversations. After scrutinizing these animations, we identified the following common elements: 1. The speaker is replaced by text-pieces; 2. Intros and outros with text are employed; 3. Lengthy and detailed text is conveyed through metaphors; 4. Music and sound effects are mood-based. The four elements co-exist as the foundation of the "no speaker" pattern. When these animations use text-pieces it is in the form of e.g. text-bubbles or a message on the animated phone instead of typical subtitles. In other words, the text was more actively integrated into the animation. Intros and outros functioned as a space to write short text-pieces that could frame the topic and sum it up in the end, such as "remember to cancel" and "The best position is the next position. Are you going to join?". The third element, the use of visual metaphors, was employed to ensure that the text did not end up too 'long'. One example of this is when the main character was feeling depressed because of an eating disorder. The animation displayed her on a complete black background accompanied with gloomy music along with keywords in a text-bubble and in the end she fell. As a result, they avoided having to explain this through long text-pieces. Furthermore, this example demonstrates how the fourth element (music and sound effects) is applied in the animation. Our interpretation of the application of these four elements is that it can be difficult to convey understandable information through strictly visual techniques (animation, pictures etc.). Therefore, when no speaker is involved, the animations need to incorporate text-pieces. This is also why the intros and outros are employed to frame the context for the viewer or to sum up the problem without a verbal explanation. Furthermore, to make sure the viewer is not overwhelmed with lengthy texts, there is also a need for visual metaphors to enhance the understanding. Finally, the use of music and sound effects is to create a mood for the viewer. We resolved to explore the strict use of only visual techniques in a visual medium such as animation. Do the visual representations rely on the often-used modality such as either a voice-over or text-pieces?

In this part of the split-test, participants were asked to watch an animation with a voice-over and no text-pieces and then the same animation with accompanying short text-pieces. This animation showed how a potential course of treatment could occur for the patient in the pain clinic. A protagonist (a woman) is shown going through this treatment, while the voiceover explains the different events and health personnel the patient can encounter. Through the animation various scene shifts happen and the voiceover explains some of the information that the patient will receive in the different courses issued from the pain clinic.

The first animation shown was without text (see Fig. 5). As feedback to this animation, it was said: "It went fast… A lot of information". Nevertheless, when asked to repeat the events in the different scenes, the participants were able to recall most of the information. This indicates that a combination of visuals and a voice-over was enough to provide the participants with at least a superficial understanding of the information. The next animation shown included short text-pieces (see Fig. 5). Three of the participants initially missed the text-pieces in the animation, except the part where the text was used to differentiate the different professions visualized (doctor, physiotherapist etc.). They liked that the differentiation was clarified through text: "… the pictures with the five professions… there I thought that it was good, that it said something above". On the other hand, one participant felt that the other text-pieces did not contribute to a better understanding or reflection on the information. This led to a discussion among the participants about the general necessity of text-pieces, whereby some found them important and not distracting.

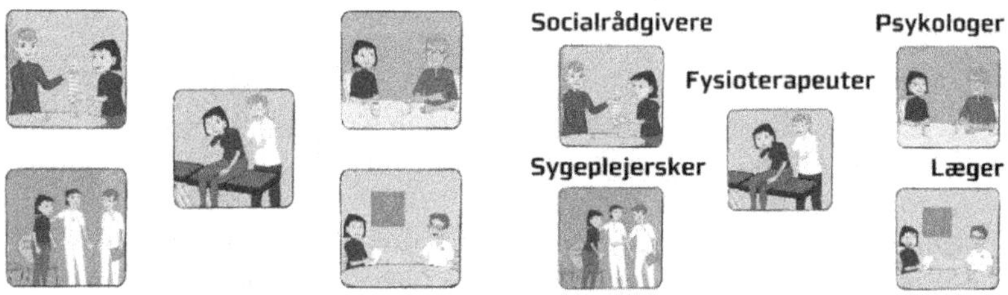

Fig. 5. Still images from the split-test with two versions of the same animated overview of sectors – one with only spoken word, and one with supplemented text overlays.

Afterwards, as a small experiment, the animation with accompanying text-pieces was shown without sound. The participants additionally agreed that the animation did not work without sound because a lot of the meaning was lost. One participant explained: "Because, it lacks a lot of coherence… you all of sudden lack… the context of what you're supposed to do". The participant would in this instance prefer if the animation had subtitles instead. This aligns with the study by Meppelink et al. [3] showing an increase in comprehension when voiceovers are used to supplement health animations. Therefore, it seems that if the animation is meant to communicate a vast amount of information, it needs either a voice-over or at least 'subtitles', otherwise important information will be lost. However, if a voiceover is already used, short text-pieces can be used to clarify rapid information such as lists of medicines, professions etc. Consequently, the result confirmed our previous interpretation – it is difficult to convey comprehensible information solely through visual techniques. However, the pattern implies no need for a voice-over if text pieces are used. This indicates that the pattern still needs to be elaborated and adjusted through future testing, as we saw in the male vs. female pattern.

4 Discussion

In this study, we explored how a design pattern language approach can support the creation and validation of design choices in health animations. As argued, traditional animation already follows several design patterns including the famous 12 animation principles. These principles explain 'how' to create realistic animation by creating the illusion of obeying the basic law of physics. In contrast, our patterns attempt to explain 'why' certain form and content is chosen and animated in a certain way. This includes answering questions such as: what drove the choice of a male voice-over or the use of a transparent body in sickness-explanation animations. Understanding the 'why' allows stakeholders to critically question the form and content decisions made by others and themselves as well as to evaluate whether the decisions are strongly substantiated or merely a subjective opinion. This further supports transparency in the development process of health animations. The ambition of applying the design pattern approach for health animation was to leverage the same strengths the approach has shown in other domains – from strengthen architectural directions, to informing user-friendly digital interfaces. In health animation, we argue that our analysis indicates that pattern languages can inform the animation process, including the discourse and not only the form. As such, the pattern library of health animations has potential to reduce future communication mistakes and as a result improve the health animation by making it clearer and more comprehensible for the citizen.

The agreement we found between the created patterns and the results from the split-test indicates the achievement of the pattern library. Additionally, the value of the 'right' design choice became evident especially in the test of the male vs female pattern, where the design pattern method enabled the ability to locate a potential relevant design choice. This not only proves the importance of pattern languages but also validates its outcomes in health animations. We argue that a split-test or other tests alone would not be able to create a strong basis for a repeated use of a particular form or content decision e.g., the use of a male voiceover. The reason for this is the many variables at play, which we experienced in our split-test. However, by using a combination of sizeable split-tests (or other tests) along with an emerging pattern library (e.g., male vs. female pattern), it is possible to validate the form and content decision by allowing us to measure the results from the test against best practices in the field. Together, they can indicate which patterns are worth being preserved and repeated even with slight changes in the future.

However, the pattern library method also comes with its limitations and problems. It has previously been reported that the validation and testing of design patterns is a difficult task due to their competing forces [15]. In the present work, we experienced this difficulty while testing the 'male vs. female voiceover' pattern. The participants preferred the male voiceover as suggested by our pattern. However, the reason for this preference is uncertain because of the challenge of isolating only one variable in an animation. We isolated the variable 'voiceover' by showing the same animation with the only change being the gender of the voiceover. Nevertheless, the feedback showed that the gender of the protagonist was important because it can create confusions when the protagonist has the same gender as the voiceover. We argue that this demonstrates the complexity of design patterns as well as their testing processes. It shows the need of a constant evolving pattern library that gets tested and adjusted over time.

5 Conclusion and Future Work

We initiated this study with a critique of previous studies for providing little-to-no rationale for the form- and content-related choices made in the design process of health animations. On that basis, we asked whether it was viable to scale the well-established design tradition of building and working with design patterns into the domain of health animation?

The initial mapping, of 72 Danish citizen-oriented health animation products, showed a broad range of animation approaches, fidelities and narrative structures being applied. Across the 23 design categories we were able to induct eight design patterns that could be described as tackling similar communicative problems, in a comparable contextual frame, and applying similar form and/or content choices as solutions. This indicates how the design pattern approach can be applied and used to create a frame of reference for health animations. However, the analysis also shows how the inducted patterns tend to blend form and content into patterns of discourse. That is, the patterns tell us more about the communicative dimension in relation to other patterns, rather than the semantics of each individual pattern alone. While this may be interpreted as the 'competing forces' of this specific application of the design pattern approach, it is also a limiting factor in our current attempts of scaling the method into the domain of animation. That being said, there is a definite potential to make the design process of this genre of animation more transparent, by utilizing this approach to articulate when we are making subjective form and content choices, and when we are leveraging past experiences through established patterns. The pilot split-test showed the potential for this by enabling a qualified hypothesis about what would work in the produced health animation variants, and what might fail when viewed by the citizens.

To further improve and develop the approach we argue that a series of further studies are required. First and foremost, the mapped health animations need to be increased from the current 72 to a substantially larger database. Furthermore, this mapping could be enriched by adding the complexity of health animations from other countries, while also creating the need for more fine-grained ways of sorting and analyzing across the categories than the current framework. Increasing the number of mapped animations will likely also produce more induced design patterns than the current eight. Additionally, an increase of mapped animations could potentially further strengthen the existing patterns with more variants, outliers, and connections among patterns.

Another important step is the implementation of this approach in both health practice and in academia. In the health care practice, a prominent issue is to identify the most suitable process of including and using pattern languages to communicate and participate in the design process. In academia, the challenge will be to effectively combine the transparency, made possible through design patterns, with the traditional effect studies most often seen in health animation studies. Combined, the two methods will be able to achieve a more precise determination of which parts of health animations work, for whom, and with what level of effect. Finally, due to the limited pilot split-test, the design patterns within this study are currently to be considered as activity patterns; not yet fully formed and validated by continuous development and testing. In conclusion, the design pattern library of health animation needs to be seen through the same lens of previous pattern libraries in their infancy: as a living 'evolving document' open to be challenged, modified, and even gradually replaced as the scope of their use is tested further by a maturing community of designers of health animations.

Appendix

Appendix 1:

Health Animation Mapping (accessed 16.6.2021)

https://docs.google.com/spreadsheets/d/1CWJwVGx7N9NTYDrIv-FOfC7zxOO p0rpqlS5p9U3SOqw/edit?usp=sharing.

Appendix 2:

Health Animation Design Pattern Library (accessed 27.1.2021)

https://docs.google.com/document/d/1eq373UTr56zNHMfxMDcLQVRHb0fu-Clz gKuCcl9iGlo/edit?usp=sharing.

References

1. Vistisen, P.: Sketching with Animation: Using Animation to Portray Fictional Realities – Aimed at Becoming Factual. Aalborg Universitetsforlag, Aalborg (2016)
2. Vistisen, P.: Science Visualization: Principles for an emerging animation community to consider. WeAnimate - Dan. Anim. Soc. (ANIS) **1**(3), 80–85 (2019)
3. Meppelink, C.S., van Weert, J.C., Haven, C.J., Smit, E.G.: The effectiveness of health animations in audiences with different health literacy levels: an experimental study. J. Med. Internet Res. **17**(1), 1–13 (2015)
4. Calderón, J.L., Shaheen, M., Hays, R.D., Fleming, E.S., Norris, K.C., Baker, R.S.: Improving diabetes health literacy by animation. Diabetes Educ. **40**(3), 361–372 (2014)
5. Alexander, C.: Notes on the Synthesis of Form (Later Pr. edition). Harvard University Press (1964)
6. Alexander, C., Ishikawa, S., Silverstein, M.: A Pattern Language: Towns, Buildings, Construction. OUP USA (1977)
7. Borchers, J.O.: A pattern approach to interaction design. In: Proceedings of the 3rd Conference on Designing Interactive Systems, Processes, Practices, Methods, and Techniques, pp. 369–378. ACM Press, New York (2000)

8. van Welie, M., van der Veer, G.C., Eliëns, A.: Patterns as tools for user interface design. In: Vanderdonckt, J., Farenc, C. (eds.) Tools for Working with Guidelines, pp. 313–324. Springer, London (2001). https://doi.org/10.1007/978-1-4471-0279-3_30
9. Gamma, E.: Design patterns – ten years later. In: Broy, M., Denert, E. (eds.) Software Pioneers, pp. 688–700. Springer, Heidelberg (2002). https://doi.org/10.1007/978-3-642-59412-0_39
10. Berkman, N.D., Sheridan, S.L., Donahue, K.E., Halpern, D.J., Viera, A., Crotty, K., et al.: Low health literacy and health outcomes: an updated systematic review. Ann. Intern. Med. **155**(2), 97–107 (2011)
11. Van Welie, M., Veer, G.: Pattern Languages in InteractionDesign: Structure and Organization (2003)
12. Kruschitz, C., Hitz, M.: Human-computer interaction design patterns: structure, methods, and tools. Int. J. Adv. Softw. **3**(1), 225–237 (2010)
13. Chang, D., Tuovinen, J.E.: Gestalt theory in visual screen design—a new look at an old subject—open research online. In: Selected Papers from the 7th World Conference on Computers in Education (WCCE 2001), Copenhagen, Computers in Education 2001: Australian Topics, vol. 8, pp. 5–12. Australian Computer Society (2002)
14. Johnston, O., Thomas, F.: The Illusion of Life: Disney Animation (Rev Sub Edition). Disney Editions, Glendale (1995)
15. Dawes, M.J., Ostwald, M.J.: Christopher Alexander's a pattern language: analysing, mapping and classifying the critical response. City Territory Architect. **4**(1), 17 (2017)
16. Norman, D.A., Draper, S.W.: User Centered System Design: New Perspectives on Human-Computer Interaction. L. Erlbaum Associates Inc., Mahwah (1986)
17. Tidwell, J.: Designing Interfaces: Patterns for Effective Interaction Design, 2nd edn. O'Reilly, Newton (2011)
18. Bayle, E., et al.: Putting it all together: towards a pattern language for interaction design: a CHI 97 workshop. ACM SIGCHI Bull. **30**(1), 17–23 (1998)
19. Wurhofer, D., Obrist, M., Beck, E., Tscheligi, M.: Introducing a comprehensive quality criteria framework for validating patterns. In: 2009 Computation World: Future Computing, Service Computation, Cognitive, Adaptive, Content, Patterns, pp. 242–247 (2009)
20. Baecker, R., Small, I.: Animation at the Interface. I B. Laurel (Red.), Art & Human- Computer Interface Design (1990)
21. Al Owaifeer, A., Alrefaie, S., Alsawah, Z., Al Taisan, A., Mousa, A., Ahmad, S.: The effect of a short animated educational video on knowledge among glaucoma patients. Clin. Ophthalmol. **12**, 805–810 (2018)
22. Chakravarthy, B., et al.: Randomized pilot trial measuring knowledge acquisition of opioid education in emergency department patients using a novel media platform. Subst. Abuse **39**(1), 27–31 (2018)
23. Gholami, M., Pakdaman, A., Montazeri, A., Jafari, A., Virtanen, J.I.: Assessment of periodontal knowledge following a mass media oral health promotion campaign: a population-based study. BMC Oral Health **14**(1), 31 (2014). https://doi.org/10.1186/1472-6831-14-31
24. Cleeren, G., Quirynen, M., Ozcelik, O., Teughels, W.: Role of 3D animation in periodontal patient education: a randomized controlled trial. J. Clin. Periodontol. **41**(1), 38–45 (2014)
25. Ferguson, M., Brandreth, M., Brassington, W., Leighton, P., Wharrad, H.: A randomized controlled trial to evaluate the benefits of a multimedia educational program for first-time hearing aid users. Ear Hear. **37**(2), 123–136 (2016)
26. Govender, R., Taylor, S.A., Smith, C.H., Gardner, B.: Helping patients with head and neck cancer understand dysphagia: exploring the use of video-animation. Am. J. Speech Lang. Pathol. **28**(2), 697–705 (2019)
27. Grigsby, T.J., Unger, J.B., Molina, G.B., Baron, M.: Evaluation of an audio-visual novela to improve beliefs, attitudes and knowledge toward dementia: a mixed-methods approach. Clin. Gerontol. **40**(2), 130–138 (2017)

28. Jones, A.S.K., Fernandez, J., Grey, A., Petrie, K.J.: The impact of 3-D models versus animations on perceptions of osteoporosis and treatment motivation: a randomised trial. Ann. Behav. Med. **51**(6), 899–911 (2017)
29. Polk, J.A., Woolridge, N., Wilson-Pauwels, L., Jenkinson, J., Mackay, M.: Improving parents' early recognition and understanding of infant cranial abnormalities through web-based 2-D animations of 3-D structures. J. Biocommun. **29**(4), 16–20 (2003)
30. Narimatsu, H., et al.: Usefulness of a bidirectional e-learning material for explaining surgical anesthesia to cancer patients. Ann. Oncol. **22**(9), 2121–2128 (2011)
31. Wells, P.: Understanding Animation. Routledge, New York (1998)
32. Betancourt, M.: The History of Motion Graphics. Wildside Press, Rockville (2013)
33. Bordwell, D., Thompson, K.: Film Art: An Introduction, 10th edn. McGraw-Hill, New York (1993)
34. Taylor, R.: Encyclopedia of Animation Techniques. Chartwell Books, New York (2003)
35. UIPatterns, User Interface Design Patterns. http://ui-patterns.com/. Accessed 30 Oct 2021

The Effect of Characters' Locomotion on Audience Perception of Crowd Animation

Wenyu Zhang(✉) and Nicoletta Adamo-Villani

Purdue University, West Lafayette, IN 47906, USA
zhan2586@purdue.edu

Abstract. A common practice in crowd animation is the use of human templates. A human template is a 3D character defined by its mesh, skeletal structure, materials, and textures. A crowd simulation is created by repeatedly instantiating a small set of human templates. For each instance, one texture is randomly chosen from the template's available texture set, and color and shape variety techniques are applied so that multiple instances of the same template appear different [1]. When dealing with very large crowds, it is inevitable to end up with instances that are exactly identical to other instances, as the number of different textures and shape modifications is limited. This poses a problem for crowd animation, as the viewers' perception of identical characters could significantly decrease the believability of the crowd simulation. A variety of factors could affect viewers' perception of identical characters, including crowd size, distance of the characters from the camera, background, movement, lighting conditions, etc. The study reported in this paper examines the extent to which the type of locomotion of the crowd characters affects the viewer's ability to perceive identical instances within a medium size crowd (20 characters). The experiment included 83 participants and compared the time participants took to spot identical characters in three different locomotion scenarios (e.g., standing, walking, and running). Findings show that the type of locomotion did not have a statistically significant effect on the time subjects took to identify identical characters within the crowd.

Keywords: Crowd animation · Virtual character · Perception

1 Introduction

Various crowd simulation techniques have been developed and are widely applied in the visual effects, animation, and video game industries. However, there is still a lack of research on the perceptual factors that could affect the audience experience of the crowd, such as the degree of realism of the characters, the level of detail, the crowd motions, etc. The work reported in the paper aims to fill this gap by examining the effects of characters' locomotion on the viewer's perception of identical characters in medium-sized crowd simulations.

M. Wölfel et al. (Eds.): ArtsIT 2021, LNICST 422, pp. 398–412, 2022.
https://doi.org/10.1007/978-3-030-95531-1_27

More specifically, this research focused on crowds which require a heterogeneous character appearance and motion. The goal of the study was to determine whether it is possible to use a lower number of entity characters depending on the particular type of locomotion performed by the crowd. The findings from the study have practical implications for real-time simulations (e.g., computer games) as well as off-line rendering simulations (e.g., crowds in film). One of the main goals in animation production is to maximize the visual quality of moving pictures with a minimal expenditure of technical resources. The results from this study can help achieve this goal. They can help those involved in animation production save time and resources by decreasing the number of different human templates that are necessary in certain crowd simulations.

2 Literature Review

2.1 Character Animation

The production process of character animation can be broken down into five basic parts: modeling, texturing, rigging, animating and rendering. Modeling, texturing, and rendering all determine the physical appearance of characters, while rigging and animating control the characters' movements and facial expressions.

Modeling is a process whereby the creator defines the shape of the characters without giving consideration to their texture. It allows the creator to display several basic properties of a character, including height, gender, age, body shape, hair style, and muscle level. Zell *et al.* [2] noted that "shape is the main descriptor for realism, and material increases realism only in case of realistic shapes."

Rigging and Animating are two key factors in crowd animation. Traditionally speaking, the character models in crowd animation are polygonal meshes rigged by bones. When the joints are rotated, the vertices cluster attached to the joints becomes deformed along a predetermined trajectory, which is how character animation is generated [3]. In real world production, character motion can be created either by animators' key-framing and adding frame interpolation, by using physics-based animation generated by computer simulation tools, by capturing real-time motion data from devices on actors (motion capture), or via any combination of the three aforementioned techniques [4]. However, certain comprehensive methods need to be adopted in creating crowd animation because moving crowds involve complicated mechanics which require algorithmics [5]. Such methods, which go beyond an individual character's locomotion, have been studied by previous researchers. For example, walking is a common mode of locomotion that can easily be produced for an individual character.

However, in the case of a group of walking characters (e.g., pedestrians on the street), factors such as collision avoidance must also be considered [6]. Additionally, many algorithms related to the motion trajectories of crowds have been developed in recent years. For instance, Yu and Terzopoulos [7] developed a novel framework for pedestrian characters, including behavioral interaction in urban settings. Guy *et al.* [8] presented a technique called Personality Trait Theory to create heterogeneous crowd motion. Sun *et al.* [9] simulated realistic crowd trajectory in an urban scenario surrounded by traffic, vehicles, intersection, etc.

In a crowd simulation, characters' locomotion and behavior inevitably rely on the nature and quality of the algorithms operating behind the scenes.

2.2 Crowd Rendering

3D renderers are tools used to output final image sequences in animation production. They generate various results depending on the shading models and numerous physical parameters inputted by the animators. Traditionally, crowd animation was rendered through animation programs. However, with the development of 3D games, many programs tend to have the capability to simulate crowds in real time. In the earlier studies on crowd rendering, limited computational speed was the main problem in the creation of 3D scenes with populated characters. Many acceleration techniques for the rendering of large environments were subsequently invented. A technique called "instance" has been frequently used in crowd simulation. In a shading API (Application Programming Interface) such as OpenGL or DirectX, a geometry shader can be used to deform the vertices and the triangle mesh of a crowd with only one call in GPU (Graphics Processing Unit) [10]. Ashraf and Zhou [11] applied a hardware-accelerated method through programmable shaders to animated crowds. In Peng *et al.* [12], developments have been made to utilize GPU in their parallel architecture to improve the performance of graphics computation. They invented a mesh simplification algorithm which can render a real-time crowd system on the GPU. Klein *et al.* [13] created an innovative method which allows instances of 3D characters with controllable parameters to be rendered on the web. In the work of [3], a novel crowd rendering system that simultaneously runs real-time on GPU and decreases the computation load on the graphics card was developed. The scene includes 30,000 instances in real-time motion.

Shading is also a significant component in rendering real-time images. Maciel and Shirley [14] implemented the LOD (Level-of-Detail) technique to create impostors to reduce the complexity of rendering. This method later evolved to an IBR (Image-based Rendering) technique which was adopted by many researchers. In a study by Tecchia and Chrysanthou [15], the Image-based Rendering (IBR) method was adopted, and the characters were pre-computed and animated. A multi-pass algorithm was first used to retouch different parts on the character, followed by the addition of efficient shading and shadow. In a study by Tecchia *et al.* [16], they used an approach whereby each character was transformed into an image-based impostor which possesses an adaptive resolution depending on one's viewing angle. Ciechomski *et al.* [17] presented a customized hardware rendering pipeline which created texture variety from a single texture in HSB color space. Millan and Rudomin [18] combined the imposter and instancing techniques and created a program which is more efficient in rendering large crowds.

2.3 Perception of Virtual Characters

Ciechomski *et al.* [17] has stated that "For a human crowd, variation can come from the following aspects: gender, age, morphology, head, kind of clothes, color of clothes and behaviors." In other words, the perception of human crowds in animation mainly depends on two aspects - appearance and behavior.

All the virtual CG characters can be classified into two categories: photo-realistic and stylized. In a study by Zell *et al.* [4], it was found that factors such the shape of a character's body and its material (especially the albedo texture) can significantly affect audience perception. These two factors have a strong influence on how realistic the characters are perceived to be.

Another factor that affects the believability of perception is the facial proportion of characters. Green *et al.* [19] concluded that facial height, jaw width, and eye separation are all considered to be important factors which can increase the appeal of animated characters.

Besides their exterior appearance, the behavior or motion of characters likewise plays an important role in creating realistic perceptions. Based on a study [20], when the characters are in motion (e.g., walking or running) as opposed to staying still, viewers can appreciate that the virtual characters resemble real world human beings, instead of perceiving them as a group of static dot-shape objects. Research by McDonnell *et al.* [21] compared the reaction times in spotting appearance-based duplicated characters versus motion-based duplicated characters. They concluded that characters cloned by appearance are more conspicuous than characters cloned by motion. Also, they discovered that the position layout of characters affected the viewers' perception - horizontal layout makes it easier for the audience to spot cloned characters compared to a vertical or diagonal layout. One limitation of their experiment is that all the testing characters were positioned facing forward, which is not considered typical in crowd animations. Pražák and O'Sullivan [22] studied the locomotion variety in crowd animation perception. They adopted motion capture techniques to capture 83 actors' real-world motion data (including both males and females) and created a virtual scene to perform the experiment. They claimed that at least three different locomotion types are needed to be displayed for each gender to achieve a realistic level of behavioral variety in a pedestrian scene. However, their character set was relatively small, with only 24 characters being shown at a time in each scene. Moreover, they did not examine the effects of the various types of motion in the experiment.

Eye tracking has become quite popular in perception studies in recent years. Using an eye tracking device, McDonnell *et al.* [23] found that head and upper body are the first part viewers tend to notice, regardless of the character's position, motion, gender, size, etc. They also found that creating more kinds of head accessories and variable top textures is more effective at increasing variety than alternating the facial geometry of characters.

When it comes to facial close-ups, the eyes tend to catch viewers' attention more than other body parts. A recent study [24] confirmed that viewers primarily maintain their glance at the virtual characters' eyes and mouth. On average, it was found that participants spend around 35% of the time looking at the eyes, while spending no more than 10% of the time focusing on other parts of the body.

Figure 1 is a screenshot of an animated commercial short for Westfield Stirling Shopping Mall [25]. Some of the CG characters are walking randomly in the mall; while some are standing still. They all have different appearance and slightly difference behavior which increases the perception fidelity of crowd animation.

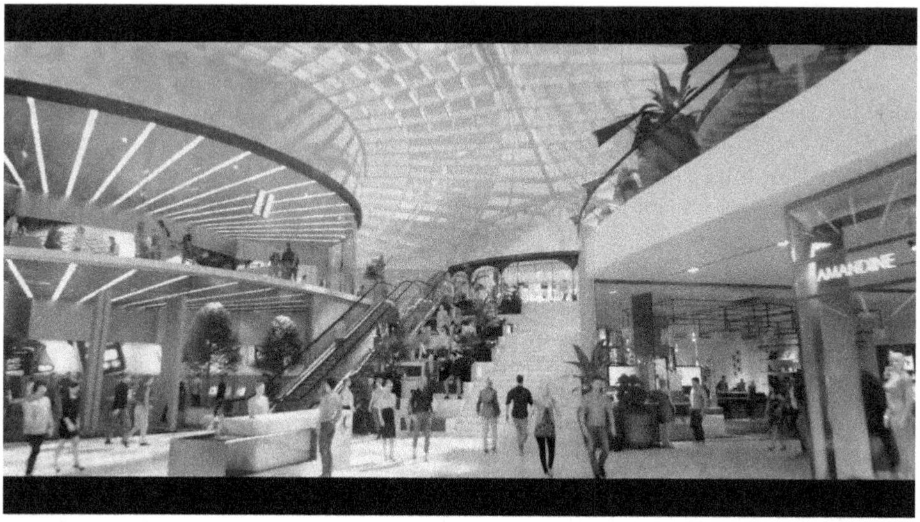

Fig. 1. A screenshot of the visualization project Westfield Stirling by New Holland Creative.

3 Methodology

The goal of this study was to determine whether different types of locomotion would affect viewers' perception of the crowd. The participants watched randomized video clips representing three scenarios and were then instructed to complete a related online survey. The study adopted a quantitative research approach that compared the length of time that participants spent on each scenario to identify identical characters. A customized Bayesian Linear Mixed Model was employed to analyze the collected data.

The independent variables in this research were the type of locomotion (standing, walking, running) of the 3D characters in the crowd and the gender of the participants. The dependent variable was the length of time subjects took to identify two identical characters in the crowd.

3.1 Hypotheses

H_{01}: Participants will spend the same amount of time to identify identical characters in all the three locomotion scenarios.

H_{a1}: Participants will spend different amount of time to identify identical characters in each of the three scenarios. Specifically, participants will spend more time to identify identical characters in the Running Scenario than in the Walking and Standing Scenarios, respectively.

H_{02}: Participants will spend the same amount of time to identify identical characters regardless of the participants' gender.

H_{a2}: The time participants will spend to identify identical characters will vary depending on the participants' gender.

3.2 Subjects

A total of 83 participants took part in this study. Thirty-three participants were students from the Computer Graphics Technology department at Purdue University. Fifty participants were selected via a survey posted on Amazon Turk. The participants were recruited without regard to gender and resulting in 46 males and 37 females in the pool. Participants' age ranged from 18 to 64 years old. Participants' familiarity with computer animation ranged from zero experience to very familiar with computer animation. All the participants could see the computer screen clearly, with or without corrective lenses.

3.3 Stimuli

The stimuli used in this study consist of three online videos demonstrating different types of character locomotion within crowd animation, along with an online survey. The crowd animation video clips were created using Maya 2016 with Golaem plugin and were rendered using Mental Ray renderer. The rendered videos contain both highlight and shadow in order to simulate realistic lighting. However, the materials on the characters do not include any other channels besides diffuse textures. All the characters' exterior, such as garment texture, is from the preset package of Golaem plugin. The characters' locomotion (e.g., walking, running) was also created using Golaem presets. The characters' moving trajectories are customized to allow the characters to have specific paths without moving out of the frame. Also, to assure all the other parameters stayed uniform, the camera angle, lighting, shadow, contrast, are set up completely identical in each video clip. The camera is positioned at one side of the scene with a tilting angle of 30° towards the ground. The lens has a view angle of 35° to capture the full scene.

In each scene, there are 18 characters with heterogeneous appearance and only two characters with homogeneous appearance, which includes skin color, hair color, color of shirt, pants and shoes. In the standing scenario, characters stand still on the ground surface and exhibit casual turning-in-place movements.

In the walking scenario, characters walk in random trajectories on the ground surface. In the running scenario, characters run around in random trajectories.

After the animation was completed, all frames were exported from the animation package. Image sequences for each scenario were processed in video editing programs and output as three 98-s video clips. Each video clip had a 10-s opener with instruction reminding viewers to be prepared for the experiment. Each video clip looped 10 times itself. All the experiment videos were in sRGB color space without any post-processing or visual effects.

The three video clips for experiment were uploaded to online video platform Vimeo which can be viewed through following link. Screenshots from the three videos are shown in Fig. 2.

- https://vimeo.com/367614577
- https://vimeo.com/367615337
- https://vimeo.com/367613352

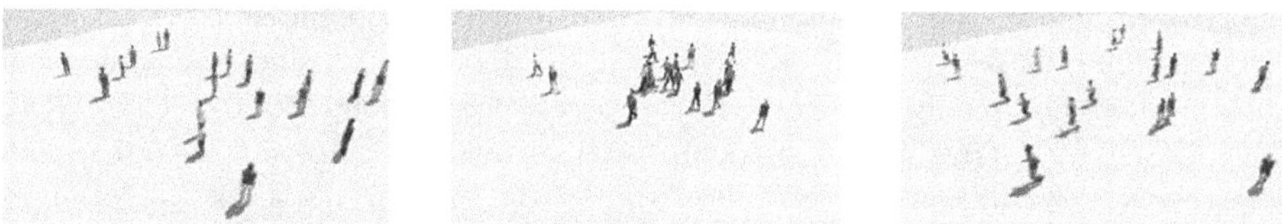

Fig. 2. Locomotion scenarios illustrating standing, walking, running, respectively.

3.4 Evaluation Instrument

This experiment required participants to view a series of animation video clips. Therefore, a laptop or personal computer with proper display and fast Internet access was required. Mobile devices were not allowed in this study given that the screen resolution on such devices and the various nuances of person-device interaction might affect the perception results.

Data collection was performed via an online survey created using the Qualtrics survey platform. The videos were embedded into the survey platform and all user interaction controls were disabled. The survey included the IRB consent form, detailed experimental instructions, a demo video, three formal testing videos, a demographics questionnaire, and optional feedback.

Since this study required quick responses from the participants, detailed instructions along with a video tutorial were displayed to participants at the beginning of the experiment to ensure they thoroughly understood the experimental procedure. In addition, a demo video was presented to allow participants to familiarize themselves with the procedure and promote reliable results. Participants were expected to adjust page zoom to a suitable resolution in the browser to allow them to watch the entire frame.

Formal video clips for testing began to play automatically as soon as the participant displayed the page. Each video clip had a text reminder stating: "Please move the cursor on the blue button. (Do not click until you have found two identical characters)." Along with each formal experiment video, there were required questions on the following page letting participants select the identical character, if found. Each question had only one correct answer out of three choices. The answer did not contain any text but only a pair of screenshots of the characters (full body front and back) appeared in the video. Thus, viewers might have had a more intuitive impression to select the character they believed they have found. Participants were forced to select an answer before they could jump to the next page.

In order to decrease potential confounds stemming from the learning effect (whereby participants' performance improves over time as they are exposed to the same stimulus), the order of the three video scenarios was randomized. We randomized the video groups into three different combinations to make sure each scenario would not always appear at the first. This greatly reduced the audience's learning effect. The order combinations were Standing-Walking-Running, Walking- Running-Standing and Running-Standing-Walking.

3.5 Procedure

For each scenario, the video clip started to play automatically and looped for 15 times. All the interaction controls were disabled on the videos. Thus, participants were not able to pause, adjust speed, download, or loop the video by themselves. Participants were asked to click on the blue button showing "CLICK ME" at the bottom right corner of each scenario page as soon as they spotted the two identical characters. The system recorded the exact response time for each participant. Figure 3 shows a screenshot of the Walking Scenario stimuli.

Next, participants were asked to select which of the three types of characters were identical in the video clip. After a selection was made, the page would progress to the next video. After viewing all the video clips and answering the pertaining questions, the participants were asked to fill out a brief demographic questionnaire. It collected participants gender, age and their familiarity of computer animation. Finally, they were given the option to share any feedback or comments they may have had regarding their experience before concluding the study.

4 Data Analysis

After the experiment was conducted, participant response times (i.e., the amount of time each participant spent to identify identical characters in each video) were collected. Since there were fixed and multiple random factors in this study, a Bayesian Linear Mixed Model was used to determine whether the response times varied significantly across the three locomotion scenarios (standing, walking, and running).

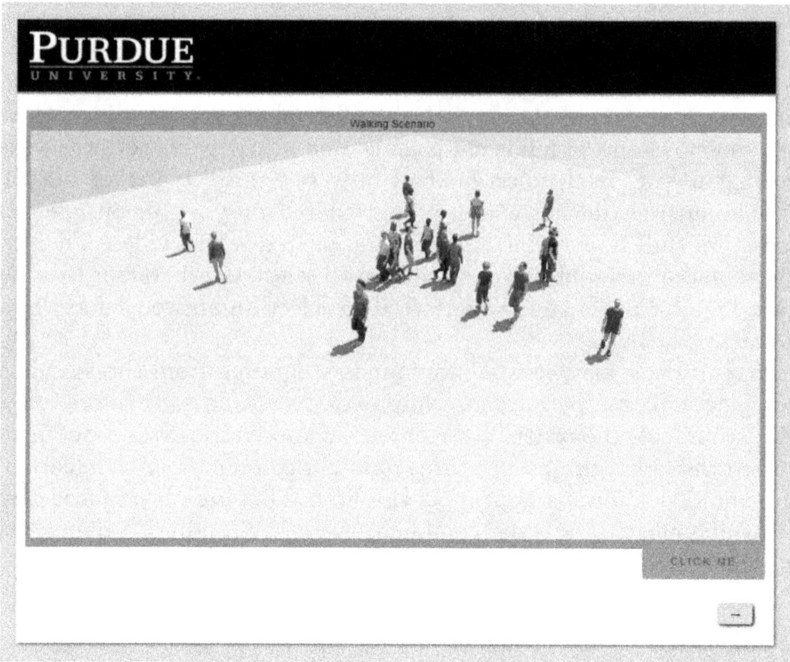

Fig. 3. A screenshot of formal testing video clip (walking scenario).

4.1 Data Pre-processing

The dependent variable in this study was the response time, or the length of time that participants spent on each scenario before clicking the mouse (indicating that they identified two identical characters). First, an accuracy check was performed to clean up the collected data; participants who selected incorrect answers were subsequently removed from data set. Standing Scenario had an accuracy of 75%; Walking Scenario had lowest accuracy of 62%; Running Scenario had highest accuracy of 87%. Figure 4 is a bar graph to visualize the accuracy result. Since the actual video would not play until the 10th second and would terminate at the 98th second, participants' who had spent less than 10 s and greater than 98 s in watching each video clip were removed from data set. After the clean-up, there were 51 available responses in the data set, 28 from males and 23 from females. The reaction times across the three video types were then analyzed using Bayesian Linear Mixed Model.

4.2 Data Model

Each participant in our study was exposed to all three video categories. Only the subjects who identified every pair correctly and responded within the acceptable range of response times (as explained above) were included in the response time

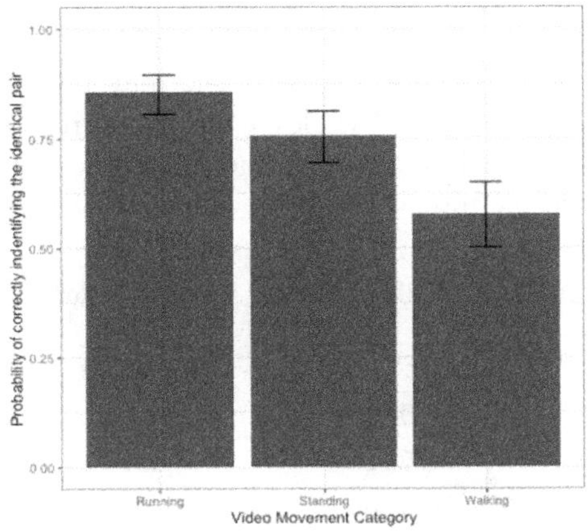

Fig. 4. Response accuracy of each locomotion scenario.

analysis. Combining all the factors which might affect the result of this study, we attempted to fit the model as below:

$$Time_{ijk} = \mu + Video_i + Subject_k + Period_j + Sequence + Gender + \epsilon_{ijk}$$

where:

1. $Time_{ijk}$ is the actual response time for subject k watching video i in time period j.
2. μ is the overall mean expected response time.
3. $Video_i$ is the effect of the ith video category (Running, Walking, Standing) on the expected response time.
4. $Subject_k \sim N(0, \sigma^2_{subj})$ is the random effect of subject k on expected response time.
5. $Period_j$ is the effect of the jth time period on the expected response time.
6. *Sequence* is the effect of video display order on the expected response time.
7. *Gender* is effect of different gender on the expected response time.
8. $\epsilon_{ijk} \sim N(0, \sigma^2)$ is the error between expected and actual response time.

The model includes fixed-effects stemming from our independent variable video category (standing, walking, running) and factors corresponding to video order, period, gender as explained above. In addition, we included random subject effect, in order to control for heterogeneity of each subject.

In this study, there were three categories under $Video_i$, 51 different individuals under $Subject_k$, three categories under $Period_j$, three different orders under *Sequence*, and two categories under *Gender*. $Video_i$ includes VideoS, VideoW, VideoR; $Period_j$ includes Period1, Period2, Period3; *Sequence* includes Sequence1, Sequence2, Sequence3; *Gender* includes GenderMale and GenderFemale.

4.3 Data Analysis

As the graph suggests, participant response accuracy was highest in the running category, followed by standing and walking scenarios, respectively.

Using Bayesian Linear Mixed Model, such result was yielded from the data model with a 95% credible interval ranging from 2.5% to 97.5% (Table 1).

Table 1. Multiple-factor credible interval

	.lower	.upper	.width	.point	.interval
VideoS	−18.0	13.9	0.95	median	qi
VideoW	−2.75	28.1	0.95	median	qi
Period2	−22.3	9.85	0.95	median	qi
Period3	−24.3	7.85	0.95	median	qi
Sequence2	−23.6	12.7	0.95	median	qi
Sequence3	−22.5	15.3	0.95	median	qi
GenderMale	−31.1	−0.44	0.95	median	qi

In this case, VideoR, Period1, Sequence1 and GenderFemale are used as baselines. The most plausible values with higher probability of representing the true estimate indicate that the mean of the intervention group VideoS and VideoW should be either lower or higher compared to the comparison group VideoR. As 0 lies within the interval, we do not have statistically significant evidence to claim that there is difference between VideoS, VideoW, and VideoR.

Credible interval for Period2 and Period3 contains 0. This indicates we do not have statistically significant evidence to claim that there is difference between Period1, Period2 and Period3. Accordingly, credible interval for Sequence2 and Sequence3 contains 0. This indicates we do not have statistically significant evidence to claim that there is difference between Sequence1, Sequence2, and Sequence3, either.

However, GenderMale has both negative lower bound and upper bound which does not contain 0. Thus, *Gender* turned out to be significant factor in this data model. Two interaction plots regarding $Video_i$ and *Gender* were generated after this interesting finding as shown in Fig. 5. In the first plot, it shows that male participants always had shorter response time than female participants across all the three video types, especially in Standing and Running Scenario. In the second plot, female participants tended to have lower variance while male participants had a higher variance. However, both genders performed worst in Walking Scenario.

4.4 Results

Results from the data analysis showed that the time participants took to identify two identical characters in the crowd were not significantly affected by different locomotion categories. Hence, we failed to reject the null hypothesis. There

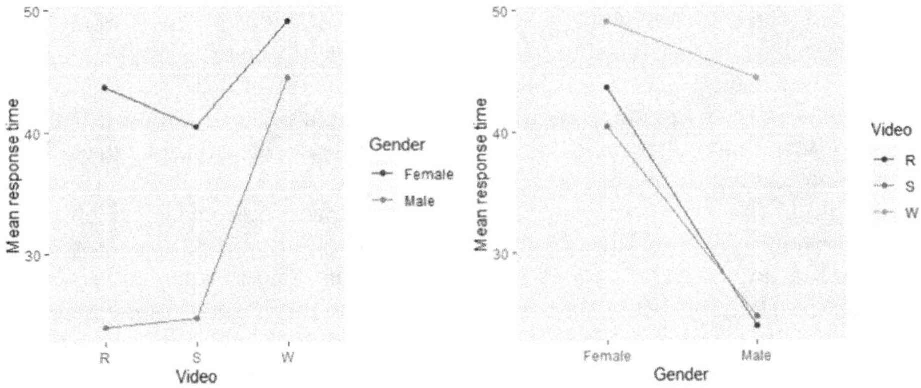

Fig. 5. Visualization of response time per gender.

was no significant difference in reaction time across the three different crowd animation scenarios. However, gender had a significant effect on participants' perception of identical characters within the crowd. Male viewers tended to be able to spot identical characters quicker than female viewers. In the three types of scenarios, male and female viewers had smaller difference in Walking Scenario while they had major difference in Standing and Running Scenario.

5 Discussion and Future Work

The results of the experiment reported in the paper indicate that the type of characters' locomotion (e.g., standing, walking, or running) in a medium size crowd animation has no significant effect on the audience's perception of the crowd. In particular, the type of locomotion exhibited by the characters in the crowd scenario does not significantly impact the time people take to spot identical characters. Findings also suggest that the gender of the participants has an impact on perception of crowd animation, with male participants being able to spot identical characters in a crowd more quickly than female participants.

The findings of the study have important practical implications for animation production. They can help animators save time and resources by optimizing the number of different human templates that are necessary in certain scenarios of crowd simulation (e.g., stadium audience, city street pedestrians, architecture visualization, etc.).

This study had several limitations and potential confounds which could be overcome in future experiments. First, a power analysis was not performed prior to the actual experiment. The researcher used as much of the subject's background characteristics and demographics in the design and analysis of the study to obtain as much power as possible under the circumstances. The pool of subjects was fairly representative of the target population, as it included participants of different ages ranging from 18 to 54 years old and with a wide range of animation experience.

Second, the position of the characters in the crowd at any given moment of time might have had an effect on participants' perception. For example, identifying two identical characters that happened to be running close to each other may have been easier than if the characters were far apart. Thus, distance between two identical characters could have been a significant factor that affected the perception in such scenarios.

Third, all the shots were static without any camera movement, which is not always true in real world films. In a case with camera movement (e.g., a top-down view with a dolly shot), the audience might not be able to focus on a specific area. Hence, the probability that viewers spot identical characters may be lower.

Fourth, the videos used in this experiment were quite rudimentary and considerably lower in quality compared to real-world commercial film productions. Visual fidelity was relatively low due to quality of character texture assets and lack of surrounding environment. The videos also lacked elements used in compositing such as smoke, fog, haze, dust, and flares - all of which are inevitably present in the real world. Further, all the testing scenarios did not include any 3D objects which might become blockers (e.g., buildings, poles, signs), but only an open space on a flat ground. As a result, the audience might be able to perceive identical characters more quickly and easily in our study as compared to real-world animated films.

Fifth, a phenomenon known as the learning effect might have also played a role in this experiment. Participants might have been able to achieve better results with more and more familiarity with the testing procedure in a short period of time. The researcher used randomization to mitigate this effect. A demo video was given at the beginning of the study, so participants could become familiar with spotting identical characters before conducting the actual experiment.

Finally, viewers' perception of the characters might have been affected by the intrinsic design features of the characters, in addition to our variable of interest (locomotion). For example, it is known that human eyes are more sensitive to certain colors of the visible spectrum (e.g., solid red and yellow) than to others, and so participants' response times might have been affected by the different colors of the characters.

In future experiments, characters' motion paths could be varied to exhibit different trajectories. For example, all the characters could be running towards the same target, or all of them could be running around in a loop. It would be interesting to see whether the moving path of the crowd as a whole would affect viewers' perception of identical characters.

In addition, certain camera angles, such as the absolute top view, could make it very difficult to spot identical characters. The difficulty of perception would also depend on the distance between the rendering camera and the characters. Further, it would be worthwhile conducting research on crowd perception under moving cameras.

Future experiments could also diversify characters' appearance, so that differences in skin color, gender, body shape, and other variables can be included and their effects on audience perception could be analyzed. Characters could also be made to wear glasses, hats, and other accessories to investigate their effects on viewers' perception.

References

1. Thalmann, D., Musse, S.R.: Crowd Simulation, 2nd edn. Springer, London (2013). https://doi.org/10.1007/978-1-84628-825-8
2. Zell, E., Zibrek, K., McDonnell, R.: Perception of virtual characters. In: SIGGRAPH 2019: ACM SIGGRAPH 2019 Courses, vol. 21, pp. 1–17 (2019). https://doi.org/10.1145/3305366.3328101
3. Dong, Y., Peng, C.: Real-time large crowd rendering with efficient character and instance management on GPU. Int. J. Comput. Games Technol. **2019**, 1792304 (2019). https://doi.org/10.1155/2019/1792304
4. Zell, E., et al.: To stylize or not to stylize? The effect of shape and material stylization on the perception of computer-generated faces. ACM Trans. Graph. **34**, 1–12 (2015). https://doi.org/10.1145/2816795.2818126
5. Lemercier, S., et al.: Realistic following behaviors for crowd simulation. Eurographics **31**, 489–498 (2012). https://doi.org/10.1111/j.1467-8659.2012.03028.x
6. Reynolds, C.W.: Flocks, herds, and schools: a distributed behavioral model. Comput. Graph. **21**(4), 25–34 (1987). https://doi.org/10.1145/37402.37406
7. Yu, Q., Terzopoulos, D.: A decision network framework for the behavioral animation of virtual humans. In: Metaxas, D., Popovic, J. (eds.) Eurographics/ACM SIGGRAPH Symposium on Computer Animation, pp. 119–128 (2007). https://doi.org/10.5555/1272690.1272707
8. Guy, S.J., Kim, S., Lin, M.C., Manocha, D.: Simulating heterogeneous crowd behaviors using personality trait theory. In: Bargteil, A., Panne, M. (eds.) Eurographics/ACM SIGGRAPH Symposium on Computer Animation, pp. 43–52 (2011). https://doi.org/10.1145/2019406.2019413
9. Sun, L., Li, X., Qin, W.: Simulating realistic crowd based on agent trajectories. Comput. Anim. Virtual Worlds **24**, 165–172 (2013). https://doi.org/10.1002/cav.1507
10. Carucci, F.: GPU Gems 2, pp. 47–67. Addison-Wesley, Boston (2005)
11. Ashraf, G., Zhou, J.: Hardware accelerated skin deformation for animated crowds. In: Cham, T.-J., Cai, J., Dorai, C., Rajan, D., Chua, T.-S., Chia, L.-T. (eds.) MMM 2007. LNCS, vol. 4352, pp. 226–237. Springer, Heidelberg (2006). https://doi.org/10.1007/978-3-540-69429-8_23
12. Peng, C., Park, S.I., Cao, Y., Tian, J.: A real-time system for crowd rendering: parallel LOD and texture-preserving approach on GPU. In: Allbeck, J.M., Faloutsos, P. (eds.) MIG 2011. LNCS, vol. 7060, pp. 27–38. Springer, Heidelberg (2011). https://doi.org/10.1007/978-3-642-25090-3_3
13. Klein, F., Spieldenner, T., Sons, K., Slusallek, P.: Configurable instances of 3D models for declarative 3D in the web. In: Proceedings of 19th International ACM Conference on 3D Web Technologies, Vancouver, pp. 71–79 (2014). https://doi.org/10.1145/2628588.2628594
14. Maciel, P.W.C., Shirley, P.: Visual navigation of large environment using textured clusters. In: 1995 Symposium on Interactive 3D Graphics, pp. 95-ff (1995). https://doi.org/10.1145/199404.199420

15. Tecchia, F., Chrysanthou, Y.: Real-time rendering of densely populated urban environments. In: Péroche, B., Rushmeier, H. (eds.) EGSR 2000. Eurographics, pp. 83–88. Springer, Vienna (2000). https://doi.org/10.1007/978-3-7091-6303-0_8
16. Tecchia, F., Loscos, C., Chrysanthou, Y.: Image-based crowd rendering. IEEE Comput. Graph. Appl. **22**, 36–43 (2002). https://doi.org/10.1109/38.988745
17. Ciechomski, P.H., Schertenleib, S., Maïm, J., Maupu, D., Thalmann, D.: Real-time shader rendering for crowds in virtual heritage. In: Mudge, M., Ryan, R.N., Scopigno, R. (eds.) The 6th International Symposium on Virtual Reality, Archaeology and Cultural Heritage, pp. 1–8 (2005). https://doi.org/10.2312/VAST/VAST05/091-098
18. Millan, E., Rudomin, I.: Impostors and pseudo-instancing for GPU crowd rendering. In: Proceedings of 4th International Conference on Computer Graphics and Interactive Techniques in Australasia and Southeast Asia, Kuala Lumpur, pp. 49–55 (2006). https://doi.org/10.1145/1174429.1174436
19. Green, R.D., MacDorman, K.F., Ho, C., Vasudevan, S.: Sensitivity to the proportions of faces that vary in human likeness. Comput. Hum. Behav. **24**, 2456–2474 (2008). https://doi.org/10.1016/j.chb.2008.02.019
20. Johansson, G.: Visual perception of biological motion and a model for its analysis. Percept. Psychophys. **14**, 201–211 (1973). https://doi.org/10.3758/BF03212378
21. McDonnell, R., Larkin, M., Dobbyn, S., Collins, S., O'Sullivan, C.: Clone attack! Perception of crowd variety. ACM Trans. Graph. **27**, 1–8 (2008). https://doi.org/10.1145/1360612.1360625
22. Pražák, M., O'Sullivan, C.: Perceiving human motion variety. In: Proceedings of ACM SIGGRAPH Symposium on Applied Perception in Graphics and Visualization, pp. 87–92 (2011). https://doi.org/10.1145/2077451.2077468
23. McDonnell, R., Larkin, M., Hernández, B., Rudomin, I., O'Sullivan, C.: Eye-catching crowds: saliency based selective variation. ACM Trans. Graph. **28**, 1–10 (2009). https://doi.org/10.1145/1531326.1531361
24. Schwind, V., Jäger, S.: The uncanny valley and the importance of eye contact. In: Mensch und Computer 2015 - Tagungsband, pp. 153–162 (2015). https://doi.org/10.1515/9783110443929-017
25. Westfield Stirling short film. https://vimeo.com/317897403

Information Presentation in Autonomous Shuttle Busses: –What and How?

Markus Linnartz(✉), Yasmin Dufner, and Nicola Fricke

Transportation Management, Hochschule Karlsruhe University of Applied Sciences, Karlsruhe, Germany

{markus.linnartz,nicola.fricke}@h-ka.de

Abstract. This paper addresses what kind of information users need when driving in an autonomously shuttle and how this information is communicated. This was investigated in two studies with participants in the age-range of 23–25 years using online focus groups. Results showed that both groups rely on the "safety driver" because it supports the feeling of security. Concerning the possibilities of transmission via different human-machine-interfaces, the participants agreed in both studies that the type of information and its transmission should be similar to that used in today's public transport. Differences between the two studies arose in the discussion about the presentation of technical information. One group preferred that technical information, including the explanation of how the shuttle works and real-time sensor data of what the autonomous shuttle is detecting, be shown by default. On the contrary, the other group only preferred this information on request by the passengers. Furthermore, participants explained that such information could increase insecurity as it could be too detailed and might overwhelm passengers. Both groups agreed that providing some extra information for reducing concerns is helpful. One aspect for overcoming negative feelings in the shuttle was the idea that more infotainment options, such as showing Points of Interest, can elicit positive feelings during the ride and this in turn can decrease potential fear or trust issues with autonomous shuttles.

Keywords: Autonomous shuttle · Human-machine-interfaces · Passenger information · Trust

1 Introduction

Research on information presentation in automotive user-interfaces of highly-automated, privately used vehicles has a long history [1]. Influential factors for the acceptance of automated vehicles, such as trust in technology that lead to acceptance of the systems [2], have already been identified. Currently, autonomously driving shuttle buses are introduced in public transportation. While the technological implementation already allows testing in specific regions [3], the conditions of operation require that a "safety-driver" is always present for intervening in specific situations and that vehicles drive very slowly. Field tests showed slightly positive feelings of safety towards autonomously

M. Wölfel et al. (Eds.): ArtsIT 2021, LNICST 422, pp. 413–423, 2022.
https://doi.org/10.1007/978-3-030-95531-1_28

driving shuttles when a safety driver is present [4–6]. Only one study showed a decreased level of acceptance [7] compared to a human-operated bus.

The safety driver was perceived as a positive factor in various studies [6–9]. However, since the specific conditions (safety-driver, slow vehicle speed) will change in the future, trust in the technology and autonomous shuttles might be lower and lead to decreases in acceptance. Possible countermeasures may include presenting passengers with more information about the operations of the shuttle or other relevant aspects with the help of different human-machine-interfaces (HMI) [10]. Users must feel comfortable in the shuttle and be able to trust the technology of the shuttle to feel safe [8]. Therefore, expectations and requirements of potential passengers concerning information presentation in future autonomous shuttles were investigated in two studies and the results are reported in this research paper.

2 Study Design

2.1 Research Questions

For the conducted studies a fully autonomous driving shuttle without safety-driver was assumed and verbally introduced to the participants. To investigate expectations and requirements from potential passengers, two studies were conducted in May 2021. These two studies occurred independently from one another. Based on addressing similar topics, their results are combined in this paper. Since the studies apply slightly different approaches, as described below, the two studies are hereinafter referred to as Study A or Study B. Research questions that are addressed by both studies and can be evaluated similarly are the following:

1. Do potential users of autonomous shuttles need specific information from an internal human-machine-interface (HMI)?
2. What information do the passengers need during an autonomous shuttle ride?
3. How should this information be displayed via HMIs?
4. Which information should be given in case of a problem with the shuttle?

In this context, it is interesting to determine what information can improve the users' feeling of safety within the shuttle during the ride and how this information can be transmitted via HMIs. The following additional research questions are specifically related to Study A:

- What are the reasons for concerns using autonomous shuttles?
- Can information reduce concerns about autonomous shuttles?

Study B also poses an additional research question:

- Which information should be continuously displayed, and which information only on request?

2.2 Materials and Methods

These questions are investigated in two online focus groups. One advantage of doing focus groups is that new creative ideas can be generated collaboratively, which might have remained hidden in individual interviews [11]. Through the group dynamics, ideas can be further discussed, extended, and directly evaluated by several potential users. Since the goal of the two studies is to get an impression of the users' requirements for the information in future autonomous shuttles, this was selected as the preferred research method. While it is true that focus group studies are not representative, they generate new ideas and findings for topics that are under-researched so that they can be further investigated in future research using quantitative methods. Of course, the implications of this study are therefore limited to the investigated sample. Overall, this research focuses on idea-generation aiming for new ideas, wishes and requirements of individual potential users.

Both group discussions took place online and in German using the platform Zoom [12]. All participants joined the online conference with audio and video. A PowerPoint presentation with prepared questions, images, and video materials from an autonomous shuttle was used to guide the group through several topics. Mural [13] was used as a digital bulletin board in which all participants interacted simultaneously to brainstorm ideas during the group discussion phase. For analyzing the results (i.e., coding the transcripts), the software MAXQDA [14] was used.

2.3 Participants

Focus Group A. The five participants in the focus group of Study A are aged between 23 and 25 and all live either in Stuttgart, Germany or in Karlsruhe, Germany. Four of them are female, one male. They mainly use public transportation, cycling or walking as a primary mode of transportation. In addition, all of them have completed a bachelor's degree in the various fields of mathematics, transportation management or public administration. Three of them are currently master's students and two participants have a full-time job. Two of the five participants have already used an autonomous shuttle in the past.

Focus Group B. The focus group in Study B also consists of five participants aged between 23 and 25. All five participants are male and live in Stuttgart, Germany or in Karlsruhe, Germany. They mainly use bicycle, public transportation, and cars as their primary modes of transportation. In addition, all of them are students of the study program transportation management. Four of the five participants have already ridden an autonomous shuttle and all five participants had some experience with autonomous vehicles.

2.4 Procedure

The online focus groups, lasting approximately two hours, were conducted in May 2021. In the two groups, the participants discussed the following aspects:

1. Introduction: This phase was the same for both groups. The participants introduced themselves to each other and were familiarized with the topic with the help of pictures and videos. In the videos, a shuttle from Monheim, Germany drives through a roundabout and through a residential street. The pictures show the interior of the shuttle without the displays, so that the participants are not already focused on the displays.
2. Experiences: During the next phase of both studies, everyone could share their experience with autonomous shuttles. Additionally, group A participants were confronted with the statement that some people have concerns about autonomous shuttles and were asked about their opinion about the underlying reasons.
3. Required Information: Study A participants were asked about what information they would need from an autonomous shuttle if there would be no safety driver on board which they could ask. They were presented with a scenario in which they are in a foreign city and want to get to a tourist attraction, and they know that an autonomous shuttle services the route. After this short introduction to the scenario, they had to use the Mural tool to cluster the different aspects which they identified into categories relating to the time at which the information should be provided. The participants of Study B also sorted relevant information into clusters but were introduced to another potential scenario of going from home to a supermarket. Another topic of Study B was how frequent information should be presented.
4. Information in problematic situations: In both groups, a different situation was introduced in which the shuttle behaves unusually for reasons that are not obvious to the passengers. Participants of Study A were confronted with the problem that the shuttle drives very slowly. An example video was shown how the autonomous shuttle in Monheim, Germany drives very slowly for no apparent reason and cars overtake the shuttle. Additionally, the scenario of a shuttle which brakes suddenly for reasons unknown to the passengers was introduced verbally and the participants had to discuss which information they expected to receive in such situations. The scenario with the sudden braking was also introduced to the participants of Study B.
5. Technical transmission: In addition to the discussion about the content of the information, participants of both studies were also asked about their ideas for transmitting the information.
6. Ending: At the end of the group discussion, Study A participants had to discuss if they think autonomous shuttles would be available in the future and whether they think the discussed information can help to reduce possible concerns against driving in an autonomous shuttle. Study B participants were encouraged to summarize the discussed topics in a short questionnaire. As an example, one of these questions was: Which of the discussed information would you like to receive continuously?

Proceedings from both focus groups were separately recorded and transcribed. Finally, through coding, the individual statements were categorized into different topics. The results of the coding and categorization into the topics are described in the following section.

3 Results

Since the described studies are qualitative in nature, they do not allow for inferential statistical analyses and therefore the results are also qualitative.

3.1 Reasons for Concerns and Insecure Feeling

In general, the participants of Study A have a positive attitude towards autonomous shuttles, even if they have some concerns. They think that the technology is not yet fully developed and that it would feel strange to sit in such a vehicle without a driver. They also note that they cannot imagine riding in an autonomous shuttle, especially in a busy city center with a lot of traffic, because they would not feel safe. Similarly, the participants of Study B say that the safety driver had to intervene in many situations during their trips with the autonomous shuttle. When asked why they think others have concerns about autonomous shuttles, participants of Study A mention the loss of control and that the passengers should have total trust in the technology. In their opinion, this aspect can be counteracted by driving with the safety driver present at the beginning or by weekly test drives. Even if they like the presence of the safety driver at the beginning, they think that when the safety driver is no longer on board, they do not want too much specific information about the technology of the shuttle.

3.2 Required Information

Concerning the question of which information the users need during the ride with an autonomous shuttle, three different types can be distinguished: basic, technical, and supplementary information. In the following, the results are clustered into these three categories.

Basic Information. For both groups, the main focus concerning basic information is on route information. They expect to see the planned route with the next stops of the shuttle. Discrepancies between the displayed route and the actual route in a situation where, for example, the planned route is unexpectedly closed, and the shuttle therefore detours, should be avoided because this could create insecurity. In this case, the participants of Study A request real-time information and display of the modified route. Other requested basic information of both groups is related to the arrival and travel times, including possible delays, the ticket prices, the rules of conduct inside the shuttle, transfer options to other modes of transport, and the current time and date. Information on changing trains should also be available during the ride based on the participants responses. Additionally, the participants request this basic information continuously during the ride. The participants of Study A specifically state that the basic information should be similar to the information currently given by public transportation systems. They explain further that they do not want to feel as if they are in a special vehicle. Instead, they prefer the feeling as if they are travelling on a normal public bus. For example, one participant of Study A said: "I think it helps people to feel a bit safer, to feel normal, when you see something like that [Information] there, because you already know it from other modes of transport" (translated from German).

Technical Information. Technical details about the autonomous shuttle should be explained, according to the participants of Study A, to non-experts in such a way that they can be easily understood by everyone and thus can also increase trust in the technology. For example, it should be shown what the sensors can detect, how fast the shuttle is travelling, etc. Accident statistics should also be shown, so that people can see that the shuttles are much safer than other modes of transport. All information should be easily available for all passenger groups including older passengers, who may not have a smartphone or be able to use a touchscreen. According to the participants of study B, the technical information should help to better understand why the shuttle slows down or even stops. Participants also request some kind of feedback for the user in the form of symbols. They would like to know which events and situations the shuttle is reacting to, including the information that the shuttle detected a person on the street and therefore slows down.

Participants from Study A state that advanced information during the ride should only be presented in such a way that interested users can inform themselves about autonomous shuttle technology. However, the other passengers who do not want this information should not be confronted with it, because this may increase concerns rather than reduce them. One participant said: "For those who want more information, more information should be made available, but not all of them should be flooded with this information. Because I think that this can also result in insecurity, if you get in and first think: 'Okay, wow I don't know what to do.' Then maybe I'd better get out again" (translated from German). On the contrary, participants of Study B say that they would like to have the information always shown, and therefore they can always get the information if they need it. This can be seen in one statement of a participant in Study B: "It does not hurt to always display it and if you need the information, you can pick it up. If not, then you just do not have to look at it (...)" (translated from German).

However, in the opinion of participants from Study A, real-time data, such as a visualization of what the sensors are currently detecting, should never be available, as this would make passengers focus too much on this information. In this case, the passengers might look for errors in the system behavior, or immediately detect the smallest mistakes of the system, which would increase the potential for fear and reduce the trust in the shuttle. This result is completely different to the opinions of Study B participants who wish for real-time data of sensors and additional information such as the speed of the shuttle.

Supplementary Information. In addition to the technical data of the shuttle, the participants of both studies would like to have tourist information or information about the ride and the route. This would also help to bring the passengers' attention to a positive topic and make autonomous driving as normal as possible. Tourist information, such as Points of Interest, should be shown. Here, the participants of Study B also add the occupancy of the Points of Interest, shown for example on a display in the shuttle, as interesting information that could be transmitted to know off-peak times to visit the supermarket or gym.

In general, according to Study A participants, the ride should be a positive event and therefore information that elicits positive feelings and possibly distracts from potential insecurities should be presented. The reasoning is that the passengers have positive feelings concerning infotainment options and this in turn can decrease negative thoughts about autonomous shuttles. The participants of Study B mention that the comparison with the infotainment display from light rail vehicles sums up well what kind of entertainment information the participants would also like to see in this case.

Participants of Study A prefer similar information to that currently displayed in public transportation vehicles so that autonomous shuttles do not feel different from a normal bus. They favor a display with information about the planned route, the next stops, the time to the next stops, the transfer possibilities, and information about important Points of Interest. Participants from Study B additionally ask for information about the date and time and the current speed of the shuttle. However, both groups would like to see a map instead of the currently displayed line path with the expected route.

3.3 Display via Human-Machine-Interfaces (HMIs)

The participants of both groups prefer displays for presenting basic information to the passengers. In their opinion, these displays should show the current route, the planned stops, the current speed, time and date, and transfer options to other modes of transport. The participants of Study A compare this to the screens in the Stuttgart Underground or Karlsruhe Light Rail and state that the information shown in autonomous shuttles should be presented similarly so that the passengers do not feel like they are sitting in a special vehicle. Furthermore, the supplementary information like place and description of Points of Interests or news should be shown on a display. Participants of Study B prefer seeing the real-time sensor data on a head-up display integrated into the front windshield of the shuttle.

Information via loudspeakers is mainly requested by the participants of Study B. They imagine auditory information at the beginning of the ride, like a start signal tone or information about still missing passengers, and the transfer possibilities via loudspeakers. The participants of Study A on the contrary do not prefer audio information, except in the event of more severe technical problems.

As described in the previous chapter, the technical information requested by the participants of Study A should only be available on request from a passenger. They think this information could be displayed via a QR Code, which is placed in the shuttle and redirects to the information page from the shuttle operator. A built-in touchscreen is another idea so that the passengers can search for the technical and other basic information, which the passengers specifically request. For older passengers, the technical details should be available on paper flyers inside the vehicle.

Other HMIs that help to reduce concerns in the opinion of the participants of Study A are, for example, emergency stop buttons that can be used to get off at the next possible opportunity and an intercom station that connects you to the control center. If possible, a "real person" should be on the other end of the line. This would ensure easy communication in case of insecurity. Again, it is important for the participants that the way in which this information is displayed is not very different from normal buses.

Summarizing the findings obtained in both studies are displayed in Table 1. Each type of information is divided into the technical requirements and in which study it was mentioned.

Table 1. Required information and transmission

Type of information	Technical requirements	Study
Route	Display	A I B
Travel- and arrival time	Display	A I B
Ticket price	Sign/Display	A
Rule of conduction	Sign	A
Transfer possibilities	Display	B
Start information	Display + Auditory	B
Date and time	Display	B
Technical information	QR-Code, touchscreen, flyer	A
Real-time sensor data	Symbol, Display	B
Real-time speed	Display	B
Traffic situation	Display	B
Points of interest	Display	A I B
News	Display	A

3.4 Information in Case of a Problem

In the case of smaller problems such as driving with slow vehicle speed, the participants of Study A prefer a message on the display that communicates something positive such as "we are currently driving with increased attention". Moreover, this information should not be too obvious and should not give the impression that something is wrong. This aspect is similar to the ideas from the previous chapter in which the participants prefer discreet information presentation. Participants of Study B prefer real-time sensor data to know why the shuttle slows down.

In the case of larger problems such as suddenly heavy braking followed by a complete stop, the participants of both groups expect more detailed information about the reason and whether and when the journey will continue. They prefer in that case announcements via the loudspeakers with an explanation what has happened, a forecast if and how long the disturbance will exist and instructions on what to do, as it is common in many trains nowadays. This information also has to appear on the display. Participants of Study A mention specifically that the person speaking via the loudspeakers should be a real human to increase the feeling of safety as it might make passengers more insecure if announcements sound like a machine.

3.5 Reducing Concerns

Participants of Study A believe that flooding the passengers with information will not reduce the concerns about autonomous shuttles. They think in addition to the basic information, some positive information should be available on request, which shows that autonomous driving is safer than riding a normal public bus nowadays. Study B participants in contrast think that too much specific information, like showing real-time sensor data on a display, is important for the passengers' feeling of safety. Therefore, it can be seen that the preference for the type of information which should be given to the passengers to reduce concerns is different in both studies even though both groups think that some selected extra information is helpful.

3.6 Continuous Information

As described in Subsect. 2.1, Study B also explored what information participants expected to receive continuously during the ride with an autonomous shuttle. This category would include information about the speed of the shuttle, the arrival time, the route, and information about Points of Interest and transfer possibilities. Route information and arrival time information is expressly mentioned and discussed several times by different participants. The study by Mirnig et al. [10] shows that the display of stops and the planned route is an important issue for information transfer in autonomous shuttles and there are different ways to design this. The participants of Study B explain these requirements by saying that they have a plan after arriving at the desired target or use it for certain source-target connections and that it would therefore be particularly important to know when they would arrive at their target and to know with which route the shuttle would try to reach the target.

3.7 Summary

In summary, the participants of the focus groups prefer similar basic information. This information should be presented using a display inside the autonomous shuttle, similar to the implementations in today's public transport. Furthermore, the participants of both studies expect supplementary information like Points of Interest. Differences between the two focus groups arose in the discussion about the presentation of technical information. The participants of Study A prefer that the information is only presented at the request of passengers. The participants of Study B on the contrary, favor that the information is shown the whole time. They also wish to see real-time sensor data, which participants of Study A believe will increase the concerns about autonomous shuttles.

4 Discussion and Outlook

Some results of the focus group studies are in line with other research findings. First, most participants state that they like the "safety driver" as it supports the feeling of security. This is similar to other studies [4, 6–9]. Nevertheless, the results of Study A suggest, that users do not prefer to be flooded with a lot of extra information. They prefer to have basic information, like the planned route and some supplementary information that give a positive driving experience and increase trust towards the shuttle. A similar aspect is also described in another study which found that the information needs to be easily accessible, but shall not be disturbing [15]. This matches the findings of this study in which the participants believe that the technical details and specifics should be provided in an easily-accessible way for all age groups that does not overload the passengers. For this reason, Study A participants believe this kind of information should only be given on request and ought not be provided continuously. On the contrary, the participants of Study B are interested in seeing real-time technical information about the shuttle (e.g., what the shuttle currently detects). Moreover, they think that such information would lower their concerns about autonomous shuttles. Participants from Study A do not wish that this information be displayed permanently. Study A participants only support providing this information to interested passengers on request (e.g., with an application on the smartphone). They explain that they prefer not to be reminded about the driverless shuttle as this would lead to lower trust. Instead, one can interpret that they pretend to be in a "normal" shuttle in order to feel more secure. This finding is partially in contrast to previous research such as [2], who conclude that all automated systems should be transparent about their system status in order to increase trust. We must acknowledge, however, that the reported studies A and B were focus groups with only five participants each and do not allow for generalizations. Any causal relationships would need to be tested in subsequent experiments.

An interesting aspect is the idea that providing entertainment features could lead to positive feelings. This was explained as possibly counteracting potential insecurities or trust issues. Another option could be using a social agent to interact with the passengers, compensating for the missing "safety driver". This could be especially helpful since the participants wished for a "human" in the shuttle (in the case of the helpline). A social agent has been found to be helpful in a study of automated vehicle driving because it increased trust in the automated driving system [16]. It would be interesting in further research to investigate whether increasing the anthropomorphic features is especially helpful in this context for mimicking an actual person.

Since the findings from the two focus groups are only an early starting point of researching which information presentation and new technological features can support trust and acceptance of autonomous shuttles, they surely cannot be generalized to the entire population. One problematic aspect of this and similar studies is that users who have not used the innovative new technologies under investigation, tend to stick to aspects which they already know. This became apparent specifically in Study A in which the participants prefer to "pretend" that they are using a normal bus and prefer the same information presentation as the one they are used to. If Study A had more participants with previous experiences riding in autonomous shuttles (such as in Study B) the results could have been very different. Furthermore, a greater variety of age groups shall be addressed in further studies since the groups in both studies were very homogeneous, that means all of them were young people who have graduated from university in the last few years. Other focus group participants may have different ideas about required information and transmission via HMIs, so conclusions drawn from this study are not generally valid for other focus groups of this topic. For this reason, future studies should incorporate specific social groups, such as the elderly, to get a better sense of the variety of preferences.

Acknowledgements. We would like to thank all participants of both studies for their committed and enthusiastic participation in the group discussion. Only with the help of the participants could the described findings be obtained.

References

1. Helldin, T., Falkman, G., Riveiro, M., Davidsson, S.: Presenting system uncertainty in automotive UIs for supporting trust calibration in autonomous driving. In: Proceedings of the 5th International Conference on Automotive User Interfaces and Interactive Vehicular Applications (AutomotiveUI 2013), Eindhoven, Netherlands, 28–30 October 2013, pp. 210–217. ACM, New York (2013)
2. Lee, J.D., See, K.A.: Trust in automation: designing for appropriate reliance. Hum. Factors **46**, 50–80 (2004)
3. Riener, A., Appel, A., Dorner, W., Huber, T., Kolb, J.C., Wagner, H. (eds.): Autonome Shuttlebusse im ÖPNV. Springer, Heidelberg (2020). https://doi.org/10.1007/978-3-662-59406-3
4. Friebel, P.: Fahrgastbefragung der Linie 708. Zwischenstand zur Akzeptanz eines automatisiert fahrenden Kleinbusses in Wusterhausen/Dosse (2019)
5. Schäfer, P., Altinsoy, P.: Autonom am Mainkai. Nutzerakzeptanz und betriebliche Herausforderungen autonomer Shuttles in Frankfurt am Main. Frankfurt University of Applied Sciences - Research Lab for Urban Transport, Frankfurt am Main (2021)
6. Zankl, C., Rehrl, K.: Digibus 2017. Erfahrungen mit dem ersten selbstfahrenden Shuttlebus auf öffentlichen Straßen in Österreich (2018)
7. Salonen, A.O.: Passenger's subjective traffic safety, in-vehicle security and emergency management in the driverless shuttle bus in Finland. Transp. Policy **61**, 106–110 (2018)
8. Mantel, R.: Akzeptanz eines automatisierten Shuttles in einer Kleinstadt Analyse anhand einer Trendstudie und Fahrgastbefragung. J. für Mobilität und Verkehr **19**, 25–35 (2021)
9. Wintersberger, P., Frison, A.-K., Thang, I., Riener, A.: Mensch oder Maschine? Direktvergleich von automatisiert und manuell gesteuertem Nahverkehr. In: Riener, A., Appel, A., Dorner, W., Huber, T., Kolb, J.C., Wagner, H. (eds.) Autonome Shuttlebusse im ÖPNV, pp. 95–113. Springer, Heidelberg (2020). https://doi.org/10.1007/978-3-662-59406-3_6
10. Mirnig, A.G., Gärtner, M., Wallner, V., Trösterer, S., Meschtscherjakov, A., Tscheligi, M.: Where does it go? A study on visual on-screen designs for exit management. In: Proceedings of the 11th International Conference on Automotive User Interfaces and Interactive Vehicular Applications: Adjunct Proceedings (Automotive UI 2019), Utrecht, Netherlands, 21–25 September 2019, pp. 233–243. ACM (2019)
11. Zwick, M., Schröter, R.: Konzeption und Durchführung von Fokusgruppen am Beispiel des BMBF-Projekts "Übergewicht und Adipositas bei Kindern, Jugendlichen und jungen Erwachsenen als systemisches Risiko." In: Schulz, M., Mack, B., Renn, O. (eds.) Fokusgruppen in der empirischen Sozialwissenschaft, pp. 24–48. VS Verlag für Sozialwissenschaften, Wiesbaden (2012)
12. Zoom Homepage (2021). https://zoom.us/. Accessed 31 Oct 2021
13. Mural Homepage (2021). https://www.mural.co/. Accessed 31 Oct 2021
14. MAXQDA Homepage (2021). https://www.maxqda.de/. Accessed 31 Oct 2021
15. Mathis, L.-A., et al.: Creating informed public acceptance by a user-centered human-machine interface for all automated transport modes. In: Proceedings of 8th Transport Research Arena (TRA 2020), Helsinki, Finland, 27–30 April 2020 (2020)
16. Kraus, J.M., Nothdurft, F., Hock, P., Scholz, D., Minker, W., Baumann, M.: Human after all: effects of mere presence and social interaction of a humanoid robot as a co-driver in automated driving. In: Proceedings of the 8th International Conference on Automotive User Interfaces and Interactive Vehicular Applications (AutomotiveUI 2016), Ann Arbor, MI, USA, 24–26 October 2016, pp. 129–134. ACM, New York (2016)

AI Assisted Design of Sokoban Puzzles Using Automated Planning

Tomáš Balyo[1(✉)] and Nils Froleyks[2]

[1] CAS Software AG, Karlsruhe, Germany
tomas.balyo@cas.de
[2] Johannes Kepler University, Linz, Austria
nils.froleyks@jku.at

Abstract. Designing interesting and challenging levels for a puzzle game is a very difficult and time consuming task. It is often possible to develop random puzzle generators that can produce solvable levels. However, in order to obtain appealing levels, usually a human designer needs to be involved. In this paper we propose a new generic method for assisting human designers to create solvable levels for a puzzle game by using Automated Planning. We will demonstrate our method on the well-known Japanese puzzle game Sokoban.

Keywords: Sokoban · Automated planning · Puzzle generation

1 Introduction

Sokoban is a puzzle game that originated in Japan. It was invented by Hiroyuki Imabayashi, and published in 1982 by Thinking Rabbit [11]. The word Sokoban is Japanese for warehouse keeper. Each puzzle represents a warehouse, where boxes are randomly placed. A warehouse keeper (the player) has to push the boxes around the warehouse so that all boxes end up in designated goal positions.

The game of Sokoban is a complicated computational problem. It was first proven to be NP-hard [6] and later PSPACE-complete [3]. While the rules are simple, even small levels can require a lot of computation to be solved. Designing interesting solvable levels is also challenging and a subject of academic research [5,14,15,18].

Automated planning [8] is one of the central techniques in artificial intelligence. The task of planning is to find a sequence of actions, i.e., a plan, that transforms the world from a given initial state to a goal state, i.e., a state that satisfies the given goal conditions. Planning is a very competitive research area and there exist multiple high performance planning tools that are constantly being developed and improved [16].

In this paper we will demonstrate how the power of automated planning tools can be utilized to design a system that can generate challenging solvable puzzles

M. Wölfel et al. (Eds.): ArtsIT 2021, LNICST 422, pp. 424–441, 2022.
https://doi.org/10.1007/978-3-030-95531-1_29

and intelligently assist a human puzzle designer. To our best knowledge, this is the first time, that automated planning has been used in this context.

We will demonstrate our technique on the example of Sokoban puzzles, since the game is widely known and well studied. Nevertheless, the technique can be used for any puzzle game that satisfies the following conditions:

- *Single player.* The game is played by a single player. There may exist helpful or adversary agents in the game as long as their behavior is fully deterministic and specified by simple rules.
- *Finite and discrete game world.* Each game state can be fully described with finitely many finite domain variables.
- *Deterministic gameplay.* Random events or random outcomes of player actions are not allowed.
- *Full observability.* There are no hidden or unknown elements that influence the gameplay.

The rest of the paper is organized as follows. In the next section we will provide the preliminary definitions of automated planning and the rules of Sokoban. Then we will review the related work in the area of procedural generation of Sokoban levels. Following that we will describe our new method and our new tool that implements it. Finally, we will present an evaluation of our tool.

2 Preliminaries

2.1 Automated Planning

As we already briefly stated in the introduction, planning is the task of finding a plan (a sequence of actions) that transforms the world from a given initial state to a goal state that satisfies the goal conditions. How to represent the world states, goal conditions, and describe the set of possible actions is defined in this Subsection.

Planning problems are modeled using the Planning Domain Definition Language (PDDL) [9], which is based on the programming language LISP [21]. PDDL is a very rich language with many features, however, we will only require a small subset of it which we will describe below.

The basic building blocks of PDDL are *Objects* and *Types*. Each object is of a certain type. For example if we define a type "city" then we can define the objects "Paris", "London", and "Madrid" of the type "city". Another type could be "person" and the objects of this type are for example "Alice" and "John". In PDDL we would express this using the following lines:

```
(:types city person - object)
(:objects
  Paris London Madrid - city
  Alice John - person
)
```

In PDDL we can refer to objects using variables. Variable names always start with a question mark "?" and each variable has a type. For example a variable "c" of the type "city" would be declared as: `(?c - city)`.

Variables appear in *Predicates*, which are atomic statements that are used to express certain conditions. For example, a predicate called "livesIn" could have two parameters, one of the type "person" and one of the type "city". In PDDL we would declare this predicate as `(livesIn ?p - person ?c - city)` and it would mean that an object of type "person" lives in an object of type "city". Using the predicate we can now declare facts about our objects by substituting variables with objects of the proper type, for example:

`(livesIn Alice Madrid), (livesIn John London).`

The last building block of PDDL that we need are operators, which can be intuitively understood as templates for actions. Actions change the world state by modifying the truth values of predicates. An action a consists of a name $name(a)$, a set of preconditions $pre(a)$ and a set of effects $eff(a)$. Both preconditions and effects are sets of grounded predicates (predicates where all variables are substituted by objects).

1. Preconditions represent the predicates that must be true in the given world state in order to execute the action. We say that an action a is applicable in a given world state s if and only if all predicates in $pre(a)$ hold true in s.
2. Effects are used to update the world state after the action is executed. Positive effects are predicates that will become true (unless they are already true) after the action is executed. Negative effects are negated predicates (wrapped in *not*) and they become false. All other predicates that are not involved in the effects of the executed actions remain unchanged.

The following is an example of an action representing moving Alice form Madrid to Paris:

```
(:action move-Alice-Madrid-Paris
  :precondition (and
    (livesIn Alice Madrid)
  )
  :effect (and
    (not (livesIn Alice Madrid))
    (livesIn Alice Paris)
  )
)
```

The precondition is that Alice lives in Madrid and the effects are that Alice does not live in Madrid anymore and she lives in Paris. If we wish to model all possible movements for both Alice and John and the three cities, we would need to write down 12 actions that are very similar to each other. A better solution is to use the already mentioned operators, i.e., action templates. Operators look like actions with the difference that they may have parameters and use predicates with variables in the preconditions and effects. An operator for the move actions would be declared as follows:

```
(:action move
  :parameters(?p - person ?from ?to - city)
  :precondition (and
    (livesIn ?p ?from)
  )
  :effect (and
    (not (livesIn ?p ?from))
    (livesIn ?p ?to)
  )
)
```

A planner would then generate all the possible actions from this template by substituting all the possible combinations of objects for the three parameters. This process is referred to as grounding.

Now we have everything we need to fully describe a planning problem in PDDL, which consists of the following elements:

1. set of used types
2. set of predicates
3. set of operators
4. list of all the objects in the problem together with their types
5. the initial state of the world in the form of grounded predicates (predicates with objects substituted for all variables)
6. the goal conditions in the form of grounded predicates

When describing a planning problem in PDDL we split the description into two files: domain.pddl and problem.pddl. The first file, domain.pddl, contains the types, predicates, and operators. The rest is written in the problem.pddl file. For our moving example the domain.pddl would be:

```
(define (domain moving)
  (:requirements :strips :typing)
  (:types city person - object)
  (:predicates
    (livesIn ?p - person ?c - city)
  )
  (:action move
    :parameters(?p - person ?from ?to - city)
    :precondition (and
      (livesIn ?p ?from)
    )
    :effect (and
      (not (livesIn ?p ?from))
      (livesIn ?p ?to)
    )
  )
)
```

and the problem.pddl would contain:

```
(define (problem moving-1)
  (:domain moving)
  (:requirements :strips :typing)
  (:objects
    Paris London Madrid - city
    Alice John - person
  )
  (:init
    (livesIn Alice Madrid)
    (livesIn John London)
  )
  (:goal (and
    (livesIn Alice Paris)
    (livesIn John Paris)
  ))
)
```

The domain file describes the general planning problem of moving people between cities, while the problem file describes the concrete problem instance of moving John and Alice from London and Madrid to Paris. An automated planner would now take these two files and find a plan, which in this case would consist of two actions: `move-alice-madrid-paris` and `move-john-london-paris`.

Since automated planning is a very competitive research field, it is easy to find well performing planning tools that are freely available on the internet. One way to choose a good planner is to look at the International Planning Competition website [16], where state-of-the-art planners are evaluated and compared in regular time intervals.

2.2 Sokoban

Each Sokoban level consists of a two dimensional rectangular grid of squares (see Fig. 2 for an example). If a square contains nothing it is called a floor. Otherwise it is occupied by one of the following entities (see Fig. 1):

- *Wall.* Walls make up the basic outline of each level. They cannot be moved and nothing else can be on a square occupied by a wall. A legal level is always surrounded by walls.
- *Box.* A box can either occupy a goal or an otherwise empty square. It can be moved in the four cardinal directions by *pushing* (see below).
- *Goal.* Goals are treated like floors for the most part. Only when each goal is occupied by a box the game is completed. In a legal level the number of goals matches the number of boxes. For the sake of simplicity, we will call a square that is either a goal or a floor square *free* since the worker and boxes can enter both.

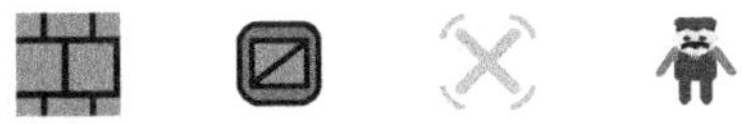

Fig. 1. The four kinds of tiles that make up a Sokoban warehouse: Wall, Box, Goal, and Worker (from left to right).

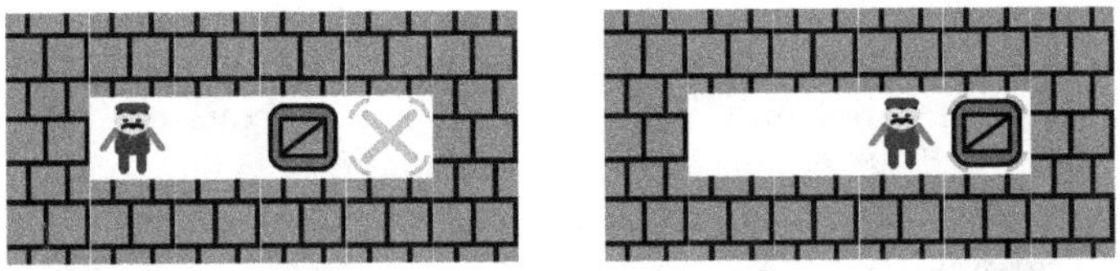

Fig. 2. A simple Sokoban level in its initial (left) and solved (right) state. The solution to this level consists of two steps: MOVE-RIGHT and PUSH-RIGHT.

- *Worker.* There must be exactly one worker in each level. It is the only element that is directly controlled by the player.

There are two kind of moves in Sokoban:

1. *Move the worker.* The worker can be moved in the four cardinal directions (up, down, left, right) by one square in each step. This movement is directly controlled by the player. The worker may be moved onto an adjacent free square.
2. *Push a box.* The worker can push a box in a certain direction if the square behind the box is free. To be precise, there are always three squares (A,B,C) involved in a push move. The first (A) contains the worker, the second (B) contains a box and the third one (C) is a free (empty or goal) square. These three squares must form a single line of adjacent squares. After the push is performed, the box occupies the free square (C) and the worker occupies the square formerly occupied by the box (B).

The goal of the game is to find a *solution*, which is a sequence of moves and pushes. Executing a solution leads to every box ending up on a goal. It does not matter which box ends up on which goal. A level may have no solution. Such a level is undesirable and should not be presented to a human player for obvious reasons.

3 Related Work

Most academic work on Sokoban focused on developing efficient solvers that find short, but not necessarily optimal solutions. Most are based on performing a heuristic search on the state space of the Sokoban puzzle. To make the search efficient multiple domain specific enhancements are used. Most notably, heuristics that recognize if the current state is already unsolvable and abort the branch of the search accordingly. The first solver to implement these techniques was

Rolling Stone [13] followed by *JSoko*[1], *YASS*[2], *Takaken*[3] and *GroupEffort* [7]. Botea et al. [2] used automatic planning but instead of a simple encoding to planning (see Sect. 4.1) they decomposed the warehouse into a set of different rooms connected by tunnels. A plan that successfully moves the boxes between the rooms is translated to actual box pushes and player movements afterwards.

Another topic related to our work was investigated more recently. Assessing the difficulty of a given level is important for designing new ones. Humans enjoy problem solving, but only if the problem is of adequate difficulty. Jarušek et al. [12] conducted an empirical study on how easily humans solve a set of Sokoban levels. They collected over 700 h of test data from different participants to establish a ground truth and presented a set of nontrivial metrics trying to predict the collected data. Ashlock et al. [1] observed artificial agents that were the result of an evolutionary learning process on randomly generated Sokoban levels. Due to the limited capabilities of the agents the metrics they present are not useful to predict the difficulty of harder levels. Van Kreveld et al. [20] developed a metric that is not specific to Sokoban but supposed to be generic enough to capture the difficulty of different grid-based puzzle games.

The first published Sokoban level generator algorithm is by Murase et al. [15]. Their approach has three phases.

1. *Generate random levels.* In this phase predefined templates of rooms are placed randomly over a prototype level consisting of only walls. The templates are placed such that they are connected by passages. Then boxes and goal tiles are placed randomly.
2. *Filter out unsolvable levels.* Phase one may generate levels that have no solution. According to the authors this happens in around half of the cases. In this phase they use a Sokoban solver to try to find a solution and filter out unsolvable levels.
3. *Evaluation.* In this phase the levels are automatically evaluated to determine whether they are interesting. The evaluation is based on simple metrics such as the length of the solution, the number of changes in directions when pushing a box and the number of detours.

The complexity of this approach is dominated by phase two – filtering out unsolvable levels. This step requires solving Sokoban problems, which is a PSPACE-complete problem [4].

The approach of Taylor and Parberry [18] is similar to Murase et al. in that they first generate a random level based on placing templates of walls. Then they randomly place goals with boxes on them in the rooms. At this point they actually have a solved Sokoban puzzle. In the following stage they "unsolve" the level by doing reverse Sokoban moves, i.e., pulling boxes away from the goals. The aim of this stage is to reach a state that is far as possible from the solved state. They do this by running an iterative deepening search of the state space.

[1] https://www.sokoban-online.de/.
[2] https://sourceforge.net/projects/sokobanyasc/.
[3] http://www.ic-net.or.jp/home/takaken/e/soko/index.html.

The complex part of this algorithm is the search for the starting state in the second stage. The process is very memory intensive, since all the visited states have to be kept in memory in order to avoid looping. On the other hand, the algorithm has the anytime property, i.e., it can be stopped at any time to return a valid solution, however, letting it run longer will yield a better solution.

In [19] an auditory Stroop test was performed to compare the engagement of players while playing hand-crafted Sokoban levels against levels generated by the approach of Taylor and Parberry [18]. The experiment showed that players found procedurally generated levels equally interesting to hand-crafted levels. This demonstrates that there is entertainment value in procedurally generated puzzles.

Kartal et al. [14] propose a Monte Carlo tree search (MCTS) based Sokoban level generator. They formulate puzzle generation as an MCTS optimization problem such that the puzzles are generated through simulated gameplay. The search process starts with a level full of walls except for one tile, which contains the player in its start position. The following actions are possible at each node of the search tree:

1. *Remove a Wall.* Choose a wall that is adjacent to an empty tile and remove it. By only removing walls adjacent to empty tiles they can ensure that no unreachable rooms are generated.
2. *Place a Box.* Choose an empty tile and put a box there.
3. *Freeze the Level.* With this action the search is changed to play mode. Removing walls and placing boxes is not allowed after this action. The current positions of walls, boxes and the player constitute the starting state of the level (without any goal positions, they will be defined later).
4. *Move the Player.* Simulate play by executing random legal moves of the player, i.e., walking around and pushing boxes.
5. *Evaluate the Level.* This is the final action of each search path. The current positions of the boxes are declared to be the goal locations and the quality of the generated level is estimated based on data driven evaluation functions.

Similarly to the previously presented method, this generator also has the anytime property. It is capable of producing a wide variety of levels thanks to its stochastic nature. Nevertheless, like all the presented approaches, it has its limitations and the generation of large puzzles remains a bottleneck as the number of possible level designs grows exponentially.

An up-to-date survey on procedural puzzle generation [5] gives an overview of the methods for generating puzzles for many games similar to Sokoban.

4 Puzzle Generation as Planning

Our proposed approach is based on the idea of using automated planners to generate solvable Sokoban levels. This means that our only task is to express the problem of level generation in PDDL and the rest is taken care of by the planning tool. We will formulate the problem of Sokoban level generation as an extension of Sokoban level solving.

4.1 Sokoban Solving as Planning

Using automated planning to solve Sokoban is not a new idea by any means. Actually, Sokoban is one of the standard benchmark problems used to evaluate new planning algorithms and tools in many academic papers and the international planning competition [16]. Nevertheless, in order to keep this paper self-contained, we will present a simple PDDL encoding of Sokoban in this Subsection.

To encode Sokoban solving we will only require one kind of objects – squares. We will have one object of type square for each location in the level that is not a wall. Additionally, we will need the following predicates:

1. `(above ?a ?b - square)` meaning that square"a" is above square "b"
2. `(left_of ?a ?b - square)` meaning that square "a" is on the left side of square "b"
3. `(box_at ?a - square)` meaning that there is a box at square "a"
4. `(worker_at ?a - square)` meaning that there is the worker at square "a"

To complete the domain description we need to specify the operators. We will need two kinds of operators – move and push and we will need 4 of each for the 4 cardinal directions (up, down, left, right). Fist, we will describe the move-up operator:

```
(:action move_up
  :parameters (?from ?to - square)
  :precondition(and
    (above ?to ?from)
    (worker_at ?from)
    (not (box_at ?to))
  )
  :effect (and
    (not (worker_at ?from))
    (worker_at ?to)
  )
)
```

The operators for moving down, left, and right are analogous, they only differ on the line with `(above ?to ?from)` where move-down has `(above ?from ?to)`, move-left has `(left_of ?to ?from)` and move-right has `(left_of ?from ?to)`. Next we describe the push-up operator:

```
(:action push_up
  :parameters (?from ?to ?box_to - square)
  :precondition(and
    (above ?to ?from)
    (above ?box_to ?to)
    (worker_at ?from)
    (box_at ?to)
```

```
    (not (box_at ?box_to))
  )
  :effect (and
    (not (worker_at ?from))
    (worker_at ?to)
    (not (box_at ?to))
    (box_at ?box_to)
  )
)
```

Like in the case of move operators, the other three push operators (push down, left, and right) only differ on the lines with the "above" predicates.

What remains is to specify the initial state and the goal conditions. For the initial state we need to declare the following predicates:

1. `(above a b)` for each pair of non-wall squares such that "a" is above "b".
2. `(left_of a b)` for each pair of non-wall squares such that "a" is on the left side of "b".
3. `(box_at a)` for each square "a" that contains a box.
4. `(worker_at a)` for the square "a" that contains the worker.

As for the goal conditions, we only need to specify that the goal squares must contain a box:

1. `(box_at a)` for each square "a" that contains a goal.

An example of a Sokoban level and its encoding in given on Fig. 3.

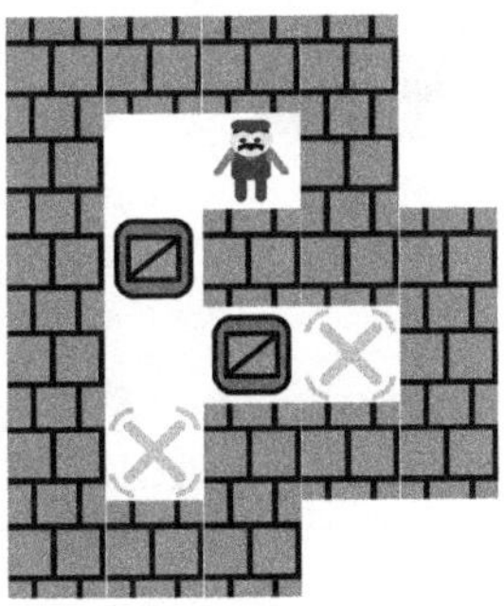

```
(:objects
  s11 s12 s21 s31 s32 s33 s41 - square
)
(:init
  (above s11 s21) (above s21 s31)
  (above s31 s41) (left_of s11 s12)
  (left_of s31 s32) (left_of s32 s33)
  (box_at s21) (box_at s32) (worker_at s12)
)
(:goal (and
  (box_at s41) (box_at s33)
))
```

Fig. 3. A Sokoban level (left) and its encoding in PDDL (right)

4.2 Level Creation as Planning

As we already mentioned in the introduction, our generator is meant to assist a human designer and not just generate fully random levels (as is the case in the related work presented in Sect. 3). Therefore we allow the user to specify the size of the puzzle by defining the outer walls. The goal positions are also set by the designer. Additional walls, boxes and even the worker position may be defined as well. Lastly, the designer specifies which of the remaining free squares may contain a wall, a box, a worker, or some combination of the three. The last thing to define is the number of walls and boxes to be added to the puzzle. If no worker position has been specified in the input then the worker will be added automatically. We will refer to this input as a *level template*. Figure 4 contains an example of a level template, the corresponding starting puzzle and final puzzle.

```
p sokogen 1 2 0   // add 1 wall and 2 boxes, do at least 0 pushes
####              // symbols: # - wall                $ - box
#2.#              //          @ - worker              . - goal
#00###            //          + - worker on goal      * - box on goal
#.333#            // The numbers specify which tokens can be placed:
# 333#            //   0 - any      3 - box              6 - wall or box
#22###            //   1 - worker   4 - worker or wall
####              //   2 - wall     5 - worker or box
```

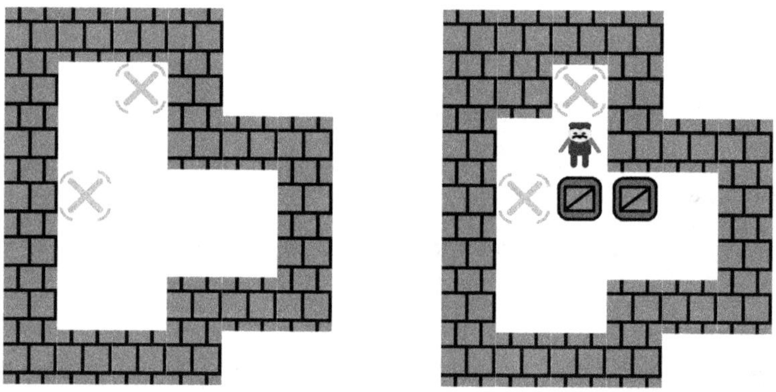

Fig. 4. A level template file for our Sokoban puzzle generator (top), the starting puzzle (down left) and the final solvable level (down right).

Transforming a level template into a solvable level will be task of the automated planner. In order to do this we must model the problem in PDDL. The PDDL model is an extension of the model used for solving Sokoban puzzles that we described in the previous Subsection. We will add four new operators:

1. *Add wall.* This operator adds a wall to one of the free squares that is allowed to contain a wall according to the level template.
2. *Add box.* Like the "add wall" operator, but for adding a box.
3. *Add worker.* Like the previous two but adds the worker.
4. *Start playing.* This operator means that we transition from the level creation phase to the playing phase of the planning problem. No more walls, boxes or workers can be added after this action is executed. Move and push actions are not allowed to happen before this action.

We also need to modify the goal conditions. For Sokoban solving we only required that all goal positions contain a box. Now we also require that the specified number of walls and boxes was placed. To model this we introduce two new types: wall and box. Then we declare as many objects of both types as we need to add according to the level template. For example, if we need to add 3 walls and 5 boxes, then 3 objects of type wall and 5 objects of type box are declared. Then with the help of two new predicates: `(wall_placed ?w - wall)` and `(box_placed ?b - box)` we can encode that all the additional walls and boxes have been placed.

In the initial state we must specify which squares may contain additional walls, boxes, or the player. For this purpose we introduce three new predicates: `(opt_wall ?s - square)` for walls, `(opt_box ?s - square)` for boxes, and `(opt_worker ?s - square)` for the worker.

To implement the *start playing* operator we will define two new predicates: `(making_level)` and `(playing)` to represent the current phase of the puzzle generation. The `(making_level)` is added to the initial state of problem definition, since we always start in this phase.

Now that we have defined all the new predicates we can model the four new operators in PDDL. We start with operators to place walls and boxes.

```
(:action place_wall
  :parameters(
    ?w - wall ?to - square
  )
  :precondition(and
    (making_level)
    (opt_wall ?to)
    (not (wall_placed ?w))
    (not (wall_at ?to))
    (not (box_at ?to))
  )
  :effect(and
    (wall_at ?to)
    (wall_placed ?w)
  )
)
```

```
(:action place_box
  :parameters(
    ?b - box ?to - square
  )
  :precondition(and
    (making_level)
    (opt_box ?to)
    (not (box_placed ?b))
    (not (wall_at ?to))
    (not (box_at ?to))
  )
  :effect(and
    (box_at ?to)
    (box_placed ?b)
  )
)
```

The "place worker" and "start playing" are defined next. Note, that "place worker" also changes the phase to playing. This way we can ensure that the worker is added last and only once. Thanks to this property the operators to place the walls and the boxes do not need to check whether a worker has been placed on the square where they wish to place their item.

```
(:action place_player_and_start
  :parameters(?to - tile)
  :precondition (and
```

```
(:action start_play
  :parameters()
  :precondition(and
```

```
    (making_level)
    (opt_player ?to)
    (not (wall_at ?to))
    (not (box_at ?to))
  )
  :effect(and
    (player_at ?to)
    (not (making_level))
    (playing)
  )
)
```

```
    (making_level)
  )
  :effect(and
    (not (making_level))
    (playing)
  )
)
```

Lastly, the move and push operators from the Sokoban solving domain need to be slightly updated. The predicate `(playing)` must be added to preconditions. With this we have described a correct and complete encoding of the Sokoban puzzle generation into PDDL. However, there is one small issue we need to address.

Planners always try to find short plans. This has an unpleasant consequence for our problem. The planner is motivated to place the walls and boxes in such a way, that the generated puzzle can be solved with as few moves and pushes as possible. This means, that the generated levels tend to be very easy to solve. In order to address this issue, we modeled a mechanism, that enforces a certain minimum amount of pushes in the solve phase. This value can be specified by the puzzle designer as the third parameter on the "p line" in the level template (see Fig. 4). This is modeled by adding a counter to the push operators, that is increased with each push action. Then in the goal conditions we can require that the counter reaches the required value.

For more details refer to the complete domain PDDL file available in the project's repository[4]. The repository also contains the tool that generates the PDDL problem files from a given level template.

4.3 Comparison to Related Work

Our method bears the most similarity to the approach of Kartal et al. [14] (see Sect. 3). They formulate the puzzle generation as an MCTS optimization problem, while we model it as a planning problem. They start with a level that contains the worker and is otherwise full of walls and then remove some walls and add some boxes. We start with a partially built level that already contains all the goals and then we add additional walls, boxes and the worker. Then both approaches have a special action that transitions the search into the playing mode. Finally, in our approach we try to solve the level and backtrack to the level building phase if it is not solvable. In Kartal et al. [14] random moves are executed for some time and then the reached state is declared to be the goal state.

[4] https://github.com/biotomas/sokoplan/blob/master/SokoGen/domain.pddl.

4.4 Generalization to Other Puzzles Than Sokoban

It easy to use our puzzle generation concept for other puzzles as well. In order to do that one must only write the appropriate PDDL domain file and a generator for the PDLL problem files. The domain file is usually written by hand and problem files are generated by a script or a small program, however, both can be written by hand. In any case, the effort is very low in comparison to designing and implementing a dedicated puzzle generator for a specific puzzle as is done in related work. Another advantage of our concept is, that as the state-of-the-art planners evolve and their performance improves, our puzzle generator improves with it automatically.

5 Experimental Evaluation

Our Sokoban puzzle generation tool is available online at GitHub[5]. The repository contains everything you need to build and use our tool and also to replicate the experimental evaluation we present in this section.

5.1 Setup

As our tool is based on planning, we will obviously need a planner. Any planner that supports PDDL[6] would work, but based on some preliminary evaluations we settled on using the well established state-of-the-art planner FastDownward [10] with the LAMA 2011 [17] configuration.

We generated 300 level templates to use as benchmarks (more on how is described in the next Subsection) and we gave the planner a time limit of 1 min to find a solvable puzzle. We run our experiments on a computer with an Intel(R) Core(TM) i7-7800X CPU @ 3.50 GHz processor and 64 GB of main memory. The used operating system was Ubuntu version 5.8.0-26-generic.

5.2 Benchmark Instances

All of the 300 level templates are based on 10 base templates, which we designed by hand (see Fig. 5). To create the benchmark templates of various complexities we reduced the number of goals in the base templates to a certain number. A benchmark template of complexity level x is defined as a template with x goal locations and the objective to add x walls and x boxes. We generated templates of complexity levels $1, 2, \dots, 6$. We created 5 templates for each of the 6 complexity levels and each of the 10 base templates, hence the 300 total benchmark templates.

To create a template of complexity level x from a base template we first mark each floor square as a potential position for adding any (wall, box, or worker) object. Then we randomly remove goal squares until exactly x goal squares

[5] https://github.com/biotomas/sokoplan.

[6] All available academic planners support PDDL.

Fig. 5. The 10 base templates we used to generate our 300 benchmark level templates. The name of the base templates are (top left to bottom right): O, L, U, H, XX, X, B, I, II, and Pi.

remain. An example of generating a template of complexity level 4 from a base template follows:

```
base temp.     p sokogen 4 4 4  // objective:
#####          #####            //     add 4 walls and 4 boxes
#...#          #.  #            //     do 4 at least pushes
#. .#          # 0.#            //
# # ####  =>   #0#0####         // symbols:
# #    .#      #0#000 #         //     # - wall
#..   ..#      #. 00. #         //     . - goal
########       ########         //     0 - add any object
```

Admittedly, these benchmarks are not exactly like the level templates a human designed would use. A human designer would start with a level template and then modify it after seeing the level produced by the tool. They would perform several iterations of these steps until a satisfactory level has been found. Nevertheless, we used the approach described above since we needed to generate a large number of templates of various sizes and complexity levels. However, we believe the generated templates are still representative enough to perform a meaningful experimental evaluation of our tool.

5.3 Experimental Results

The results of the experimental evaluation are presented in Table 1. Not solving a level template can either mean that it is impossible to place the given amount of walls and boxes such that a solvable level is created or that the planner could not find a solution in the given time limit (of 1 min). Unfortunately, in most cases, we cannot distinguish between these two scenarios, since planners are not very good at proving non-existence of plans.

For most of the base templates we could solve around 20 of the 30 level templates, except for X and B, which seem to be too tight to add more than 2 walls and 2 boxes in most of the cases. On the large base templates (XX, II, and Pi) we failed to solve most of the higher complexity templates. We believe that this is not due to the not existence of solutions, rather it is caused by the inability of the planner to find a solution within the given time limit. We could add 6 boxes and walls only for base templates U and I, which with 18 and 20 free squares represent middle sized levels. This seems to be the sweet spot between being to tight to place enough objects and too large to find a solution within the time limit.

Overall, the experimental evaluation showed that our approach works and we can rapidly generate levels of various shapes and complexities.

Table 1. The table contains experimental results on our benchmarks grouped by base templates and complexity levels. The first column contains the names of the base templates, see Fig. 5 for their definitions. The values in the second column are the number of free squares in the corresponding templates that can be used to place walls, boxes and the player. Columns 3 to 8 contain the number of solved instances within a time limit of 1 min for each complexity level. The final column contains the total number of solved instances within 1 min across all complexity levels.

Base Template	Free Squares	Solved for Complexity Level 1	2	3	4	5	6	Total Solved
O	10	5	5	5	5	0	0	20
L	12	5	5	5	5	1	0	21
U	18	5	5	5	2	1	1	19
H	20	5	5	5	0	2	0	17
XX	30	5	5	5	1	1	0	17
X	18	3	3	1	1	0	0	8
B	9	5	5	0	0	0	0	10
I	20	5	5	4	3	2	1	20
II	28	5	5	3	4	0	0	17
Pi	35	5	5	5	1	0	0	16

6 Conclusion

We presented a method to assist human level designers to generate solvable Sokoban puzzles using automated planners. Our method has several advantages. Firstly, it based on a very generic principle (using planners) so it can be easily modified and used to generate puzzles other than Sokoban. Secondly, it is using a constantly evolving search technology (automated planning) so the generator will automatically improve with time as planners get more and more performant. Thirdly, it is very simple and easy to implement and customize.

6.1 Future Work

As for future work we would like to improve the performance of our tool by tuning the PDDL encoding and adjusting the configuration parameters of the used planner or evaluate other available planners.

We plan to develop a user friendly graphical user interface (GUI) for our generator to make it easy to use for less technical users.

Finally, we would like to test our general method on other puzzles than Sokoban. As described in the paper, this mostly only amounts to formulating new PDDL models for the given puzzles.

References

1. Ashlock, D., Schonfeld, J.: Evolution for automatic assessment of the difficulty of sokoban boards. In: IEEE Congress on Evolutionary Computation, pp. 1–8, July 2010. https://doi.org/10.1109/CEC.2010.5586239

2. Botea, A., Müller, M., Schaeffer, J.: Using abstraction for planning in Sokoban. In: Schaeffer, J., Müller, M., Björnsson, Y. (eds.) CG 2002. LNCS, vol. 2883, pp. 360–375. Springer, Heidelberg (2003). https://doi.org/10.1007/978-3-540-40031-8_24
3. Culberson, J.: Sokoban is PSPACE-complete. In: Proceedings in Informatics, vol. 4, pp. 65–76. Citeseer (1997)
4. Culberson, J.: Sokoban is PSPACE-complete. Technical reports (Computing Science) (1997)
5. De Kegel, B., Haahr, M.: Procedural puzzle generation: a survey. IEEE Trans. Games **12**(1), 21–40 (2019)
6. Dor, D., Zwick, U.: Sokoban and other motion planning problems. Comput. Geom. **13**(4), 215–228 (1996)
7. Froleyks, N., Balyo, T.: Using an algorithm portfolio to solve Sokoban. In: Tenth Annual Symposium on Combinatorial Search, June 2017
8. Ghallab, M., Nau, D., Traverso, P.: Automated Planning and Acting. Cambridge University Press, Cambridge (2016)
9. Haslum, P., Lipovetzky, N., Magazzeni, D., Muise, C.: An introduction to the planning domain definition language. Synth. Lect. Artif. Intell. Mach. Learn. **13**(2), 1–187 (2019). https://doi.org/10.2200/S00900ED2V01Y201902AIM042
10. Helmert, M.: The fast downward planning system. J. Artif. Intell. Res. **26**, 191–246 (2006)
11. Imabayashi, H.: Sokoban Official. https://sokoban.jp/title.html
12. Jarušek, P., Pelánek, R.: Human Problem Solving: Sokoban Case Study. Fakulta informatiky, Masarykova univerzita, Brno, Technická zpráva (2010)
13. Junghanns, A., Schaeffer, J.: Sokoban: evaluating standard single-agent search techniques in the presence of deadlock. In: Mercer, R.E., Neufeld, E. (eds.) AI 1998. LNCS, vol. 1418, pp. 1–15. Springer, Heidelberg (1998). https://doi.org/10.1007/3-540-64575-6_36
14. Kartal, B., Sohre, N., Guy, S.: Data driven Sokoban puzzle generation with monte Carlo tree search. In: Proceedings of the AAAI Conference on Artificial Intelligence and Interactive Digital Entertainment, vol. 12 (2016)
15. Murase, Y., Matsubara, H., Hiraga, Y.: Automatic making of Sokoban problems. In: Foo, N., Goebel, R. (eds.) PRICAI 1996. LNCS, vol. 1114, pp. 592–600. Springer, Heidelberg (1996). https://doi.org/10.1007/3-540-61532-6_50
16. Pommerening, F., Torralba, A., Balyo, T., Vallati, M., Chrpa, L., McCluskey, L.: The international planning competition (1998–2018). https://www.icaps-conference.org/competitions/
17. Richter, S., Westphal, M., Helmert, M.: LAMA 2008 and 2011. In: International Planning Competition, pp. 117–124 (2011)
18. Taylor, J., Parberry, I.: Procedural generation of Sokoban levels. In: Proceedings of the International North American Conference on Intelligent Games and Simulation, pp. 5–12 (2011)
19. Taylor, J., Parberry, I., Parsons, T.: Comparing player attention on procedurally generated vs. hand crafted Sokoban levels with an auditory Stroop test. In: Proceedings of the Foundations of Digital Games (2015)
20. van Kreveld, M., Löffler, M., Mutser, P.: Automated puzzle difficulty estimation. In: 2015 IEEE Conference on Computational Intelligence and Games (CIG), pp. 415–422 (2015). https://doi.org/10.1109/CIG.2015.7317913
21. Winston, P.H., Horn, B.K.: LISP, 2nd edn. Osti.gov, United States (1986)

Logo Generation Using Regional Features: A Faster R-CNN Approach to Generative Adversarial Networks

Aram Ter-Sarkisov(✉) and Eduardo Alonso

CitAI Research Center, Department of Computer Science City, University of London, London, UK
alex.ter-sarkisov@city.ac.uk

Abstract. In this paper we introduce Local Logo Generative Adversarial Network (LL-GAN) that uses regional features extracted from Faster R-CNN for logo generation. We demonstrate the strength of this approach by training the framework on a small style-rich dataset of real heavy metal logos to generate new ones. LL-GAN achieves Inception Score of 5.29 and Frechet Inception Distance of 223.94, improving on state-of-the-art models StyleGAN2 and Self-Attention GAN .

Keywords: Deep learning · Generative adversarial networks · Logo generation.

1 Introduction

Generative Adversarial Networks (GANs) were first introduced in [7]. They have gained a wide recognition in the Artificial Intelligence community due to their ability to approximate the distribution of real data by generating fake data. Recent advances include Progressive-Growing GANs, StyleGAN and StyleGAN2 that learn styles at different resolutions [14–16], Self-Attention GANs (SAGANs) that learn the connections between different spatial locations [29], CycleGANs and Pix2Pix GANs for unpaired style transfer [12, 30] and Wasserstein loss function [1].

Faster R-CNN and Mask R-CNN [6, 9, 24] are state-of-the-art open-source deep learning algorithms for object detection and instance segmentation that work in multiple stages, unlike single-shot models like YOLO [23].

Faster R-CNN first predicts regions containing objects based on overlaps (Intersect over Union, IoU) between fixed-size rectangles known as anchors and ground truth bounding boxes using Region Proposal Network (RPN). Then, it pools features from these areas by cropping and resizing corresponding areas in features maps. This is done using Region of Interest Pooling (RoIPool) to construct fixed-size Regions of Interest (RoIs) containing rescaled regional features for each object (later replaced by more accurate Region of Interest Align,

M. Wölfel et al. (Eds.): ArtsIT 2021, LNICST 422, pp. 442–456, 2022.
https://doi.org/10.1007/978-3-030-95531-1_30

RoIAlign [9]). These local features are fed through fully connected (`fc`) layers to independently predict the object classes and refine bounding box prediction. In addition to this, Mask R-CNN segments objects' masks.

One of the new and challenging areas in GANs and neural style transfer is the creation of logos and fonts. This area includes style and shape transfer between fonts [3,4], logo synthesis [19,21,26], transfer of style to font [2] and font generation [8]. A specific challenge in this area is disentaglement of content and style learning, often done through training of two different encoders and feature concatenation, as in [4], and separation of transfer of shape and texture (ornamentation), done through pretraining of the shape model and ornamentation model that takes the shapes and adds ornamentation [3]. Logo synthesis (style transfer), as in [19,21,26], also uses conditional input (random vector + sparse vector for the class).

We address the shortcomings of the state-of-the-art models, such as the size of the output, which in most cases is limited to 64×64 pixels. This size is sufficient for separate characters/glyphs or small logos, as readability does not suffer. For larger logos or words, model output must be upsampled. Another limitation we address is the size of the training data: we leverage Faster R-CNN's capacity to sample a batch of regional features in a single image to overcome the need for a large dataset.

In this paper we present a GAN model for generating logos of heavy metal bands. To the best of our knowledge, it would be the first GAN study that is focused on the generation of band logos. With respect to specifically heavy metal logos, recently, there were two related publications: in [28] style transfer model based on [5] was used to fuse the style of heavy metal bands logos, e.g. Megadeth and the content of corporate logos, e.g. Microsoft. In [25] the styling of heavy metal logos and its association with genre and readability are investigated.

Measured by Frechet inception distance [11], Inception score [27] and detection accuracy, the presented model confidently outperforms the state-of-the-art StyleGAN2 and SAGAN frameworks. Our contribution consists of the following:

- Local Logo GAN (LL-GAN) framework: training the Generator by comparing regional features extracted from the fake and real data using RoIAlign module in Faster R-CNN. Since loss is computed only on regional features, the Generator's parameters receive updates only from the region containing the logo in the real data. This model augments the baseline GAN framework, serving as an additional source of gradients for the Generator's parameters. Ground truth bounding box is used to determine positive RoIs in the fake image, therefore the Generator learns to output spatially-aware logos. A number of RoIs is sampled from each image using RPN and RoIAlign modules, which compensates for the sparsity of the data,
- Logo generator. The model is capable of generating style-rich heavy metal logos consisting of glyph-like structures that closely resemble real-life band logos without suffering from the mode collapse. This includes an augmenta-

tion of the DCGAN's model architecture [22] that allows for creation of large images (282 × 282),
- Style-rich metal band logos dataset. Images with heavy metal band logos were scraped from the internet and labelled at text level (bounding box around the band's logo). Each image contains a single-word logo, with a simple background (e.g. black or white) across 10 bands selected for the style of the logo. The dataset consists of 923 images and an equal number of bounding box coordinates of the logo.

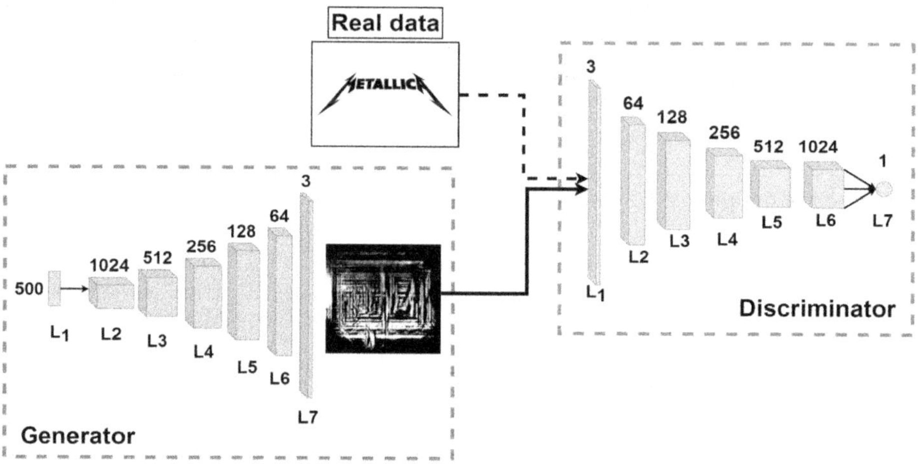

Fig. 1. DCGAN+ framework. Details of the architecture of both models is presented in Table 2. Values in each module in the number of feature maps in the Convolution (Discriminator) or Transposed Convolution (Generator) models. Normal arrows: features and fake data, broken arrow: real data.

2 Our Approach

Model sizes and structures are compared in Table 1.

2.1 DCGAN+ Framework

DCGAN+ is an augmentation of the DCGAN architecture [22] that enables generation of larger images in a single shot. The main idea behind the architecture is the selection of the right rate of upsampling and downsampling of feature maps in each model (kernel size, stride, padding). Figure 1 and Table 2 provide a summary of the models' architectures. This solution successfully addresses the problem of the size of the generated logo, as we increase it from at most 64 × 64, as in [26] to 282 × 282.

Table 1. Comparison of sizes of the frameworks. G: generator, D: discriminator, F: Faster R-CNN.

Framework	Number of Parameters	Structure of the framework
DCGAN+	43.83M + 3.93M	G + D
LL-GAN	43.83M + 3.93M + 41.43M	G + D + F
StyleGAN2 [16]	84.69M (Total)	G + D
StyleGAN2 w/attention [16]	85.87M (Total)	G + D
SAGAN [29]	8.1M + 4.92M	G + D
DCGAN [22]	3.5M + 2.7M	G +D
Faster R-CNN [10]	41.80M	F

Table 2. DCGAN+ framework. G: Generator, D: Discriminator

Model	Block	Depth	Kernel	Stride	Pad
G	L_1(Input)	500	0	0	1
	L_2	1024	8	2	0
	L_3	512	4	2	0
	L_4	256	4	2	1
	L_5	128	4	2	1
	L_6	64	2	2	1
	L_7(`tanh`)	3	2	2	1
D	L_1(Input)	3			
	L_2	64	4	2	1
	L_3	123	3	2	1
	L_4	256	3	2	1
	L_5	512	3	2	1
	L_6	1024	3	2	1
	L_7(`fc`)	1	-	-	-

2.2 LL-GAN Framework

Overall framework is presented in Fig. 2. Generator and Discriminator are the same as in DCGAN+. One of the key contributions of this paper is the use of local features from the RoIAlign stage in Faster R-CNN to compute style loss. We use the ground truth bounding box around the band logo to extract one RoI from the real data, skipping the RPN stage. For the fake data, RPN predicts raw boxes passed on to RoIAlign that uses these predictions to extract RoI features and outputs B positive predictions (i.e. RoI box predictions that have IoU with the ground truth box greater than a pre-defined threshold), each of fixed size $H \times W \times C$. Each RoI's height and width are hyperparameters, and depth C is determined by the depth of the FPN feature map, see [18].

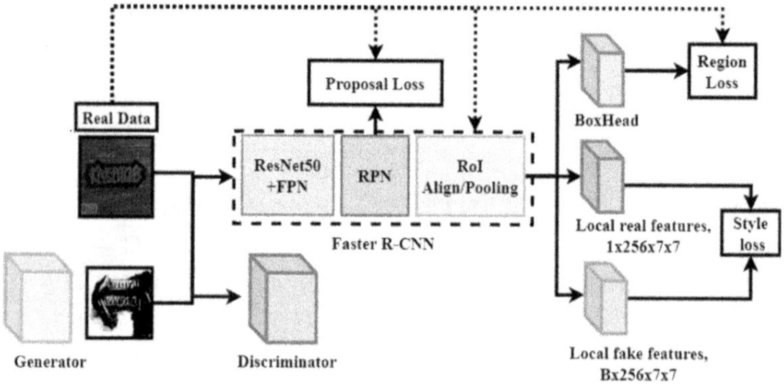

Fig. 2. LL-GAN framework. Normal arrows: features, dotted arrows: box coordinates, broken line box: Faster R-CNN.

Feature loss is computed between B positive RoIs from the fake and the single RoI from the real data (ground truth region). The number of RoIs varies from image to image, but on the average grows as the fake data increasingly resembles the real data.

Each of C feature maps extracted from the real data is vectorized, i.e. an ith feature map is converted into a vector with $H \cdot W = HW$ elements which we refer to as $\mathcal{F}_i^r$. Dot-product is computed between each (i, j) pair of vectorized feature maps to obtain matrix $\mathcal{G}^r$ with dimensionality $C \times C$ (i.e. each (i, j) element in $\mathcal{G}^r$ is a dot product of the vectors $\mathcal{F}_i^r$ and $\mathcal{F}_j^r$), see Eq. 1.

$$\mathcal{G}_{i,j}^r = \mathcal{F}_i^r \otimes \mathcal{F}_j^r \tag{1}$$

For each kth RoI extracted from the fake data, we also compute Gram matrix $\mathcal{G}^{k,f}$, Eq. 2, where $\mathcal{F}_i^{k,f}$ is an ith vectorized feature map in the kth RoI. Therefore $\mathcal{G}_{i,j}^{k,f}$ is the dot-product between each (i, j) pair of vectorized feature maps in k^{th} RoI, $\mathcal{F}_i^{k,f} \otimes \mathcal{F}_j^{k,f}$.

$$\mathcal{G}_{i,j}^{k,f} = \mathcal{F}_i^{k,f} \otimes \mathcal{F}_j^{k,f} \tag{2}$$

Equations 1 and 2 compute correlation between regional features, which represents the style. The normalized style loss of kth RoI, D_k is computed using L_2 distance between $\mathcal{G}^r$ and $\mathcal{G}^{k,f}$ elementwise, Eq. 3. Finally, we sum B normalized RoI losses, Eq. 4.

$$D_k = \frac{\sum_{i=1}^{C} \sum_{j=1}^{C} \left(\mathcal{G}_{i,j}^r - \mathcal{G}_{i,j}^{k,f}\right)^2}{(2 \times H \times W)^2} \tag{3}$$

$$L^S = \frac{\sum_{k=1}^{B} D_k}{B} \tag{4}$$

The main idea of computing style loss using Eqs. 1–4 is to train the Generator to evolve features that approximate the distribution of the real logos, and in the same region as in the real data. The first requirement (style) is satisfied by Eqs. 1 and 2, the second one (spatial awareness) by the RoIAlign functionality: by backpropagating loss extracted from a region in the fake data, Generator learns to evolve region-aware logos. Total loss in this framework is computed using Eq. 7.

$$L^D = \mathbf{E}_{x\sim p(x)} \log D(x) + \mathbf{E}_{z\sim p(z)} \log(1 - D(G(z))) \tag{5}$$

$$L^G = \mathbf{E}_{z\sim p(z)} \log D(G(z)) \tag{6}$$

$$L_{Total} = L^G + L^D + L^S \tag{7}$$

Equations 5 and 6 are the usual Discriminator and Generator losses, both computed using binary cross-entropy, for the real data x and fake data z, except that Generator loss maximizes the loss function instead of minimizing it, see Sect. 4 for details. L^S is the style loss in Eq. 4.

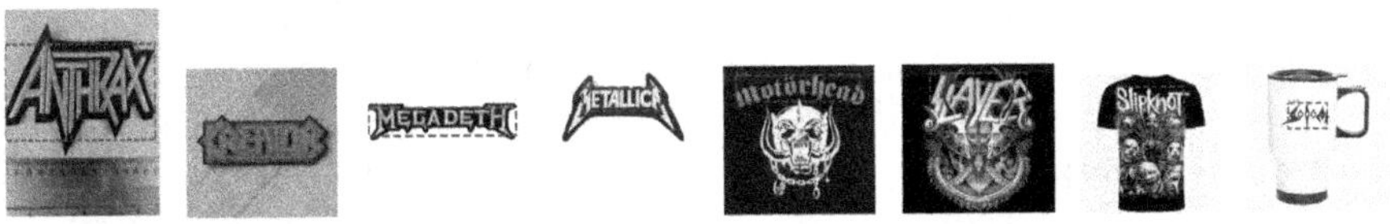

Fig. 3. Examples of logos used in the training data overlaid with bounding box and score predictions by Faster R-CNN. Best viewed in color.

3 Dataset Construction and Labeling

To train LL-GAN models, dataset must have labels consisting of bounding boxes around logos (one box per image). Therefore, dataset construction consists of three stages: first, we scrape the logos from the internet and manually labelled a small portion of it. Next, we train Faster R-CNN on a labelled text and logo ICDAR dataset, to predict boxes around words, and finetuned it to the labelled portion of the metal logo data. Finally, we use this model on the remaining scraped data to label each metal logo with the bounding box.

3.1 Raw Dataset

Our real dataset consists of 923 images of varying sizes. Each image contains a heavy metal band's logo, predominantly with a neutral (e.g. black or white) background. This was done in order to prevent the generator from learning background features and instead focus on the logo style and semantics. Ten bands were selected purely for the style of their logos: Anthrax, Kreator, Manowar, Megadeth, Metallica, Motorhead, Sepultura, Slayer, Slipknot, Sodom. The sizes of images vary between 50×50 and 512×1024 pixels, with the majority about

200×200. Examples with the overlaid bounding boxes are presented in Fig. 3. This is a very challenging dataset, for two reasons: it is very small, and it is rich in style (specific styles of heavy metal logos/fonts) and weak in content, because each image contains only a single logo, there's a limited number of observations for each logo. As we explained in Sect. 2 and show in Sect. 4, the ability of Faster R-CNN to learn and extract regional features from a single image addresses this challenge.

3.2 Faster R-CNN Logo Detector

To detect boxes around text in logos, we finetuned the out-of-the-box Faster R-CNN model from Torchvision v0.3.0 library with ResNet50 backbone feature extractor and FPN pretrained on MS COCO 2017 to ICDAR Focused Scene Text (ICDAR-FST2013), [13] dataset that contains 223 images of street signs for 100 epochs. This model was trained to detect separate words in various contexts. Next, we fintetuned it for 500 epochs to a portion of the metal logo dataset. The model predicts only two classes (object vs background) per RoI, and we capped the number of candidates in RPN stage at 1024 and also used a slightly larger RPN anchor generator (5 anchor sizes between 16 and 256 and 5 scales, between 0.25 and 2, a total of 25/location), learning rate of $1e{-}5$, regularization hyperparameter (weight decay) of $1e{-}2$ and Adam optimizer with $\beta_1 = 0.9, \beta_2 = 0.99$. Other important hyperparameters (positive/negative box thresholds, RoI dimensions, RoI batch size, heads sizes) were the same as in the baseline Torchvision model. First, this model was used to label the rest of the metal logo data for experiments in Sect. 4.Then, in Sect. 5, this model was used to detect logos produced by generators in all LL-GAN frameworks and to evaluate the accuracy of outputs of all generators and produce results in Table 4.

4 Experiments

4.1 DCGAN+ Framework

We trained both Generator and Discriminator in the DCGAN+ framework from scratch with a learning rate of $1e{-}4$ and weight regularization coefficient of $1e{-}3$ for both models using Adam optimizer [17], batch size of 128 and binary cross-entropy loss for 1000 epochs. This took about 6 h on a GPU with 8Gb VRAM. Following the recommendations in [7] and Pytorch GAN tutorial, Discriminator is updated using real and fake data (1 iteration). Then, the fake data is relabelled as real and the Generator is updated by computing loss using real labels. This is done to avoid premature convergence.

4.2 LL-GAN Framework

For LL-GAN we used the pretrained weights and the same architecture for the Generator and Discriminator from DCGAN+. Only Generator and Discriminator were trained, all Faster R-CNN weights trained in Sect. 3 remained frozen, since the logo detector model was specifically trained to detect single logos anywhere. Real and fake data is processed differently by the logo detector. From the real data, only single RoI regional features with dimensions $C \times H \times W$ is extracted and vectorized, Eq. 1, using ground truth bounding box, hence RPN stage is skipped, and no gradients are computed. Fake data is fed forward through the whole framework (see Fig. 2), RoI features are extracted and vectorized, Eq. 2 for the loss, Eqs. 3–7 and gradient computation.

Also, RoI module, during processing of fake images, always appends the ground truth bounding box coordinates to the list of RoIs. The reason for that is that early in training, Generator cannot output high-quality logos, and therefore Faster R-CNN will not be able to find good RoIs anywhere in the fake data. As a result, the number of positive RoIs (B in Eq. 4) varied from image to image, but overall increased due to the improvement in the work of the Generator. In addition to the baseline LL-GAN framework that uses Eq. 7 loss function, we experimented with a number of tricks:

- In addition to style loss in Eq. 4, we added detection loss from fake data. Ground truth bounding box coordinates were taken from the real logo that was used to train the Generator. This added two more loss functions: raw boxes in RPN and refined boxes in RoI,
- Extend ground truth bounding boxes around logos to add more context when computing the Generator's loss. We experimented with different values and found 20 pixels in each direction the optimal number for the tradeoff between context and background noise.
- Compute L_2 loss between backbone features extracted from real and fake data, similar to content loss in neural style transfer [5]. Features were taken from all outputs of FPN layers. Therefore, in addition to B RoIs from which we compute L^S, we add the loss from features extracted from the whole image. The objective of adding this loss is to improve the Generator's ability to output a more neutral, e.g. black, background.
- Full model: we combine base model and all three extensions

We trained in total five frameworks (baseline + three augmentations + full model). Each framework was trained for 500 epochs, using Adam optimizer($\beta_1 = 0.9, \beta_2 = 0.999$), regularization parameter (weight decay) of $1e{-3}$. Hyperparameters of Faster R-CNN logo detector were the same across all frameworks, and shared most of them with the pre-trained logo detector, including the size of the RoIs, $H = 7, W = 7, C = 256$. Since logo generation is a very spatially sensitive task, we used different thresholds for positive and negative candidates both at RPN and RoIAlign stages: the positive threshold was 0.9 and negative 0.1.

4.3 StyleGAN2

StyleGAN [15] and StyleGAN2 [16] are the state-of-the art GANs that can learn different styles and generate high-quality large images, this includes training on small dataset (<5000 images). We trained StyleGAN2 on our data to generate images size 256×256, using high truncation $\psi = 1$ coefficient(no gradient averaging), augment the data by 25%, with the learning rate of $1e{-}4$ for both Generator and Discriminator, Adam optimizer ($\beta_1 = 0.5, \beta_2 = 0.999$), self-attention mechanism [29] and batch size of 4 (maximum possible for this image size on the GPU with 8 Gb of VRAM. We trained each model (with and without attention modules) for 100000 steps (~100 epochs), which took about 72 h, but we noticed that after about 20000 steps the model starts to overfit and exhibits a strong mode collapse. We therefore report the best result for each model (20000 steps for the StyleGAN2 with attention and 15000 for StyleGAN2 without attention).

4.4 Self-Attention GANs

We also train SAGAN, [29], with spectral normalization [20] and Hinge loss function. We used the recommended hyperparameters: latent dimension size 128, batch size of 64, Generator learning rate $1e{-}4$, Discriminator learning rate $4e{-}4$ and Adam optimizer ($\beta_1 = 0, \beta_2 = 0.9$). Generator's architecture consists of 7 modules ($ConvTranspose2D + BatchNorm + ReLU$, each equipped with a spectral transformer. Self-attention module is added to block 3 with 256 feature maps and map size of 16×16. The model outputs images size 256×256. SAGAN framework was trained for 300000 iterations (~330 epochs). Training was stopped due to the obvious mode collapse.

5 Evaluation of Results

Examples of outputs of all models are presented in Fig. 5. In Table 3 we report FID and IS scores, in Table 4 we report quality and detection results for all models. The best results are bold+italicized, second best bold and third-best italicized. For FID score, we used the layer with 2048 maps, for IS scores we split the sample into either 1 or 10 subsets. Each model generates 512 images which are processed by Faster R-CNN logo detector. If it predicts a logo with confidence score exceeding the pre-defined threshold of 0.75, the detection is considered to be a True Positive (TP), otherwise it is a False Positive (FP). The assumption of this test is that a good Generator would output images that contain exactly single identifiable logo. If the detector predicts more than one logo in a single image with confidence exceeding this threshold, all predictions other than the best-scored one are counted as FPs. If it predicts no logos at all, it is also counted as an FP. Detection rate is defined as $\frac{TP}{TP+FP}$, average confidence is averaged over all detections, including those below the threshold.

5.1 DCGAN+ and LL-GAN

DCGAN+ achieves the best FID score of 220.155, in which it confidently outperforms far more sophisticated state-of-the-art models. It also achieves the third-best results across all other scores. The baseline model is capable of producing high-quality realistic logos in the style of heavy metal bands without overfitting to any particular feature. Among its weaknesses are the inconsistency in glyph stlye, both in terms of color and background noise, see Figs. 4 and 5. In particular, some logos are red and yellow and consist of thin vertical lines. Vanilla LL-GAN model achieves the best IS scores of 6.339 and 5.292 and outputs highly detectable logos with high confidence.

Table 3. Comparison of models' performance-Quality. Italicized+bold: best, bold: second-best, italicized: third-best

Framework name	FID	IS(1)	IS(10)
DCGAN+	***220.155***	*6.023*	*5.105*
LL-GAN	**223.948**	***6.339***	***5.292***
+ FRCNN loss	271.030	5.705	4.947
+ extended boxes	247.181	5.753	4.901
+ backbone features	*237.752*	4.590	4.095
Full	249.694	**6.232**	**5.150**
StyleGAN2 ($\psi = 0.6$)	329.026	2.840	2.766
StyleGAN2 ($\psi = 1.0$)	354.873	2.497	2.433
+attention	328.859	2.356	2.298
SAGAN	283.554	3.581	3.394

Most logos generated by the vanilla model are very realistic, resemble real glyphs, are consistent in colors (mostly red and white, as in the training data), and do not experience mode collapse. Also LL-GAN with all three augmentations perform well, producing IS scores of 6.232 and 5.150. In Fig. 4 we placed outputs from DCGAN+ and different LL-GAN models that output logos with similar features side-by-side to highlight the advantages of our approach. The same features produced by LL-GAN generators are more homogeneous in color and shape, the background contains fewer geometric artefacts and is more consistent and neutral. Metrics discussed in this section confirm that this consistency does not come at the cost of lower variance in the output.

5.2 State-of-the-Art Models

StyleGAN2 is capable of producing logos with very consistent structures, but due to the size of the dataset suffers from mode collapse. This is reflected in the highest detection score of 0.687 and low FID and IS scores: the generated structures are consistent enough to be classified as a logo, but do not resemble the training data and are very similar. SAGAN also suffers from mode collapse.

By comparing results in Tables 3 and 4 and Fig. 5 to the models' architectures and sizes in Table 1, LL-GAN models are comparable in size to StyleGAN2, but their Generators output more interesting logos.

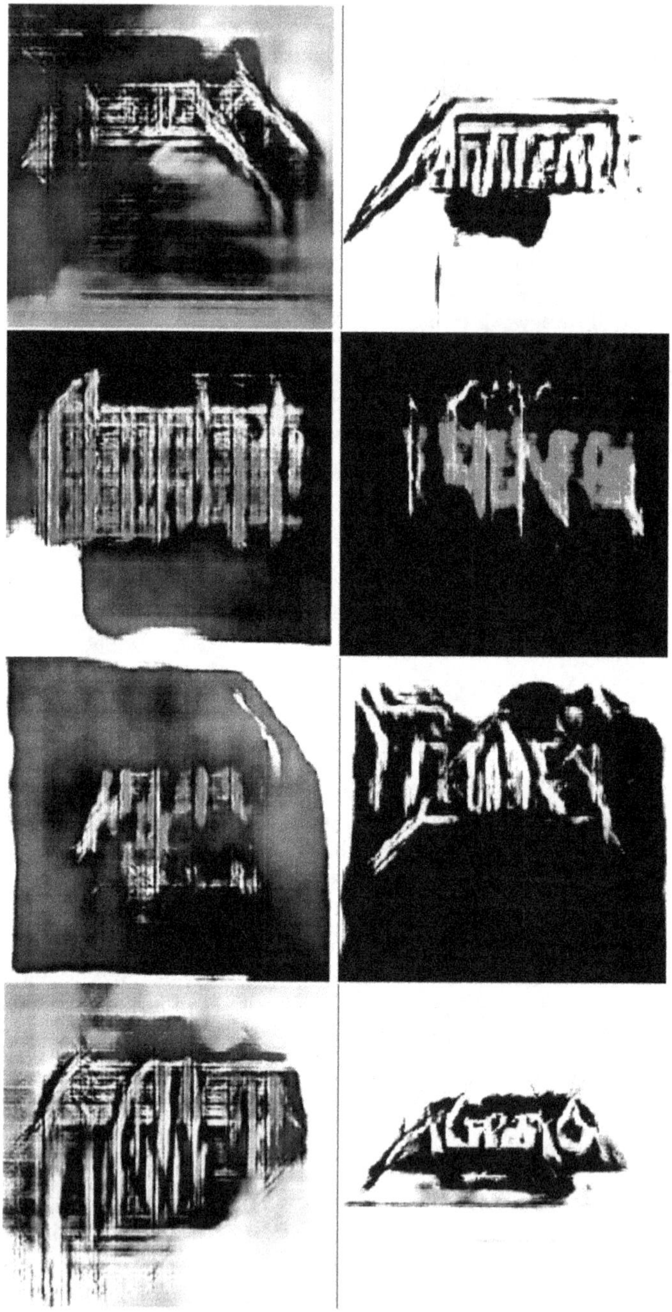

Fig. 4. Comparison of DCGAN+ (left) and LL-GAN output (right). First row: DCGAN+ vs LL-GAN, second row: DCGAN+ vs LL-GAN(+backbone features), third row: DCGAN+ vs LL-GAN (full), fourth row: DCGAN+ vs LL-GAN(+FRCNN losses). The obvious weakness of DCGAN+ that LL-GAN fixes is the lack of shape (glyphs are made up of thicker, shorter features without gaps) and color (all glyphs in the logo have the same color) consistency. Each row used the same Generator input. Best viewed in color. (Color figure online)

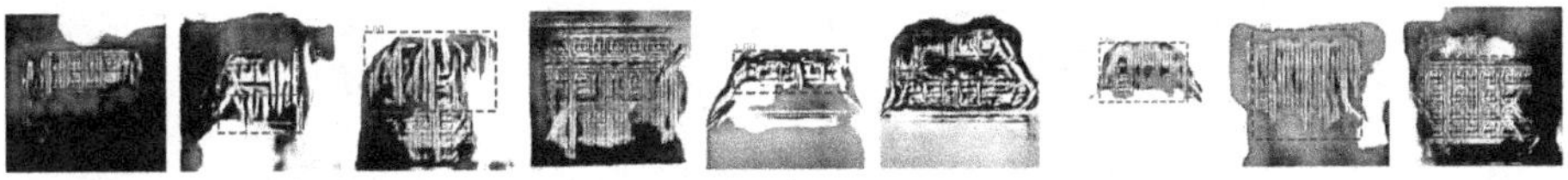

(a) DCGAN+

(b) LL-GAN

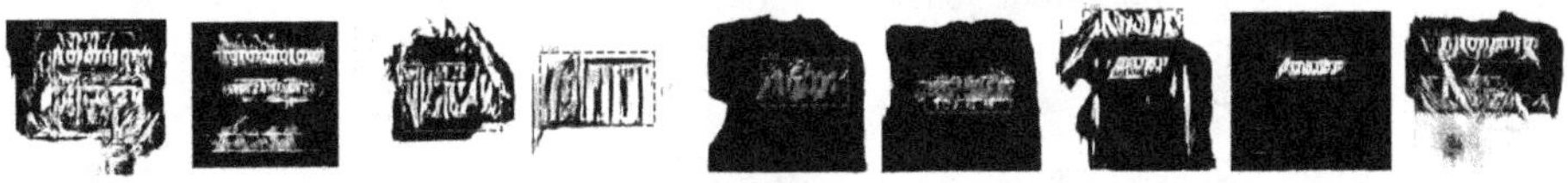

(c) LL-GAN + extended boxes

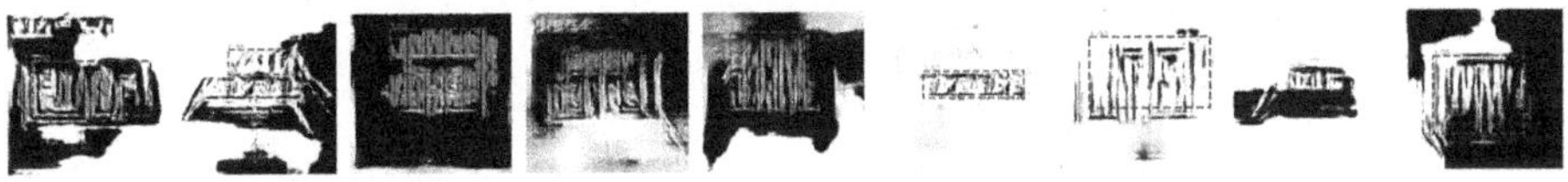

(d) LL-GAN + Faster R-CNN loss

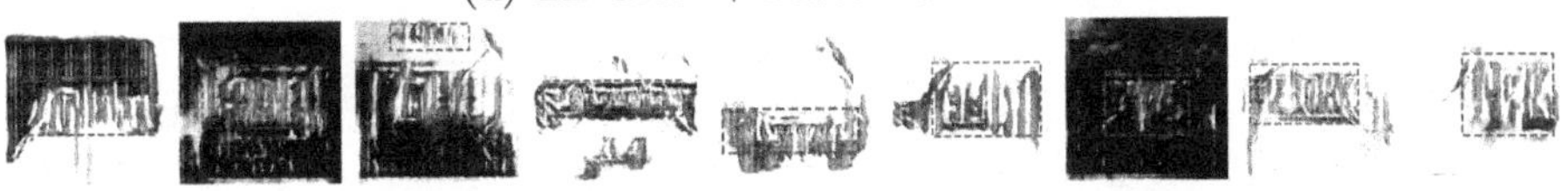

(e) LL-GAN + backbone features

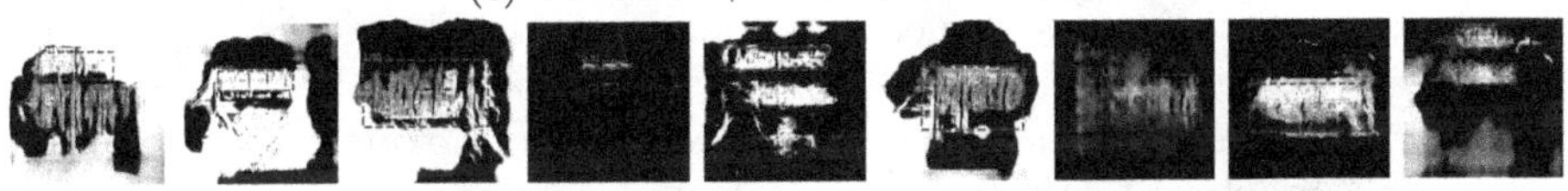

(f) LL-GAN (full)

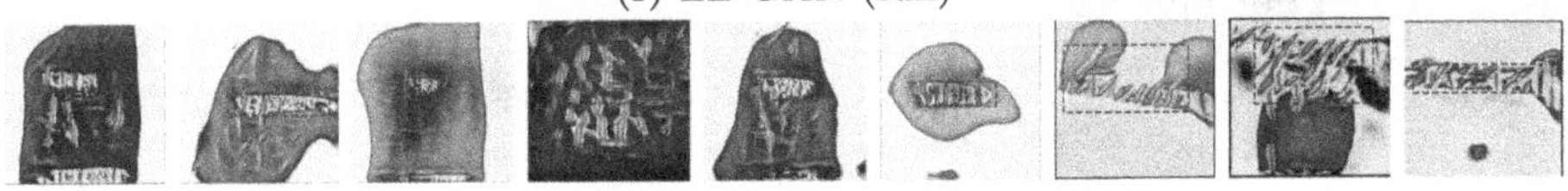

(g) StyleGAN2 ($\psi = 1$)

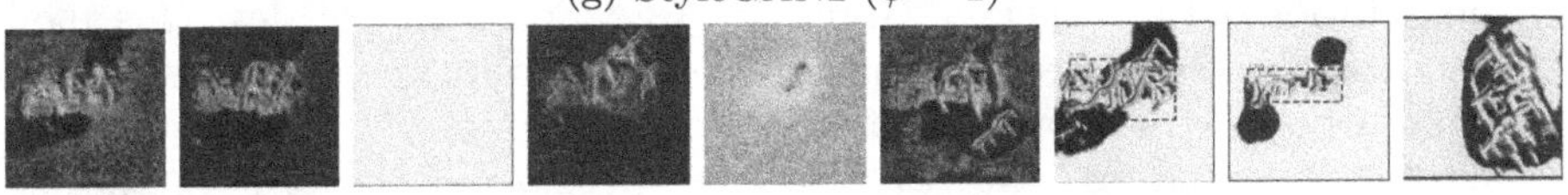

(h) StyleGAN2 ($\psi = 1$) + attention module

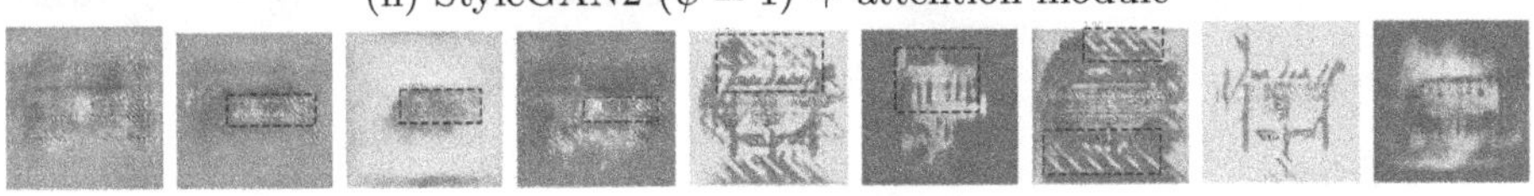

(i) SAGAN

Fig. 5. Examples generated by the models presented in the paper overlaid with bounding boxes predicted by the Faster R-CNN logo detection (+confidence score). Three last images for StyleGAN2 and StyleGAN2+Attention models were obtained using mixing regularities, see [16] for details. All DCGAN+ and LL-GAN images are 282×282, all other models are 256×256. Best viewed in color. (Color figure online)

Table 4. Comparison of models' performance-Detection. Italicized+bold: best, bold: second-best, italicized: third-best

Framework name	Detection rate	AvgConf
DCGAN+	*0.670*	*0.739*
LL-GAN	**0.674**	**0.746**
+ FRCNN loss	0.640	***0.827***
+ extended boxes	0.666	0.707
+ backbone features	0.622	0.701
Full	0.590	0.638
StyleGAN2($\psi = 0.6$)	0.554	0.670
StyleGAN2($\psi = 1.0$)	***0.687***	0.684
+attention	0.578	0.569
SAGAN	0.561	0.600

6 Conclusion

Generation of logos is a challenging problem that is becoming increasingly more popular in deep learning community. In this paper we presented a novel framework that fuses Faster R-CNN and GANs for generating large (282×282) heavy metal logos. The model was trained on a small style-rich dataset of real-life band logos. Results achieved by LL-GAN confidently outperform the state-of-the-art models trained on the same dataset, and we intend to explore the capacity of Faster R-CNN detector to extract and learn from regional features further. The advantages of our approach include:

- The novel idea of training the Generator using losses extracted from regional features in the real and fake data using Faster R-CNN.

- Computation of the style loss (Gram matrix) on regional features. This allows to use correlation between features in the fake and real data to transfer style from real to fake data, and construct samples from every image.
- The use of bounding boxes to determine the size of the RoIs in the fake data. Changing this size can improve results, e.g. by creating a more stable background.

Also, we would like to address certain limitations of the presented solution:

- Dataset and scope. All models were trained on a small dataset collected specifically to create logos in a particular style. We are confident this approach can be scaled to more general problems (e.g. logo stylization, style transfer, conditional logo creation) and larger datasets.
- Disentaglement and fusion of style and content. Disentanglement of style from content is active area of research in the font generation community [3,4]. In this paper we only used a single Generator for the logo generation. This result

can be improved both by augmenting the architectures, and fusing the style and content datasets.

References

1. Arjovsky, M., Chintala, S., Bottou, L.: Wasserstein gan (2017). arXiv preprint arXiv:1701.07875
2. Atarsaikhan, G., Iwana, B.K., Uchida, S.: Contained neural style transfer for decorated logo generation. In: 2018 13th IAPR International Workshop on Document Analysis Systems (DAS), pp. 317–322. IEEE (2018)
3. Azadi, S., Fisher, M., Kim, V.G., Wang, Z., Shechtman, E., Darrell, T.: Multi-content gan for few-shot font style transfer. In: Proceedings of the IEEE Conference on Computer Vision and Pattern Recognition, pp. 7564–7573 (2018)
4. Gao, Y., Guo, Y., Lian, Z., Tang, Y., Xiao, J.: Artistic glyph image synthesis via one-stage few-shot learning. ACM Trans. Graph. (TOG) **38**(6), 1–12 (2019)
5. Gatys, L.A., Ecker, A.S., Bethge, M.: Image style transfer using convolutional neural networks. In: Proceedings of the IEEE Conference on Computer Vision and Pattern Recognition, pp. 2414–2423 (2016)
6. Girshick, R., Donahue, J., Darrell, T., Malik, J.: Rich feature hierarchies for accurate object detection and semantic segmentation. In: Proceedings of the IEEE Conference on Computer Vision and Pattern Recognition, pp. 580–587 (2014)
7. Goodfellow, I., et al.: Generative adversarial nets. In: Advances in Neural Information Processing Systems, pp. 2672–2680 (2014)
8. Hayashi, H., Abe, K., Uchida, S.: Glyphgan: style-consistent font generation based on generative adversarial networks (2019). arXiv preprint arXiv:1905.12502
9. He, K., Gkioxari, G., Dollár, P., Girshick, R.: Mask R-CNN. In: Proceedings of the IEEE international conference on computer vision. pp. 2961–2969 (2017)
10. He, K., Zhang, X., Ren, S., Sun, J.: Deep residual learning for image recognition. In: Proceedings of the IEEE Conference on Computer Vision and Pattern Recognition, pp. 770–778 (2016)
11. Heusel, M., Ramsauer, H., Unterthiner, T., Nessler, B., Hochreiter, S.: Gans trained by a two time-scale update rule converge to a local nash equilibrium. In: Advances in Neural Information Processing Systems, pp. 6626–6637 (2017)
12. Isola, P., Zhu, J.Y., Zhou, T., Efros, A.A.: Image-to-image translation with conditional adversarial networks. In: Proceedings of the IEEE Conference on Computer Vision and Pattern Recognition, pp. 1125–1134 (2017)
13. Karatzas, D., et al.: ICDAR 2013 robust reading competition. In: 2013 12th International Conference on Document Analysis and Recognition, pp. 1484–1493. IEEE (2013)
14. Karras, T., Aila, T., Laine, S., Lehtinen, J.: Progressive growing of GANS for improved quality, stability, and variation (2017). arXiv preprint arXiv:1710.10196
15. Karras, T., Laine, S., Aila, T.: A style-based generator architecture for generative adversarial networks. In: Proceedings of the IEEE Conference on Computer Vision and Pattern Recognition, pp. 4401–4410 (2019)
16. Karras, T., Laine, S., Aittala, M., Hellsten, J., Lehtinen, J., Aila, T.: Analyzing and improving the image quality of stylegan. In: Proceedings of the IEEE/CVF Conference on Computer Vision and Pattern Recognition, pp. 8110–8119 (2020)
17. Kingma, D.P., Ba, J.: Adam: a method for stochastic optimization (2014). arXiv preprint arXiv:1412.6980

18. Lin, T.Y., Dollár, P., Girshick, R., He, K., Hariharan, B., Belongie, S.: Feature pyramid networks for object detection. In: Proceedings of the IEEE Conference on Computer Vision and Pattern Recognition, pp. 2117–2125 (2017)
19. Mino, A., Spanakis, G.: Logan: Generating logos with a generative adversarial neural network conditioned on color. In: 2018 17th IEEE International Conference on Machine Learning and Applications (ICMLA), pp. 965–970. IEEE (2018)
20. Miyato, T., Kataoka, T., Koyama, M., Yoshida, Y.: Spectral normalization for generative adversarial networks (2018). arXiv preprint arXiv:1802.05957
21. Oeldorf, C., Spanakis, G.: Loganv2: Conditional style-based logo generation with generative adversarial networks (2019). arXiv preprint arXiv:1909.09974
22. Radford, A., Metz, L., Chintala, S.: Unsupervised representation learning with deep convolutional generative adversarial networks (2015). arXiv preprint arXiv:1511.06434
23. Redmon, J., Divvala, S., Girshick, R., Farhadi, A.: You only look once: unified, real-time object detection. In: Proceedings of the IEEE Conference on Computer Vision and Pattern Recognition, pp. 779–788 (2016)
24. Ren, S., He, K., Girshick, R., Sun, J.: Faster R-CNN: Towards real-time object detection with region proposal networks. In: Advances in Neural Information Processing Systems, pp. 91–99 (2015)
25. Rijken, G.J., Cutura, R., Heyen, F., Sedlmair, M., Correll, M., Dykes, J., Smit, N.: Illegible semantics: exploring the design space of metal logos (2021). arXiv preprint arXiv:2109.01688
26. Sage, A., Agustsson, E., Timofte, R., Van Gool, L.: Logo synthesis and manipulation with clustered generative adversarial networks. In: Proceedings of the IEEE Conference on Computer Vision and Pattern Recognition, pp. 5879–5888 (2018)
27. Salimans, T., Goodfellow, I., Zaremba, W., Cheung, V., Radford, A., Chen, X.: Improved techniques for training GANS. In: Advances in Neural Information Processing Systems, pp. 2234–2242 (2016)
28. Ter-Sarkisov, A.: Network of steel: Neural font style transfer from heavy metal to corporate logos (2020). arXiv preprint arXiv:2001.03659
29. Zhang, H., Goodfellow, I., Metaxas, D., Odena, A.: Self-attention generative adversarial networks. In: International Conference on Machine Learning, pp. 7354–7363 (2019)
30. Zhu, J.Y., Park, T., Isola, P., Efros, A.A.: Unpaired image-to-image translation using cycle-consistent adversarial networks. In: Proceedings of the IEEE International Conference on Computer Vision, pp. 2223–2232 (2017)

User Study on the Effects Explainable AI Visualizations on Non-experts

Sophia Schulze-Weddige(✉) and Thorsten Zylowski

Future Labs, CAS Software AG, CAS-Weg 1-5, 76131 Karlsruhe, Germany
sophia.schulze-weddige@gmx.de, thorsten.zylowski@cas.de

Abstract. Artificial intelligence is drastically changing the process of creating art. However, in art, as in many other domains, algorithms and models are not immune from generating discriminatory and unfair artifacts or decisions. Explainable Artificial Intelligence (XAI) makes it possible to look into the "black box" and to identify biases and discriminatory behaviour. One of the main problems of XAI is that state-of-the-art explanation tools are usually tailored to AI experts. This paper evaluates how intuitively understandable the same tools are to laypeople. By using the prototypical use case of predictive sales, and testing the results with users, the abstract ideas of XAI are transferred to a real-world setting to study its understandability.

Based on our analysis, it can be concluded that explanations are easier to understand if they are presented in a way that is familiar to the users. A presentation in natural language is favorable because it presents facts unambiguously. All relevant information should be accessible in an intuitive manner that avoids sources of misinterpretations. It is desirable to design the system in an interactive way that allows the user to request further details on demand. This makes the system more flexible and adjustable to the use case. The results presented in this paper can guide the development of explainability tools that are adapted to a non-expert audience.

Keywords: Explainable AI · Human-centric AI · User study

1 Introduction

Many facets of art can be created by artificial intelligence, including paintings and literary works, as well as audio and video art. However, these systems can contain biases and show discriminatory behavior. For example, biases were found in AI-based generated art [17]. In addition, there is a large body of work dealing with the classification of art, especially paintings [2,19]. For training such classifiers, datasets are collected that may be biased (e.g. eurocentristic bias, gender bias, etc.). This would result in classifications favoring certain regions or groups. One can easily imagine a classifier that systematically rates European paintings higher (in price, in quality) than paintings from regions that are less strongly

M. Wölfel et al. (Eds.): ArtsIT 2021, LNICST 422, pp. 457–467, 2022.
https://doi.org/10.1007/978-3-030-95531-1_31

represented internationally, simply because the representation of the different regions is unequally distributed in the data. In cases like these, where the decisions affect important aspects of our lives, it is indispensable to understand the underlying decision process to control for fairness and safety. Explainable Artificial Intelligence (XAI) is one way to make the decisions of automated systems transparent. It has been used to uncover cases in which algorithms had been unfair, for example towards protected minorities or women [3]. Detecting these biases is a first and necessary step to eliminate them.

XAI has become very popular in recent years. Many large software companies such as Microsoft[1] and IBM[2] develop tools to access the decision process of AI systems. Novel papers about the topic are being published regularly. But many explanation methods currently available are tailored for people with prior knowledge in machine learning. This makes them difficult to understand for laypeople. The need for human-centric explanations has been reported frequently and attempts have been made to provide them [16].

This paper aims to evaluate out-of-the-box explainability methods in the context of non-expert users. That is, evaluate whether the tools available are suited to explain automated decision processes to people that have no proficiency in AI. A user study is conducted based on a real-world example from the sales industry which leads to precise recommendations for the development of human-centered XAI. There are excellent summaries about XAI [15] and human-centered AI [5], but they fail to provide detailed suggestions because many decisions depend heavily on the use case. By generalizing the insights from the results of this specific use case, this paper aims to provide valuable information which is intended to help other practitioners to make informed choices for their explanation systems.

2 Related Work

XAI aims to clarify how an automated decision is generated. Hereby, many XAI approaches focus on the systems that generate the decision. While it is imperative to accurately portray the decision process, correctness is not sufficient to make explanations understandable to humans [11]. The notion of human-centered XAI puts the human back in the focus of attention. The goal is to provide explanations that appeal to the person using the system. This means the explanations are easy to understand and not misleading. As the users cannot be expected to have prior knowledge about AI, the explanation should be adjusted to the target audience [13].

Explainability includes the ability of humans to understand the explanation. When designing a human-centered system, the first step is to define precisely whom the explanation is aimed at. Then, the goal of the explanation needs to be determined. Some guidelines help in designing such a system [9]. But they state

[1] See Microsoft's toolkit at https://github.com/interpretml/interpret.
[2] See IBM's toolkit at https://github.com/Trusted-AI/AIX360.

that many use case-specific decisions need to be made. As the use cases can be distinct, it is difficult to give general best practices.

One way to design an ideal system for a specific use case follows the sociotechnological approach described by Ehsan and Riedl [5]. They state that the social and the technical part of a human-centered system co-evolve in an iterative process. They describe a cycle of altering the system to the needs of the user and evaluating the effect. One example for a use case-specific implementation is called "Glass Box" [16]. They developed a chat- or voice-based interactive dialogue for the loan application data set [4]. In their study, participants are presented with counterfactual explanations for why their loan application was rejected. If they are not contemptuous with the answer they can ask follow-up and what-if questions. More research on use case-specific implementations for human-centered XAI is needed to develop a deeper understanding of how to make explanations understandable for humans.

3 The User Study

A user study is conducted to evaluate different explainability tools based on a real-world example in predictive sales which aims to learn from past sales outcomes to predict customer behavior in the future. The explanation methods evaluated in this study are based on a classifier that predicts the status of a lead. A lead is the collection of data about an individual customer. The data set contains 440 variables and comprises roughly 10000 leads. A random forest classifier was used for the classification task. An averaged accuracy of 0.9 was yielded and a precision, recall, and F1-score of 0.91.

An online application is developed in which the participants can choose between three different example leads. One of the example leads is lost, the second is won and the third is very close to the decision boundary. The examples are selected randomly in the interval of their prediction, namely [0, 0.5), [0.49, 0.5), [0.5, 1]. For each example, the participant can further choose between five visualizations which are generated with the SHAP [10], LIME [12] and alibi explain [18] python packages. They are chosen based on popularity and type of explanation. Only instance-based explanations are included, as users are usually more interested in understanding information regarding themselves, rather than aim to comprehend the underlying model behavior in detail [1].

3.1 The Explanation Tools

LIME provides local explanations for individual predictions by fitting interpretable models to input-output pairs. These pairs are generated by randomly perturbing the instance at question to depict its local surroundings. The perturbed instances are then used as an input to the model to calculate their outputs. The generated samples are weighted by their distance to the instance in question. LIMEs visualization is divided into three parts. On the left, the prediction probability for the different classes is depicted. In the center, the feature importance of the ten most important features is shown. On the right, the corresponding values of these features can be found. Colors indicate whether the influence of that feature is positive (orange) or negative (blue).

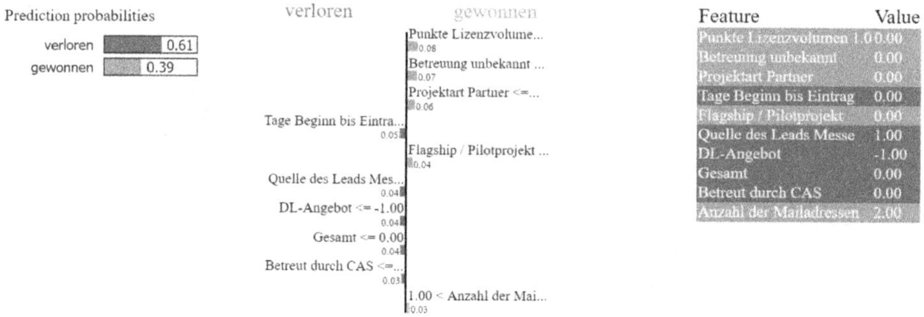

Fig. 1. Sample explanation from the custom application with LIME. (Color figure online)

Lundberg and Lee developed an explanation method that is based on Shapley values from cooperative game theory [10]. Shapley values distribute the surplus in a cooperative game between the contributors depending on their influence on the outcome [8]. In the context of XAI, the features can be interpreted as the players and the prediction as the outcome. Hence, Shapley values are a measure for the contribution of the individual features to the overall prediction.

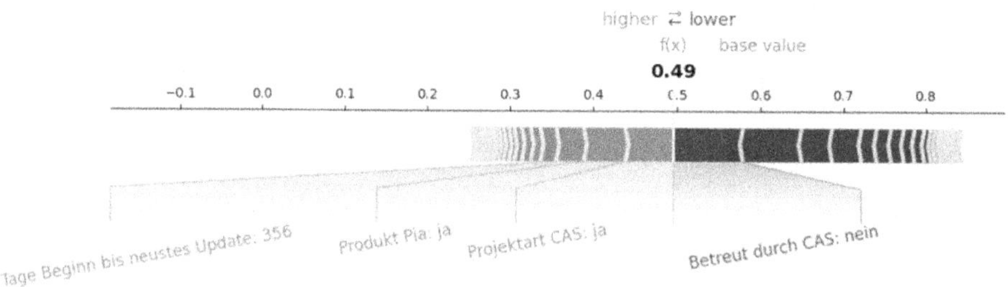

Fig. 2. Sample force plot with SHAP.

Three visualizations from the SHAP package are used in this study. All of them show the features with the highest Shapley values. The first one uses the analogy to a force that is pushing the prediction to its final value to visualize the effect of individual variables (Fig. 2). Variables with a positive effect push the prediction higher up the number strip, which is indicated by arrows to the right. Variables with a negative effect push the prediction to the left, which means lowering the prediction value. The forces are at an equilibrium in the final prediction value. The width of the arrows indicates the strength of the effect. Variables with an effect of at least 5% are written out together with the corresponding value.

In the second visualization, Shapley values are expressed in a bar plot (Fig. 3). The bar plot is bi-directional, which means the bars start at 0 in the center and point either to the left for negative values or to the right for positive ones. Additionally, the bars are color-coded with blue for negative and red for positive effects. The length of the bar indicates the magnitude of the Shapley value. The y-axis shows the names of the variables and the corresponding values of the instance.

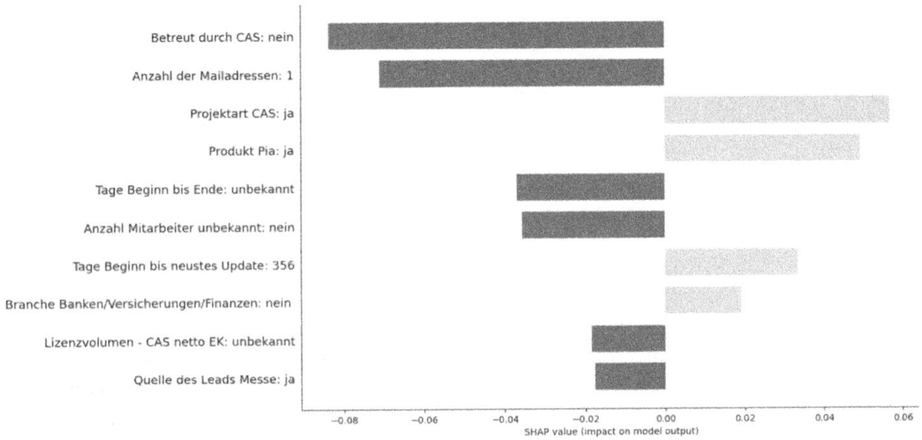

Fig. 3. Sample bar plot with SHAP.

Thirdly, the same information can be depicted in a decision plot (Fig. 4). This type of plot shows a decision path that can be followed from bottom to top, where it ends at the final prediction. Which variable is being considered can be seen on the y-axis. The x-axis shows the current prediction value. When the line moves to the left, it denotes a negative effect in the corresponding variable. When it moves to the right, the effect is positive.

Lastly, explanations are presented as counterfactual examples. By making explicit what would have to change in the input in order to yield a different output, they not only provide information on the reasons behind a decision but also on how to alter it in the future. This information is highly valuable for humans who usually ask for why something happened *rather* than something else [11]. Moreover, counterfactual explanations are easy to understand because they are presented in natural language which is "the most accessible modality of explanation" [5]. The changes are presented in a bullet point list stating what could be done to improve the prediction of the instance. Further, the magnitude of the change is shown in brackets. A bullet point list could look like this:

- Source of the lead is **not** trade fair (0.1)
- Project type **is** partner (0.07)

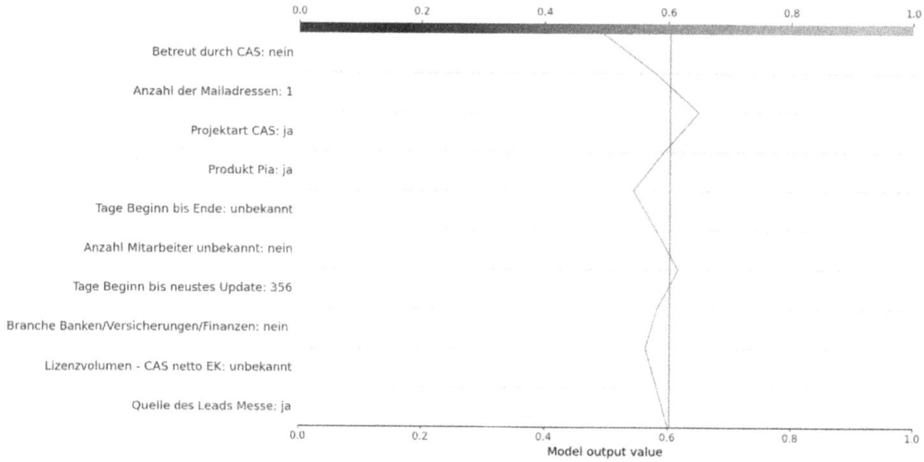

Fig. 4. Sample decision plot with SHAP.

3.2 Study Design

In the user study, participants first make themselves comfortable with the use of the online application by exploring it freely. Then they answer six questions. The necessary information to answer them can be found in the application. The participants are not instructed to search for the answers in a specific way but are free to use the application to their liking. This includes choosing which and how many methods to consult before answering each question. On the one hand, the questions aim to evaluate whether the participants can extract the relevant information from the application. On the other hand, the questions help to see which explainability methods are preferably used to find the information. After each question, participants indicate which methods they used for their answer. It is possible to select multiple methods if more than one were considered. During the whole study, participants are asked to describe their train of thought and opinion about the methods. At the end, five statements from the system causability scale (SCS) [7] are used to inquire the opinion about the different explainability methods. The agreement to those statements is measured on a five-item Likert scale [14].

1. I understood the explanations within the context of my work.
2. I did not need support to understand the explanations.
3. I was able to use the explanations with my knowledge base.
4. I think that most people would learn to understand the explanations very quickly.
5. I did not need more references in the explanations: e.g., medical guidelines, regulations.

4 Results

Fifteen employees from CAS Software AG participated in the user study, five of which work in the sales department, five in the research department and the remaining five in consulting, development, or product management. Neither of the participants is an expert in AI or has seen the explainability tools before, except for one who briefly encountered SHAP. The interviews lasted between 20 and 50 min with an average of 30 min per participant. The results of the Shapiro Wilk normality test show that the data from the SCS is not normally distributed for the bar plot ($p = 0.0003$) and the counterfactual explanation ($p = 0.01$). Thus, a non-parametric method for the evaluation of the ratings is used. The Wilcoxon signed rank test for dependent samples is conducted pair-wise for all explainability methods.

Table 1. Table of summary statistics. This table shows the means, standard deviations and the SCS scores for the five explainability methods.

	Bar plot	Decision plot	Force plot	LIME	Counterfactuals
Mean	4.41	3.49	3.29	3.07	4.08
Std	1.05	1.28	1.47	1.29	1.21
SCS score	88	70	66	61	82

The mean values and standard deviations can be found in Table 1. As the mean rankings for the bar plot ($\mu = 4.41$) and the counterfactual explanations ($\mu = 4.08$) are unequivocally higher than for the decision plot ($\mu = 3.49$), the force plot ($\mu = 3.29$) and the LIME plot ($\mu = 3.07$), a one-sided hypothesis test is conducted. For the decision, force, and LIME plot the effect is not as clear. Therefore, a two-sided test was conducted to compare these three among each other.

Nine out of the fifteen participants gave the bar plot the highest rating and another four participants gave it the second-highest rating. From the remaining six participants, four gave the highest score to the counterfactual explanations and two to the decision plot. All participants except for two have a mean rating higher than 4.0 for the bar plot. For the counterfactual explanations, it is all but four.

The results show that the bar plot has a significantly higher rating than the decision plot ($p = 0.0062$), the force plot ($p = 0.0011$), and the LIME plot ($p = 0.0003$). The counterfactual explanations have a significantly higher rating than the LIME plot ($p = 0.002$). No significant differences can be found between the ratings of the decision, force, and lime plot. For an overview of all the results see Table 2.

After each question, the participants indicated which methods they used to find the relevant information. The bar plot was used most frequently. It was

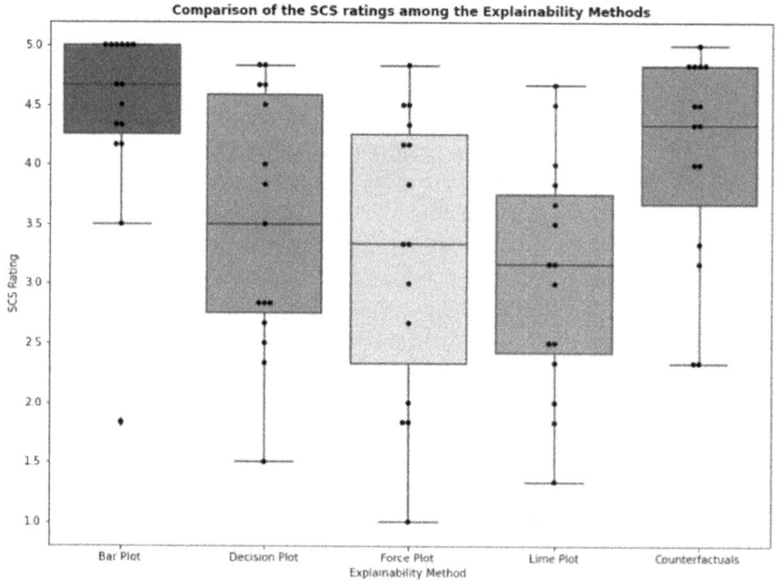

Fig. 5. Box plots for the SCS ratings. The box plots show the median and the first and third quartile, as well as the minimum and maximum values. The ratings of each participant for each plot are displayed as a black dot.

Table 2. Results of the Wilcoxon signed-rank test. This table shows the p-values calculated by the Wilcoxon signed-rank test. A p-value lower than 0.01 is considered significant and shown in bold.

	Counterfactuals	Decision plot	Force plot	LIME
Bar plot	0.103	**0.006**	**0.001**	**0.0004**
Counterfactuals		0.069	0.028	**0.002**
Decision plot			0.599	0.256
Force plot				0.495

involved in answering one of the questions in 44 cases. The second most frequently used method was the counterfactual explanations with 34 cases, followed by the decision plot with 27 cases, the LIME plot with 24 cases, and lastly the force plot with 16 cases.

The results for the SCS ratings match with the statement of the participant during the user study. All of the participants talked positively about the bar plot and five explicitly stated it as their favorite method. Ten participants noted that they had trouble understanding the decision plot and six said that the force plot has too little information as it only shows a small number of variables. Moreover, nine participants said that they had trouble understanding the variables.

5 Discussion and Conclusion

The analysis of the SCS ratings clearly shows that the bar plot is the favorite explainability method in this study. It has a significantly higher rating than the decision, force, and LIME plot and it received the highest overall ranking for almost two-thirds of the participants. Further, it has been used most frequently during the study. It appears that many users are familiar with bar plots, which shows that most people prefer visualizations that they already know. This can be confirmed by the second highest ranking method, the counterfactual examples. These are easy to understand as they are presented in text form. The familiarity helps to focus on the relevant aspects. It makes sense to pursue using simple, familiar plots if possible. In cases where a more complex display of information is needed, precise and easy clarifications should be provided. In many applications, it is not feasible to carry out a tutorial or lengthy introduction. Hence, the application must be self-sufficient in its usage.

Although simplicity is favorable to see all information at a glance, relevant information such as the direction of the plot (decision plot and force plot) and the values on the axes (bar plot) should be made clearly visible. Relevant clarifications about the interpretation of the plot should be integrated directly into the plot. Important aspects of the plot should be highlighted. This particularly holds for parts of the plots that might lead to misinterpretations. For example, the effect size of the bar plot might be overestimated because the bars are scaled by the width of the plot. This makes it difficult to grasp the real effect size from the plot. Directing the focus to the numeric value of the effect by increasing the font size of the x-axis or adding the values next to the plots, reduces the chances of misinterpretation. The counterfactual explanations could be enhanced by transforming the bullet points into more natural sentences and presenting precise actionable suggestions for improvement. In general, user studies are highly relevant to detect pitfalls like these and should be conducted to build an explanation system that matches the target group.

Moreover, the possibility for interaction between the user and the application is highly favorable. One way to allow for interaction is to provide details on demand. Following Grice's maxims, only information that is relevant to the situation and not known yet should be displayed [6]. This can vary between instances. Thus, it facilitates understandability if the user can ask for clarification if necessary but is not overwhelmed with details at the beginning. Further, integrating interactions to the application makes it more flexible and adjustable to the user's needs. That means the same system can be used by different people in a way that suits them. One way to anticipate interactions is in the form of interactive plots. For example, hovering over variable names to receive further elaborations or linking sources where additional help can be found.

In addition, participants asked for more details about the variables. Especially possible value instantiations are of interest. Nine participants explicitly stated that they had trouble understanding the variables. An explanation regarding the variable names and possible values could facilitate the understanding. The variables are the basis of the explanations. If they are not clear to the user it makes the interpretation of the explanations difficult even if the effect was understood correctly.

All in all, in order to yield higher understandability the explanation methods should be adjusted to the use case and the target audience. But some general considerations can guide the design choices. Simple and familiar visualizations are preferable over complex and detailed ones. Possible sources of misinterpretation should be detected and the visualizations should aim to direct the focus to relevant information to avoid them. Moreover, allowing for interaction between the user and the system increases flexibility and enhances the user experience. Following these suggestions, explanation tools can be used to reveal the underlying decision process of an algorithm to non-experts in AI art, as well as various other domains.

References

1. Arya, V., et al.: One Explanation Does Not Fit All: A Toolkit and Taxonomy of AI Explainability Techniques (2019). arXiv preprint arXiv:1909.03012
2. Cetinic, E., Grgic, S.: Genre classification of paintings. In: 2016 International Symposium ELMAR, pp. 201–204 (2016). https://doi.org/10.1109/ELMAR.2016.7731786
3. Chiusi, F.: Report: automated society 2020. J. Chem. Inf. Model. **110**(9), 1689–1699 (2017)
4. Dua, D., Graff, C.: UCI machine learning repository (2017). https://archive.ics.uci.edu/ml/datasets/statlog+(german+credit+data)
5. Ehsan, U., Riedl, M.O.: Human-centered explainable AI: towards a reflective sociotechnical approach. In: International Conference on Human-Computer Interaction, pp. 449–466 (2020). http://arxiv.org/abs/2002.01092
6. Grice, H.P.: Logic and conversation. In: Cole, P., Morgan, J.L. (eds.) Speech Acts, Syntax and Semantics, vol. 3, pp. 41–58. Academic Press, New York (1975)
7. Holzinger, A., Carrington, A., Müller, H.: Measuring the quality of explanations: the system causability scale (SCS): comparing human and machine explanations. KI - Kunstliche Intelligenz **34**(2), 193–198 (2020)
8. Kuhn, H.W., Tucker, A.W.: Contributions to the Theory of Games (AM-28), Vol. II. Annals of Mathematics Studies, Princeton University Press (2016). https://books.google.de/books?id=Pd3TCwAAQBAJ
9. Liao, Q.V., Gruen, D., Miller, S.: Questioning the AI: informing design practices for explainable AI user experiences. Conf. Human Factors Comput. Syst. - Proc. (2020). https://doi.org/10.1145/3313831.3376590
10. Lundberg, S.M., Lee, S.I.: A unified approach to interpreting model predictions. Adv. Neural Inf. Process. Syst. **2017**(Section 2), 4766–4775 (2017)
11. Miller, T.: Explanation in artificial intelligence insights from the social sciences. Artif. Intell. **267**, 1–38 (2019). arXiv:1706.07269
12. Ribeiro, M.T., Singh, S., Guestrin, C.: "Why should I trust you?": explaining the predictions of any classifier (2016)
13. Ribera, M., Lapedriza, A.: Can we do better explanations? A proposal of user-centered explainable AI. In: CEUR Workshop Proceedings, Vol. 2327 (2019)
14. Robinson, J.: Likert Scale, pp. 3620–3621. Springer, Netherlands, Dordrecht (2014). https://doi.org/10.1007/978-94-007-0753-5

15. Samek, W., Montavon, G., Lapuschkin, S., Anders, C.J., Müller, K.R.: Explaining deep neural networks and beyond: a review of methods and applications. Proc. IEEE **109**(3), 247–278 (2021). https://doi.org/10.1109/JPROC.2021.3060483
16. Sokol, K., Flach, P.A.: Glass-box: explaining AI decisions with counterfactual statements through conversation with a voice-enabled virtual assistant. In: IJCAI, pp. 5868–5870 (2018)
17. Srinivasan, R., Uchino, K.: Biases in generative art - a causal look from the lens of art history (2021)
18. Van Looveren, A., Klaise, J.: Interpretable counterfactual explanations guided by prototypes (2019). arXiv preprint arXiv:1907.02584
19. Zujovic, J., Gandy, L., Friedman, S., Pardo, B., Pappas, T.N.: Classifying paintings by artistic genre: an analysis of features classifiers. In: 2009 IEEE International Workshop on Multimedia Signal Processing, pp. 1–5 (2009). https://doi.org/10.1109/MMSP.2009.5293271

Author Index

CPSIA information can be obtained
at www.ICGtesting.com
Printed in the USA
LVHW060433150222
711162LV00004B/99

9 783030 955304